SELECTED ASPECTS OF CANCER PROGRESSION: METASTASIS, APOPTOSIS AND IMMUNE RESPONSE

Cancer Growth and Progression

Volume 11

Series Editors

Hans E. Kaiser, D. Sc.
Professor,
Department of Pathology,
School of Medicine, University of Maryland, Baltimore, MD, U.S.A.
& Department of Clinical Pathology, University of Vienna, Austria

Aejaz Nasir, M.D., M.Phil., FCAP
Department of Interdisciplinary Oncology-Pathology,
Moffit Cancer Center & Research Institute,
Tampa, FL, U.S.A.

Production Assistant:

Yasmin Qayumi, B.S.

Selected Aspects of Cancer Progression: Metastasis, Apoptosis and Immune Response

Edited by

Hans E. Kaiser

Department of Pathology,
School of Medicine,
University of Maryland,
Baltimore, MD, U.S.A
and
Department of Clinical Pathology,
University of Vienna, Austria

and

Aejaz Nasir

Moffitt Cancer Center & Research Institute,
Tampa, FL, U.S.A

Associate Editor:

Nelly Adriana Nasir, MD

Department of Pathology,
Sir Mortimer Jewish General Hospital,
McGill University, Montreal, Canada

 Springer

Hans E. Kaiser
Department of Pathology, School of Medicine
University of Maryland, Baltimore, MD,
U.S.A
and
Department of Clinical Pathology
University of Vienna
Austria

Aejaz Nasir
Moffitt Cancer Center & Research Institute
Tampa, FL
U.S.A

ISBN 978-1-4020-6728-0   e-ISBN 978-1-4020-6729-7

Library of Congress Control Number: 2008920057

Printed on acid-free paper

9 8 7 6 5 4 3 2 1

springer.com

Contents

H.E. Kaiser and A. Nasir (eds.), Selected Aspects of Cancer Progression:
Metastasis, Apoptosis and Immune Response, 1–10.
© *Springer Science + Business Media B.V. 2008*

CHAPTER ONE

Metastasis: a current perspective

David T. Denhardt[1], Ann F. Chambers[2], and Danny R. Welch[3]

Abstract: Topics included in this brief review of the metastatic process include recent (2004) progress in our understanding of tumor cell-host cell interactions and issues concerning the establishment of metastases. This includes the critical steps of intravasation and extravasation as well as establishment and evolution of micrometastases. Also summarized are details of metastasis-suppressing and metastasis-including genes and how they impact on signal transduction pathways. Finally, two proteins that play complex roles in the metastatic process, osteopontin and issue inhibitor of metastasis-1, are discussed in more depth.

Keywords: tumor-host interactions, metastasis-suppressing genes, metastasis-inducing genes, osteopontin, TIMP-1.

In this brief overview of metastasis we have sought to summarize the major concepts currently in play and to highlight some of the recent advances in our understanding of tumor metastasis. The complexity of the metastatic process precludes, in the limited space available, a comprehensive overview and citation of important primary references; for this we apologize. An excellent and more extensive review of metastasis was published recently by Bogenrieder and Herlyn (2003). We leave to others in these volumes a detailed discussion of the many genes in which mutations can occur, contributing to the formation of a primary tumor. This includes genes that allow cancer cells to subvert the mechanisms regulating the differentiation, proliferation and survival of primary tumor cells (Hanahan and Weinberg 2000).

Tumor cell–host cell interactions

Solid tumors that form in a tissue as the consequence of the genesis, proliferation and evolution of a transformed cell are typically a mix of descendents of the initiating tumor cell, normal cells that were present in the tissue, and cells that have invaded the emerging tumor, macrophages for example. Arguably the most important of the tumor cell–host cell interactions is that which results in the development of a tumor vasculature, angiogenesis. The formation of new blood vessels is essential if the nascent tumor is to grow in size. Thus cells within the tumor secrete proteases and chemotactic factors such as FGF and VEGF that stimulate the angiogenic process by stimulating capillary endothelial sprouting from pre-existing blood vessels or by recruiting circulating endothelial

[1] Department of Cell Biology and Neuroscience,
Rutgers University, Piscataway, NJ 08854 USA
[2] London Regional Cancer Centre, London, Ontario,
N6A 4L6 Canada
[3] Department of Pathology and Comprehensive
Cancer Center, University of Alabama at Birmingham,
Birmingham, AL 35294 USA

cells (Rafii 2000). Proteases, including matrix metal-loproteinases, urokinase, and thrombin, facilitate the penetration of the tumor by the growing capillary by degrading extracellular matrix barriers. Inhibitors of these proteinases, tissue inhibitors of metalloprotein-ases (TIMPs) and urokinase-like plasminogen activa-tor inhibitors (uPAIs) for example, will oppose tumor angiogenesis and act as tumor suppressors. A good overview of the involvement of proteases in cancer is that by Wall et al. (2003). Important points made in this review included the role of the uPA receptor in controlling cell migration and proliferation, and the fact that cells can penetrate the extracellular matrix (ECM) by an amoeboid process independent of known proteolysis.

The stability of epithelial cells is maintained in large part by their interactions with neighboring stromal cells. Disruption of these interactions, mediated by cytokines and proteinases or by altered expression of genes supporting cell–cell interactions, results in the epithelial–mesenchymal transition, a step in the pro-gression of epithelial tumors. The interactions, direct or indirect, between tumor cells and host cells leads to emerging properties of the growing tumor not without parallels to a developing organ, albeit a less well-organized organ (Bissell and Radisky 2001; Wiseman and Werb 2002; Wang and Tetu 2002). De Wever and Mareel (2003) have reviewed many of the aspects of tumor cell–stromal cell interactions with a focus on TGF-β and the possible origin of myofibroblasts from fibroblasts/myocytes. TGF-β and PDGF, which normally regulate wound healing, can be produced by the cancer cell and act on the stromal compartment to elicit expression of genes that can stimulate tumor progression. An example of this is the production of TIMP-1 by stromal cells in response to a tumor cell stimulus (Nakopoulou et al. 2002). This is discussed further below. There is an increasing appreciation of the fact that secreted proteases do not degrade ECM proteins solely to overcome physical barriers to cell migration (Hojilla et al. 2003). Cleavage of ECM pro-teins can generate peptides that have profound effects on cell behavior, as exemplified by endostatin, an inhibitor of angiogenesis produced by the cleavage of collagen XVIII. Products of fibrin(ogen) cleavage also stimulate metastasis (Palumbo et al. 2002).

Finally, there are also proteases on the cell surface, adamalysin-like and disintegrin metalloproteinases

(ADAMs) for example, that may cleave cell sur-face proteins and thereby modulate cell activities. As one example, tumor necrosis factor alpha-convert-ing enzyme (TACE) processes the membrane-bound precursors of both TGF-α and TNF-α to their mature forms (Moss et al. 2001). TGF-α, like other members of the EGF-like peptide family, generally stimulates tumor growth (Normanno et al. 2001). Although in some contexts TNF-α stimulates tumor destruction by inducing vascular collapse, in other contexts it acts as a tumor promoter (Szlosarek and Balkwill 2003). Similarly, TGF-β may either stimulate or inhibit tum-origenesis/metastasis (Wakefield and Roberts 2002). For both these latter cytokines, it is the intensity of the signaling that regulates the resulting response of the cell, and this is a function of both the level of active cytokine produced and the abundance of active recep-tors on the target cell.

Genetic instability and the resulting clonal heteroge-neity are major forces driving the evolution of a tumor. Set within an underlying pattern of gene expression characteristic of the tissue origin, an increasing number of genes are up- or down-regulated as the cancer progresses. It is from within this witches' brew of cells expressing different characteristics that the metastatic cell evolves, aided by host stromal cells. Unexpectedly, it appears that, at least in mammary carcinomas, seem-ingly normal stromal cells (fibroblasts) exhibit genetic alterations, loss of heterozygosity, that accompany the neoplastic alteration of the epithelial cells (Moinfar et al. 2000). Similarly Dong-Le Bourhis et al. (1997) reported that fibroblasts from normal breast tissues, but not putatively normal fibroblasts from cancerous breast tissues, could inhibit the proliferation of MCF-7 cells. This has led to the recognition that drugs targeted to cancer cells, Herceptin for example, may suppress tumor cell growth indirectly via an action on stromal cells (Corsini et al. 2003).

Establishment of metastases

The physical process of metastasis can be bro-ken down into three steps: Escape of a tumor cell from the primary tumor and entrance (intravasation) into the circulatory (hematopoietic, lymphatic) sys-tem; Survival of physical trauma and immunological attack in the circulation; Exit (extravasation) from

the circulation and establishment at a secondary site. Intravasation presumably involves both the abrogation of cell–cell contacts that hold the tumor cell in the tumor and the expression of proteases that enable the tumor cell to penetrate the wall of the capillary or lymphatic vessel. During invasion the migrating cell ceases proliferation as the result, at least in basal cell carcinomas, of the up-regulation of p16^{INK4a}, an inhibitor of CDK4/6, which together with cyclin D is responsible to phosphorylating Rb and relieving inhibition of E2F. This was shown by an immunohistochemical study of p16^{INK4a} levels and Rb phosphorylation, comparing cells at the invasive front with cells in the center of the tumor mass (Svensson et al. 2003). In tumors in which the tumor suppressor p16^{INK4a} is down-regulated, migration and cell cycle progression may not be mutually exclusive. Although it was long thought that metastasizing cells would have a low probability of surviving in the circulation, intravital videomicroscopy studies have shown the contrary, that at least some varieties of cancer cells survive very well in the circulation and can extravasate efficiently (MacDonald et al. 2002).

It is believed that in individuals with a healthy immune system any tumor cell expressing an unfamiliar antigen on the cell surface is likely to be recognized by cytotoxic immune cells and killed (Platsoucas et al. 2003). Although good arguments can be made that many occult malignant conversions arise and are eliminated by immunological mechanisms, experimental evidence bearing directly on this question is limited. The ability of the immune system to detect and then to eliminate tumors depends on many factors, not the least being where the tumor develops and the novel antigens it presents. If novel antigens are not presented during inflammatory processes of tumor development, e.g. during angiogenesis, the immune system will be tolerant of the cancer cell. If a tumor epitope is recognized as foreign by the immune system, then the tumor cells will be killed unless the growing tumor finds a way to subvert the recognition process, for example by down-regulating MHC class-I expression, inhibiting the antigen-processing machinery, or incapacitating the response of dendritic or other antigen-presenting cells (Pardoll 2001; Vicari et al. 2002; Garcia-Lora et al. 2003). Unfortunately, this seems to happen all too often.

Figure 1 illustrates various aspects of the metastatic process in experimental model systems. Recent research has shown that although the extravasation process is reasonably efficient, it is the subsequent proliferation and establishment of a secondary tumor that is very inefficient (Chambers et al. 2002; MacDonald et al. 2002). Circulating tumor cells are frequently trapped in the first capillary bed they encounter after entering the circulation, thus accounting in part for why cancers in certain organs preferentially metastasize to a limited number of locations in the body. As an alternative to extravasation, the trapped (or adherent) cancer cell may proliferate intravascularly for an initial period (Wong et al. 2002). Attachment of a tumor cell to the endothelium depends on the interaction of mutually complementary cell surface molecules (cadherins) or receptors able to engage different domains of the same ligand, for example osteopontin, which in theory could engage an integrin on one cell and a CD44 variant on the other cell. Once a cell has extravasated, studies employing intravital videomicroscopy have shown that the extravasated cell may remain quiescent for extended periods before beginning to proliferate. These experiments suggest that, in contrast to earlier beliefs, it may be this last step, the ability to proliferate in a new environment, that is apparently most difficult to accomplish. Whether, or how soon, a subset of extravasated tumor cells begin to proliferate and form a tumor is determined by its interactions with cells in the new location. If the new location provides the appropriate support structure (soluble factors, a suitable ECM, and friendly interactions with neighboring cells) then rapid growth of the secondary tumor will occur. If on the other hand conditions are not conducive for growth the extravasated tumor cell may remain dormant for months or years until it begins to proliferate.

Metastasis-suppressing and metastasis-inducing functions

Several groups have searched for metastasis suppressor genes using various procedures, some based on the presumption that the mRNA encoding a metastasis suppressor gene would be expressed at lower levels in metastatic cells as compared with a non-metastatic but otherwise similar tumorigenic cell (Welch et al. 2000; Steeg 2003). The metastasis-suppressing action of the putative suppressor can be confirmed by the demonstration that when expressed in a metastatic

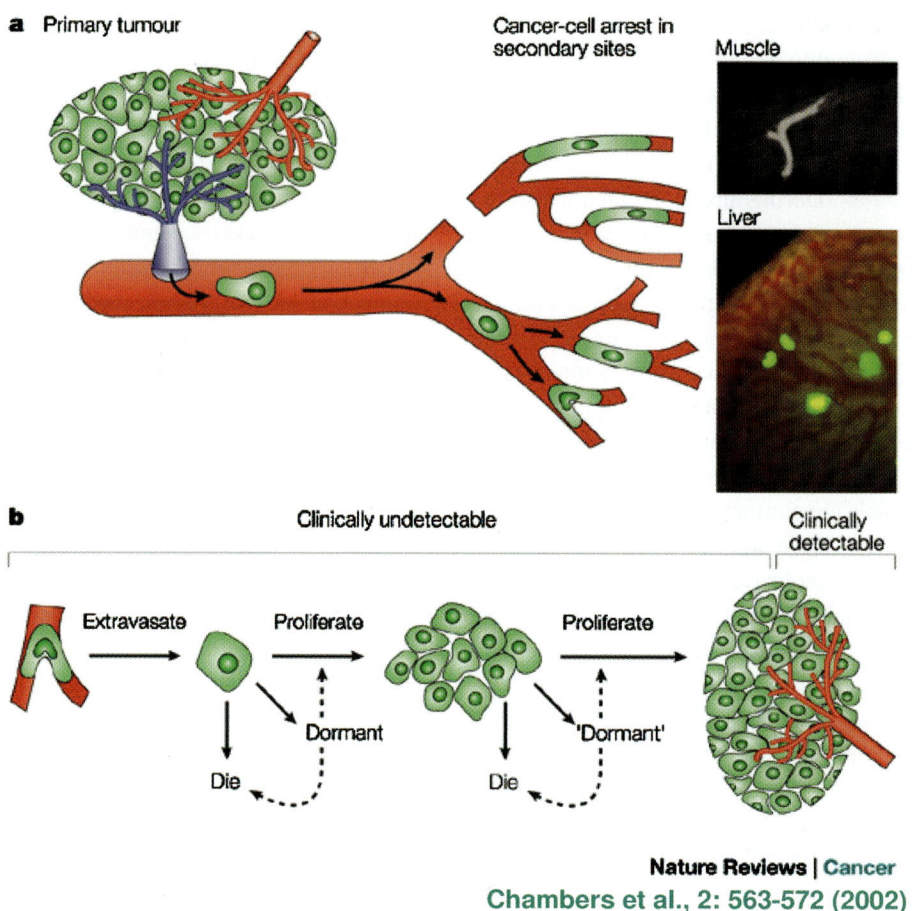

Nature Reviews | Cancer
Chambers et al., 2: 563-572 (2002)

Fig. 1 The Metastatic Process. (a) Escape of tumor cells from the primary tumor (intravasation) and their dissemination in the circulation throughout the body. Cancer cells tend to arrest in the small capillary beds of those organs immediately downstream of the tumor. This is one reason why breast tumors tend to metastasize to lung, whereas tumors secondary to colon cancers are often found in the liver. However, in order for breast cancer cells to metastasize to bone, they need to pass through the lung microcirculation and enter the arterial circulation; once they arrive in bone, evidence suggests that their growth in that microenvironment may be favored (Chambers et al. 2002). (b) Once a tumor cell arrives in a secondary organ, it has a number of fates, which depend on interactions of the cell with its microenvironment. Solitary cells may die, become dormant, or proliferate. Similarly, micrometastases may die, become "dormant" (in which proliferation and apoptosis are balanced), or grow progressively and become vascularized. The proportion of cells that pass thorough each of these steps determines the overall metastatic burden. Dormant cells that subsequently become activated, or "dormant" micrometastases in which the balance between apoptosis and proliferation becomes tipped to favor growth, may be responsible for late recurrences seen in patients after long periods of clinical tumor dormancy. (Reprinted from Chambers AF, Groom AC, MacDonald IC, Dissemination and growth of cancer cells in metastatic sites, Nature Reviews Cancer 2:563–72. Copyright (2002), with permission from Macmillan Magazines Ltd.)

cell it reduces the metastatic efficiency. NM23 was the first to be described and was subsequently found to be a histidine protein kinase. It phosphorylates the scaffold protein KSR, leading to reduced ERK activation. Several additional metastasis suppressor genes (MKK4, RhoGDI2, KAI1) also modulate signal transduction pathways. MKK4 activates the p38 and JNK kinases; RhoGDI2 regulates Rho and Rac function (which stimulate motility and invasiveness); and KAI1 (CD82) is a tetraspannin that interacts with cer-

tain integrins and the epidermal growth factor receptor (Steeg 2003).

Genomic analysis (loss of heterozygosity, comparative genomic hybridization, accompanied by microcell-mediated chromosome transfer) is an alternative strategy for locating potential metastasis suppressor genes. Genes identified using these strategies include KISS-1, BRMS1, CRSP3 and TXNIP (Shevde and Welch 2003). KISS-1, which has a very short half-life, is a ligand for a G-protein coupled receptor. CRSP3,

a cofactor that interacts with the transcription factor SP1, and TRN, which interacts with and presumably regulates thioredoxin activity, are both involved in regulating KISS-1 expression. BRMS1 may hinder metastasis by increasing connexin 43 expression and augmenting gap junctional intercellular communication. BRMS1 shares homology with mSds3 and is capable of forming a transcription-inhibiting complex with histone deacetylase and retinoblastoma binding protein 1 (Meehan et al. 2004).

Metastasis is a complex process, requiring the expression of many genes to execute the process. Thus it should not be surprising that anything that can reduce expression of these genes can hinder metastasis. That many of the identified genes are involved in signal transduction or cell–cell interactions is to

be expected. Because their identification is largely based on a loss-of-function strategy, it has been more of a challenge to identify them than to identify genes (oncogenes) that confer a gain-of-function, tumorigenicity, on normal cells. There is also the fact that we have a good cell culture model for tumorigenicity (growth in soft agar), but not for metastatic ability, which can only be conclusively shown in an animal. It is likely that more metastasis suppressor genes remain to be discovered, each of which holds the promise of providing an approach to blocking or slowing the metastatic spread of a cancer.

As illustrated in Fig. 2 (Shevde and Welch 2003), many proteins participate in one or another aspect of metastasis. These include cell adhesion proteins, proteinases, cytoskeletal proteins and cell signaling

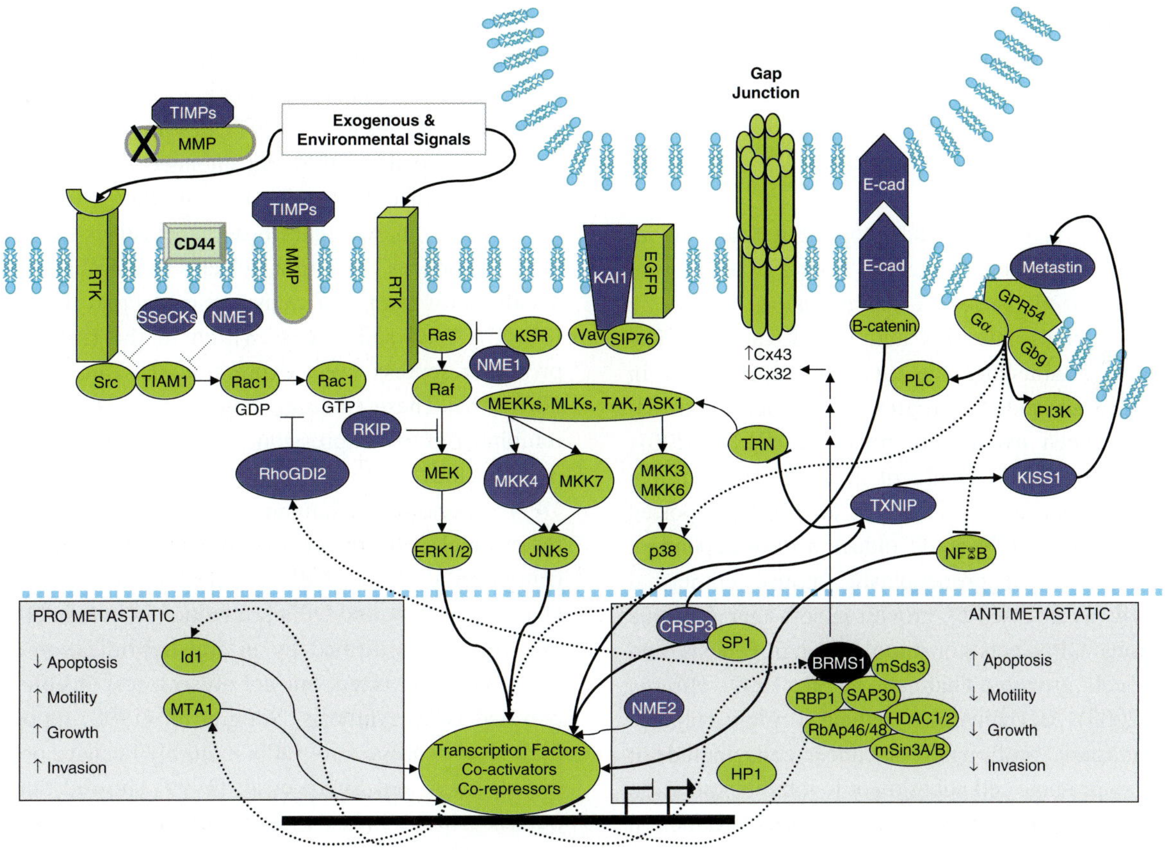

Fig. 2 An illustration of possible mechanisms of action of metastasis suppressor genes. Shown in cartoon form is a portion of mammalian cell with the plasma membrane at the top and the nucleus at the bottom. Known metastasis suppressor proteins are represented within open shapes; putative metastasis suppressor proteins are in stippled boxes. Solid black lines illustrate known connections, whereas the dotted lines illustrate connections that remain to be confirmed. Shadowed boxes in the nucleus contain pro- and anti-metastatic genes involved in transcription. (Reprinted from Shevde LA, Welch DR, Metastasis suppressor pathways – an evolving paradigm, Cancer Letters 198:1–20. Copyright (2003), with permission from Elsevier.)

proteins (receptors and their ligands, intracellular signal transduction intermediates, transcription factors) (Engers and Gabbert 2000). To cite a few examples: Van Golen et al. (2002) reported that the RhoC GTPase was overexpressed in inflammatory breast cancer cells and apparently responsible for the increased motility and invasiveness of these cells. Oft et al. (2002) described evidence for a multistep model of tumor progression entailing stepwise increases in the levels of Smad2 and H-ras activity that drive the epithelial–mesenchymal transition of squamous carcinoma cells. In the MCF10A human breast cell line Smad2/3 signalling appears to be inhibitory in the early steps of tumorigenesis but stimulatory in the later stages of metastasis, at least to the lung after tail vein injection (Tian et al. 2003). Because they are less well known as metastasis stimulators, we conclude this review with a brief discussion of two genes, osteopontin and TIMP-1, that illustrate the difficulties in reaching a satisfying understanding of how cells metastasize.

Osteopontin

OPN is a glycosylated phosphoprotein found in all body fluids and bone; it acts as both a cytokine and (in mineralized tissues) a cell adhesion protein by binding various integrins and CD44 variants. The literature establishing the involvement of OPN in tumorigenesis and the progression of metastatic disease has been extensively reviewed (Weber 2001; Furger et al. 2001; Denhardt et al. 2003; Yeatman and Chambers 2003). Chang et al. (2003) reported that exogenous OPN could induce a dose-dependent transformation of preneoplastic mouse epidermal JB6 cells as assessed by growth in soft agar. In some situations OPN acts to stimulate angiogenesis and tumor cell growth (Thalmann et al. 1999; Hirama et al. 2003). Mechanisms by which it might enhance the metastatic proficiency of cancer cells include it ability to promote cell adhesion or to inhibit expression of inducible nitric oxide synthase (OPN production by metastasizing cancer cells might protect them from being killed by NO produced by cytotoxic macrophages). There is also growing evidence that OPN may facilitate metastases to bone (Nemoto et al. 2001; Kang et al. 2003; Allan et al. 2004). OPN has been shown in several contexts to inhibit apoptosis, perhaps

thereby allowing higher levels of expression of the *ras* oncogene. Another mechanism by which OPN might foster metastasis is by promoting the migratory and invasive properties of the cells. In an elegant series of studies Tuck et al. (2000, 2001, 2003) found that OPN enhances the migratory and invasive properties of mammary epithelial cells, apparently by up-regulating expression of urokinase-type plasminogen activator (uPA) and enhancing the activity of various growth factor receptor kinases including hepatocyte growth factor receptor (Met) and epidermal growth factor (EGF). Shijubo et al. (2000) reported that OPN could synergize with VEGF to stimulate endothelial cell migration, and Leali et al. (2003) showed that OPN could enhance FGF-2-mediated angiogenesis.

Rudland and colleagues have shown that if OPN expression is specifically upregulated in a benign rat mammary cell line, RAMA37, either by transfection with certain "metastasis-inducing sequences" or a plasmid engineered to express OPN, the cells acquire a malignant phenotype. This and other research, reviewed by Oates et al. (1997), has provided compelling evidence that OPN can enhance the metastatic proficiency of cancer cells. Recent studies from this group have shown that these "metastasis-inducing DNA sequences" can bind to Tcf-4, a member of the T cell factor family of transcription factors that bind to a CAAAG sequences, which in the (rat) OPN promoter repress OPN expression (El-Tanani et al. 2001a, b). These findings suggest that TCF-4 is an inhibitor of OPN expression.

Substantial data have accumulated documenting OPN expression in human cancers, produced either by stromal cells or by the tumor cells themselves (Furger et al. 2001). Gillespie et al. (1997) and Tuck et al. (1998) reported OPN production in human breast cancers and confirmed by in situ hybridization that tumor cells are often, but not always, responsible for elevated OPN synthesis. Sung et al. (1998) reported high level expression of OPN in highly invasive breast cancer lines, and Sharp et al. (1999) showed that in human primary breast cancers OPN is expressed predominantly by the cancer cells. One mechanism by which OPN might influence the metastatic process is via its interaction with CD44 variants that are associated with metastasis (Katagiri et al. 1999). Whether this involves stimulating cell adhesion and/or activation of signal transduction pathways is not known.

Research has also demonstrated an association between OPN levels in tumors and/or blood and patient outcome, suggesting that OPN may play an important functional role in human cancer (Furger et al. 2001). Western blotting analysis revealed that OPN was present at elevated levels in patients with metastatic cancers. Chambers et al. (1996) reported that OPN was over-expressed in series of human lung cancers, relative to adjacent normal tissue, with a statistically significant association between OPN presence in the tumors and patient survival. In a series of 154 cases of lymph node-negative breast cancers, tumor cell OPN (as detected by immunohistochemistry using the anti-OPN mAB53) was found to be associated with poor patient survival (Tuck et al. 1998). OPN mRNA overexpression is correlated with high-grade, late-stage hepatocellular carcinoma, early tumor recurrence, and unfavorable prognosis (Pan et al. 2003). Weber (2001) has reviewed the importance of OPN as a promising target for cancer therapy. Should it turn out to be a significant target for clinical intervention, the intriguing question arises: which among the various documented or suggested roles for OPN in the metastatic process are critically relevant?

Tissue inhibitor of metalloproteinases-1

Our second example of a protein whose contribution to the metastatic process is puzzling is TIMP-1, a glyco-sylated 28-kD protein tightly structured by six disulfide bonds (Brew et al. 2000). It is identical to EPA (eryth-roid potentiating activity), which promotes an increase in erythroid progenitor cells. TIMP-1 inhibits many members of the metalloproteinase (MP) superfamily, which includes some of the adamalysin-like disintegrin and metalloproteinases (ADAMs) and many of the matrix metalloproteinases (MMPs) (Baker et al. 2002). These are Ca^{2+}-dependent, Zn^{2+}-containing enzymes that are usually activated extracellularly by proteolytic cleavage. Many of the MMPs effectively degrade struc-tural proteins such as collagens, laminin, elastin, and fibronectin in the ECM (Nagase and Woessner 1999). Many of the ADAMs on the other hand focus their cleavage activity on proteins at the cell surface, pro-ducing effector molecules such as TGFα or the HER2 ectodomain (Codony-Servat et al. 1999). Proteinases, including some MMPs and ADAMs, known to cleave cell surface molecules, have been termed "sheddases"

(Kheradmand and Werb 2002). Via these extracellular activities MPs play roles in various physiological func-tions including embryogenesis, angiogenesis, wound healing, ovulation, inflammation and fibrosis. Because of their potent proteolytic activities they are closely regulated, in part by TIMP-1.

MPs, especially MMPs and the membrane-type-MMPs, enable the metastasis of cancer cells by facilitating the passage of malignant cells through basement membranes and the extracellular matrix (Westermarck and Kahari 1999; Hotary et al. 2000; Itoh and Nagase 2002). They also enable the growth of tumors as a result of their role in angiogenesis, enabling the growing tumor to develop a vasculature (Folkman 2002). The ability of TIMP-1 to inhibit both angiogenesis and metastasis would appear to make TIMP-1, and other TIMPs for that matter (there are 4), effective anti-cancer agents (Nii et al. 2000). This has spawned commercial efforts to utilize MMP inhibitors as anti-cancer agents (Coussens et al. 2002). A discon-certing finding, however, was that an elevated level of expression of TIMP-1 mRNA in primary breast carcinomas was associated with metastasis (Ree et al. 1997). Moreover, a higher level of TIMP-1 expres-sion predicts poorer patient prognosis with respect to the length of the disease-free interval and survival rate (McCarthy et al. 1999; Yoshikawa et al. 1999; Ylisirnio et al. 2001; Holten-Andersen et al. 2002; Nakopoulou et al. 2002).

It is well established that TIMP-1, produced by the tumor cells, is inhibitory to tumor progression by virtue of its ability to inhibit metalloproteinases and block angiogenesis and tumor cell invasion. It is less well appreciated that TIMP-1 can also stimulate growth and/or DNA replication of tumor cells, including MG63 (human osteosarcoma), K562 (human erythro-leukemia), MCF-7 (human mammary adenocarcinoma) and BC-61 (human breast carcinoma) (Hayakawa 1994; Yamashita et al. 1996). In MG63 cells TIMP-1 stimulated the Ras-MAPK pathway and DNA synthesis (Yamashita et al. 1996; Wang et al. 2002). This growth promoting activity was downregulated by tyrosine phosphorylation inhibitors. Yan and Moses (2001) dis-cussed the biphasic effects (stimulation of tumor cell proliferation followed by tumor necrosis as the result of inhibition of angiogenesis) of TIMP-1 on tumor progression in the context of Burkitt's lymphoma. Zeng et al. (1995) reported that the correlation between

increased TIMP-1 mRNA levels and colorectal cancer progression reflected a growth-promoting function for human TIMP-1. Rat breast carcinoma lines engineered to overexpress TIMP-1 exhibited increased VEGF expression and more aggressive tumor development (Yoshiji et al. 1998). Porter et al. (2004) reported data suggesting that TIMP-1 stimulates proliferation of MDA-MB-435 cells by inhibiting a putative cell surface proteinase, possibly an ADAM.

As noted above, a number of studies suggest strongly that TIMP-1 fosters metastasis. For example, Nakopoulou et al. (2002) reported a significant correlation in 117 invasive human breast carcinomas between elevated TIMP-1 mRNA levels in stromal cells at the tumor-stromal interface and lymph node metastases, c-erbB-2 expression and adverse patient prognoses. TIMP-1 mRNA was not detected in tumor cells and adjacent normal breast tissue. Additionally, the positive correlation of high TIMP-1 mRNA expression with estrogen receptor levels suggested an association with tumor cell differentiation. One obvious approach to elucidating the action of TIMP-1 would be to demonstrate its metastasis-promoting action directly, for example by engineering a tumorigenic, non-metastatic cell line to over-express TIMP-1. This is unlikely to succeed however, because TIMP-1 is strongly inhibitory to the growth of a primary tumor; it inhibits angiogenesis and suppresses tumor cell invasion (i.e. is anti-metastatic). But how then does one explain that in a significant number of tumors elevated TIMP-1 expression strongly correlates with progression to an invasive, malignant cancer? There is a considerable literature that interactions among stromal cells and epithelial cells ("bidirectional stromal–epithelial crosstalk") play an important role in the progression of breast cancer (Moinfar et al. 2000; Wang and Tetu 2002; LeBedis et al. 2002). It may be that the production of TIMP-1 by stromal cells is the key to understanding the correlation between elevated TIMP-1 expression in tumors and enhanced aggressiveness of the tumor cells. If so, the puzzle remains why TIMP-1 produced by tumor cells is inhibitory to tumor growth and metastasis whereas TIMP-1 produced by tumor stromal cells stimulates tumor progression. In both cases it is believed that the protein is secreted.

Acknowledgements Research in the authors' laboratories has been supported by the National Institutes of Health, the Cancer Federation Inc., the Canadian Institutes of Health Research, Canadian Breast Cancer Research Alliance, and the National Foundation for Cancer Research Center for Metastasis Research.

References

Allan AL, Tuck AB, Bramwell VHC, Vandenberg TA, Winquist EW, Chambers AF (2004) Contribution of osteopontin to the development of bone metastasis. In: Singh G, Rabbani SA (eds) Bone cancer metastasis. Humana, Totowa, NJ

Baker AH, Edwards DR, Murphy G (2002) Metalloproteinase inhibitors: biological actions and therapeutic opportunities. J Cell Sci 115:3719–3727

Bissell MJ, Radisky D (2001) Putting tumours in context. Nat Rev Cancer 1:46–54

Bogenrieder T, Herlyn M (2003) Axis of evil: molecular mechanisms of cancer metastasis. Oncogene 22:6524–6536

Brew K, Dinakarpandian D, Nagase H (2000) Tissue inhibitors of metalloproteinases: evolution, structure and function. Biochim Biophys Acta 1477:267–283

Chambers AF, Wilson SM, Kerkvliet N, O'Malley FP, Harris JF, Casson AG (1996) Osteopontin expression in lung cancer. Lung Cancer 15:311–323

Chambers AF, Groom AC, MacDonald IC (2002) Dissemination and growth of cancer cells in metastatic sites. Nat Rev Cancer 2:563–572

Chang PL, Cao M, Hicks P (2003) Osteopontin induction is required for tumor promoter-induced transformation of preneoplastic mouse cells. Carcinogenesis 24:1749–1758

Codony-Servat J, Albanell J, Lopez-Talavera JC, Arribas J, Baselga J (1999) Cleavage of the HER2 ectodomain is a pervanadate-activable process that is inhibited by the tissue inhibitor of metalloproteases-1 in breast cancer cells. Cancer Res 59:1196–1201

Corsini C, Mancuso P, Paul S, Burlini A, Martinelli G, Pruneri G, Bertolini F (2003) Stroma cells: a novel target of herceptin activity. Clin Cancer Res 9:1820–1825

Coussens LM, Fingleton B, Matrisian LM (2002) Matrix metalloproteinase inhibitors and cancer: trials and tribulations. Science 295:2387–2392

Denhardt DT, Mistretta D, Chambers AF, Krishna S, Porter JF, Raghuram S, Rittling SR (2003) Transcriptional regulation of osteopontin and the metastatic phenotype: evidence for a Ras-activated enhancer in the human OPN promoter. Clin Exp Metastasis 20:77–84

De Wever O, Mareel M (2003) Role of tissue stroma in cancer cell invasion. J Pathol 200:429–447

Dong-Le Bourhis X, Berthois Y, Millot G, Degeorges A, Sylvi M, Martin PM, Calvo F (1997) Effect of stromal and epithelial cells derived from normal and tumorous breast tissue on the proliferation of human breast cancer cell lines in co-culture. Int J Cancer 71:42–48

El-Tanani M, Barraclough R, Wilkinson MC, Rudland PS (2001a) Regulatory region of metastasis-inducing DNA is the binding site for T cell factor-4. Oncogene 20:1793–1797

El-Tanani M, Barraclough R, Wilkinson MC, Rudland PS (2001b) Metastasis-inducing DNA regulates the expression of the osteopontin gene by binding the transcription factor Tcf-4. Cancer Res 61:5619–5629

Engers R, Gabbert HE (2000) Mechanisms of tumor metastasis: cell biological aspects and clinical implications. J Cancer Res Clin Oncol 126:682–692

Folkman J (2002) Role of angiogenesis in tumor growth and metastasis. Semin Oncol 29 (Suppl 16):15–18

Furger KA, Menon RK, Tuck AB, Bramwell VH, Chambers AF (2001) The functional and clinical roles of osteopontin in cancer and metastasis. Curr Mol Med 1:621–632

Garcia-Lora A, Algarra I, Garrido F (2003) MHC class I antigens, immune surveillance, and tumor immune escape. J Cell Physiol 195:346–355

Gillespie MT, Thomas RJ, Zhou PU, Martin TJ, Findlay DM (1997) Calcitonin receptors, bone sialoprotein and osteopontin are expressed in primary breast cancers. Int J Cancer 10:812–815

Hanahan D, Weinberg RA (2000) The hallmarks of cancer. Cell 100:57–70

Hayakawa T (1994) Tissue inhibitors of metalloproteinases and their cell growth-promoting activity. Cell Struct Funct 19:109–114

Hirama M, Takahashi F, Takahashi K, Akutagawa S, Shimizu K, Soma S, Shimanuki Y, Nishio K, Fukuchi Y (2003) Osteopontin overproduced by tumor cells acts as a potent angiogenic factor contributing to tumor growth. Cancer Lett 198:107–117

Hojilla CV, Mohammed FF, Khokha R (2003) Matrix metalloproteinases and their tissue inhibitors direct cell fate during cancer development. Br J Cancer 89:1817–1821

Holten-Andersen MN, Christensen IJ, Nielsen HJ, Stephens RW, Jensen V, Nielsen OH, Sorensen S, Overgaard J, Lilja H, Harris A, Murphy G, Brunner N (2002) Total levels of tissue inhibitor of metalloproteinases 1 in plasma yield high diagnostic sensitivity and specificity in patients with colon cancer. Clin Cancer Res 8:156–164

Hotary K, Allen E, Punturieri A, Yana I, Weiss SJ (2000) Regulation of cell invasion and morphogenesis in a three-dimensional type I collagen matrix by membrane-type matrix metalloproteinases 1, 2, and 3. J Cell Biol 149:1309–1323

Itoh Y, Nagase H (2002) Matrix metalloproteinases in cancer. Essays Biochem 38:21–36

Kang Y, Siegel PM, Shu W, Drobnjak M, Kakonen SM, Cordon-Cardo C, Guise TA, Massague J (2003) A multigenic program mediating breast cancer metastasis to bone. Cancer Cell 3:537–549

Katagiri YU, Sleeman J, Fujii H, Herrlich P, Hotta H, Tanaka K, Chikuma S, Yagita H, Okumura K, Murakami M, Saiki I, Chambers AF, Uede T (1999) CD44 variants but not CD44s cooperate with beta1-containing integrins to permit cells to bind to osteopontin independently of arginine-glycine-aspartic acid, thereby stimulating cell motility and chemotaxis. Cancer Res 59:219–226

Kheradmand F, Werb Z (2002) Shedding light on sheddases: role in growth and development. Bioessays 24:8–12

Leali D, Dell'Era P, Stabile H, Sennino B, Chambers AF, Naldini A, Sozzani S, Nico B, Ribatti D, Presta M (2003) Osteopontin (Eta-1) and fibroblast growth factor-2 cross-talk in angiogenesis. J Immunol 171:1085–1093

LeBedis C, Chen K, Fallavollita L, Boutros T, Brodt P. (2002) Peripheral lymph node stromal cells can promote growth and tumorigenicity of breast carcinoma cells through the release of IGF-I and EGF. Int J Cancer 100:2–8

MacDonald IC, Groom AC, Chambers AF (2002) Cancer spread and micrometastasis development: quantitative approaches for in vivo models. Bioessays 24:885–893

McCarthy K, Maguire T, McGreal G, McDermott E, O'Higgins N, Duffy MJ (1999) High levels of tissue inhibitor of metalloproteinase-1 predict poor outcome in patients with breast cancer. Int J Cancer 84:44–48

Meehan WJ, Samant RS, Hopper JE, Carrozza MJ, Shevde LA, Workman JL, Eckert KA, Verderame MF, Welch DR (2004) The BRMS1 metastasis suppressor forms complexes with RBP1 and the mSin3 histone deacetylase complex and represses transcription. J Biol Chem 279:1562–1569

Moinfar F, Man YG, Arnould L, Bratthauer GL, Ratschek M, Tavassoli FA (2000) Concurrent and independent genetic alterations in the stromal and epithelial cells of mammary carcinoma: implications for tumorigenesis. Cancer Res 60:2562–2566

Moss ML, White JM, Lambert MH, Andrews RC (2001) TACE and other ADAM proteases as targets for drug discovery. Drug Discov Today 6:417–426

Nagase H, Woessner F (1999) Matrix metalloproteinases. J Biol Chem 274:21491–21494

Nakopoulou L, Giannopoulou I, Stefanaki K, Panayotopoulou E, Tsirmpa I, Alexandrou P, Mavrommatis J, Katsarou S, Davaris P (2002) Enhanced mRNA expression of tissue inhibitor of metalloproteinase-1 (TIMP-1) in breast carcinomas is correlated with adverse prognosis. J Pathol 197:307–313

Nemoto H, Rittling SR, Yoshitake H, Furuya K, Amagasa T, Tsuji K, Nifuji A, Denhardt DT, Noda M (2001) Osteopontin deficiency reduces experimental tumor cell metastasis to bone and soft tissues. J Bone Miner Res 16:652–659

Nii M, Kayada Y, Yoshiga K, Takada K, Okamoto T, Yanagihara K (2000) Suppression of metastasis by tissue inhibitor of metalloproteinase-1 in a newly established human oral squamous cell carcinoma cell line. Int J Oncol 16:119–124

Normanno N, Bianco C, De Luca A, Salomon DS (2001) The role of EGF-related peptides in tumor growth. Front Biosci 6:D685–707

Oates AJ, Barraclough R, Rudland PS (1997) The role of osteopontin in tumorigenesis and metastasis. Invasion Metastasis 17:1–15

Oft M, Akhurst RJ, Balmain A (2002) Metastasis is driven by sequential elevation of H-ras and Smad2 levels. Nat Cell Biol 4:487–494

Palumbo JS, Potter JM, Kaplan LS, Talmage K, Jackson DG, Degen JL (2002) Spontaneous hematogenous and lymphatic metastasis, but not primary tumor growth or angiogenesis, is diminished in fibrinogen-deficient mice. Cancer Res 62:6966–6972

Pan HW, Ou YH, Peng SY, Liu SH, Lai PL, Lee PH, Sheu JC, Chen CL, Hsu HC (2003) Overexpression of osteopontin is associated with intrahepatic metastasis, early recurrence, and poorer prognosis of surgically resected hepatocellular carcinoma. Cancer 98:119–127

Pardoll D (2001) T cells and tumours. Nature 411:1010–1012

Platsoucas CD, Fincke JE, Pappas J, Jung WJ, Heckel M, Schwarting R, Magira E, Monos D, Freedman RS (2003) Immune responses to human tumors: development of tumor vaccines. Anticancer Res 23:1969–1996

Porter JF, Shen S, Denhardt DT (2004) Tissue inhibitor of metalloproteinase-I stimulates proliferation of human cancer cells by inhibiting a metalloproteinase. Brit J Cancer 90:463–70

Rafii S (2000) Circulating endothelial precursors: mystery, reality, and promise. J Clin Invest 105:17–19

Ree AH, Florenes VA, Berg JP, Maelandsmo GM, Nesland JM, Fodstad O (1997) High levels of messenger RNAs for tissue inhibitors of metalloproteinases (TIMP-1 and TIMP-2)

in primary breast carcinomas are associated with development of distant metastases. Clin Cancer Res 3:1623–1628

Sharp JA, Sung V, Slavin J, Thompson EW, Henderson MA (1999) Tumor cells are the source of osteopontin and bone sialoprotein expression in human breast cancer. Lab Invest 79:869–877

Shevde LA, Welch DR (2003) Metastasis suppressor pathways – an evolving paradigm. Cancer Lett 198:1–20

Steeg PS (2003) Metastasis suppressors alter the signal transduction of cancer cells. Nat Rev Cancer 3:55–63

Shijubo N, Uede T, Kon S, Nagata M, Abe S (2000) Vascular endothelial growth factor and osteopontin in tumor biology. Crit Rev Oncog 11:135–146

Sung V, Gilles C, Murray A, Clarke R, Aaron RD, Azumi N, Thompson EW (1998) The LCC15-MB human breast cancer cell line expresses osteopontin and exhibits invasive and metastatic phenotype. Exp Cell Res 241:273–284

Svensson S, Nilsson K, Ringberg A, Landberg G (2003) Invade or proliferate? Two contrasting events in malignant behavior governed by p16(INK4a) and an intact Rb pathway illustrated by a model system of basal cell carcinoma. Cancer Res 63:1737–1742

Szlosarek PW, Balkwill FR (2003) Tumour necrosis factor alpha: a potential target for the therapy of solid tumours. Lancet Oncol 4:565–573

Thalmann GN, Sikes RA, Devoll RE, Kiefer JA, Markwalder R, Klima I, Farach-Carson CM, Studer UE, Chung LW (1999) Osteopontin: possible role in prostate cancer progression. Clin Cancer Res 5:2271–2277

Tian F, DaCosta Byfield S, Parks WT, Yoo S, Felici A, Tang B, Piek E, Wakefield LM, Roberts AB (2003) Reduction in Smad2/3 signaling enhances tumorigenesis but suppresses metastasis of breast cancer cell lines. Cancer Res 63:8284–8292

Tuck AB, Elliott BE, Hota C, Tremblay E, Chambers AF (2000) Osteopontin-induced, integrin-dependent migration of human mammary epithelial cells involves activation of the hepatocyte growth factor receptor (Met). J Cell Biochem 78:465–475

Tuck AB, Hota C, Chambers AF (2001) Osteopontin(OPN)-induced increase in human mammary epithelial cell invasiveness is urokinase (uPA)-dependent. Breast Cancer Res Treat 70:197–204

Tuck AB, Hota C, Wilson SM, Chambers AF (2003) Osteopontin-induced migration of human mammary epithelial cells involves activation of EGF receptor and multiple signal transduction pathways. Oncogene 22: 198–1205

Tuck AB, O'Malley FP, Singhal H, Harris JF, Tonkin KS, Kerkvliet N, Saad Z, Doig GS, Chambers AF (1998) Osteopontin expression in a group of lymph node negative breast cancer patients. Int J Cancer 79:502–508

van Golen KL, Bao LW, Pan Q, Miller FR, Wu ZF, Merajver SD (2002) Mitogen activated protein kinase pathway is involved in RhoC GTPase induced motility, invasion and angiogenesis in inflammatory breast cancer. Clin Exp Metastasis 19:301–311

Vicari AP, Caux C, Trinchieri G (2002) Tumour escape from immune surveillance through dendritic cell inactivation. Semin Cancer Biol 12:33–42

Wakefield LM, Roberts AB (2002) TGF-beta signaling: positive and negative effects on tumorigenesis. Curr Opin Genet Dev 12:22–29

Wall SJ, Jiang Y, Muschel RJ, DeClerck YA (2003) Meeting report: Proteases, extracellular matrix, and cancer: an AACR Special Conference in Cancer Research. Cancer Res 63:4750–4755

Wang CS, Tetu B (2002) Stromelysin-3 expression by mammary tumor-associated fibroblasts under in vitro breast cancer cell induction. Int J Cancer 99:792–799

Wang T, Yamashita K, Iwata K, Hayakawa T (2002) Both tissue inhibitors of metalloproteinases-1 (TIMP-1) and TIMP-2 activate Ras but through different pathways. Biochem Biophys Res Commun 296:201–205

Welch DR, Steeg PS, Rinker-Schaeffer CW (2000) Molecular biology of breast cancer metastasis. Genetic regulation of human breast carcinoma metastasis. Breast Cancer Res 2:408–416

Wiseman BS, Werb Z (2002) Stromal effects on mammary gland development and breast cancer. Science 296:1046–1049

Weber GF (2001) The metastasis gene osteopontin: a candidate target for cancer therapy. Biochim Biophys Acta 1552:61–85

Westermarck J, Kahari V-M (1999) Regulation of matrix metalloproteinase expression in tumor invasion. FASEB J 13:781–792

Wong CW, Song C, Grimes MM, Fu W, Dewhirst MW, Muschel RJ, Al-Mehdi AB (2002) Intravascular location of breast cancer cells after spontaneous metastasis to the lung. Am J Pathol 161:749–753

Yamashita K, Suzuki M, Iwata H, Koike T, Hamaguchi M, Shinagawa A, Noguchi T, Hayakawa T (1996) Tyrosine phosphorylation is crucial for growth signaling by tissue inhibitors of metalloproteinases (TIMP-1 and TIMP-2). FEBS Letts 396:103–107

Yan L, Moses MA (2001) A case of tumor betrayal: biphasic effects of TIMP-1 on Burkitt's lymphoma. Am J Pathol 158:1185–1190

Yeatman TJ, Chambers AF (2003) Osteopontin and colon cancer progression. Clin Exp Metastasis 20:85–90

Ylisirnio S, Hoyhtya M, Makitaro R, Paaakko P, Risteli J, Kinnula VL, Turpeenniemi-Hujanen T, Jukkola A (2001) Elevated serum levels of type I collagen degradation marker ICTP and tissue inhibitor of metalloproteinase (TIMP) 1 are associated with poor prognosis in lung cancer. Clin Cancer Res 7:1633–1637

Yoshiji H, Harris SR, Raso E, Gomez DE, Lindsay CK, Shibuya M, Sinha CC, Thorgeirsson UP (1998) Mammary carcinoma cells over-expressing tissue inhibitor of metalloproteinases-1 show enhanced vascular endothelial growth factor expression. Int J Cancer 75:81–87

Yoshikawa T, Saitoh M, Tsuburaya A, Kobayashi O, Sairenji M, Motohashi H, Yanoma S, Noguchi Y (1999) Tissue inhibitor of matrix metalloproteinase-1 in the plasma of patients with gastric carcinoma. A possible marker for serosal invasion and metastasis. Cancer 86:1929–1935

Zeng ZS, Cohen AM, Zhang ZF, Stetler-Stevenson W, Guillem JG (1995) Elevated tissue inhibitor of metalloproteinase 1 RNA in colorectal cancer stroma correlates with lymph node and distant metastases. Clin Cancer Res 1:899–906

H.E. Kaiser and A. Nasir (eds.), Selected Aspects of Cancer Progression:
Metastasis, Apoptosis and Immune Response, 11–19.
© *Springer Science + Business Media B.V.* 2008

CHAPTER TWO

Control of cell motility during tissue invasion

James Varani

Abstract: Invasion of the surrounding tissue by tumor cells is a complex process and involves tissue destruction and active tumor cell movement. Regulation of cell motility is complex. There are cell type-specific differences (for example, differences between fibroblasts and epithelial cells) and differences between transformed cells and their normal counterparts. Under normal circumstances, fibroblasts exist in the tissue as single cells. Reflecting this, fibroblasts migrate as single cells. Epithelial cells exist as sheets of cells or three-dimensional masses. Epithelial cell migration involves groups of cells moving together as well as movement by individual cells. For both fibroblasts and epithelial cells, *in vitro* motility requires a substratum to which the cells can adhere. Components of the extracellular matrix including fibronectin and laminin provide a suitable substrate. Differences between normal and transformed cells in elaboration of fibronectin/laminin and differences in interaction with these motility-inducing matrix molecules contribute to differences in motility. When fibroblasts interact with fibronectin (or other motility-supporting matrix molecules), binding to cell surface adhesion receptors initiates a cascade of signaling events that results in organization of the cell for movement. In epithelial cells, interaction with the matrix is also critical to cell movement. However, significant movement does not occur unless there is also interaction with soluble motility-stimulating factors (i.e. growth factors, chemokines and complement components). While the process of tumor cell motility has been extensively studied in simple two-dimensional or three-dimensional assay systems, recent studies using intact tissue in organ culture have validated many of the observations resulting from *in vitro* studies in simple two-dimensional or three-dimensional assay systems.

Keywords: Fibroblasts, Keratinocytes, Tumor cells, Fibronectin, Laminin, Motility, Invasion, Organ culture

Introduction

Motility in higher eukaryotic cells is a fundamental, though complex, biological process. In development, embryonic cells travel long distances through the tissues before eventually taking up the positions they will occupy in the adult organism. Wound-healing is another biological process in which motility is critical. During wound healing, epithelial cells, fibroblasts and endothelial cells, which are quiescent in the healthy adult tissue, become activated as a result of wounding. Active movement in two and three dimensions ensues. Wound-healing in the skin has been most well-studied,

The Department of Pathology, University of Michigan,
Ann Arbor, MI 48109 USA

for obvious reasons, but healing of wounds in other tissues and organs is thought to proceed along similar lines. In pathological wound healing conditions (for example, fibrosis) events that underlie motility in a normal wound-repair situation also occur, but regulation of the process is defective. Motility is also thought to play a role in local spread of malignant tumor cells though adjacent tissues as well as to the formation of distant metastasis.

While cell motility is assumed to be critical to a number of *in vivo* biological processes, direct evidence is, in most cases, lacking. This is due to the nature of motility and the limitations of studying *in vivo* a dynamic process that occurs over hours to days. Our ability to make measurements that unequivocally demonstrate the importance of motility in, for example, tumor cell invasion is limited. Thus, we have had to rely on surrogate markers and indirect evidence. While indirect and largely circumstantial, this evidence is, nonetheless, quite convincing. What is the evidence that indicates a critical role for cell motility in tumor invasion? First, there is morphological and ultrastructural evidence. Precise examination of motile fibroblasts and epithelial cells *in vitro* has demonstrated morphological features of actively moving cells that are not observed in stationary cells. Specifically, *in vitro* motile cells demonstrate a wide ruffled membrane with multiple focal adhesions at the leading edge and a much narrower membrane at the tail. Cytoskeletal elements are attached at focal adhesion sites and oriented in the plane of movement (Brakebusch et al. 2002; Friedl and Wolf 2003; Hood and Cheresh 2002). These same morphological features are observed in invasive tumor cells at the invasive front *in vivo* (Carr 1983).

Second, while it is difficult to directly observe motility occurring *in vivo* under most circumstances, there are exceptions. The process of neutrophil egress from the vasculature into extravascular tissue space can be observed by time-lapse photography. Using transparent tissues such as rodent mesenteric vessels, one can actually watch as leukocytic cells attach to the endothelial cell surface and then move by a process known as diapedesis between endothelial cells, out of the vascular bed and into the tissue space (Devreotes and Zigmond 1988). Such studies have been instrumental in demonstrating the role of various cytokines, chemokines and adhesion factors in the migratory

process. Such studies have also been instrumental in demonstrating that factors regulating cell motility *in vivo* have a dramatic (motility-inducing) effect on the same cells in sophisticated *in vitro* motility assays. Experimental studies of a similar nature have demonstrated analogous behavior by circulating tumor cells (Woods 1958; Farina 1998). While demonstrating motility in such models is clearly not the same as demonstrating motility during tissue invasion *in situ*, these studies provide direct evidence that motility can occur *in vivo*.

Additional evidence that motility is important for tumor invasion comes from correlative studies. Numerous studies over the past 25 or more years have demonstrated that experimental tumors contain mixtures of cells with different phenotypic characteristics. From many different tumors, subpopulations of cells that differ in their capacity for motility *in vitro* have been selected. When such cells are examined for their capacity to form invasive tumors upon injection into animals, there is virtually a 100% correlation between motility and invasive capacity (Hart 1979; Varani et al. 1985). Among the same populations, there is essentially no correlation between motility and tumorigenicity.

In summary, it is well accepted that cell motility plays an important role in many biological processes, including tumor cell invasion and metastasis. This is in spite of the fact that, with rare exceptions, there is little direct proof *in vivo* to implicate motility. Given the assumed importance of motility in various biological processes, it is not surprising that extensive efforts have been made to assess and understand the process of cell movement and how it is regulated. Most of the effort has involved *in vitro* experimentation. This review focuses on what is known about the role of motility in the invasion process, primarily as it is carried out by cells from fibroblastic and epithelial tumors.

In vitro regulation of motility in fibroblasts and epithelial cells

Motility in malignant cells is not uniquely a function of the malignant phenotype. Quite the opposite, malignant cells mimic their normal counterparts in their motility response. Mesenchymal cells exist in

nature primarily as single cells. Both normal mesen-chymal cells and their malignant counterparts migrate as single cells *in vitro*. Fibroblasts migrate over and through a variety of different substrates, and do so quite robustly in the absence of soluble motility-stimulating factors. The key is a substratum that either contains a matrix component with which the cells can interact or on which the cells can deposit their own matrix. In contrast, epithelial cells form cohesive units. Epithelial cell migration reflects this. Migration of epithelial cells occurs as sheets of cells, as when cells at a wound edge migrate to close a wound, or as finger-like projections of cells during epithelial tumor invasion. Single cell migration of epithelial cells also occurs, but this is usually found in advanced tumors, where cohesion between cells has been substantially disrupted. Given these multiple patterns of migra-tion, are there underlying principles? The following sections of this review focus on two issues – (i) the nature and role of the migration-supporting substrate and (ii) the role of soluble migration-promoting factors – in supporting mesenchymal and epithelial cell movement.

Motility in normal and malignant fibroblasts

Neither fibroblasts nor epithelial cells migrate effi-ciently *in vitro* unless the migration substrate contains one or more extracellular matrix components with which the cells can interact. For fibroblasts, fibronec-tin appears to be among the most potent migration-stimulating factors (Yamada et al. 1978; Roberts et al. 1988; Ryseck et al. 1989), although other matrix components including structural proteins (collagens and elastin) also support motility (Stetler-Stevenson et al. 1993). When fibroblasts are plated onto a sub-strate that contains fibronectin or laminin, migration is rapidly initiated. However, if there is insufficient matrix to stimulate migration efficiently, cell move-ment still occurs – only more slowly. Synthesis of the supportive matrix and deposition of it into the substrate is a necessary pre-requisite for movement.

Fibronectin was discovered in the late 1960s to be a major cell-surface protein on fibroblasts. In early stud-ies, this protein was of interest because it (i) mediated cell–substrate attachment and (ii) was lost from the surface of transformed cells (Hynes 1976; Olden and Yamada 1977; Vahari and Mosher 1978). Based on these observations, it was easy to see how the malig-nant phenotype, characterized by a loss of adhesive functions, might reflect the lack of fibronectin on the surface of transformed cells. Loss of tumor adhesive function and the capacity for invasion through tis-sues was interpreted to reflect loss of this adhesion protein.

Efforts to understand how fibronectin was lost from the cell surface during transformation revealed that fibronectin receptor expression (identified as the α5β1 integrin) was lacking on transformed cells (Plantefaber and Hynes 1989). The absence of the α5β1 integrin not only correlated with a lack of surface fibronectin but was also directly implicated in decreased binding of exogenous fibronectin by transformed cells. The response to exogenous fibronectin was not absolute and this was revealed to be due to the presence of other (though less efficient) fibronectin-binding receptors (such as α4β1 and α3β1) on the transformed cell surface (Plantefaber and Hynes 1989). Subsequent studies revealed a more complex relationship between surface fibronectin and fibronectin receptor expression. Cells lacking the α5β1 fibronectin receptor were shown to down-regulate fibronectin biosynthesis (Senger et al. 1983; Chakrabarty et al. 1989a; Chakrabarty et al. 1987; Varani and Chakrabarty 1990). When cells expressing a conditional "malignant phenotype", were induced to express normal characteristics, fibronectin recep-tor expression and fibronectin biosynthesis were induced in parallel (Senger et al. 1983; Chakrabarty et al. 1989a; Chakrabarty et al. 1987; Varani and Chakrabarty 1990). Likewise in normal cells, factors that induced fibronectin expression on the cell sur-face induced both increased receptor expression and increased ligand biosynthesis (Roberts et al. 1988; Ryseck et al. 1989; Shumaker et al. 1994).

It might seem intuitive that if fibronectin were an important motility-supporting substrate for fibroblasts and transformed cells were less efficient than their normal counterparts in binding fibronectin, they would also be less motile. Several factors contribute to mak-ing this not the case. First, while fibroblast motility is a function of cell–substrate adhesion, the two proc-esses are not the same. Motility requires adhesion to a substratum, but if the attachment is too strong, cells cannot break the attachment and, therefore, do not move efficiently. By reducing adhesion to fibronectin

(but not eliminating it altogether) transformation could allow the cells to express a motile phenotype under conditions in which the normal counterpart cells remain sessile. Second, fibronectin is a large, complex molecule with several distinct domains. Fragments of the intact molecule have biological effects (including motility-inducing activity) not found in the intact molecule. Third, and, perhaps most important, motility is a complex process with a unique set of intracellular signals. Interaction with fibronectin through $\alpha5\beta1$ results in generation of a different set of intracellular signals than interaction through other fibronectin-binding integrins (Hynes 1992).

Fibronectin is not the only component of the extracellular matrix to influence fibroblast motility. Another matrix component, laminin, also supports migration efficiently. In early studies it was demonstrated that purified laminin from a basement membrane-producing rodent tumor supported attachment and motility in transformed fibroblasts as well as (or better than) their normal counterpart cells (Terranova et al. 1982; Rao et al. 1983; Fligiel et al. 1986). Laminin receptors, primarily the $\alpha6\beta1$ integrin, but including other integrin and non-integrin moieties, were demonstrated to be expressed on cells that migrated in response to laminin (Brakebusch et al. 2002; Hood and Cheresh 2002; Terranova et al. 1982; Rao et al. 1983; Friedl et al. 1998). Since laminin is a major component of basement membranes, it could be understood how the ability of a cell to interact with this basement membrane component could promote invasion by malignant cells. Unlike fibronectin, however, laminin was not lost from the surface of malignant fibroblasts. Quite the contrary, several studies demonstrated that laminin expression was often increased on transformed cells (Chakrabarty et al. 1989a; Chakrabarty et al. 1987; Varani et al. 1983; Bober et al. 1987; Ramlpoldi et al. 1985). Since there was a correlation between laminin expression on the cell surface (as opposed to unoccupied laminin receptors) and motility, this suggested that migration on laminin might involve the utilization of newly synthesized laminin laid down as part of the extracellular matrix to provide the substrate for migration of the advancing cells.

In summary, both fibronectin and laminin support fibroblast migration, and malignant as well as normal fibroblasts are capable of utilizing either matrix component as a migration substrate. With regard to fibronectin, there is a loss of the matrix component from the surface of transformed cells associated with a loss of receptor. However, while the major fibronectin-binding receptor is lost from transformed cells, other receptors with fibronectin-binding capacity remain. In contrast, laminin is not lost from the cell surface in transformed fibroblasts, and in many cases, increased cell surface expression is observed. How can these disparate findings be tied into a cogent picture of how migration is influenced by the matrix? The simplest interpretation is that both fibronectin and laminin provide a supportive matrix for fibroblast migration. Fibronectin is plentiful in the extracellular matrix – presumably in sufficient amount to support migration of transformed fibroblasts without continued elaboration. Laminin is present in smaller amounts. However, fibroblastic cells – and in particular, transformed fibroblasts – synthesize the supportive matrix as they crawl along.

Motility in normal and malignant epithelial cells

Whereas fibroblasts exist as single cells in most circumstances, epithelial cells normally reside in cell sheets or as three-dimensional masses. At one extreme are squamous epithelial cells, consisting of large, flattened cells that adhere tenaciously to one another as well as to the underlying basement membrane and form three-dimensional, multi-layered structures. At the other extreme are secretory epithelial cells that are relatively less adhesive and form sheets of cells consisting of a single cell layer of columnar or cuboidal cells. Not surprising, adhesion in epithelial cells is a complex process and involves cell–cell as well as cell–substrate interactions (Clark et al. 1982; Varani et al. 1987; Nickoloff et al. 1988). Adding a further dimension to the complexity of epithelial cell adhesive function is the fact that progressive changes in adhesive properties accompany the process of differentiation. Finally, as epithelial tumors progress, adhesive properties undergo still more changes. In advanced carcinomas, the epithelial cells are often described as have a "spindle-shape" morphology (epithelial–mesenchymal transformation). Such cells appear similar to fibroblasts and exist as single cells. Cell–cell adhesion functions are lost in such cells.

Given the complexity in epithelial cell adhesive functions, it is not surprising that regulation of motility

in these cells should also be complex. Epithelial cells can migrate in two-dimension (for example, across the surface of the basement membrane or provisional matrix during wound-repair) as well as in three-dimension (during tumor invasion). In spite of these complexities, there are some unifying principles. First (like fibroblasts), epithelial cell migration is dependent on the presence of a substrate to which the cells can adequately adhere. While a number of different matrix components are able to support epithelial cell movement, fibronectin provides an optimal substrate, either for migration in two dimensions or three dimensions (Varani et al. 1987; Nickoloff et al. 1988).

Studies conducted in our laboratory have examined fibronectin-mediated adhesion and motility in human squamous epithelial tumor cells (Frenette et al. 1988; Varani et al. 1991; Chakrabarty et al. 1989b). Initially, we utilized emerging data from mesenchymal cell studies as a guidepost. It quickly became apparent however (not surprising) that while fibronectin provided an optimal migration substrate for epithelial cells as well as fibroblasts, cell–substrate interactions were not identical. In squamous epithelial cells, fibronectin was not lost from the cell surface during transformation. Quite the contrary, fibronectin biosynthesis and fibronectin receptor ($\alpha5\beta1$) expression were both highly expressed on these cells and this was correlated with utilization of the matrix component for both attachment and migration (i.e. antibodies that interfered with fibronectin interactions decreased adhesion and motility in parallel). Normal human epithelial cells (epidermal keratinocytes) as well as malignant squamous epithelial cells synthesized large amounts of fibronectin and utilized the endogenously produced fibronectin as an adhesion and migration-supporting substrate. When normal epithelial cells were induced to differentiate, responsiveness to fibronectin decreased. This was associated with a reduction in fibronectin biosynthesis and a corresponding reduction in surface $\alpha5\beta1$ integrin (Chakrabarty et al. 1989b). When malignant squamous epithelial cells were treated with the same differentiation inducing protocol, differentiation failed to occur in some tumor lines and was incomplete in others (Varani et al. 1991; Chakrabarty et al. 1989b; Chakrabarty et al. 1990). Fibronectin production and fibronectin receptor expression remained high and responses to fibronectin were sustained in these cells.

We next turned our attention to adenocarcinoma cells – i.e. colon carcinoma cell – thinking that findings from studies with squamous epithelial cells would be helpful. Again it was found that behavior in one cell type (in this case, squamous epithelial cells) was not predictive of behavior in another (i.e. in adenocarcinoma cells). Specifically, it was observed that adenocarcinoma cells synthesized small amounts of fibronectin (and laminin) and did not attach or spread well on either matrix component (Nabeshima et al. 1998; Shimao et al. 1999; Inoue et al. 2001). Likewise, motility was deficient on these substrates. To make matters more perplexing, when these cells were induced to differentiate, matrix biosynthesis increased along with a corresponding increase in responsiveness (Nabeshima et al. 1998; Shimao et al. 1999; Inoue et al. 2001). Thus, the behavior of the colonic adenocarcinoma cells was almost the opposite of what was anticipated, based on findings squamous epithelial cells. Studies by other investigators with squamous carcinoma cells or adenocarcinoma cells from a number of different tumor types have demonstrated similar relationships between fibronectin production and cellular motility (Clark et al. 1982; Varani et al. 1987; Nickoloff et al. 1988; Smith et al. 1981; Taylor-Papadimitriou et al. 1985; Brown and Parkinson 1985; Akiyama et al. 1990; Clark et al. 1985).

Motility in normal and malignant epithelial cells: role of soluble motility-supporting factors

While mesenchymal cells and epithelial cells are both capable of migration *in vitro*, a major difference between the two cell types is that, *in vitro* at least, fibroblast movement occurs in a serum-free, growth factor-free environment while epithelial cell movement does not. Even when a supportive substrate is present, there is essentially no movement of epithelial cells in the absence of a soluble motility-promoting factor. While numerous motility-promoting factors have been identified (Barrandon and Green 1987; Jeffers et al. 1996), not all of these factors are equal. For example, studies in our laboratory examined a series of epithelial cell growth factors for ability to stimulate motility in normal human epidermal keratinocytes. Each of four factors was examined under conditions in which the growth-promoting activity was equivalent. Two of the factors, epidermal growth factor (EGF) and

hepatocyte growth factor (HGF; also known as scatter factor) proved to be potent stimulators of motility. Two other epithelial cell growth factors – insulin-like growth factor-1 (IGF-1) and keratinocyte growth factor (KGF) – were less effective, even when used at concentrations that were equipotent to EGF and HGF in terms of proliferation (Zeigler et al. 1996b). In the same studies it was demonstrated that although fibroblasts migrated actively in the absence of soluble factors, growth factors to which they respond (including EGF and IGF-1) promoted an incremental increase in migration. Thus, it is not that mesenchymal cells do not respond to motility-inducing factors; it is, rather, that they can mount a substantial response even in their absence. Epithelial cells, on the other hand, appear to be dependent on such signals.

Invasion = Motility + Tissue Destruction

There is another component to invasion – i.e. tissue destruction. Some degree of tissue destruction accompanies invasion with all solid tumors. In some cases, the degree of tissue destruction is minimal, but in other cases, it may be extensive. Basal cell tumors of the skin, which rarely metastasize to distant sites, often produce extensive destruction of the surrounding tissue – including cartilage and bone (Brodland 1998). Research over the past 2 decades has shown that tissue destruction during invasion is dependent on proteolytic enzymes – including members of the serine proteinase family (i.e. tissue plasminogen activator and urokinase-type plasminogen activators), certain of the cysteine proteinases (cathepsins) and, most important, various members of the matrix metalloproteinase (MMP) family (Koblinski et al. 2000). Studies using *in vitro* invasion assays (for example, matrix-coated filters in transwell- or Boyden-type chambers) have demonstrated inhibition of invasion with proteolytic enzyme inhibitors under conditions in which migration through uncoated filters or migration in two-dimensions is unaffected (Horine et al. 2001). This suggests that the two processes – i.e. active motility in the invading cell population and tissue destruction – are distinct biological events. Nevertheless, both are critical to the invasion process. Furthermore, while distinct, they intersect at several points in the invasion process. For example, damage to the connective tissue exposes matrix components

that are able to stimulate motility. Additionally, an overlapping cascade of intracellular signaling events leads to motility and, concomitantly, to the up-regulation of tissue-destructive enzymes of the MMP family (following section).

Intracellular signaling through mitogen-activated protein (MAP) kinase pathways: relations to motility and MMP elaboration during tissue invasion

We have used the differential motility response of human epidermal keratinocytes to various growth factors to help elucidate the signaling events that underlie epithelial cell motility (Zeigler et al. 1996a, b, 1999). To summarize, factors that effectively stimulate epithelial cell motility (as well as proliferation) including EGF and HGF induce sustained signaling through the extracellular signal-related kinase (Erk) and jun-N-terminal kinase (Jnk) MAP kinase pathways, while factors that are effective in stimulating proliferation but not motility in the same cells (i.e. IGF-1 and KGF) induce transient signaling only. The same growth factors that stimulate motility and proliferation also induce production of MMP-9, while factors that are ineffective in inducing motility are, likewise, ineffective in promoting MMP-9 production. Of interest, while sustained signaling through both the Erk and Jnk pathways accompany induction of motility and MMP-9 production, it appears that sustained stimulation of the Erk pathway alone is sufficient to induce motility. This is based on the finding that transfection of a Jnk-dominant-negative construct into epithelial cells, has no effect on the motility response in spite of profound inhibition of Jnk signaling (Zeigler et al. 1999). In contrast, induction of MMP-9 production appears to depend on both pathways since in either Jnk-dominant-negative transfected epithelial cells or in cells treated with Erk activation inhibitors, MMP-9 production is blocked (Zeigler et al. 1999). Our suggestion based on these findings (and consistent with past observations; Brakebusch et al. 2002; Friedl and Wolf 2003; Hood and Cheresh 2002) is that Erk activation of myosin light chain kinase (MLCK) (a key step in organizing the acto-myosin contractile system) is involved in motility-induction while stimulation of MMP-9 occurs via an effect on AP-1-mediated gene transcription (Zeigler et al. 1999).

Relationship between *in vitro* motility, tissue destruction and invasion

In vitro studies are of value in that experimental conditions can be readily manipulated to provide information on biological potential. It is imperative, however, that findings made *in vitro* be critically evaluated in light of *in vivo* events. The question which can be legitimately asked is whether *in vitro* motility in simple two-dimensional or three-dimensional assays have anything to do with how tumor cells penetrate the adjacent connective tissue during the process of invasion. To help address this issue, experiments in organ culture were carried out.

For this, we have utilized an "*in situ*" organ culture model to study the invasion process. In this model, 2 mm punch biopsies of healthy human skin are maintained in organ culture under serum-free, growth factor-free conditions or, alternatively, are maintained in the same medium but supplemented with an exogenous growth factor such as EGF. In the absence of exogenous growth factors, normal histological structure and biochemical function are maintained for several days (Fig. 1A and Varani et al. 1993; Varani et al. 1994). However, when a high level of EGF is included in the culture medium, either alone or in conjunction with other epithelial growth-promoting agents, several changes occur in the tissue. The epithelial cells undergo a proliferative response. Strands of epithelial cell grow down into the dermal space (Figs. 1B and 1C). Erosion of the dermal–epidermal basement membrane occurs, and in some areas, the basement membrane is completely lacking (Figs. 1D and 1E). In other areas, isolated epithelial cells are completely surrounded by dermal elements (Fig. 1F). Of particular interest to the present discussion, where epithelial cells are in the process of penetrating the underlying stromal tissue, extensive filopodia are present at the front of the advancing cells (Fligiel and Varani 1993). The morphological and ultrastructural features of the cells are indistinguishable from motile cells in collagen lattice cultures or in other two-dimensional or three-dimensional motility assays.

Additionally, there is up-regulation of several MMPs including MMP-9 under invasion-promoting conditions, and damage to the underlying connective tissue can be seen (Figs. 2C and 2D and Zeigler et al. 1996a; Fligiel and Varani 1993; Varani et al. 1995).

In this model, MMP activity, matrix degradation and tumor penetration into the dermis are all inhibited in the presence of TIMP-2 (Zeigler et al. 1996a). Also of interest, there is a substantial induction of fibronectin in the organ-cultured tissue under conditions in which epithelial cell invasion occurs (Fligiel and Varani 1993). Thus, we conclude that epithelial cell invasion of the stroma in this model (i) is associated with MMP induction and connective tissue breakdown; (ii) depends on elaboration of a motility-supporting matrix; (iii) is associated with morphological evidence of motility in the invading epithelial cells;

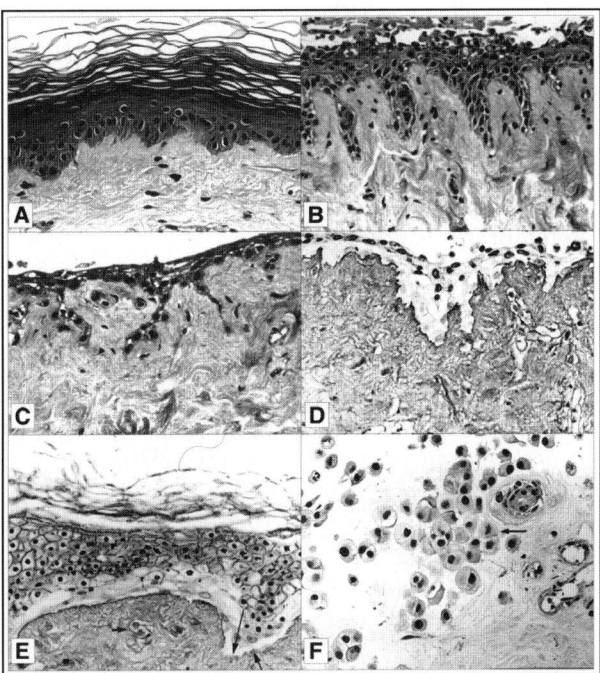

Fig. 1 Stromal invasion by human epidermal keratinocytes in organ-cultured human skin. **A**: Histological appearance of human skin after 8 days in organ culture under serum-free, growth factor-free conditions. **B** and **C**: Histological appearance of human skin after 8 days in organ culture in the presence of EGF. Note the strands of epidermal keratinocytes pushing into dermal space and the abnormalities in epidermal differentiation. Panels **A**-**C** are hematoxylin and eosin-stained. **D** and **E**: Histological appearance of human skin after 8 days in organ culture in the presence of EGF. Panels **D** and **E** were stained with the trichrome stain to enhance features of the basement membrane. It can be seen from panel **D** that the dermal – epidermal basement membrane is rough and pitted. In panel **E**, it can be seen that in places, the basement membrane has been eroded entirely (arrows). **F**: Histological appearance of human skin after 8 days in organ culture in the presence of EGF. Isolated epithelial cells can be seen in the dermis in places. These cells appear to be completely separated from the overlying epidermis. A mitotic figure can be seen (arrow)

and (iv) depends on the presence of epithelial cell growth factors in the culture medium. Taken together, these findings provide a strong correlation between epithelial cell movement in simple two-dimensional and three-dimensional *in vitro* migration assays and capacity of epithelial cells to penetrate the surrounding connective tissue in an organ culture setting. To the extent that organ culture invasion by growth factor-stimulated normal epithelial cells mimics the invasion process as carried out by malignant tumor cells *in vivo*, we can suggest similar mechanistic events. Finally, since abnormalities in the EGF receptor system are among the most common defects in epithelial cell tumors (Klapper et al. 2000; O-charoenrat et al. 2002), we hypothesize that events triggered through EGF receptor family members are central to the invasion process in intact tissue.

References

Akiyama SK, Larjava H, Yamada K. (1990) Differences in the biosynthesis and localization of the fibronectin receptor in normal and transformed cultured human cells. Cancer Res 50:1601–1607

Barrandon Y, Green H (1987) Cell migration is essential for sustained growth by transforming growth factor-a and epidermal growth factor. Cell 50:1131–1137

Bober FJ, Birk DE, Raska K (1987) Expression of varying portions of the adenovirus 12 early region 1 in transformal cells affects tumorigenicity and interaction with extracellular matrix components. Lab Invest 56:39–43

Brakebusch C, Bouvard D, Stanchi F, Sakai T, Fassler R (2002) Integrins and invasive growth. J Clin Invest 109:999–1006.

Brodland DG (1998) Basal cell carcinoma: features associated with metastasis. In: Miller SJ, Maloney ME (eds) Cutaneous oncology: pathophysiology, diagnosis and management.. Blackwell, Malden MA, pp 657–663

Brown KW, Parkinson EK (1985) Alteration of the extracellular matrix of cultured human keratinocytes by transformation and during differentiation. Int J Cancer 35:799–807

Carr I (1983) Experimental lymphatic metastasis. J Microsc 131:211–220

Chakrabarty S, Brattain MG, Ochs RL, Varani J (1987) Modulation of fibronectin, laminin and cell adhesion in the transformation and differentiation of murine AKR fibroblasts. J Cell Physiol 133:415–425

Chakrabarty S, Fan D, Varani J (1990) Modulation of differentiation and proliferation in human colon carcinoma cells by transforming growth factor β_1 and β_2. Int J Cancer 46:493–499

Chakrabarty S, Jan Y, Brattain MJ, Tobon A, Varani J (1989a) Transforming growth factor-β elicits diverse cellular responses from human colon cancer cells. Cancer Res 49:2112–2117.

Chakrabarty S, Jan Y, Levine A, McClenic B, Varani J (1989b) Fibronectin/laminin and their receptors in aberrant growth control in Ha-ras oncogene transformed and epidermal growth factor gene transformed FR3T3 cells. Int J Cancer 44:325–331

Clark RAF, Lanigan JM, Della Pelle P, Manseau E, Dvorak HF, Colvin RB (1982) Fibronectin and fibrin provide a provisional matrix for epidermal cell migration during wound healing. J Invest Dermatol 79:264–269

Clark RAF, Nielson LD, Howell SE, Folkvord JM (1985) Human keratinocytes that have not terminally differentiated synthesize laminin and fibronectin but deposit only fibronectin in the pericellular matrix. J Cell Biochem 28:127–141

Devreotes PN, Zigmond SH (1988) Chemotaxis is eukaryotic cells: a focus on leukocytes and Dictyostelium. Ann Rev Cell Biol 4:649–686

Farina K (1998) Cell motility of tumor cells visualized in living intact primary tumors using green fluorescent protein. Cancer Res 58:2528–2532

Fligiel SEG, Laybourn KA, Peters BP, Ruddon RW, Hiserodt JC, Varani J (1986) Laminin production by murine melanoma cells: possible involvement in cell motility. Clin Exp Metastasis 4:259–272

Fligiel SEG, Varani J (1993) *In situ* epithelial cell invasion in organ culture. Invasion Metastasis 13:225–233

Frenette GP, Carey TE, Varani J, Schwartz DR, Fligiel SEG, Ruddon RW, Peters BP (1988) Biosynthesis and secretion of laminin and laminin-associated glycoproteins by nonmalignant and malignant human keratinocytes: a comparison of cell lines from primary and secondary tumors in the same patient. Cancer Res 48:5193–5202

Friedl P, Danker KS, Brocker EB (1998) Cell migration strategies in 3-D extracellular matrix: differences in morphology, cell-matrix interactions and integrin function. Micros Res Tech 43:369–378

Friedl P, Wolf K (2003) Tumor cell invasion and migration: diversity and escape mechanisms. Nature Rev (Cancer) 3:362–374

Hart IR (1979) The selection and characterization of an invasive variant of the B16 melanoma. Am J Pathol 97:585–600

Hood JD, Cheresh DA (2002) Role of integrins in cell invasion and migration. Nature Revs (Cancer) 2:91–100

Horine K, Kindezelskii AL, Elner VM, Hughes BA, Petty HR (2001) Tumor cell invasion of 3-dimensional matrices: demonstration of migratory pathways, collagen disruption and intercellular cooperation. FASEB J 15:932–939

Hynes RO (1976) Cell surface proteins and malignant transformation. Biochem Biophys Acta 458:73–78

Hynes RO (1992) Integrins: Versatility, modulation and signaling in cell adhesion. Cell 69:11–25

Inoue T, Nabeshima K, Shimao Y, Meng JY, Koono M (2001) Regulation of fibronectin expression and splicing in migrating epithelial cells: migrating MDCK cells produce lesser amounts of, but more active fibronectin. Biochem Biophys Res Commun 280:1262–1268

Jeffers M, Rong S, Vande WG (1996) Enhanced tumorigenicity and invasion/metastasis by hepatocyte growth factor/scatter factor-met signaling in human tumor cells with induction of urokinase proteolysis network. Mol Cell Biol 16:1115–1125

Klapper LN, Kirschbaum MH, Sela M, Yarden Y (2000) Biochemical and clinical implications of the ErbB/HER signaling network of growth factor receptors. Adv Cancer Res 77:25–79

Koblinski JE, Ahram M, Sloane BF (2000) Unraveling the role of proteinases in cancer. Clin Chim Acta 291:113–135

Nabeshima K, Inoue T, Shimao Y, Kataoka H, Koono M (1998) TPA-induced cohort migration of well-differentiated human renal adenocarcinoma cells: cells move in a RGD-dependent manner on fibronectin produced by cells. Virchows Arch 433:243–253

Nickoloff BJ, Mitra RS, Riser BL, Dixit VM, Varani J (1988) Modulation of keratinocyte motility; correlation with production of extracellular matrix molecules in response to growth promoting and anti-proliferative factors. Am J Pathol 132:543–551

O-charoenrat P, Rhys-Evans PH, Modjtahedi H, Eccles SA (2002) The role of c-erbB receptors and ligands in head and neck squamous cell carcinoma. Oral Oncol 38:627–640

Olden K, Yamada KM (1977) Mechanisms of the decrease in the major cell surface protein of chick embryo fibroblasts after transformation. Cell 11:957–969

Plantefaber LC, Hynes RO (1989) Changes in integrin receptor expression on oncogenically transformed cells. Cell 56: 281–290

Ramlpoldi E, Larizza L, Doneda L, Barlato S (1985) Fibronectin and laminin in hybrids of rous sarcoma virus – transformal and normal mouse fibroblasts. Tumor 71:419–423

Rao CN, Barsky SH, Terranova VP, Liotta LA (1983) Isolation of a tumor cell laminin receptor. Biochem Biophys Res Commun 111:804–808

Ryseck RP, McDonald-Bravo H, Zerial M, Bravo R (1989) Coordinate expression of fibronectin, fibronectin receptor, tropomyosin and actin genes in serum-stimulated fibroblasts. Exp Cell Res 180:537–545

Senger DR, Destree AT, Hynes RO (1983) Complex regulation of fibronectin synthesis by cells in culture. Amer J Physiol 245:144–150

Shimao Y, Nabeshima K, Inoue T, Koono M (1999) Role of fibroblasts in HGF/scatter factor-induced cohort migration of human colorectal cells: fibroblasts stimulate migration associated with increased fibronectin production via upregulated TGF-beta 1. Int J Cancer 82:449–458

Shumaker DK, Sklar MD, Prochownik EV, Varani J (1994) Increase cell-substrate adhesion accompanies conditional reversion to the normal phenotype in ras-oncogene-transformed NIH-3T3 cell. Exp Cell Res 214:440–446

Smith HS, Riggs JL, Mosessan MW (1981) Production of fibronectin by human epithelial cells in culture. Cancer Res 39:4138–4144

Stetler-Stevenson WG, Aznavoorian S, Liotta LA (1993) Tumor cell interactions with the extracellular matrix during invasion and metastasis. Ann Rev Cell Biol 9:541–573

Taylor-Papadimitriou J, Burchell J, Hurst J (1985) Production of fibronectin by normal and malignant human mammary epithelial cells. Cancer Res 41:2491–2450

Terranova VP, Liotta LA, Russo RG, Martin GR (1982) Role of laminin in the attachment and metastasis of murine tumor cells. Cancer Res 42:2265–2269

Vahari A, Mosher DF (1978) High molecular weight cell surface protein (fibronectin) lost in malignant transformation. Biochim Biophys Acta 516:1–9

Varani J, Chakrabarty S (1990) Modulation of fibronectin synthesis and fibronectin binding during transformation and differentiation of mouse AKR fibroblasts. J Cell Physiol 143:445–454

Varani J, Carey TE, Fligiel SEG, McKeever PE, Dixit V (1987) Tumor type-specific differences in cell-substrate adhesion. Int J Cancer 39:397–403

Varani J, Fligiel SEG, Perone P (1985) Directional motility in malignant murine tumor cells. Int J Cancer 35:559–564

Varani J, Fligiel SEG, Schuger L, Perone P, Inman DR, Griffiths CEM, Voorhees JJ (1993) Effects of all-trans retinoic acid and Ca^{2+} on human skin in organ culture. Am J Pathol 142:189–198

Varani J, Lovett EJ, McCoy JP, Shibata S, Maddox DE, Goldstein IJ, Wicha M (1983) Differential expression of a laminin-like substance by high and low metastatic tumor cells. Am J Pathol 111:27–34

Varani J, Perone P, Griffith CEM, Inman DR, Fligiel SEG, Voorhees JJ (1994) All-trans-retinoic acid (RA) stimulates events in organ cultured human skin that underlie repair. J Clin Invest 94:1747–1756

Varani J, Perone P, Inman DR, Burmeister W, Schollenberger SB, Sitrin RJ, Johnson KJ (1995) Human skin in organ culture: Elaboration of proteolytic enzymes under growth factor-free and growth factor-containing conditions. Am J Pathol 146:210–217

Varani J, Schuger L, Fligiel SEG, Inman DR, Chakrabarty S (1991) Production of fibronectin by human tumor cells and interaction with exogenous fibronectin. Comparison of cell lines obtained from colon adenocarcinomas and squamous carcinomas of the upper aerodigestive tract. Int J Cancer 47:421–425

Woods S (1958) Pathogenesis of metastasis formation observed in the rabbit ear chamber. Arch Pathol 66: 550–558

Yamada KM, Olden K, Pastan I (1978) Transformation sensitive cell surface protein: isolation, characterization and role in cellular morphology and adhesion. Ann NY Acad Sci 312:256–267

Zeigler ME, Chi Y, Schmidt T, Varani J (1999) The role of ERK and JNK signaling pathways in growth factor-simulated cell invasion. J Cell Physiol 180:271–284

Zeigler ME, Dutcheshen NT, Gibbs DF, Varani J (1996a) Growth factor-induced epidermal invasion of the dermis in human skin organ culture: expression and role of matrix metalloproteinases. Invasion Metastasis 16:11–18

Zeigler ME, Krause S, Karmiol S, Varani J (1996b) Growth factor-induced epidermal invasion of the dermis in human skin organ culture: dermal invasion correlated with epithelial cell motility. Invasion Metastasis 16:3–10

H.E. Kaiser and A. Nasir (eds.), Selected Aspects of Cancer Progression:
Metastasis, Apoptosis and Immune Response, 21–32.
© Springer Science + Business Media B.V. 2008

CHAPTER THREE

Cell adhesion and invasion during secondary tumor formation: interactions between tumor cells and host organs

Peter Gassmann[1], Jörg Haier[1,2], and Garth L. Nicolson[2]

Abstract: To form clinically evident metastases, the main cause of death in cancer patients, cancer cells must complete a highly complex series of steps called the metastatic cascade. This includes local invasion, intravasation, transport in the circulation, adhesion and extravasation, survival, proliferation and angiogenesis. Since failure to complete any one of these steps results in metastatic failure, understanding the steps of the metastatic cascade may allow us to develop new therapeutic approaches for the treatment of cancer. Here we review the role of specific tumor cell (TC) adhesion and migration processes in organ-selective metastases formation. TC adhesion in the microvasculature of host organs is a specific and highly regulated process. Important adhesion molecules are: selectins in TC-endothelial cell adhesion and integrins in TC–extracellular matrix interactions. Defined expression of adhesion molecules and their corresponding ligands can govern this non-random process of organ-selective metastasis formation. Once TC adhesion in the microcirculation occurs, early extravasation of TCs, which is regulated by chemotactic stimuli of growth factors and/or chemokines in an organ-specific manner, can be observed.

[1]Molecular Biology Laboratory, Department General Surgery, University Hospital Münster, Germany
[2]Institute for Molecular Medicine, Huntington Beach, CA, USA

Keywords: Cell adhesion, Invasion, Migration, Metastasis, Chemotaxis, Host organs

Introduction

The ability to metastasize is the most fearsome aspect of cancer and most cancer deaths are the sequel of metastatic diseases rather than primary tumor growth. In many cases each metastatic focus appears to originate from a single tumor cell (TC) and is therefore monoclonal, at least at the beginning of the metastatic process. In order to form overt metastases, a cell must complete the metastatic cascade, a series of well-defined steps including local invasiveness, cell detachment, vascular invasion (hematogenous or lymphogenic), circulation or transport, cell arrest, extravasation, survival, proliferation and angiogenesis. The inability of a TC to complete any one step of this cascade results in metastatic failure (Nicolson 1989, 1988c). Despite the fact that even small tumors can release a high number of TCs into the circulation (Glaves et al. 1988; Glaves et al. 1986), the number of evident metastases remains unexpectedly low and several studies indicate that less than 0.01% of circulating TCs will form metastatic lesions (Fidler 1970; Liotta et al. 1974; Weiss 1986). Furthermore, the pattern of metastasis formation is not random. The reasons for the organ-specific distribution of metastases have been a matter of debate for more than a century since Stephen Paget postulated his 'seed-and-soil' theory (Paget 1889) that was later challenged by James Ewing in 1928 (Ewing 1928), proposing the pattern of metastatic growth being a result of blood supply and mechanical TC arrest.

The metastatic cascade, its inefficiency and its organ selectivity raise the question of potential thera-peutic targets for future therapies precluding the formation of overt metastases by circulating TCs. The most promising therapeutic targets may be those that are most highly regulated and most specific for TCs. These considerations have triggered further research to understand the mechanisms of each step of the metastatic cascade. Here we will review the role and mechanisms of TC adhesion and migration in meta-static target organs and their role in the formation of metastatic lesions.

Mechanisms of metastasis formation

Tumor cell arrest at distant organ sites

During the last decade various groups have reported diverging results using intravital microscopy tech-niques for the observation of the early steps of metas-tasis formation. Mechanical arrest was observed by different groups in experimental studies where human TCs were injected intra-venously in different animal models (Ding et al. 2001; Steinbauer et al. 2003; Mook et al. 2003). While multi-cellular emboli do arrest in the microcirculation in experimental metas-tases models, parental tumors usually release single TCs into the circulation that can easily pass through the microvasculature of different organs (Nicolson 1988c; Haier et al. 2003) (and our own unpublished observations). Circulating TCs must therefore estab-lish specific adhesions to the vascular wall of target organs as observed in different experimental settings (Nicolson 1988c; Glinsky et al. 2003; Glinkii et al. 2003). This hypothesis is in line with clinical obser-vations correlating the expression pattern of adhesion molecules on TCs and in metastatic lesions to differ-ent host tissues and to the clinical course of cancer patients (Haier et al. 2000).

The first step for successful circulating TCs in the blood is to establish adhesive interactions with the microvascular vessel wall of host organs (Nicolson 1988c). Successful cell arrest, however, is dependent on the balance between adhesive and anti-adhesive forces, and the rate at which adhesive interactions are formed and broken (Weis 1992). For example, the adhesion to the extracellular matrix (ECM) usually provides stronger bonds for the TCs than the transient interactions with the endothelium, mostly mediated by selectins and their ligands (Weiss et al. 1988). Once TCs attach to the endothelium, morphological and functional changes in the endothelial cells (EC) can be induced that enable direct interactions with underlying ECM structures. Subsequently, retraction of EC can lead to exposure of the basal membrane that is com-posed of ECM proteins arrayed in an organ-dependent manner. Furthermore, some parenchymatous organs, such as the liver, contain an incomplete EC lining within their microvessels that leaves gaps between the EC where subendothelial ECM is directly accessible to circulating cells (Haier and Nicolson 2001).

Cell–cell and cell–matrix adhesive interactions are complex processes mediated by adhesion molecules and their ligands. Binding of a ligand to its receptor can induce functional and morphological changes of the cellular participants, and even modifications of the expression pattern of adhesion molecules themselves. However, these alterations in the availability, activity and avidity of cell adhesion molecules are usually tightly regulated and can interfere with other cellular signalling processes, such as cell migration, invasion, secretion of degradative enzymes, or cell motility, among others. Moreover, the microenviroment can be influenced by the secretion of degradative enzymes triggered by cell–ECM binding (Westermarck and Kähäri 1999).

Cell adhesion and the endothelium

Although EC from different organs show morphologic similarities, there are distinctive regional and molecular–structural differences in the vascular endothelium (Nicolson 1988c; Auerbach 1991; Pauli et al. 1990; Fenyves et al. 1993). These differences seem to play a role in the organ-specific dissemination of metastatic TCs. For example, the endothelia of different organs display different antigenic patterns (e.g. expression of EC-specific molecules, such as CD31, von Willebrandt-factor, among others) and organ-specific functional

behaviours (e.g. secretion of specific growth factors) (Nicolson 1988c).

Initial contacts between circulating TC and EC ('docking') show similarities to the early events of leukocyte–endothelium interactions. These early contacts are weak and transient and appear to be mediated mainly by selectins and their ligands expressed on leukocytes as well as on cancer cells. Initial adhesions can result in stimulation of the endothelium and TC through cytokines, free radicals, bioactive substances, or growth factors, among others. These mediators themselves can cause specific modulation or activation of adhesion molecules expressed at the surfaces of ECs or TCs, resulting in 'locking' or stabilisation of the initially temporary bonds between cells. This adhesion stabilization is mainly mediated by different members of the integrin adhesion family. Integrin–ECM interactions provide approximately tenfold stronger binding forces compared to selectin-mediated bonds. In addition, they are able to regulate integrin-binding and stabilization processes, such as conformational changes and cross-linking of the adhesion molecules, referred to as inside-out signalling. After successful establishment of cell adhesions, integrin-mediated signal transduction pathways can initiate further endothelial cell retraction, followed by migration of the cancer cell through the underlying basement membrane ECM and invasion into the host organ parenchyma (Crissman et al. 1991).

Selectins and tumor cell adhesion

The selectins are a small group of cell adhesion molecules consisting of E-, P-, and L-selectins. Under physiological conditions E-selectins mediate the "rolling" of neutrophils, T-memory cells and monocytes on the endothelium preceding the formation of stronger adhesions. They can serve as mediators of specific transient attachment at sites of inflammation and modulate the extravasation of leucocytes toward the injured tissue (reviewed in Bird MI [1997]).

All members of the selectin family display similar molecular structures containing an N-terminal domain, a lectin-domain, an EGF-like domain and a varying numbers of complement-binding domains. Each selectin molecule is completed by a trans-membrane region and a short cytoplasmic domain.

The expression profiles and specific interactions of selectins and their main sialylated ligands sLe-a and sLe-x on TC of different entities and EC of potential target organs, respectively, appear to be important for the organ-specific distributions of metastases and correlate with the metastatic potential of several types of cancers.

One of the first lines of evidences for a significant role of selectin-mediated adhesion mechanisms in the metastatic cascade was established when it was shown that the development of experimental liver metastases from human colorectal carcinoma cells was dependent on E-selectins (Brodt et al. 1997). The expression of sLe-x was up-regulated in colon carcinomas compared to normal colon mucosa, and the expression in liver metastasis specimens was higher than in the corresponding parental tumors (Hoff et al. 1989). In two other studies the level of sLe-a and sLe-x expression paralleled the increasing potential for liver metastasis formation of colon cancer cells (Yamada et al. 1997; Sato et al. 1997). Clinical records of patients suffering from colon carcinomas indicated that the disease-free survival of patients with sLe-x-positive tumors was significantly poorer compared to patients with sLe-x-negative tumors (Nakamori et al. 1997).

Interestingly, there seems to be a bi-directional signalling between the TC 'seeds' and the surrounding microenvironmental 'soil'. For example, EC of small vessels adjacent to cancer cell nests, both in primary human colorectal carcinomas and metastatic lesions, displayed increased levels of E-selectin expression. In addition, the rates of selectin expression were inversely correlated with the distance of blood vessels to the cancer cell nests. Furthermore, EC adjacent to metastatic lesions expressed more E-selectin than EC adjacent to the primary tumor. Therefore, selectin expression appears to be inducible by some stimuli from cancer cells (Ye et al. 1995).

Although the results of some studies concerning the involvement of these receptors remain controversial, there are several other carcinomas where the metastatic dissemination appears to be affected by selectin-mediated interactions. For example, MCF-7 and T-47D breast carcinoma cells accumulated and rolled under flow conditions on TNF-stimulated, but not on unstimulated EC monolayers. These adhesive interactions were diminished or completely blocked by treatment of stimulated EC with an anti-E-selectin

antibody (Tozeren et al. 1995). In another study a similar effect was observed on adhesion of MCF-7 breast cancer cells (Lafrenie et al. 1994). In support of the specific role of selectins for the metastatic process the comparison of the expression level of E- and P-selectins and their ligands sLe-a and sLe-x between tissue specimens from metastatic lesions, primary tumors and normal epithelium of patients with breast cancer showed higher selectin levels in metastases compared to primary tumor lesions (Renkonen et al. 1997).

In addition, the importance of selectin-mediated cell adhesion mechanisms for distant organ metastasis formation was demonstrated for gastric (Nakashio et al. 1997; Alexiou et al. 2003), bladder (Numahata et al. 2002) and pancreatic (Iwai et al. 1993) carcinomas, among others. For example, the in vivo investigations of Scherbarth et al. (1997) showed that pre-treatment of B16 melanoma cells with IL-1 (and thereby induction of E-selectin) prior to intravenous injection into the blood circulation of mice induced earlier TC arrest and higher numbers of adherent cells compared to untreated cells (Scherbarth and Orr 1997). Further evidence for the ability of E-selectin to direct TCs to specific sites of the body came from the work of Biancone et al. (1996) The distributions of organ metastases from B16 melanoma cells were compared between normal mice and two transgenic mice that expressed E-selectin either constitutively in every cell or in a soluble truncated form only in the liver, respectively. The injection of TC that were negative or positive for E-selectin ligands resulted in development of lung tumor nodules. In both transgenic groups, however, only E-selectin positive TC resulted in formation of liver metastasis (Biancone et al. 1996).

One of the most interesting characteristics of E-selectin-mediated TC–EC interactions appears to be the influence of TCs on the expression of EC adhesion molecules via induction of inflammatory cytokines. For example, human MDA-MB-231 breast cancer cells were able to induce E-selectin expression on cultured HUVEC cells. This induction was increased in the presence of peripheral blood monocytes and blocked by anti-interleukin 1-β (Scherbarth and Orr 1997). In addition, the in vivo secretion of TNF-alpha and interleukin-1 was enhanced within 30–60 min after injection of cancer cells into the portal vein (Khatib et al. 1999). Furthermore, CEA-producing

cancer cells have been shown to stimulate cytokine production by hepatic Kupffer cells (Gangopadhyay et al. 1998). A more recent study from Kitaka et al. (2002) suggested that, in particular, the TNFα-p55-receptor may play a critical role during formation of liver metastases. After intrasplenic administration of colon carcinoma cells, less than 50% of TNF-Rp55 knockout mice developed liver metastases compared with 90% in wild-type mice (Kitakata et al. 2002).

Integrin-mediated tumor cell adhesion

Mechanical anchorage of cells to the ECM is a physiological prerequisite for tissue structure and function, cellular immunity and various other functions that participate in maintenance of an organism's integrity. Such cell-ECM adhesions in multicellular organisms are mostly mediated by integrins and some other cell adhesion molecules.

Integrins are a group of transmembrane proteins composed of two non-covalently bound α- and β-subunits. Combinations of 15 α- and 9 β chains form more than 25 integrin heterodimers, but β1 and β3 appear to be the most important integrins expressed on cancer cells (Hynes 1987). Both chains are required to determine the specificity of adhesive interactions, and integrin heterodimers are either monospecific, or they can bind to several ECM components. Besides their transmembrane linkage to the actin-cytoskeleton by adaptor proteins that provide the physical strength for adhesive interactions, the intra-cytoplasmic domains trigger intracellular signalling cascades. Integrin-mediated cell adhesion events can thereby regulate various cellular functions like apoptosis, anoikis and cell migration, proliferation and differentiation (Miranti and Brugge 2002).

The crucial role of integrins in cell biology involves a complexity of signalling pathways and cross-talk with other signalling cascades like growth factor or chemokine receptor signalling pathways (Ben-Ze'ev 1994).

Integrins are not only capable of mediating signals into the cell, depending on the extracellular microenvironment, but their interactions can also be modified by signals arising from inside the cell. This "inside-out"-signalling, affects integrin-mediated adhesiveness by modulating the affinity and avidity of integrins for their ligands, and therefore influence cell adhesion and motility. The binding of cytoplasmatic proteins to the intracellular part of integrins has been suggested to induce a shift in integrins' quaternary structure, resulting in the head groups becoming more "open" and available for ligand binding (Loftus and Liddington 1997; Liddington and Bankston 2000). Another mechanism for the modulation of cellular binding ability to the ECM seems to be in the relocalization of integrins to the adhesive structures by lateral mobility and shift of the integrin heterodimers within the cell membrane.

During the formation of metastases integrin-mediated interactions of carcinoma cells with ECM components appear to be among the most important determinants for organ-specific metastasis formation (Nicolson 1989). One characteristic phenomenon during the progression of TCs to the metastatic phenotype is a dramatic change in the repertoire of expressed integrins as well as their affinity and avidity for ECM binding. In immunohistochemical analyses carcinomas, in general, tend to show weaker integrin staining than found in adenomas or normal cells (Bosman et al. 1993; Koukoulis et al. 1993). But during the development of the invasive/metastatic phenotype the patterns of expressed integrins change significantly. Whereas normal colonic mucosa cells do not significantly express $\alpha 5$-integrins, this subunit seems to be up-regulated in invasive colorectal carcinomas (Koretz et al. 1994; Gong et al. 1997). In some systems these changes have been found earlier during carcinogenesis, such as in the transformation of hepatocytes (Volpes et al. 1993; Scoazec et al. 1996), among others. Differences in the surface expression and distribution of integrins in malignant tumors compared with preneoplastic tumors of the same type may be responsible for mediating the distinctive changes in adhesive properties to EC/ECM (Mizejewski 1999) in particular host organs. For example, it was demonstrated that adhesion of highly metastatic cell lines to EC or ECM can occur at higher rates than found with comparable poorly metastatic cells (Nicolson et al. 1989; Haier et al. 1999).

The repertoire of a tumor's integrins can predict the site of metastatic implantation and growth. For example, while liver metastases of gastric carcinomas have been found to be correlated with $\alpha 2\beta 1$- and $\alpha 3\beta 1$-integrin expression, $\alpha 3\beta 1$-integrin expression correlated with the development of peritoneal metastases, and $\alpha 2\beta 1$-integrin expression appeared to

mediate the formation of lymphatic metastases (Hideki et al. 1998). The $\alpha v\beta 3$-integrin (vitronectin receptor) appears to play an important role in metastatic cell growth, and $\alpha v\beta 3$-integrin activation seems to be a requirement for the metastatic phenotype of breast cancer and melanoma cells (Fielding-Habermann et al. 2001). Blocking $\alpha v\beta 3$-integrins on these cells can decrease metastatic cell growth in vivo (Trikha et al. 2002). In cooperation with bone matrix components this integrin may also govern the metastatic colonization of bones as a specific target for metastases of breast and prostate carcinoma cells by mediation of their adhesion and migration (Liapsi et al. 1996; Byzova et al. 2000). In contrast, melanoma cells seem to prefer $\alpha v\beta 3$-mediated cell adhesion to vitronectin in order to colonize lymph nodes (Nip et al. 1992). On the other hand, $\alpha v\beta 3$-integrins were found overexpressed in tumor vasculature (Max et al. 1997), which could indicate an interaction between cancer cells and host organs that promotes tumor angiogenesis.

Moreover, there is growing evidence that plasma proteins and other blood components, such as platelets, cooperate with integrins to mediate TC adhesion in target organs (Fielding-Habermann et al. 1996).

In addition to adhesion, integrins are essential elements of cell signalling. The key participant of integrin-mediated intracellular signal transduction is focal adhesion kinase (FAK). This tyrosine kinase can mediate signal transduction from the areas at the cell surface contact sites (focal adhesions and focal contacts), that interact with ECM and where integrin clustering occurs (Lu et al. 2001). Integrin clustering in focal adhesions induces FAK activation by (auto-) phosphorylation. Activated FAK recruits other signalling molecules and some oncogene products, such as Src-family kinases, to the focal adhesions. The involvement of Src has been shown to be required for integrin-mediated cell motility in various cell types, such as fibroblasts and Chinese hamster ovary (CHO) cells (Sieg et al. 1999; Cary et al. 1996; Owen et al. 1999). For example, Src inhibition resulted in impairment of the cell motility, caused by transgenic FAK expression in FAK-deficient fibroblasts. Additionally, activation of Src led to specific growth factor and ECM signals by recruiting integrin receptors into FAK-containing complexes in response to growth factor receptor ligation (Eliceiri et al. 2002). Moreover, a recent

study by Kohana et al. (2002) demonstrated, that FAK is constitutive activated in metastatic melanoma cells, but not in non-transformed melanocytes. This mechanism may play an important role in constitutive activation of integrins, and thus it may be a significant oncogenic mechanism for acquisition of an aggressive (metastatic) phenotype.

Migration and invasion into host organs

Tumor cell extravasation and migration into host organs

Following the initial adhesive interactions with ECs, TCs have to actively change their shapes by spreading or flattening to increase contact areas and initiate cell migration through the vessel wall (Menter et al. 1995). Active extravasation enables TCs to escape the high shear forces within the center of the parabolic curve of fluid flow in the blood vessel (Haier and Nicolson 2001; Nicolson 1988a). Once the TCs extravasate, they have escaped the usually toxic environment of the blood circulation and are also less amenable to immunologic attacks.

Cell motility is a crucial cell property for various physiologic and pathophysiologic processes. During embryologic development cell migration is required for translocation of cells to their final destination. Leukocyte migration has long been studied in immunologic processes, and inducible, directed migration of fibroblasts, osteoblasts and other epithelial cell types in wound healing is well established. Although there are differences between cell types regarding their actual mechanisms of locomotion, the principles of migration appear to be similar in a wide variety of cell types.

Mechanisms of cell migration

There have been several excellent reviews published recently that address the molecular and morphologic mechanisms of migration in different cell types (Sanchez-Madrid and del Pozo 1999; Entschladen and Zänker 2000; Hood and Cheresh 2002; Maghazachi 2000) so here we will give only an overview of the principles of polarisation, migration and chemotaxis in cancer cells.

The migration of a cell requires its polarisation, which defines a leading and a trailing edge. While new cell adhesions are formed at the leading edge, cell–ECM and cell–cell adhesions have to be cleaved at the rear of the cell. The actin-containing cytoskeleton must be reorganized dynamically, and the tubulin system has to be orientated towards the moving direction of the cell. These processes require clustering and redistribution of membrane proteins like cell adhesion and signalling receptors as well as compartmentation in the cytoplasm of different intracellular signalling pathways at the front and the rear of the cell.

Cells may respond by chemotaxis and orientate along a diffusion gradient of a soluble chemoattractant, or they can migrate along insoluble ECM components, at the site of cell attachment, referred to as haptotaxis.

There are a considerable number of signalling molecules and corresponding membrane receptors that can induce and govern cell migration by induction of cell spreading, polarization and reorganisation of the cytoskeleton. The leading edge of a migrating cell is represented by the lamellipodium, a protruding flat structure composed of cross-linked F-actin in a meshwork. From these lamellipodias protrusions of filopodias, also composed of F-actin bundles, may be found at the cell's edge. Furthermore, at the sites of the lamellipodias membrane ruffles are usually observed that represent actin-rich regions of the cell's leading edge that are dynamically alternating between adherent and non-adherent states.

Besides the actin-containing cytoskeleton, the microtubule system is also orientated towards the direction of cell migration. These tubulin-containing structures are involved in the turnover of cell adhesions (Kirchner et al. 2003; Kaverina et al. 1999), targeting of signalling molecules like c-src to focal adhesions, and redistribution and clustering of membrane and cytoplasmatic proteins. At the leading front of the cell integrins cluster and cooperate with growth factor receptors, such as IGR-IR, HGF-R (Met-R), EGF-R or PDGF-R, to form new adhesions as well as modulate intracellular signalling pathways to promote cell motility (Andre et al. 1999; Trusolino et al. 2001; Sieg et al. 2000). This cooperation is partially due to direct interaction of growth factor receptors with integrins, and it is partially coordinated by FAK and its associated proteins, such as src, and p130cas

(Cary et al. 1998; Moro et al. 2002).Inhibition of FAK can abolish virtually any chemotactic factor-mediated migration of invasive carcinoma cells (Hauck et al. 2001; Hsia et al. 2003).

In addition to integrins and receptor tyrosine kinases, such as growth factor receptors, other components can activate focal adhesion proteins like FAK. G-protein coupled receptors (GPCR), such as chemokine receptors, can induce phosphorylation of FAK and its associated proteins, and thus they can act as chemoattractant receptors for cancer cells, combining chemosensitivity and cell motility (Needham and Rozengurt 1998; Müller et al. 2001).

The nascent focal adhesions exert the driving propulsive forces for migrating cells (Balaban et al. 2001; Beningo et al. 2001). Integrins, growth factors and other chemoattractants can also activate members of the small GTPases, of which the best studied members are Rho A, Rac and Cdc 42. These small GTPases are the main regulators of cell motility, adhesion and shape, and they act by organizing the actin-containing cytoskeleton and cytoskeleton-associated proteins, such as Paxillin, Vinculin and MRN proteins. Rac1 promotes the formation of lamellipodias that drive cell motility in many cell types through PI(3)K signalling mechanisms (Keely et al. 1997). RhoA and RhoB can regulate the formation of stress fibres and generation of contractile forces via ROCK (Rho kinase), whereas cdc42 promotes the formation of actin-rich microspikes that can sense chemoattractant gradients (Nobes and Hall 1995; Ridley and Hall 1992). Some cancer cells may change their shapes and types of cell movement depending on the activation of these proteins. While signalling via Rho and ROCK results in a rather amoeboid type of cell movement, Rac signalling seems to modulate the formation of large protrusions and a flat cell shape (Sahai and Marshall 2003).

Growth factor signalling can promote adhesion formation at the leading front of the cell, but it can also promote de-adhesion at the rear of the moving cell. For example, EDGF signalling is involved in the cleavage of focal adhesions at the rear of migrating cells, and Impairment of focal adhesion disassembly can result in loss of cell migration (Lu et al. 2001). The tubulin system (Kaverina et al. 1999; Ballestrem et al. 2000), v-src (Fincham and Frame 1998), and calpains (Glading et al. 2002; Carragher and Frame 2002), among other components, seem to be regulators of focal adhesion turnover at the rear of migrating cells.

Role of tumor cell migration in clinical metastasis formation

A number of in vivo and human studies indicating a key role of cell motility regulation in the formation of distant organ metastases have been published.

For example, MIA is a protein expressed by malignant melanomas and is associated with tumor stage and prognosis (Bosserhoff et al. 1997). Transfection of melanoma cells with MIA results in increased motility and invasiveness in vitro as well as enhanced extravasation, invasiveness and metastatic potential in vivo without effecting cell proliferation or tumor growth at the primary tumor site (Guba et al. 2000). This indicates that the enhanced metastatic potential found in the transfectants is mainly associated with the increased motility of the TCs. Another example is AKT2, a downstream protein of the PI-3K pathway. Overexpression of AKT2 in breast and ovarian cancer cells leads to up-regulation of $\beta1$ integrins, increased motility and invasiveness through binding to collagen 4 in vitro. This is paralleled by enhanced lymphogenic and hematogenic metastases formation in vivo (Arboleda et al. 2003).

In metastatic MDA-MD-435 breast carcinoma cells the metastatic potential has been related to IGF-I-induced chemotaxis. Metastatic cells show chemotaxis to IGF-I that involves the insulin receptor substrate 2 (IRS-2). In contrast, non-metastatic parental TCs do not show chemotactic potential (Jackson et al. 2001). In a related cell line Müller et al. (2001) demonstrated that metastases formation and malignant cell growth are also governed by chemokines. Human breast cancer samples expressed high levels of the chemokine receptor CXCR4, and its ligand SDF-I was found to govern formation of distant organ metastases in vivo by a chemotaxis mechanism (Müller et al. 2001). SDF-I/CXCR4 signalling has also been demonstrated to be essential for lymphoma cell tissue invasion and lymphoma metastasis formation (Zeelenberg et al. 2001). In contrast to lymphomas, blocking of CXCR4 function in colon carcinoma cells did not block initial invasion of the lung parenchyma by TCs but did affect outgrowth of micrometastases (Zeelenberg et al. 2003).

Another important family of highly conserved membrane proteins is the transmembrane 4 superfamily. The members of this family are motility-related protein-1 (MRP-1), KAI1/CD82 and CD 151 (also called SFA-1 or PETA-3). MRP-1 and KAI1/CD82 are negative regulators of TC motility (Ikeyama et al. 1993), and thus they seem to act as metastases suppressor genes. Negative MRP-1 and KAI1/CD82 expression are associated with poor prognosis in breast (Huang et al. 1998), lung (Higashiyama et al. 1998) and pancreatic (Sho et al. 1998) cancers. Additionally there is clinical and experimental evidence for a metastasis suppressor function of KAI1/CD82 in prostate cancer (Dong et al. 1996). On the other hand, CD 151 is a positive regulator of cell adhesion and migration. CD151 is a metastasis-related antigen and appears to promote formation of metastases of lung (Adachi et al. 1996) and colon (Hashida et al. 2003) cancers by increasing cancer cell motility. Therefore, the transmembrane 4 superfamily is a clinically relevant group of antigens that control metastasis formation by regulating TC motility.

The importance of TC migration and invasion in metastasis formation is also supported by previous findings that TCs cultured from metastases have higher invasive potentials in vitro. This was usually assessed by their abilities to invade ECM protein layers in Boyden-chamber assays and show increased motilities in transwell-migration assays compared to cells cultured from parental tumors. For example, when lymphoma cells cultured from parental tumors were compared to cells isolated from lung and liver metastases, not only was organ selective adhesion demonstrated, but cells from metastatic lesions also exhibited higher invasive potentials and motilities than cells from primary tumors (Nicolson et al. 1988b). Similar observations were made for murine mammary adenocarcinoma cells (Nicolson 1988b). In addition, high and low metastatic variants of a renal carcinoma cell line as well as melanoma variants with different metastatic potentials (Hendrix et al. 1987) showed good correlations between the metastatic potentials and cell motility and invasiveness.

Tumor cell migration and invasiveness

Finally, it should be mentioned that invasive growth requires not only cell motility but also local and tightly regulated proteolysis mediated by locally secreted matrix metalloproteinases (MMPs), heparanase and other degradative enzymes (Nicolson 1989). The invasive capacity of TCs into ECM layers seems to be dependent on cell motility and proteolysis in many experimental settings (Bernhard et al. 1994; Hua and Muschel 1996). Furthermore, it has recently been shown that MMPs can process chemotactic activity on TCs and that they cooperate with integrins to promote cell migration (Gianelli et al. 2001; Morini et al. 2000). It has also been demonstrated that TCs may adopt an amoeboid type of motility allowing them to penetrate three-dimensional ECM matrixes without significant proteolysis (Sahai and Marshall 2003; Wolf et al. 2003). This may partly account for disappointing clinical results with MMP-inhibitors. Invasion in many systems seems to be mediated by multiple degradative enzymes, so inhibiting only one class of enzymes may not be sufficient to completely block invasion.

Conclusions

Reviewing the available literature concerning clinical indicators for the early steps of blood-borne tumor metastasis, there are several studies indicating that specific TC adhesion events are implicated in the organ selectivity of metastasis formation. The important adhesion events appear to be dependent on the expression patterns of TC adhesion molecules. TC motility seems to be a prerequisite for the metastatic phenotypes of tumors, and chemotaxis may be involved in the specific and selective invasion of TCs into distinct target organs. Nonetheless, metastatic dissemination and growth are multi-step and "multiplayer" processes. Our understanding of the mechanisms of metastasis formation will likely be important as a future guide for the development of new concepts in cancer therapy.

References

Adachi M, Taki T, Ieki Y, Huang C, Higashiyama M, Miyake M (1996) Correlation of KAI1/CD82 gene expression with good prognosis in patients with non-small cell lung cancer. Cancer Res 56:1751–1755

Alexiou D, Karayiannakis AJ, Syrigos KN, Zhar A, Sekara E, Michail P, Rosenberg T, Diamantis T (2003) Clinical significance of serum levels of E-selectin, intercellular

adhesion molecule-1, and vascular cell adhesion molecule-1 in gastric cancer patients. Am J Gastroenterol 98:478–485

Andre F, Rigot V, Thimonier J, Montixi, C, Parat F, Pommier G, Marvaldi J, Luis J (1999) Integrins and E-cadherin cooperate with IGF-I to induce migration of epithelial colonic cells. Int J Cancer 83:497–505

Arboleda MJ, Lyons JF, Kabbinavar FF, Bray MR, Snow BE, Ayala R, Danino M, Karlan BY, Slamon DJ (2003) Overexpression of AKT2/protein kinase Bβ leads to up-regulation of β1 integrins, increased invasion, and metastasis of human breast and ovarian cancer cells. Cancer Res 63:196–206

Auerbach R (1991) Vascular endothelial cell differentiation: organ-specificity and selective affinities as the basis for developing anti-cancer strategies. Int J Radiat Biol 60:1–10

Balaban NQ, Schwarz US, Riveline D, Goichberg P, Tzur G, Sabanay I, Mahalu D, Safran S, Bershadsky A, Addadi L, Geider B (2001) Force and focal adhesion assembly: a close relationship studied using elastic micropatterned substrates. Nat Cell Biol 3:466–472

Ballestrem C, Wehrle-Haller B, Hinz B, Imhof BA (2000) Actin-dependent lamellipodia formation and microtubule dependent tail-retraction control directed cell migration. Mol Biol Cell 11:2999–3012

Beningo K, Dembo M, Kaverina I, Small JV, Wang Y (2001) Nascent focal adhesions are responsible for the generation of strong propulsive forces in migrating fibroblasts. J Cell Biol 153:881–888

Ben-Ze'ev A (1994) Colorectal cancer and integrin family of cell adhesion receptors: current status and future directions. Eur J Cancer 30:2166–2170

Bernhard EJ, Gruber SB, Muschel RJ (1994) Direct evidence linking expression of metalloproteinase 9 (92-kDa gelatinase/collagenase) to the metastatic phenotype in transformed rat embryo cells. Proc Natl Acad Sci USA 91:4293–4297

Biancone L, Araki M, Araki K, Vassalli P, Stamenkovic I (1996) Redirection of tumor metastasis by expression of E-selectin in vivo. J Exp Med 183:581–587

Bird MI (1997) Selectins: physiological and pathophysiological roles (Review). Biochem Soc Trans 25:1199–1206

Bosman FT, de Bruine A, Flohil C, van der Wurff A, ten Kate J, Dinjens WW (1993) Epithelial–stromal cell interactions in colon cancer. Int J Dev Biol 37:203–211

Bosserhoff AK, Golob M, Buettner R, Landthaler M, Hein R (1997) [MIA (melanoma inhibitory activity). Biological functions and clinical relevance in malignant melanoma]. Hautarzt 49:762–769

Brodt P, Fallavollita L, Bresalier RS, Meterissian S, Norton CR, Wolitzky BA (1997) Liver endothelial E-selectin mediates carcinoma cell adhesion and promotes liver metastasis. Int J Cancer 71:612–619

Byzova TV, Kim W, Midura RJ, Plow EF (2000) Activation of integrin alpha (V) beta (3) regulates cell adhesion and migration to bone sialoprotein. Exp Cell Res 254:229–308

Carragher NO, Frame MC (2002) Calpain: a role in cell transformation and migration. Int J Biochem Cell Biol 34:1539–1543

Cary LA, Chang, JF, Guan JL (1996) Stimulation of cell migration by overexpression of focal adhesion kinase and its association with Src and Fyn. J Cell Sci 109:1787–1794

Cary LA, Han DC, Polte TR, Hanks SK, Guan JL (1998) Identification of p130cas as a mediator of focal adhesion kinase-promoted cell migration. J Cell Biol 140:211–221

Crissman JD, Cerra RF, Sarka F (1991) The morphology of cancer cell arrest and extravasation. In: Orr FW, Buchanan MR, Weiss L (eds) Microcirculation and cancer metastasis. CRC, Boca Raton, FL, pp 205–215

Ding LD, Sunamura M, Kodama T, Yamauchi J, Duda DG, Shimamura H, Shibuya K, Takeda K, Matsuno S (2001) In vivo evaluation of the early events associated with liver metastases of circulating cancer cells. Br J Cancer 85:431–438

Dong JT, Suzuki H, Pin SS, Bova GS, Schalken JA, Isaacs WB, Barrett JC, Isaacs JT (1996) Down-regulation of the KAI1 metastases suppressor gene during the progression of human prostatic cancer infrequently involves gene mutation or allelic loss. Cancer Res 56:4387–4390

Eliceiri BP, Puente XS, Hood JD, Stupack DG, Schlaepfer D, Huang XZ, Sheppard D, Cheresh DA (2002) Src-mediated coupling of focal adhesion kinase to intergin alpha(v)beta5 in vascular endothelial growth factor signaling. J Cell Biol 157:149–160

Entschladen F, Zänker KS (2000) Locomotion of tumor cells: a molecular comparison to migrating pre- and postmitotic leukocytes. J Cancer Clin Oncol 126:671–681

Ewing J (1928) Neoplastic diseases, 6 edn. W. B. Sanders, Philadelphia, PA

Fenyves AM, Behrens J, Spanel-Borowski K (1993) Cultured microvascular endothelial cells (MVEC) differ in cytoskeleton, expression of cadherins and fibronectin matrix. A study under the influence of interferon-gamma. J Cell Sci 106:879–890

Fidler IJ (1970) Metastasis: quantitative analysis of distribution and fate of tumor emboli labelled with 125 I-5-iodo-2′-deoxyuridine. J Natl Cancer Inst 773–782

Fielding-Habermann B, Habermann R, Saldivar E, Ruggeri ZM (1996) Role of beta3 integrins in melanoma cell adhesion to activated platelets under flow. J Biol Chem 271:5892–5900

Fielding-Habermann B, O'Toole TE, Smith JW Fransvea E, Ruggeri ZM, Ginsberg MH, Hughes PE, Pampori N, Shattil SJ, Saven A, Mueller BM (2001) Integrin activation controls metastasis in human breast cancer. Proc Natl Acad Sci USA 98:1853–1858

Fincham VJ, Frame MC (1998) The catalytic activity of Src is dispensable for translocation to focal adhesions but controls the turnover of these structures during cell motility. EMBO J 17:81–92

Gangopadhyay A, Lazure DA, Thomas P (1998) Adhesion of colorectal carcinoma cells to the endothelium is mediated by cytokines from CEA stimulated Kupffer cells. Clin Exp Metastasis 16:703–712

Gianelli G, Bergamini C, Fransvea E, Marinosci F, Quaranta V, Antonaci S (2001) Human hepatocellular carcinoma (HCC) cells require both α3β1 integrin and matrix metalloproteinases activity for migration and invasion. Lab Invest 81:613–627

Glading A, Lauffenburger DA, Wells A (2002) Cutting to the chase: calpain proteases in cell motility. Trends Cell Biol 12:46–54

Glaves D, Huben RP, Weiss L (1988) Haematogenous dissemination of cells from human renal adenocarcinomas. Br J Cancer 57:32–35

Glaves D, Ketch DA, Asch BB (1986) Conservation of epithelial cell phenotypes during haematogenous metastasis from mammary carcinomas. J Natl Cancer Inst 76:933–938

Glinkii OV, Huxley VH, Turk JR, Deutscher SL, Quinn TP, Pienta KJ, Glinsky VV (2003) Continuous real time ex vivo epifluorescent video microscopy for the study of metastatic cancer cell interactions with the microvascular endothelium. Clin Exp Metastasis 20:451–458

Glinsky VV, Glinsky GV, Glinskii OV, Huxley VH, Turk JR, Mossine VV, Deutscher SL, Pienta KJ, Quinn TP (2003) Intravascular metastatic cancer cell homotypic aggregation at the sites of primary attachment to the endothelium. Cancer Res 63:3805–3811

Gong J, Wang D, Sun L, Zborowska E, Willson JK, Brattain MG (1997) Role of $\alpha5\beta1$ in determining malignant properties of colon carcinoma cells. Cell Growth Differ 8:83–90

Guba M, Bosserhoff AK, Steinbauer M, Abels M, Anthuber M, Buettner R, Jauch KW (2000) Overexpression of melanoma inhibitory activity (MIA) enhances extravasation and metastases of A-mel 3 melanoma cells in vivo. Br J Cancer 83:1216–1222

Haier J, Korb T, Hotz B, Spiegel HU, Senninger N (2003) An intravital model to monitor steps of metastatic Tumor cell adhesion within the hepatic microcirculation. J Gastrointest Surg 7:507–514

Haier J, Nasralla M, Nicolson GL (1999) Different adhesion properties of highly and poorly metastatic HT-29 carcinoma cells with extracellular matrix components: role of integrin expression and cytoskeletal components. Br J Cancer 80:1867–1874

Haier J, Nasralla M, Nicolson GL (2000) Cell surface molecules and their prognostic values in assessing colorectal carcinomas. Ann Surg 231:11–24

Haier J, Nicolson GL (2001) The role of tumor cell adhesion as an important factor in the formation of distant colorectal metastasis. Dis Colon Rectum 44:876–884

Haier J, Nicolson GL (2001) Tumor cell adhesion adhesion under hydrodynamic conditions of fluid flow. APMIS 109:241–262

Hashida H, Takabayashi A, Tokuhara T, Hattori N, Taki T, Hasegawa H, Satoh S, Kobayashi N, Yamaoka Y, Miyake M (2003) Clinical significance of transmembrane 4 superfamily in colon cancer. Br J Cancer 89:158–167

Hauck CR, Sieg DJ, Hsia DA, Loftus JC, Gaarde WA, Monia BP, Schlaepfer DD (2001) Inhibition of focal adhesion kinase expression or activity disrupts epidermal growth factor-stimulated signaling promoting the migration of invasive human carcinoma cells. Cancer Res 61:7079–7090

Hendrix MJ, Seftor EA, Seftor RE, Fidler IJ (1987) A simple quantitative assay for studying the invasive potential of high and low human metastatic variants. Cancer Lett 38:137–147

Hideki U, Koichi H, Yamaguchi KY, Yasoshima T (1998) Separate functions of $\alpha2\beta1$ and $\alpha3\beta1$ integrins in the metastatic process of human gastric carcinoma. Surg Today 28:1001–1006

Higashiyama M, Taki T, Ieki Y, Adachi M, Huang C, Koh T, Kodama K, Doi O, Miyake M (1998) Negative motility related protein-1 (MRP-1/CD9) gene expression as a factor of poor prognosis in non-small cell lung cancer. Cancer Res 55:6640–6644

Hoff SD, Matsushita Y, Ota DM, Cleary KR, Yamori T, Hakamori S, Irimura T (1989) Increased expression of sialyl-dimeric LeX antigen in liver metastases of human colorectal carcinomas. Cancer Res 49:6883–6888

Hood JD, Cheresh DA (2002) Role of integrins in cell invasion and migration. Nat Rev Cancer 2:91–100

Hsia DA, Mitra SK, Hauck CR, Strebolow DN, Nelson JA, Ilic D, Huang S, Li E, Nemerow GR, Leng J, Spencer KSR, Cheresh DA, Schlaepfer DD (2003) Differential regulation of cell motility and invasion by FAK. J Cell Biol 160:753–767

Hua J, Muschel RJ (1996) Inhibition of metalloproteinase 9 expression by a ribozyme blocks metastasis in a rat sarcoma model system. Cancer Res 56:5279–5284

Huang C, Kohno N, Ogawa E, Adachi M, Taki T, Miyake M (1998) Correlation of reduction in MRP-1/CD9, and KAI1/CD82 with recurrences in breast cancer patients. Am J Pathol 153:973–983

Hynes RO (1987) Integrins: a family of cell surface receptors. Cell 48:549–554

Ikeyama S, Koyama M, Yamaoka M, Sasada R, Miyake M (1993) Suppression of cell motility and metastases by transfection with human motility related protein-1 (/MRP-1/CD9) DNA. J Exp Med 177:1231–1237

Iwai K, Ishikura H, Kaji M, Sugiura H, Ishizu A, Takahashi C, Kato H, Tanabe T, Yoshiki T (1993) Importance of E-selectin (ELAM-1) and sialyl Lewis(a) in the adhesion of pancreatic carcinoma cells to activated endothelium. Int J Cancer 54:972–977

Jackson JG, Zhang X, Yoneda T, Yee D (2001) Regulation of breast cancer cell motility by insulin receptor susbtrate-2 (IRS-2) in metastatic variants of human breast cancer cell lines. Oncogene 20:7318–7325

Kahana O, Micksche M, Witz IP, Yron I (2002) The focal adhesion kinase (P125FAK) is constitutively active in human malignant melanoma. Oncogene 21:3969–3977

Kaverina I, Krylyshkina O, Small JV (1999) Microtubule targeting of substrate contacts promotes their relaxation and dissociation. J Cell Biol 146:1033–1043

Keely PJ, Westwick JK, Whitehead IP, Der CJ, Parise LV (1997) Cdc42 and Rac1 induce integrin-mediated cell motility and invasiveness through PI(3)K. Nature 390:632–636

Khatib AM, Kontogianea M, Fallavollita L, Jamison B, Meterissian S, Brodt P (1999) Rapid induction of cytokine and E-selectin expression in the liver in response to metastatic tumor cells. Cancer Res 59:1356–1361

Kirchner J, Kam, Z, Bershadsky AD, Geiger B (2003) Live-cell monitoring of tyrosine phosphorylation in focal adhesions following microtubule disruption. J Cell Sci 116:975–986

Kitakata H, Nemoto-Sasaki Y, Takahashi Y, Kondo T, Mai M, Mukaida N (2002) Essential roles of tumor necrosis factor receptor p55 in liver metastasis of intrasplenic administration of colon 26 cells. Cancer Res 62:6682–6687

Koretz K, Brüderlein S, Henne C, Fietz T, Laque M, Möller P (1994) Comparative evaluation of integrin alpha- and beta-chain expression in colorectal carcinoma cell lines and there tumors. Virchows Arch 425:229–236

Koukoulis GK, Virtanen I, Moll R, Quaranta V, Gould VE (1993) Immunolocalisation of integrins in the normal and neoplastic colonic epithelium. Arch B Cell Pathol Incl Mol Pathol 63:173–184

Lafrenie RM, Gallo S, Podor TJ, Buchanan MR, Orr FW (1994) The relative roles of vitronectin receptor, E-selectin and alpha4-beta1 in cancer cell adhesion to interleukin-1-treated endothelial cells. Eur J Cancer 30A:2151–5158

Liapsi H, Flath A, Kitazawa S (1996) Integrin alpha V beta 3 expression by bone-residing breast cancer metastases. Diagn Mol Pathol 5:127–135

Liddington RC, Bankston LA (2000) The structural basis of dynamic cell adhesion: heads, tails, and allostery. Exp Cell Res 261:37–43

Liotta LA, Kleinermann J, Saidel GM (1974) Quantitative relationships of intravascular tumor cells, tumor vessels, and pulmonary metastases following TC implantation. Cancer Res 34:997–1004

Loftus JC, Liddington RC (1997) New insights into the integrin-ligand interaction. J Clin Invest 100 (Suppl 11):77–81

Lu Z, Jiang G, Blume-Jensen P, Hunter T (2001) Epidermal growth factor-induced tumor cell invasion and metastasis initiated by dephosphorylation and downregulation of focal adhesion kinase. Mol Cell Biol 21:4016–4031

Maghazachi AA (2000) Intracellular signaling events at the leading edge of migrating cells. IJBCB 32:931–943

Max R, Gerritsen RR, Nooijen PT, Goodman SL, Sutter A, Keilholz U, Ruiter DJ, De Waal RM (1997) Immunohistochemical analysis of integrin $\alpha v\beta 3$ expression on tumor-associated vessels of human carcinomas. Int J Cancer 71:320–324

Menter DG, Fitzgerald L, Patton JT, McIntire LV, Nicolson GL (1995) Human melanoma integrins contribute to arrest and stabilisation potential while flowing over extracellular matrix. Immunol Cell Biol 73:575–583

Miranti CK, Brugge JS (2002) Sensing the environment: a historical perspective on integrin signal transduction. Nat Cell Biol 4:83–90

Mizejewski GJ (1999) Role of integrins in cancer: survey of expression patterns. Proc Soc Exp Biol Med 222:124–138

Mook ORF, Marle J, Vreeling-Sindelarova H, Jongens R, Frederiks WM, Noorden CJK (2003) Visualisation of early events in tumor formation of eGFP-transfected rat colon cancer cells in liver. Hepatology 38:295–304

Morini M, Mottolese M, Ferrari N, Ghiorzo F, Buglioni S, Mortarini R, Noonan DM, Natali PG, Albini A (2000) The alpha 3 beta 1 integrin is associated with mammary carcinoma cell metastasis, invasion, and gelatinase B (MMP-9) activity. Int J Cancer 87:336–342

Moro L, Dolce L, Cabodi S, Bergatto E, Erba EB, Smeriglio M, Turco E, Retta SF, Giuffrida MG, Venturino M, Godovac-Zimmermann J, Conti A, Schaefer E, Beguinot L, Tacchetti C, Gaggini P, Silengo L, Tarone G, Defilippi P (2002) Integrin-induced epidermal growth factor (EGF) receptor activation requires c-Src and p130Cas and leads to phosphorylation of specific EGF receptor tyrosines. J Biol Chem 277:9405–9414

Müller A, Homey B, Soto H, Ge N, Catron D, Buchanan ME, McClanahan T, Murphy E, Yuan W, Wagner S, Barrera JL, Mohar A, Verastegui E, Zlotnik A (2001) Involvment of chemokine receptors in breast cancer metastases. Nature 410:50–56

Nakamori S, Kameyama M, Imaoka S, Furukawa H, Ishikawa O, Sasaki Y, Izumi Y, Irimura T (1997) Involvement of carbohydrate antigene sialyly Lewis(x) in colorectal cancer metastasis. Dis Colon Rectum 40:420–431

Nakashio T, Narita T, Sato M, Akiyama S, Kasai Y, Fujiwara M, Ito K, Takagi H, Kannagi R (1997) The association of metastasis with the expression of adhesion molecules in cell lines derived from human gastric cancer. Anticancer Res 17:193–299

Needham LK, Rozengurt E (1998) $G\alpha 12$ and $G\alpha 13$ stimulate Rho-dependent tyrosine phosphorylation of focal adhesion kinase, Paxilin, and p130 Crk-associated substrate. J Biol Chem 273:14626–14632

Nicolson GL (1988a) Cancer metastasis: TC and host organ properties important in metastasis to specific secondary sites. Biochem Biophys Acta 948:175–224

Nicolson GL (1988b) Differential organ tissue adhesion, invasion, and growth properties of metastatic rat mammary adenocarcinoma cells. Breast Cancer Res Treat 12: 167–176

Nicolson GL (1988c) Organ specificity of tumor metastasis: role of preferential adhesion, invasion and growth of unique malignant cells at specific organ sites. Cancer Metastasis Rev 7:143–188

Nicolson GL (1989) Metastatic tumor cell interactions with the endothelium, basement membrane and tissue. Curr Opin Cell Biol 1:1009–1019

Nicolson GL, Belloni PN, Tressler RJ, Dulski K, Inoue T, Canaugh PG (1988) Adhesive, invasive, and growth properties of selected metastatic variants of a murine large cell lymphoma. Invasion Metastasis 9:102116

Nicolson GL, Belloni PN, Tressler RJ, Dulski K, Inoune T, Cavanaugh PG (1989) Adhesive, invasive, and growth properties of selected metastatic variants of a murine large-cell lymphoma. Invasion Metastasis 9:102–116

Nip J, Shibata H, Loskutoff DJ, Cheresh DA, Brodt P (1992) Human melanoma cells derived from lymphatic metastases use integrin alpha v beta 3 to adhere to lymph node vitronectin. J Clin Invest 90:1406–1413

Nobes CD, Hall A (1995) Rho, Rac and Cdc42 GTPases regulate the assembly of multimolecular focal complexes associated with actin stress fibers, lamellipodia, and filopodia. Cell 81:53–62

Numahata K, Satoh M, Handa K, Saito S, Ohyama C, Ito A, Takahashi T, Hoshi S, Orikasa S, Hakamori SI (2002) Sialyl-Le(x) expression defines invasive and metastatic properties of bladder carcinoma. Cancer 94:673–685

Owen JD, Ruest PJ, Fry DW, Hank SK (1999) Induced focal adhesion kinase (FAK) expression in FAK-null cells enhances cell spreading and migration requiring both auto- and activation loop phosphorylation sites and inhibits adhesion-dependent tyrosin phosphorylation of Pyk2. Mol Cell Biol 19:4806–4818

Paget S (1889) The distribution of secondary growth in cancer of the breast. Lancet 1:571–573

Pauli BU, Augustin-Voss HG, el-Sabban ME, Johnson RC, Hammer DA (1990) Organ-preference of metastasis. The role of endothelial cell adhesion molecules. Cancer Metastasis Rev 9:175–189

Renkonen J, Paavonen T, Renkonen R (1997) Endothelial expression of sialyl Lewis(x) and sialyl Lewis(a) in lesions of breast carcinoma. Int J Cancer 74:296–300

Ridley AJ, Hall A (1992) The small GTP-binding protein Rho regulates the assembly of focal adhesion and actin stress fibers in response to growth factors. Cell 70:389–399

Sahai E, Marshall CJ (2003) Differing modes of tumor cell invasion have distinct requirements for Rho/ROCK signalling and extracellular proteolysis. Nat Cell Biol 5:711–719

Sanchez-Madrid F, del Pozo MA (1999) Leukocyte polarisation in cell migration and immune interactions. EMBO 18:501–511

Sato M, Narita T, Kimura N, Zenita K, Hashimoto T, Manabe T, Kannagi R (1997) The association of sialyl Lewis(a) antigen with the metastastic potential of human colon cancer cells. Anticancer Res 17:3505–3511

Scherbarth S, Orr FW (1997) Intravital videomicroscopic evidence for regulation of metastasis by the hepatic microvasculature: effects of interleukin-1 alpha on metastasis and the location of B16F1 melanoma cell arrest. Cancer Res 57:4105–4110

Schlaepfer DD, Hauck CR, Sieg DJ (1999) Signaling through focal adhesion kinase. Prog Biophys Mol Biol 71:435–478

Scoazec JY, Flejou JF, D'Errico A, Bringuier AF, Fiorentino M, Zamparelli A, Feldmann G, Grigioni WF (1996) Fibrolamellar carcinoma of the liver: composition of the extracellular matrix and expression of cell adhesion molecules. Hepatology 24:1128–1136

Sho M, Adachi M, Taki T, Hashida H, Konishi T, Huang C, Ikeda N, Nakajima Y, Kanehiro H, Hisanga M, Nakano H, Miyake M (1998) Transmembrane 4 superfamily as a prognostic factor in pancreatic cancer. Int J Cancer 79:509–517

Sieg DJ, Hauck CR, Ilic D, Klingbeil CK, Schaefer E, Damsky CH, Schlaepfer DD (2000) FAK integrates growth-factor and integrin signals to promote cell migration. Nat Cell Biol 4:249–257

Sieg DJ, Hauck CR, Schlaepfer DD (1999) Requirement role of focal adhesion kinase (FAK) for integrin-stimulated cell migration. J Cell Sci 112:2677–2691

Steinbauer M, Guba M, Cernaianu G, Köhl G, Cetto M, Kunz-Schughart LA; Geissler EK, Falk W, Jauch KW (2003) GFP-transfected tumor cells are useful in examining early metastasis in vivo, but immune reaction precludes long-term development studies in immunocompetent mice. Clin Exp Metastasis 20:135–141

Tozeren A, Kleinmann HK, Grant DS, Morales D, Mercurio AM; Byers SW (1995) E-selectin-mediated dynamic interactions of breast cancer cells with endothelial-cell monolayers. Int J Cancer 60:426–431

Trikha M, Zhou Z, Timar J, Raso E, Kennel M, Emmell E, Nakada MT (2002) Multiple Roles for platelet GPIIb/IIIa and $\alpha v \beta 3$ integrins in tumor growth, angiogenesis, and metastases. Cancer Res 62:2824–2833

Trusolino L, Bertotti A, Comoglio M (2001) A signaling adaptor function for $\alpha 6 \beta 4$ integrin in the control of HGF-dependent invasive growth. Cell 107:643–654

Volpes R, Van den Oord JJ, Desmet VJ (1993) integrins as differential cell lineage markers of primary liver tumors. Am J Pathol 142:1483–1492

Weis L (1992) Biomechanical interactions of cancer cells within the microvasculature during haematogenous metastases. Cancer Metastasis Rev 11:227–235

Weiss L (1986) Metastatic inefficiency: causes and consequences. Cancer Metastasis Rev 3:1–24

Weiss L, Orr FW, Honn KV (1988) Interactions of cancer cells with the microvasculature during metastasis. FASEB J 2:12–21

Westermarck J, Kähäri VM (1999) Regulation of metalloproteinase expression in tumor invasion. FASEB J 13:781–792

Wolf K, Mazo I, Leung H, Engelke K, von Adrian UH, Deruygina EI, Strongin AY, Brockner EB, Friedl P (2003) Compensation mechanism of tumor cell migration: mesenchymal-amoeboid transition after blocking of pericellular proteolysis. J Cell Biol 160:267–277

Yamada N, Chung YS, Takasuka S, Arimoto Y, Sawada T, Dohi T, Sowa M (1997) Increased sialyl LeA expression and fucosyltransferase activity with acquisition of a high metastatic capacity in a colon cancer cell line. Br J Cancer 76:582–587

Ye C, Kiriyama K, Mistuoka C, Kannagi R, Ito K, Watanabe T, Kondo K, Akiyama S, Takagi H (1995) Expression of E-selectin on endothelial cells of small veins in human colorectal cancer. Int J Cancer 61:455–460

Zeelenberg IS, Ruuls-Van Stalle L, Roos E (2001) Retention of CXCR4 in the endoplasmatic reticulum blocks dissemination of a T-cell hybridoma. J Clin Invest 108:269–277

Zeelenberg IS, Ruuls-Van Stalle L, Roos E (2003) The chemokine receptor CXCR4 is required for outgrowth of colon carcinoma micrometastases. Cancer Res 63:3833–3839

H.E. Kaiser and A. Nasir (eds.), Selected Aspects of Cancer Progression:
Metastasis, Apoptosis and Immune Response, 33–58.
© Springer Science + Business Media B.V. 2008

CHAPTER FOUR

Genes and metastasis: experimental advances and clinical implications

Alison L. Allan and Ann F. Chambers

Abstract: Metastasis is the spread of cancer from a primary site resulting in the establishment of secondary tumors in distant locations. It is often difficult to determine if the metastatic process has begun at the time of removal of a primary tumor, creating uncertainty in patients with regards to possible cancer recurrence. Moreover, the majority of cancer deaths occur due to the physiological effects of metastasis rather than from the consequences of the primary tumor. The identification and characterization of molecular factors which contribute to metastasis therefore provides the potential for developing novel prognostic and/or therapeutic strategies to combat cancer mortality. The development of metastasis requires the coordinated expression and function of many different genes. Dysregulation of the balance between the expression of genes which promote metastasis and those which suppress metastasis can influence a cell's ability to successfully complete all the steps in the metastatic cascade. In this chapter, we summarize what is known about the metastatic process and discuss some of the many studies which have investigated different classes of molecules known to be associated with and/or contribute to metastasis. These factors include oncogenic proteins, growth and angiogenic factors, cellular receptors, factors important for cell attachment, enzymes which degrade the extracellular matrix, molecules important for intracellular signalling, chemoattractant molecules, and factors associated with gene transcription. We also consider the emerging role of DNA microarray technology in helping to define the relationship between genes and metastasis both within and between different tumor types. Finally, we discuss how this evolving knowledge about metastasis-associated genes can be translated into clinical strategies for the management, treatment, and cure of metastatic disease.

Keywords: metastatic process, Metastasis promoting genes, Metastasis suppressor genes, Molecular signatures, Prognostic tools, Therapeutic targets

Introduction

Over the last several decades, tremendous research advances have been made in the realms of cancer prevention, detection, and management of primary tumors. Despite this, cancer still remains among the leading causes of death worldwide, primarily due to the failure of effective clinical management of metastatic disease. When the primary tumor is detected and removed before metastasis occurs, prognosis is good and the chance of cancer-free survival is high. However, if cancer cells have already begun to dissemi-

London Regional Cancer Program, London, Ontario, Canada;
Department of Oncology, University of Western Ontario,
London, Ontario, Canada

nate from the primary tumor and spread to other parts of the body, current therapeutic strategies largely depend on the use of systemic cytotoxic therapies such as chemotherapy which frequently result in severe side-effects for the patient and, in many cases, often do not yield long-term success. Moreover, metastatic spread after the removal of a primary tumor can be difficult to identify, creating uncertainty in patients with regards to possible cancer recurrence. The identification and characterization of molecular factors which contribute to metastasis will therefore provide the potential for developing novel prognostic and/or therapeutic strategies to combat cancer. This chapter will review what is currently known about the metastatic process, discuss the growing number of molecular determinants of metastasis that are currently under investigation, and speculate about the clinical importance and implications of this research.

The metastatic process

Metastasis is the spread of cancer cells from a primary site resulting in the establishment of secondary tumors in distant locations (Chambers 1999; Chambers et al. 2000; Chambers et al. 2002; Fidler 1991). In the past several years, significant inroads have been made towards elucidating the physical steps involved in metastasis. It is widely accepted that the metastatic process is comprised of a series of sequential steps, and cancer cells must successfully complete each step in order to give rise to a metastatic tumor (Chambers et al. 2001; Chambers et al. 2002; Fidler 1991, 1999; Folkman 1992; MacDonald et al. 2002; Woodhouse et al. 1997).

Steps in the metastatic cascade

During cancer progression, the malignant population of cells in a primary tumor may reach a size $(1–2 mm^3)$ which necessitates the growth of new blood vessels (angiogenesis) in order to supply the tumor with factors required for its metabolism and continued proliferation (Folkman 1986, 1992). This vascularization of the primary tumor enhances the opportunity for tumor cells to enter the bloodstream, a process called intravasation. Intravasation requires heterotypic interactions between tumor cells and endothelial cells which in turn lead to tumor cell adhesion, migration,

and invasion through the extracellular matrix into the vasculature (Chambers et al. 2001; Chambers et al. 2002; MacDonald et al. 2002). Since these newly formed blood vessels often lack an intact endothelial cell wall, they can provide a relatively accessible escape route for tumor cells to enter the circulation (Butler and Gullino 1975). In addition, cells may also disseminate from the primary tumor through the lymphatic system. However, because there is no direct flow from the lymphatic system to other organs, cancer cells that escape via this route (i.e. during breast cancer metastasis) must still enter the venous system in order to be distributed to remote parts of the body (Chambers et al. 2002; Swartz and Skobe 2001).

Tumor cells that disseminate from the primary tumor and survive the challenges of host anti-tumor immune responses and hemodynamic shear stresses in the circulation are then carried to the capillary beds of secondary organs. Tumor cells are usually arrested by size restriction in the first capillary bed that they encounter, although some may pass through the first capillary bed and travel on to other secondary sites depending on regional blood pressure and deformability of the cells. (Al-Mehdi et al. 2000; Chambers et al. 2002; Morris et al. 1993; Scherbarth and Orr 1997). Arrested tumor cells can escape from the circulation (extravasate) by invading and migrating back through the endothelial cell wall, through the extracellular matrix, and into the secondary organ (Koop et al. 1994; Koop et al. 1996). Once in the new site, cells must initiate and maintain growth to form micrometastases, and (as with the primary tumor) these micrometastases require angiogenesis for nutrition and growth in order to grow into macroscopic tumors that are sufficiently large enough to cause physiological effects on the host (Folkman 1992; Holmgren et al. 1995; Luzzi et al. 1998; Naumov et al. 2001).

The successful metastatic cell must therefore negotiate a number of different steps, including dissemination from the primary tumor via the blood or lymphatic system, survival within the circulation, extravasation into a secondary site, initiation of growth into micrometastases, and maintenance of growth as a vascularized macrometastases. Considering the onerous nature of this multistep process, it is not surprising that several lines of experimental and clinical evidence indicate that metastasis is an inherently inefficient process (Weiss 1990). Experimental studies have

shown that early steps in hematogenous metastasis (i.e. survival in the bloodstream, extravasation) are remarkably efficient, with greater that 80% of cells successfully completing the metastatic process to this point. However, only a small subset of these cells (i.e. ~2%, depending on the experimental model) can initiate growth as micrometastases, and an even smaller subset (i.e. ~0.02%, depending on the experimental model) are able to persist and grow into macroscopic tumors (Cameron et al. 2000; Chambers et al. 2000; Chambers et al. 2001; Luzzi et al. 1998). It is thus the later steps in metastasis involving growth at the secondary site that have been shown, at least in experimental models, to be highly inefficient. Indeed, these findings are supported by clinical observations that, despite the fact that patients diagnosed with cancer may have hundreds to thousands of single disseminated cancer cells that can be detected in the bloodstream and sites remote from the primary tumor, only a very small percentage progress to form overt macrometastases (Weiss 1990, 1992).

Organ-specific metastasis

The movement of tumor cells within and between secondary organs is not random; rather, it depends to a large extent on the location of the primary tumor relative to the body's natural pattern of blood flow. For example, tumor cells that enter the circulation from most parts of the body (i.e. liver, breast, bone) are carried by the systemic venous system directly to the heart and then circulated to all organs of the body via the systemic arterial system. In contrast, cells entering the bloodstream from splanchnic organs such as the colon are circulated first through the liver via the hepatic portal vein, and then into the venous system (Chambers et al. 2000; Chambers et al. 2001; MacDonald et al. 2002; Weiss 1992). It is well established that certain types of cancers have organ-specific preferences for metastatic growth. For example, colorectal cancers preferentially metastasize to the liver, prostate cancer often metastasizes to bone, and breast cancer favors metastasis to regional lymph nodes, lungs, liver, bone, and brain (Chambers et al. 2002).

In the 1920s, James Ewing proposed that blood flow patterns alone were sufficient to account for both the physical delivery of tumor cells to secondary organs and for patterns of organ-specific metastasis (Ewing 1928). However, a number of different theories have challenged this idea by proposing that there are additional, molecular-level mechanisms which explain why and how different types of cancer cells may arrest or grow in "favorite" sites of metastasis development. The most central of these theories is the "seed and soil" theory of metastasis, first proposed in 1889 by Stephen Paget (Paget 1889, 1989). Paget predicted that a cancer cell (the "seed") can survive and proliferate only in secondary sites (the "soil") that produce growth factors appropriate to that type of cell, and this theory has largely withstood the test of time (Fidler 2001; Fidler et al. 2002; Hart 1982). A second idea suggests that the endothelial cells in the vascular beds of certain organs express adhesion molecules specific for particular types of cancer cells (Orr and Wang 2001). A third concept, often called the homing theory, proposes that different organs produce chemotactic factors which can attract specific types of tumor cells to "home" to and arrest in a particular organ (Moore 2001; Muller et al. 2001; Murphy 2001; Wang et al. 1998). It is unlikely that these theories are mutually exclusive, rather, the increasing number of molecular players that are thought to be involved in the metastatic cascade (Tables 1 and 2, discussed below) instead suggests that all of these various factors cooperate with mechanical influences such as blood flow in order to contribute to the ability of specific types of cancer cells to establish themselves as metastases in various target organs.

Experimental models of metastasis

The current knowledge about steps in the metastatic cascade and organ-specific metastasis has arisen from both clinical and experimental studies. The process of metastasis can be modelled experimentally using "spontaneous" or "experimental" metastasis assays (reviewed in Welch 1997). In spontaneous metastasis assays, cancer cells are injected to form a primary tumor, preferentially in an orthotopic site (the correct anatomical site for primary tumor growth; i.e. the mammary fat pad for breast cancer cells), and the endpoint of the assay is the detection of metastases in sites distant from the primary tumor. In experimental metastasis assays, cancer cells are injected directly into the circulation in order to model the later steps in the metastatic process, and the endpoint of the

Table 1 Metastasis promoters*

Molecule	Type of molecule	Proposed function(s)	Possible metastasis step(s) influenced
Ras	Oncogenic protein	Stimulation of tumor cell growth and survival	Growth of primary tumor and growth of metastasis, has the potential to influence most other steps via stimulation of signal transduction pathways
VEGF	Growth factor	Stimulation of endothelial and tumor cell growth, protection from anti-tumor immune responses	Growth of primary tumor, angiogenesis, tumor cell survival, growth of metastases
HGF and Met	Growth factor and growth factor receptor	Stimulation of endothelial and tumor cell growth, motility, invasion	Growth of primary tumor, angiogenesis, intravasation, extravasation, growth of metastases
EGFR	Growth factor receptor	Stimulation of endothelial and tumor cell growth, adhesion, motility, invasion, inhibition of apoptosis	Growth of primary tumor, angiogenesis, intravasation, extravasation, tumor cell survival, growth of metastases
Integrins	Transmembrane receptors	(via signalling interactions): stimulation of endothelial and tumor cell growth, adhesion, migration, survival, tumor cell invasion	Growth of primary tumor, angiogenesis, intravasation, extravasation, tumor cell survival, growth of metastases
OPN	Integrin-binding secreted phosphoprotein	Adhesion, endothelial and tumor cell migration, invasion, survival	Angiogenesis, intravasation, extravasation, tumor cell survival
MMPs	Proteases	Stimulation of tumor cell growth, extracellular matrix degradation, migration, invasion	Growth of primary tumor, angiogenesis, intravasation, extravasation, growth of metastases
uPA and uPAR	Protease and protease receptor	Endothelial and tumor cell adhesion and migration, extracellular matrix degradation, tumor cell invasion and proliferation	Growth of primary tumor, angiogenesis, intravasation, extravasation, growth of metastases
CXCR4	Chemokine	Tumor cell homing or growth in secondary organs	Determination of organ-specific metastasis

*References pertaining to each specific molecule can be found in the appropriate sections in the text. This table shows only a representative selection of the many metastasis-promoting molecules discovered to date. For a more comprehensive review, see Price et al. 1997.

assay is, again, the formation of visible metastases in secondary sites. Furthermore, the use of a technique called intravital videomicroscopy (IVVM) facilitates direct observation of the microcirculation *in vivo* and allows for detection and tracking of tumor cells as they move into or out of the circulation and migrate into the host tissue. The combined use of experimental metastasis assays, IVVM, and quantitative approaches that follow the fate of cancer cells has been valuable in identifying and characterizing the different steps in the metastatic cascade (Chambers et al. 1995; MacDonald et al. 2002). It has been suggested that spontaneous and experimental metastasis assays may

measure different properties (Glaves 1983; Nicolson and Custead 1982; Stackpole 1981), and thus the use of several different assays as well as different cell types is required in order to fully understand the various aspects of the metastatic process.

Although experimental models of metastasis have allowed for the identification of a number of mechanical factors and molecular changes which contribute to a cell's ability to metastasize, these endpoint assays have not been useful in clarifying the biological relevance of particular molecules to specific steps in the metastatic cascade. However, the implementation of molecular tools which allow for the overexpression or

Table 2 Metastasis suppressors*

Molecule	Type of molecule	Proposed function(s)	Possible metastasis step(s) influenced
Nm23	Histadine kinase	Disruption of Ras/Raf/ERK-MAPK pathway	???
KAI1	Tetraspanin	Integrin interactions, EGFR desensitization	???
MKK4	MAP kinase kinase	Phosphorylates and activates p38 and JNK, potentially resulting in increased apoptosis	Growth of metastases???
RHOGDI2	Inhibitor of guanine nucleotide dissociation for Rho GTPases	Modulation of JNK, p38, and ERK pathways via interactions with Rho proteins, potentially resulting in effects on cell morphology, motility, and adhesion	Intravasation, extravasation???
BRMS1	Part of a histone deacetylase complex	Transcriptional regulation of gap-junctional communications between cells	???
KiSS1	G-protein coupled ligand	Phosphorylation of ERK and p38 pathways, potentially resulting in inhibition of cell motility and invasion	Intravastion, extravasation, growth of metastases???
CRSP3	Transcriptional co-activator	Transcriptional regulation	Growth of metastases???
TXNIP	Thioredoxin (TRX) inhibitor	Regulation of DNA synthesis, signaling cascade, and transcriptional activity (via TRX)	Growth of metastases???

*References pertaining to each specific molecule can be found in the appropriate sections in the text. For a more comprehensive review, see Debies and Welch 2001; Kauffman et al. 2003; Yoshida et al. 2000; Steeg 2003.

inhibition of specific molecules of interest, combined with the use of animal models to assess the functional consequences of such changes has proven to be a more successful strategy. The remainder of this chapter focuses on recent experimental advances towards understanding the relationship between genes and metastasis, and the implications that this research has for clinical prognosis and treatment of metastasis.

Molecular mechanisms of metastasis

A number of events in the metastatic cascade (i.e. angiogenesis, adhesion, migration, invasion, and growth) also play an important role in tightly regulated normal biological processes such as reproduction, embryogenesis, wound healing, smooth muscle cell migration, and liver regeneration. The fundamental difference between normal and malignant cells is thought to lie in their ability to be regulated at the biochemical level: in malignant cells, the molecules that start, maintain, or stop these cellular processes

are often expressed at inappropriate levels or at an inappropriate time and place (Kohn and Liotta 1995; Liotta et al. 1991; Nicolson 1991).

How does this escape from normal physiological regulation occur? The metastatic cascade is largely thought to be a continuation of the multistep process of tumorigenesis, which has been likened to a "Darwinian-type" cellular evolution. The progression of an altered population of cells towards malignancy is characterized by increasing genomic instability as a result of the accumulation of both gain-of-function and loss-of-function changes such as mutations, translocations, chromosomal aberrations, gene deletions, and gene amplifications. These molecular changes manifest themselves at the cellular level, resulting in increased susceptibility to genotoxic insult, altered cell cycle control, dedifferentiation, and acquisition of the ability to proliferate autonomously (Vogelstein and Kinzler 1993). With this increasing genetic instability, the prevailing theory of metastasis predicts that some cancer cells may either acquire one or more subsequent molecular changes or simply succumb to

the collective effect of accumulated "advantageous" mutations such that they can become invasive, escape from the primary tumor, and travel to secondary sites or organs (Poste and Fidler 1980).

A growing number of genes or gene products have been identified as being upregulated or downregulated during metastasis (Price et al. 1997; Ramaswamy 2003). However, when determining the relative importance of such molecules in regulating metastasis, it is extremely important to identify not only the expression pattern of these genes in tumors and tumor cells, but also the functional association between altered gene expression and the acquisition or maintenance of a metastatic cell phenotype. Experimentally, the relationship between expression and metastatic function can be examined by manipulating gene expression and observing the resultant biological effects using *in vitro* cell culture systems and *in vivo* animal models. Upregulation or overexpression of a particular gene can be achieved by stably transfecting the gene of interest into normal or cancer cells using a variety of methods, including calcium phosphate co-precipitation, electroporation, polycationic liposomes (lipofection), microinjection, and viral vectors (Watson et al. 1992). Gene expression can be downregulated or inhibited using various antisense targeting strategies, such as treatment or transfection with antisense oligonucleotides, antisense vectors, ribozymes, or small inhibitory RNA (siRNA) molecules (Crooke 2001; Paul et al. 2002). Furthermore, depending on the type of gene product produced, the expression of some proteins can be inhibited by treatment with neutralizing or blocking antibodies and peptides. Experimental or spontaneous metastasis assays can then be used to compare the metastatic ability of the transfected cells relative to control cells in which expression of the gene of interest has not been manipulated. Alternatively, transgenic or knockout animals can be generated, although this approach is generally more time-consuming and often prone to complications because of the fact that many metastasis-associated genes are also critical for embryogenesis, development, and normal cellular processes (Kohn and Liotta 1995; Liotta et al. 1991; Nicolson 1991).

The use of experimental strategies such as these has allowed for the identification of a number of molecules which, through gain or loss of expression, can drive the metastatic potential of a cell (reviewed in detail in Price et al. 1997; Steeg 2003). As discussed above, the metastatic process is comprised of a series of sequential steps which requires the coordinated interplay of a variety of cell types, including tumor cells, endothelial cells, stromal cells, and immune cells. These critical cell–cell and cell–matrix communications are regulated by molecular signals which facilitate the successful completion of various cellular events such as adhesion, migration, invasion, survival, and growth.

Metastasis-promoting genes

Metastasis-promoting genes are genes whose increased expression and/or activation (usually as result of abnormal regulation) can contribute to the acquisition of a metastatic cellular phenotype. These genes encode a variety of different types of proteins, including oncogenic proteins, growth and angiogenic factors, cellular receptors, chemoattractant molecules, factors important for cell attachment, and enzymes which degrade the extracellular matrix. While a thorough review of all of these molecules is beyond the scope of this chapter, a selection (based on involvement in a number of different metastasis steps, and/or involvement in a large number of human cancers) are summarized in Table 1 and will be discussed in greater detail below.

Ras

Oncogenes code for proteins that are important for the regulation of cellular growth control and responsiveness to extracellular signals. When these control mechanisms are constitutively upregulated in the absence of external growth signals, the uncontrolled proliferation that is typical of cancer cells can occur (Hanahan and Weinberg 2000). Members of the Ras oncogene family provide an example of how activated oncogenes can lead to phenotypic changes that promote metastatic behavior through alterations in downstream gene expression (Chambers and Varghese 2002).

Ras oncogenes encode a family of membrane-bound small GTP binding proteins that act as cellular transducers, relaying signals from the cell surface to the cytoplasm which in turn activate signalling cascades that regulate the expression of a number of different genes. In addition to activating the Ras/Raf/ ERK-MAPK (extracellular signal-regulated kinase/

mitogen-activated protein kinase) pathway, Ras proteins can also associate with effector molecules involved in other signalling pathways, such as the phosphatidylinositol 3-kinase (PI3K) pathway. Depending on cell type and molecular cross-talk between signalling pathways, Ras proteins can regulate the expression of downstream genes which are important for various cellular responses, including growth, differentiation, and apoptosis (reviewed in Campbell et al. 1998; Frame and Balmain 2000).

Ras genes have been shown to be activated by mutation in a large percentage of human cancers, and in many other tumors Ras is activated by other mechanisms (Bos 1989; Clark and Der 1995). Functionally, the H-*ras* oncogene was one of the first oncogenes to be shown to be able to morphologically transform NIH 3T3 cell *in vitro* (Parada et al. 1982; Shih et al. 1979; Shih and Weinberg 1982; Tabin et al. 1982), and these *ras*-transformed cells were also able to form tumors in nude mice (Blair et al. 1982; Chang et al. 1982; Fasano et al. 1984). Shortly afterwards, it was shown that *ras*-transformed NIH 3T3 cells, in addition to being tumorigenic, were also highly metastatic *in vivo* (Bondy et al. 1985; Muschel et al. 1985; Thorgeirsson et al. 1985). Although it was initially difficult to comprehend how changes in the expression of a single gene could be sufficient to induce the complex steps required for tumorigenicity and metastasis, the subsequent identification of Ras as a key signal transduction molecule suggested that Ras-induced malignancy was a result of the coordinated activity of multiple downstream molecules which in turn affected multiple cellular events. Oncogenes such as Ras thus have the potential to contribute to all steps in the metastatic process via activation of factors such angiogenic factors, adhesion molecules, proteases, growth-related genes, or survival factors. A number of Ras-induced genes have been shown to be functional contributors to metastasis, including the angiogenic factor vascular endothelial growth factor (VEGF), the secreted integrin-binding protein osteopontin (OPN), the cysteine proteases Cathepsins L and B, and calcyclin (Chambers and Tuck 1993). Some of these genes are discussed in the sections below.

Although the particular step(s) of metastasis which are influenced by Ras remain to be fully elucidated, studies using IVVM and quantitative cell accounting analysis revealed that Ras activation is important for regulating the ability of tumor cells to initiate and maintain growth of micrometastases once they have arrived at the secondary site (Koop et al. 1996; Varghese et al. 2002). It was observed that the balance between proliferation and apoptosis was tipped in favor of growth in micrometastases with activated Ras, and in favor of death in the control micrometastases (Varghese et al. 2002). These findings therefore suggest that disruption of the proliferation:apoptosis equilibrium to favor progressive growth is one mechanism by which oncogenic signalling can influence metastatic potential.

Vascular endothelial growth factor

The expression of vascular endothelial growth factor (VEGF) has been shown to upregulated in a number of different human tumor types, and has been correlated with poor clinical outcome (Berse et al. 1992; Chung et al. 1996; Gasparini 2000; Junker 2001; Shen et al. 2000). Through paracrine and autocrine interactions with endothelial cells, tumor cells, and immune cells, VEGF has been shown to play a variety of roles in the tumor microenvironment, including stimulation of angiogenesis, protection from host anti-tumor immune responses, and promotion of tumor and endothelial cell growth (Gabrilovich et al. 1996; Herold-Mende et al. 1999; Masood et al. 2001). It has been demonstrated that human VEGF can exist as several different isoforms which are generated by alternative slicing of a single primary transcript of the VEGF gene (Robinson and Stringer 2001). While the importance of the various isoforms is not fully understood, overexpression of the secreted isoforms VEGF121 and VEGF165 has been shown to be important for tumor angiogenesis in mouse models (Grunstein et al. 2000; Kondo et al. 2000; Zhang et al. 2002). Although VEGF is most commonly studied in relation to the development of primary tumors, several studies have also identified it as being important for metastasis. Overexpression of VEGF in melanoma or colon cancer cells can lead to increased metastasis as assessed by both spontaneous and experimental metastasis assays (Claffey et al. 1996; Kondo et al. 2000). Furthermore, inhibition of VEGF by treatment with antisense molecules or anti-VEGF blocking antibodies results in a significant decrease in the metastatic ability of a number of different cancer cell types (Claffey et al. 1996; Melnyk et al. 1996; Warren et al. 1995; Yano et al. 2000).

Hepatocyte growth factor and Met

Hepatocyte growth factor (HGF) was originally identified in the liver as a paracrine and autocrine growth factor (Fausto 1991). Also known as scatter factor (SF), HGF is expressed in a variety of other organs and has been shown to have significant effects on cell growth, morphology, and motility of both epithelial and endothelial cells. HGF is synthesized as an inactive precursor protein, and must undergo extracellular proteolytic cleavage in order to become active. The cellular effects of HGF are mediated by way of signal transduction through its cognate receptor Met, a transmembrane tyrosine kinase that is the protein product of the *c-met* proto-oncogene (Jiang and Hiscox 1997; Stella and Comoglio 1999). Overexpression of both HGF and Met has been observed in many types of epithelial cancers, including breast carcinoma, and high levels of HGF in breast tissue has been shown to correlate with poor survival (To and Tsao 1998; Tuck et al. 1996; Yamashita et al. 1994). A number of studies have provided strong evidence that HGF and Met contribute to tumor progression and metastasis via a multiplicity of functional mechanisms. For example, HGF is a potent angiogenic factor, and can stimulate endothelial cell growth, motility, and organization into new vessels (Rosen and Goldberg 1995). Transfection of HGF and/ or Met into different epithelial cell types can promote an invasive cellular phenotype *in vitro* and increased tumorigenicity and metastatic ability *in vivo* (Firon et al. 2000; Lamszus et al. 1997; Lin et al. 1998; Meng et al. 2000). Furthermore, downregulation of Met activity by transfection with either a dominant negative form of Met or a Met-specific hammerhead ribozyme results in a reduction in invasiveness, tumorigenicity and metastasis (Firon et al. 2000; Jiang et al. 2001). Preliminary studies by Meng et al. (2000) indicate that HGF may even play a role in decreasing sensitivity to chemotherapeutic agents, although the *in vivo* relevance of these results still needs to be examined. Taken together, these studies indicate that HGF and Met play an important role during cancer progression and metastasis by stimulating angiogenesis, motility, invasion, and growth.

Epidermal growth factor receptor

The epidermal growth factor receptor (EGFR) is another example of a receptor tyrosine kinase that can modulate signalling pathways during metastasis.

EGFR is a member of a subfamily of four closely related receptors: EGFR (also known as ErbB-1), HER-2/*neu* (ErbB-2), HER-3 (ErbB-3), and HER-4 (ErbB-4). These receptors can homodimerize or heterodimerize following autocrine or paracrine interactions with ligands such as epidermal growth factor (EGF) and transforming growth factor α (TGFα). After ligand binding, the tyrosine kinase intracellular domain of the receptor is activated by autophosphorylation, which in turn initiates a cascade of downstream intracellular events via the Ras/Raf/MAPK signalling pathway (Ullrich and Schlessinger 1990; Wells 1999). Experimental and clinical studies have demonstrated that EGFR functionally contributes to a variety of processes that are crucial to cancer progression and metastasis, including cytoskeletal reorganization, adhesion, motility, invasion, angiogenesis, cell survival, and tumor cell growth (Chakravarti et al. 2002; Chan et al. 2002; Ciardiello et al. 2001; Naramura et al. 1993; Price et al. 1996; Riedel et al. 2002; Segall et al. 1996; Shibata et al. 1996; Turner et al. 1996). It has been suggested that the increased activity of the EGFR pathway in cancer cells may occur as a result of several mechanisms, including overexpression of EGFR, decreased receptor turnover, increased ligand concentrations, and/or EGFR gene alterations which result in the presence of abnormal receptors. Not surprisingly, EGFR has been implicated in a wide variety of human cancers, and the presence of EGFR expression is usually associated with advanced disease and poor prognosis (Ciardiello and Tortora 2001; Salomon et al. 1995; and references therein).

Integrins

The integrins are family of dimeric transmembrane receptors comprised of α and β subunits. At least 24 different heterodimers can be formed by non-covalent associations between 18 α and 8 β subunits, and each heterodimer can bind a wide variety of ligands, including extracellular matrix (ECM) proteins (i.e. laminin, vitronectin), E-cadherin, and the secreted phosphoprotein osteopontin (OPN). Integrin-ligand interactions can induce activation and clustering of the focal adhesion complex (FAC), which serves to assemble structural and regulatory proteins such that they can mediate cytoskeletal shape and migration and create a framework for the association of signalling molecules. Integrins can coordinate with growth factor pathways (i.e. VEGF, EGFR, HGF/Met) to activate a number of signalling cascades such as the focal adhesion kinase (FAK)-Src cascade, the PI3K cascade, and the Ras/Raf/ERK cascade. Therefore, although originally identified as important cell-surface adhesion receptors that mechanically link the cytoskeleton to the ECM or to other cells during cellular migration and invasion, integrins have also more recently been recognized as key signalling receptors capable of influencing a dynamic range of cellular processes, including migration, proliferation, differentiation, survival, and angiogenesis (reviewed in Giancotti and Ruoslahti 1999; Newham and Humphries 1996).

Various integrins have been shown to have prognostic value for a number of human cancers, and depending on the stage of tumor progression, some integrins may be up or down-regulated (Mizejewski 1999; Varner and Cheresh 1996). In particular, it is believed that there is significant involvement of integrins that contain αv, α6, and/or β1 subunits. For example, increased expression of the αvβ3 integrin has been associated with metastatic cancers of the lung, breast, colon, and skin (Albelda et al. 1990; Clarke et al. 1997; Felding-Habermann et al. 2001; Vonlaufen et al. 2001). Expression of the α6β1 integrin has been shown to increase invasion and/or metastasis of pancreatic carcinoma cells, breast carcinoma cells, and oral squamous cell carcinoma cells (Shaw et al. 1996; Vogelmann et al. 1999; Zhang et al. 1996), while α6β4 can influence invasion and metastasis of breast cancer cells in association with various signalling pathways (Abdel-Ghany et al. 2001; Shaw et al. 1997; Trusolino et al. 2001). In addition, activated vascular endothelial cells express a number of integrins (most notably αvβ3) which have been found to be important for regulating tumor angiogenesis (Eliceiri and Cheresh 1999). A large number of other experimental studies support a functional role for integrins by demonstrating that inhibition of specific integrins via antibody or peptide antagonists can lead to a reduction in primary tumor growth, angiogenesis, cell survival, and the development of distant metastases (reviewed in Tucker 2002). The involvement of integrins in so many aspects of cancer progression underlines the importance of these signalling receptors in regulating the metastatic process.

Osteopontin

Osteopontin (OPN) is a secreted phosphoprotein that has been implicated in a number of normal and pathologic conditions. Clinical studies have demonstrated that OPN is overexpressed in many human cancers, including breast, ovarian, prostate, colon, lung, liver, esophageal, and gastric cancers, and in some cases OPN is associated with cancer progression (Agrawal et al. 2002; Brown et al. 1994; Fedarko et al. 2001; Furger et al. 2001; Ye et al. 2003). In particular, it has been shown that there is a strong correlation between elevated OPN levels in patients with breast or prostate cancer and increased tumor aggressiveness, increased tumor burden, and poor prognosis/survival rates (Hotte et al. 2002; Singhal et al. 1997; Tuck et al. 1997; Tuck et al. 1998).

OPN can be produced in the tumor microenvironment by both tumor cells and other surrounding cells, such as fibroblasts, inflammatory cells, or endothelial cells (Brown et al. 1994). The OPN protein contains several conserved structural elements including heparin- and calcium-binding domains, a thrombin-cleavage site, a CD44 binding site, and an RGD (Arg-Gly-Asp) integrin-binding amino acid sequence (Sodek et al. 2000). Based on the presence of these domains, it is not surprising that experimental studies have shown that the importance of OPN lies in its ability to interact with a diverse range of factors, including cell surface receptors (integrins, CD44), growth factor/receptor pathways (VEGF, HGF/Met, TGFα/EGFR), and secreted proteases (urokinase plasminogen activator, matrix metalloproteinases). These complex signalling interactions can result in changes in gene expression which ultimately lead to increased malignant cell behavior such as adhesion, migration, invasion, tumor angiogenesis, enhanced tumor cell survival, and metastasis (Denhardt and Chambers 1994; Geissinger et al. 2002; Philip et al. 2001; Shijubo et al. 2000; Tuck et al. 1999; Tuck et al. 2000; Tuck et al. 2003; Xuan et al. 1995; Weber et al. 1997). In addition, downregulation of OPN by antisense treatment results in decreased tumorigenicity of ras-transformed fibroblasts (Behrend et al. 1994), while OPN-deficient mice injected with melanoma cells show reduced experimental metastasis compared to wild-type controls (Nemoto et al. 2001). Taken together with the clinical observations, these experimental studies indicate that OPN is not

merely associated with cancer, but that it actually plays a multi-faceted functional role in malignancy.

Matrix metalloproteinases

Matrix metalloproteinases (MMPs) are a family of secreted or transmembrane metalloenzymes that require the metal ion Zn^{2+} to carry out their functional role in the degradation of extracellular matrix and basement membrane components during normal and abnormal physiological processes (Brinckerhoff and Matrisian 2002; Egeblad and Werb 2002). MMPs are produced as proenzymes that require extracellular activation through proteolytic cleavage of an amino terminal domain. Although all MMPs are thought to contain a number of conserved functional and structural domains that result in similar mechanisms of latency, activation, and proteolysis, members of the MMP family differ from each other based on structural domains that determine cellular location, matrix binding, inhibitor binding, and/or substrate specificity for each MMP (reviewed in Nagase and Woessner 1999). Under normal physiologic conditions, the net activity of MMPs is tightly regulated by maintaining an equilibrium between levels of activated MMPs and levels of their endogenous inhibitors, known as tissue inhibitors of metalloproteinases (TIMPs). However, during cancer progression, this balance is often altered in favor of enhanced cellular invasiveness via increased production of MMPs by a number of cell types, including fibroblasts, infiltrating immune cells, endothelial cells, and tumor cells. The overexpression of MMPs has been positively correlated with increasing tumor stage in many types of human cancer, and this is reflected by both an increase in the relative expression levels of individual MMPs as well as an increase in the number of different MMP family members that are expressed (Stetler-Stevenson et al. 1996).

Based on the inherent degradative activity of MMPs, it was originally believed that the major contribution of these enzymes to the metastatic process occurred during the steps of intravasation and extravasation. Experimental studies using *in vivo* endpoint metastasis assays have shown that transfection of specific MMP family members such as gelatinase A (MMP-2) and gelatinase B (MMP-9) can increase the metastatic potential of bladder cancer cells or rat embryonic fibroblasts (respectively) (Bernhard et al. 1994; Kawamata et al. 1995), while melanoma cells injected into MMP-2 knockout mice show reduced tumorigenicity and metastatic ability as well as decreased tumor-associated angiogenesis (Itoh et al. 1998). Additional studies have demonstrated that transfection or treatment with different TIMPs or synthetic inhibitors of MMPs can result in a reduction of *in vitro* cellular invasiveness through basement membranes such as Matrigel (Albini et al. 1991; Khokha et al. 1992; Taraboletti et al. 1995), supporting a central role for MMPs in facilitating of the breakdown of physical barriers between a primary tumor and distant sites of metastases development. However, a number of contradictory studies have reported inconsistent effects of MMPs in promoting metastasis, suggesting that their role in cancer progression is more complex than first hypothesized (reviewed in Chambers and Matrisian 1997; McCawley and Matrisian 2000). For example, overexpression of matrilysin (MMP-7) in colon cancer cells (Witty et al. 1994) or stromelysin-3 (MMP-11) in breast cancer cells (Noel et al. 1996) results in increased tumorigenicity without a corresponding increase in metastatic ability, suggesting that MMPs may be involved in growth of the primary tumor. Furthermore, studies which utilized IVVM to delineate the steps at which MMPs affect metastasis indicate that these enzymes also play a critical role in survival of disseminated tumor cells and post-extravasation growth of metastases (Koop et al. 1994). Although the detailed mechanisms by which MMPs can regulate the tumor growth environment remain to be elucidated, it has been proposed that they may facilitate the release of sequestered growth factors in the extracellular matrix surrounding the growing tumor, either directly or via a proteolytic cascade. Therefore, MMPs provide important functional contributions to cancer progression and metastasis through their roles in primary tumor growth, angiogenesis, intravasation, extravasation, and growth of distant metastases.

Urokinase plasminogen activator (uPA) and uPAR

Plasminogen activators (PA) are serine-specific proteases that convert inactive plasminogen to active plasmin, a degradative enzyme that has broad specificity for various ECM proteins. Plasminogen activators exist in two forms: tissue plasminogen activator (tPA), which is present in plasma and is important for the degradation of fibrin-rich blood clots, and urokinase plasminogen activator (uPA), which is primarily involved in cell-mediated proteolysis in physiological situations such as tissue remodeling, embryogenesis, and metastasis (Collen 1999). The uPA receptor (uPAR) can be upregulated by a number of growth factors, including HGF and EGF, and uPAR binds uPA with high affinity. Interestingly, uPAR does not internalize bound uPA, thus providing cells with the ability to generate localized protease activity at their cell surface in order to facilitate cell invasion and migration through the ECM. As well as playing a direct enzymatic role in the degradation of ECM components, uPA and uPAR have also been shown to have a number of indirect effects via the activation of latent growth factors (i.e. HGF, transforming growth factor β), the release of sequestered angiogenic factors (i.e. VEGF, basic fibroblast growth factor) from the ECM, the activation of MMPs, and the inactivation of TIMPs. Additionally, uPA and uPAR have been implicated in integrin-mediated activation of various signalling pathways such as the MAPK pathway and the PI3K pathway, resulting in downstream effects on cellular adhesion, migration, invasion, and proliferation (Blasi and Carmeliet 2002; and references therein).

Overexpression of both uPA and uPAR has been observed in many kinds of human cancer, and is frequently associated with tumor progression, risk of tumor recurrence, and poor prognosis (Brunner et al. 1999; Duffy et al. 1990; Hasui et al. 1996; Look et al. 2002; Oka et al. 1991; Stephens et al. 1999). In the tumor microenvironment, uPAR has been shown to be expressed by a number of cell types, including tumor cells and stromal cells (Blasi and Carmeliet 2002). The functional importance of uPAR in tumorigenicity and metastasis was demonstrated by Yu et al. (1997), who showed that human epidermoid carcinoma cells transfected with uPAR produced rapidly growing, highly metastatic tumors when inoculated on chick embryonic chorioallantoic membranes (CAMs). Furthermore, reduced expression of surface uPAR caused these cells to go into a prolonged state of dormancy. Additional experimental studies using mouse models and a variety of tumor cell types showed a reduction in tumor growth, angiogenesis, and metastasis following treatment with antisense uPAR or inhibitors/antagonists which interfered with uPA/uPAR

interactions (Crowley et al. 1993; Lakka et al. 2001; Min et al. 1996). Therefore, by utilizing both direct proteolytic and indirect non-proteolytic mechanisms, the interactions between uPA and its receptor uPAR play an important role in potentiating invasive cellular events which contribute to the metastatic process.

Chemokines and chemokine receptors

Chemokines are molecules which are functionally and structurally similar to growth factors. Their activity is mediated by interactions with G-protein-coupled receptors which, through signal transduction, can then induce cellular adhesion and directional migration (Zlotnik and Yoshie 2000). Previously shown to be involved primarily in trafficking and homing of haematopoietic cells such as leukocytes, studies by Muller et al. (2001) indicate that chemokines may also play a role in organ-specific metastasis. These studies showed that, relative to normal breast tissue, the chemokine receptor CXCR4 was overexpressed in breast tumors, and that its chemokine ligand CXCL12 was overexpressed in tissues to which breast cancer cells often metastasize, such as lymph nodes, bone marrow, liver, and lung. Functional studies revealed that treatment of breast carcinoma cells with exogenous CXCL12 could induce invasive cellular behavior *in vitro*, and that inhibition of *in vivo* interactions between CXCL12 and CXCR4 by treatment with a CXCR4-neutralizing antibody resulted in blockage of both experimental and spontaneous metastasis (Muller et al. 2001). The overexpression of CXCR4 has also been shown to occur in a variety of other cancers, including prostate cancer (Taichman et al. 2002), pancreatic cancer (Koshiba et al. 2000), and B-cell lymphoma (Arai et al. 2000), suggesting that this phenomenon is probably not limited to breast cancer.

The study by Muller et al. (2001) is the first of its kind to show the importance of chemokines in promoting organ-specific metastasis. The authors suggest that chemokines contribute to metastasis by causing cancer cells to "home" to specific secondary sites. But how can this proposed mechanism be reconciled with the idea that physical and mechanical factors (such as blood flow patterns) have been found to play an important role in determining where circulating cancer cells arrest? This apparent discrepancy can be easily resolved by considering that the ability of

chemokine-receptor interactions to initiate signalling pathways provides them with a means to influence cellular processes such as cytoskeletal rearrangement, migration, invasion, and growth (Zlotnik and Yoshie 2000). The involvement of chemokines in metastasis therefore illustrates the idea that the coordinated activity of physical, mechanical, and molecular factors is critical for the metastatic process (Chambers et al. 2002).

Although only a small selection of the known metastasis-promoting molecules have been presented here, the studies discussed in this section representatively illustrate the important concept that the development of metastasis requires the coordinated expression and function of many different genes. Conversely, it can be theoretically predicted that only one gene might be required to block metastasis, since the failure to complete any particular step in the metastatic process could result in the loss of a cell's metastatic potential. An emerging field of study involving metastasis suppressor genes is aimed at testing this theory.

Metastasis suppressor genes

Metastasis suppressor genes can be broadly defined as genes which suppress the ability of metastases to form, without affecting the growth of the primary tumor (Steeg 2003). These genes have been identified by their reduced or absent expression in metastatic tumor cells relative to tumorigenic but non-metastatic cells. Similar to the paradigm of tumor suppressors such as p53 and Rb (Picksley and Lane 1994), it is thought that the loss of metastasis suppressor gene function can lead to an escape from normal cellular control and the development of metastasis. Accordingly, the restoration of metastasis suppressor gene function should lead to suppression or interruption of the metastatic cascade.

Within the last 2 decades, at least eight such metastasis suppressor genes have been identified (Table 2). Although the molecular mechanisms by which these genes can suppress metastasis are not yet fully understood, growing evidence suggests that their functional effect may be to regulate the ability of tumor cells to initiate and maintain growth of metastases at the secondary site via mediation of important signal transduction pathways (Kauffman et al. 2003; Steeg 2003; Table 2). Relative to the study of metastasis-promoting genes, the field of metastasis suppressor

genes is very much in its infancy. Current research efforts have been focused on transfecting the proposed suppressors (either as chromosomal segments or single genes) into metastatically competent cells and testing the ability of these genes to reduce metastasis formation without affecting the growth of the primary tumor. Ongoing studies are working towards dissecting the detailed molecular mechanisms utilized by these genes, as well as the particular steps in the metastatic process which may be affected. In addition, two of the most well-characterized metastasis suppressor genes (Nm23 and KAI1) as well as MKK4 have been reported to have low expression in some human tumor types, and in many cases this reduced expression can be correlated with advanced tumor stage and/or poor patient prognosis (Freije et al. 1998; Guo et al. 1996; Guo et al. 1998; Kim et al. 2001; Miyazaki et al. 2000; Yang et al. 1997; Yu et al. 1997). Based on *in vitro* and *in vivo* studies, all eight of the genes listed in Table 2 fulfill the definition of a metastasis suppressor, and the following discussion will briefly summarize what is currently hypothesized and/or known about the molecular activity of these genes.

Nm23

Nm23 was the first metastasis suppressor gene to be discovered (Leone et al. 1991), and it has been demonstrated to suppress metastasis in a number of different tumor cell types, including melanoma, breast, colon, and oral squamous cell carcinoma (Leone et al. 2001, Leone et al. 1993; Miyazaki et al. 1999; Tagashira et al. 1998). Nm23 has recently been shown to be a histadine kinase which can phosphorylate the KSR (kinase suppressor of Ras) protein (Hartsough et al. 2002). KSR is known to be a scaffold protein for the Ras/Raf/ERK-MAPK pathway, and as such can provide a docking site for the assembly of important signalling molecules (Morrison 2001). It is hypothesized that the phosporylation of KSR by Nm23 could result in an altered intracellular location of KSR and/or a disruption in the assembly of signalling proteins, resulting in reduced ERK activation (Hartsough et al. 2002). Given the importance of the Ras/Raf/ERK-MAPK cascade to the metastatic process, disruption of this pathway would be a successful cellular strategy towards the suppression of metastasis.

KAI1

KAI1 has been shown to suppress metastasis in prostate and breast cancer cells (Dong et al. 1995; Yang et al. 2001), and loss of KAI1 expression has been widely correlated with the progression of many types of human cancer (Guo et al. 1996; Guo et al. 1998; Miyazaki et al. 2000; Yang et al. 1997; Yu et al. 1997). Conserved domains in the KAI1 protein backbone suggest that it is a member of the tetraspanin family of proteins, which function as membrane adaptors that coordinate large complexes of proteins, including cell adhesion molecules, integrins, kinases and phosphatases (Maecker et al. 1997). KAI1 can form distinct complexes with both integrins and EGFR, and KAI1-EGFR interactions have been shown to reduce the effects of downstream signalling of EGFR in response to stimulation with EGF (Bienstock and Barrett 2001; Odintsova et al. 2000). Thus, similar to Nm23, interference with signalling pathways important to metastases development (in this case EGFR and integrin-mediated pathways) has the potential to inhibit a cell's ability to successfully complete the metastatic process.

MKK4

Reduced expression of MKK4 has been observed in highly metastatic prostate and ovarian carcinoma cells (Yamada et al. 2002; Yoshida et al. 1999). MKK4 is a MAPK kinase (MAPKK) protein involved in the p38 and JNK (Jun terminal kinase) arms of the MAPK signalling cascade. Located directly upstream of JNK and p38, MKK4 can activate these proteins via phosphorylation (Cuenda 2000). The JNK pathway is most commonly associated with the induction of apoptosis in response to cellular stress, leading to the hypothesis that MKK4 may be involved in the suppression of metastasis by tipping the cellular proliferation:apopotosis balance in favor of apoptosis, thus negatively influencing the growth of metastases at secondary sites (Chekmareva et al. 1998; Yoshida et al. 1999).

RHOGDI2

The RHOGDI2 gene was recently identified as a potential metastasis suppressor by way of cDNA microarray comparison between two bladder cancer cell lines of varying metastatic aggressiveness. Subsequent transfection of RHOGDI2 into the more aggressive cell line demonstrated that this gene could reduce metastatic potential without affecting primary tumor growth, thus validating it as a *bona fide* tumor suppressor gene (Gildea et al. 2002). RHOGDI2 is an inhibitor of guanine nucleotide dissociation for Rho GTPases, a family of proteins that play an important role in regulation of the cytoskeleton resulting in downstream cellular effects on cell morphology, motility, and adhesion (Schmitz et al. 2000). Interactions between Rho GTPases and RHOGDI2 can result in modulation of the JNK, p38, and ERK signalling pathways (Schmitz et al. 2000), indicating that decreased expression of RHOGDI2 during metastasis could potentially play a functional role in the extensive dysregulation of intracellular signalling.

BRMS1

The BRMS1 gene has been identified as a metastasis suppressor in breast and melanoma cells (Seraj et al. 2000; Shevde et al. 2002). Although it is not currently known what type of protein the BRMS1 gene encodes, nuclear localization of the protein combined with the presence of coiled-coil domains, glutamate-rich domains, and a leucine zipper domain suggests that it could be part of a transcriptional complex (Debies and Welch 2001). Functionally, a study by Saunders et al. (2001) indicates that BRMS1 may be involved in gap-junction communications between cells. Gap-junctions are membrane channels comprised of clusters of connexin proteins which facilitate the transfer of signalling molecules from cell to cell, either through homotypic (between the same cell type) or heterotypic (between different cell types) interactions (Yamasaki et al. 1995). Saunders et al. (2001) demonstrated that breast cancer cells transfected with BRMS1 showed increased gap-junctional communication relative to control cells, possibly due to BRMS1-mediated changes in connexin gene expression (Debies and Welch 2001). The authors propose that these increased cell–cell communications between non-metastatic cells (expressing BRMS1) and metastatic cells (not expressing BRMS1) could suppress metastasis through the directional distribution of signalling molecules important for controlling the metastatic process. This hypothesis suggests that, in addition

to regulating signalling proteins (as is predicted for Nm23, KAI1, MKK4, and RHOGDI2), metastasis suppressor genes may also facilitate the cell–cell exchange of these signalling molecules.

KiSS1, CRSP3, and TXNIP

Welch et al. (1994) showed that microcell-mediated transfer of chromosome 6 into metastatic melanoma cells could result in decreased metastasis with no effect on tumorigenicity. Subsequent studies by this group identified a metastasis suppressor gene, KiSS1, as being responsible for this effect (Lee et al. 1996). KiSS1 is a precursor for the secreted neuropeptide Metastin/Kisspeptin, a ligand for a G-protein coupled receptor. Metastin/Kisspeptin can induce phosphorylation of components of the ERK and p38 signalling pathways resulting in inhibition of cell motility and invasion, potentially via differential regulation of MMPs (Kotani et al. 2001; Yan et al. 2001). Interestingly, the KiSS1 gene was mapped not to chromosome 6, but to chromosome 1, suggesting that KiSS1 was regulated by another gene on chromosome 6. Indeed, this group recently identified the transcriptional co-activator CRSP3 (mapped to chromosome 6) as being capable of upregulating KiSS1 expression. Another gene, TXNIP (mapped to chromosome 1) was also shown to be upregulated by CRSP3 and determined to be involved with suppression of melanoma cell metastasis (Goldberg et al. 2003). TXNIP is a vitamin D3 regulated protein which binds to the redox-active state of thioredoxin (TRX), a protein thought to play a role in cellular growth control via regulation of DNA synthesis, signalling cascades, and transcription factor activity (Nakamura et al. 1997). Independent of KiSS1, both CRSP3 and TXNIP were determined to be metastasis suppressor genes in their own right, leading to the theory that perhaps that CRSP3 can regulate a common signalling pathway which involves both KiSS1 and TXNIP, and reduced expression of any of these genes can result in disruption of transcriptional regulation in favor of metastasis development (Goldberg et al. 1999; Goldberg et al. 2003).

In addition to the eight *bona fide* metastasis suppressor genes discussed here, several other genes have been implicated in metastasis suppression. For example, protease inhibitors such as mapsin and TIMPs as well as the membrane proteins CD44 and E-cadherin have all been shown to have some involvement in metastasis suppression (Christofori and Semb 1999; Gao et al. 1997; Toi et al. 1998; Zhang et al. 2000). However, conflicting studies have shown that (depending on tumor type and/or cellular situation) E-cadherin, different TIMPs, and different CD44 variants can also function as metastasis promoters (De Marzo et al. 1999; Kauffman et al. 2003; Ree et al. 1997), indicating that further studies are clearly needed in order to fully delineate the role that these molecules play in the metastatic process.

Summary

Dysregulation of the balance between the expression of genes which promote metastasis and the expression of genes which suppress metastasis can influence a cell's ability to successfully complete all the steps in the metastatic cascade (Fig. 1). Combined with the growing evidence that cell–cell and cell–matrix interactions are absolute requirements for metastasis development, the observation that both tumor cells and host cells (i.e. endothelial cells, stromal cells, immune cells) can produce these and other metastasis-associated factors continues to validate Paget's classical theory of "seed and soil" at both the cellular and molecular level. Signals relayed from host cells in the tumor microenvironment (the "soil") can be transmitted to tumor cells (the "seed") via cell surface receptors which initiate signalling cascades in order to regulate cellular behavior. The inappropriate response of tumor cells to normal host signalling and/or the inability of the host to appropriately regulate cellular behavior can result in cancer progression and metastasis. The ability of tumor cells to develop and maintain a metastatic phenotype is therefore reliant on factors intrinsic to each tumor cell (genetic influences), factors intrinsic to the host environment (epigenetic influences), and the complex interactions between the two.

Studies involving the experimental manipulation of individual metastasis-associated genes have led to tremendous advances in elucidating specific molecular mechanisms involved in the various steps of metastasis. However, these findings must be carefully interpreted in order to avoid a reductionist or oversimplified view of the metastatic process, particularly considering the variability between different types of cancer

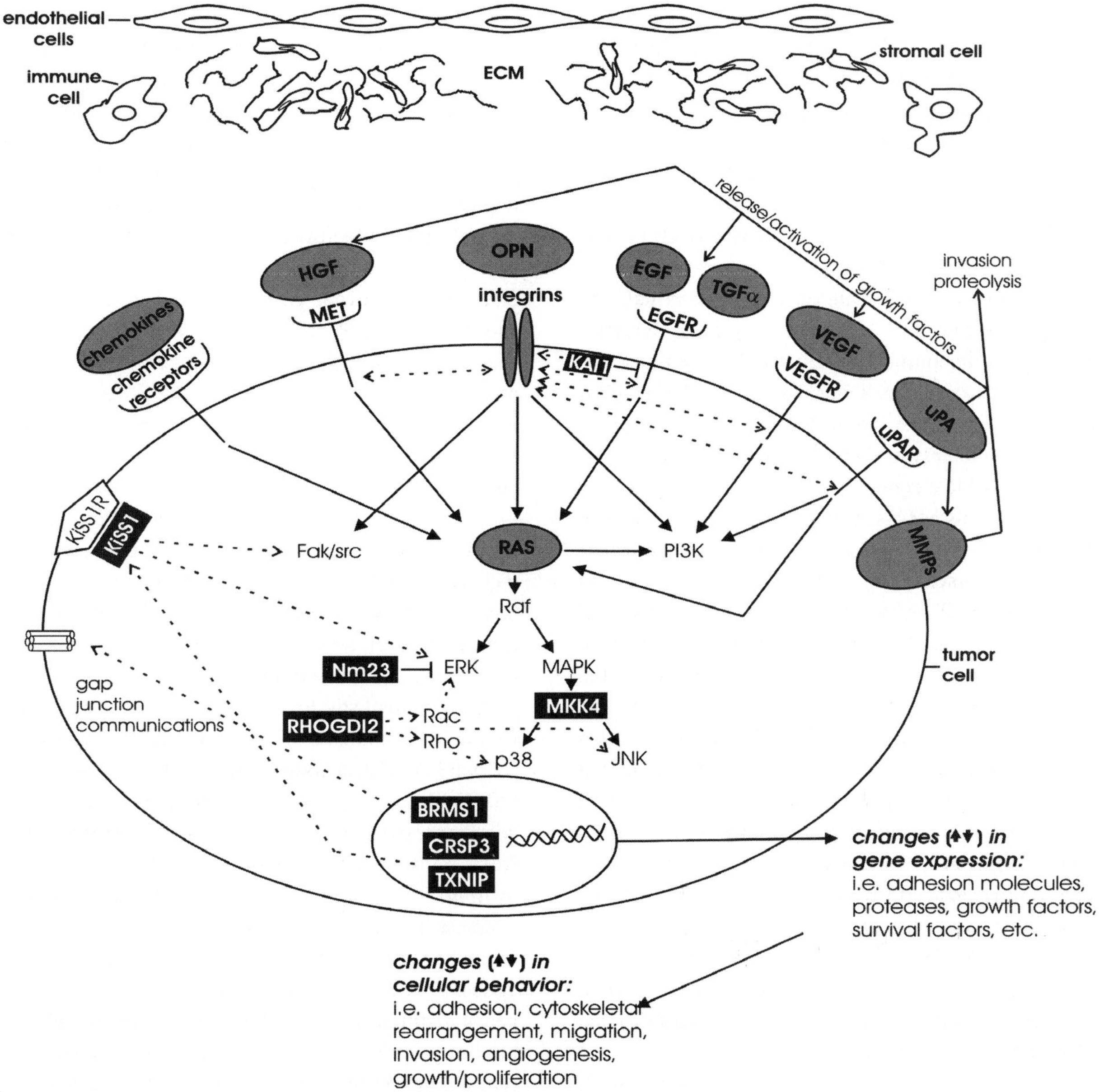

endothelial — cells

immune cell

ECM

stromal cell

release/activation of growth factors

invasion proteolysis

HGF

OPN

integrins

EGF

TGFα

MET

EGFR

VEGF

VEGFR

uPA

chemokines chemokine receptors

KAI1

uPAR

KiSS1R

KiSS1

Fak/src

RAS

PI3K

MMPs

tumor cell

Raf

gap junction communications

Nm23

ERK

MAPK

RHOGDI2

Rac

MKK4

Rho

p38

JNK

BRMS1

CRSP3

TXNIP

changes (▲▼) in gene expression: i.e. adhesion molecules, proteases, growth factors, survival factors, etc.

changes (▲▼) in cellular behavior: i.e. adhesion, cytoskeletal rearrangement, migration, invasion, angiogenesis, growth/proliferation

Fig. 1 Molecular mechanisms of metastasis. By using the various metastasis-associated molecules discussed in the text as examples, this schematic diagram is a hypothetical representation of the cellular and molecular cross-talk which is required for metastasis progression. Metastasis-associated molecules are produced by both the tumor cell and host cells in the tumor microenvironment. Progression of metastasis is therefore reliant on host/tumor interactions, the coordinated activity of multiple genes, and a dysregulation of the balance between metastasis suppressors (black rectangles) and metastasis promoters (gray ovals)

and the inherent genetic instability associated with tumor progression. The ongoing development of high-throughput molecular analysis tools such as DNA microarray technology may provide a more accurate picture of the relationship between genes and metastasis both within and between different tumor types.

Microarrays: tools for solving the metastasis puzzle?

DNA microarrays are microchips which contain tens of thousands of individual nucleic acid samples, each representing a known gene or a defined expressed

sequence tag (EST). The value of DNA microarrays to researchers and clinicians lies in the ability to simultaneously observe the expression patterns of multiple genes in response to a given experimental treatment, thus leading to information about transcriptional changes involved in specific signalling pathways and the identification of novel genes which may be functionally important for specific cellular processes. In addition, DNA microarrays can be used to identify broad-spectrum changes in gene expression patterns between normal and abnormal tissues, potentially defining a "fingerprint" or "signature" of the transcriptional status of the cell during the development of diseases such as metastatic cancer. The use of DNA microarrays to answer discovery-based research questions (i.e. which genes are expressed differently in metastatic cells versus normal cells?) has enhanced traditional hypothesis-driven research (i.e. can a particular gene functionally contribute to metastasis?) by facilitating the discovery of new metastasis-associated genes. A number of studies involving different tumor cell types have used microarrays to compare the expression profiles of cell lines with varying degrees of metastatic potential, leading to the identification of novel genes which can subsequently be examined in greater detail in order to determine their functional influence on the metastatic process (Chakrabarti et al. 2002; Cooper et al. 2003; Dong et al. 2001; Euer et al. 2002; Gildea et al. 2002; Goldberg et al. 2003; Otsuka et al. 2001; Sakakura et al. 2002; Scholl et al. 2000; Wang et al. 2002).

Concurrent to these experimental studies, several large-cohort clinical studies have illustrated the value of microarray-mediated gene profiling as a tool for predicting patient survival for different types of cancer, including B-cell lymphoma and cancers of the breast, lung, and brain (Beer et al. 2002; Pomeroy et al. 2002; Shipp et al. 2002; van de Vijver et al. 2002; van't Veer et al. 2002). Of these cancers, the poor outcome of breast cancer patients in particular is almost invariably a result of metastatic disease. Two recent studies (van de Vijver et al. 2002; van't Veer et al. 2002) report that that a 70-gene "signature" of primary breast tumors can more accurately predict the likelihood of metastases development and patient outcome than current clinical and histological criteria such as age, lymph node status, histological tumor grade, and estrogen-receptor (ER) status (McGuire 1991).

Not surprisingly, this molecular signature includes changes in the expression patterns of several genes which encode functional classes of molecules known to be associated with metastatic cell behavior, such as signal transduction molecules, cell cycle control proteins, growth and angiogenic factors, and molecules important for cell motility and invasion. However, several genes which have previously been correlated with breast cancer metastasis, such as uPA, cyclin D1, estrogen ER-α, HER2/*neu* and c-myc (Bieche and Lidereau 1995; Janicke et al. 2001; van Diest et al. 1997) were noticeably absent from the 70-gene profile, indicating that further studies are required in order to clarify how this gene signature relates to functional mechanisms which influence cell behavior during breast cancer metastasis.

Based on the findings of van't Veer and colleagues, it is reasonable to suppose that molecular signatures of metastasis might also exist in primary tumors of other highly metastatic cancers, such as prostate and colon carcinomas. However, given the intrinsic genetic heterogeneity both within and between tumors, it is probably unrealistic to assume that such a metastasis signature exists that is common to all types of cancer. Or is it? A controversial study by Ramaswamy et al. (2003) suggests otherwise. This study analyzed the gene expression profile of adenocarcinoma metastases from multiple solid tumor types (breast, uterine, ovarian, colon, prostate, and lung) relative to the gene expression profile of primary tumors from the same tumor spectrum. A 17-gene molecular signature comprised of 8 upregulated genes and 9 downregulated genes distinguished the metastases from the primary tumors, and this signature was common across tumor types. Furthermore, it was also observed that this metastasis signature was present in some primary tumors, and that these tumors were most likely to be associated with metastasis development and poor clinical outcome. The authors conclude that these results collectively support the presence of generic metastasis-determining molecular programs for all tumors rather than distinct mechanisms of metastasis for different tumor types.

This study is the first of its kind to provide evidence for a molecular signature that is biologically informative in multiple tumor types. While these findings are intriguing and potentially very important for the field of cancer metastasis research, there are several

aspects of the study which are unsettling and in direct opposition to many traditional views of the metastatic process. For example, none of the 17 genes identified in this study as being predictive for metastasis correspond to any confirmed metastasis-promoting or metastasis-suppressing genes, or even to functional classes of molecules shown to be associated with metastasis. Instead, these genes mostly encode for proteins which are predicted to be important for the regulation of overall and/or critical cell machinery, i.e. components of the translational apparatus and proteins involved in mitosis. The specific molecular mechanisms by which these genes (alone or in combination) can influence metastasis are not known, nor are they really even speculated upon by the authors. However, the generality of the gene products suggests that they could affect diverse downstream molecular targets which most likely differ from tumor to tumor because of genetic instability.

How can these observations be reconciled with the hundreds of studies that have defined specific molecules as being functional contributors to the metastatic process? The answer might be found, once again, by examining the mechanisms of host–tumor interaction (Liotta and Kohn 2003). Because the authors did not microdissect the tumor cells to separate them from host cells before analyzing gene expression, it is impossible to separate cause from consequence. That is to say, it is not clear whether the 17-gene expression profile is the *cause* of metastasis (i.e. tumor cell influence on the host), the *consequence* of metastasis (i.e. host influence on the tumor cell), or a combination of both. Another intriguing idea to take into account is the suggestion that host genetic background may influence metastatic potential, perhaps via allelic differences in metastasis suppressor genes (Hunter et al. 2001). It is entirely possible that, rather than indicating that a given tumor has a higher likelihood of becoming metastatic, the 17-gene metastasis signature may instead be more reflective of a host's genetic predisposition to develop metastatic disease. Clarification of the role that host-tumor interactions and host genetic background play in dictating this 17-gene molecular signature would probably help to put the findings of this study into context with current knowledge about the metastatic process.

Notwithstanding the issues of functional mechanisms, host–tumor interactions, and the contribution of host genetic background, the observation that primary tumors can express molecular signatures predictive of their likelihood to metastasize has sparked new debate about the origin of the metastatic cell (Bernards and Weinberg 2002; Edwards 2002; Gatenby and Maini 2002; Sherley 2002). The prevailing theory of metastasis predicts that some cells within the primary tumor become metastatic by acquiring one or more molecular changes (in addition to changes acquired during tumorigenicity) and/or by simply succumbing to the collective effect of accumulated advantageous mutations (Poste and Fidler 1980). Furthermore, the acquisition of a metastatic cell phenotype is believed to be a rare, stochastic event that occurs late in multistep tumorigenicity, providing the cell with the ability to become invasive, escape from the primary tumor, travel through the circulation, and initiate and maintain growth at a distant site (Poste and Fidler 1980; Weiss 1990). However, in light of the findings of microarray studies (Ramaswamy et al. 2003; van de Vijver et al. 2002; van't Veer et al. 2002), it has been suggested instead that some primary tumors are "pre-configured" by their molecular signature to metastasize, and that this proclivity occurs as a result of an early event (rather than a late event) in tumorigenicity (Bernards and Weinberg 2002; Ramaswamy et al. 2003; van de Vijver et al. 2002).

Bernards and Weinberg (2002) identified several potential implications of this early acquisition of metastatic potential. For example, they hypothesized that the same molecular changes that can confer a selective growth advantage early in tumorigenicity can then confer the ability of that cell to escape from the primary tumor later in tumorigenicity. This scenario would suggest that genes which are "specifically and exclusively" involved in regulating metastasis do not exist. Certainly, genes that *exclusively* regulate metastasis probably do not exist: functional studies examining metastasis-promoting and metastasis-suppressing genes indicate that molecules involved in metastasis are virtually always, under normal circumstances, involved in mediating key cellular processes and/or in maintaining physiological homeostasis. However, the idea that there are no genes which are capable of *specifically* regulating the metastasis process contradicts the findings of decades of molecular-based cancer metastasis research. Bernards, Weinberg, and others (Ramaswamy et al. 2003; van de Vijver et al. 2002)

also suggest that the early acquisition of a metastatic phenotype calls into question the value of early cancer detection and treatment, since even very small primary tumor cell populations theoretically may already have the ability to disseminate to distant sites. Ironically, the most inconsistent aspect of this debate about the origin of the metastatic cell is the aspect which is at its very core: the definition of an "early" versus a "late" event in multistep tumorigenicity. It is well established that, by the time a tumor is clinically detectable, it is invariably well into the tumorigenic process (Liotta et al. 1991; Nicolson 1991). Therefore, without the ability to mathematically define the exact time at which molecular changes are acquired relative to the timing of particular steps in tumor development and metastasis, it is difficult to draw appropriate conclusions about the validity of (and the relationship between) each theory (Gatenby and Maini 2002).

These issues aside, the undeniable power of DNA microarrays as tools for high-throughput gene expression analysis is reflected by the contributions that they continue to make towards the elucidation of molecular aspects of metastasis. Moreover, the chance that these valuable tools could be used in the clinical setting for the benefit of patients is an exciting prospect. The final part of this chapter will discuss how our evolving knowledge about metastasis-associated genes (both as molecular signatures and as functional entities) can be translated into clinical strategies for the management of metastatic disease.

Metastasis genes and the clinic: prognostic and therapeutic implications

By striving to gain an understanding of the molecular details of metastasis, the studies that have been presented herein have the common goal of ultimately contributing to improved cancer patient cure rates and quality of life. Any therapy that prevents metastasis from developing to a point where it can do physiological harm to a patient has potential for clinical utility. Each of the individual steps in metastasis (intravasation, survival in the circulation, extravastation, and growth in the secondary site) therefore presents a possible target for therapeutic intervention. Many of the most promising anti-cancer drugs currently in clinical trials are those which target molecules involved in

multiple steps in the metastatic process across various tumor types, i.e. MMP inhibitors (Coussens et al. 2002; Overall and Lopez-Otin 2002), VEGF inhibitors (Zhu et al. 2002), and small molecule antagonists which interfere with critical ligand interactions of cell signalling receptors such as EGFR (Ciardiello and Tortora 2001) and the $\alpha v \beta 3$ integrin (Tucker 2002). However, the fact that the metastatic process is known to be inherently inefficient suggests that there may be particular steps which are either more clinically accessible than others and/or which are better biological targets for the prevention of metastasis. Specifically, it has been proposed that the final broad step of initiation and maintenance of metastases growth in secondary sites could be successfully inhibited by targeting components of oncogene-activated pathways (i.e. the Ras/Raf/MAPK pathway) (Chambers and Varghese 2002), by limiting the growth of metastases via inhibition of angiogenesis (Folkman 2002), or perhaps even by trying to pharmaceutically upregulate metastasis suppressor gene function (Kauffman et al. 2003; Steeg 2003). In addition, the accumulated evidence that implicates "seed and soil" interactions as critical components of the metastatic process suggests that interference with host–tumor cell compatibility (i.e. by inhibiting organ-specific metastatic growth mediated by chemokine receptors) (Muller et al. 2001) might also provide a basis for useful therapeutic strategies.

The same major obstacles that have thus far prevented researchers from identifying an absolute molecular program that defines metastasis – namely the complex nature of the metastatic process and the inherent heterogeneity within and between tumor types – have also presented significant clinical challenges in the successful management of metastatic disease. Therefore, while therapeutics that target individual metastasis-promoting or metastasis-suppressing molecules are promising, each drug (if used on its own) will probably only benefit that subset of patients which have dysregulated expression of the particular target molecule(s). The use of microarray studies to identify the association between multiple genes in a given metastatic tumor will hopefully be valuable for identifying novel signalling pathways which could be targeted for the development of anticancer therapies more broadly applicable to a greater number of patients. However, until this occurs, the most successful therapeutic strategies to combat metastatic

cancer will probably prove to be those that use specific molecular therapies (in appropriate subsets of patients) in combination with more traditional approaches such as chemotherapy and hormone therapy.

A potentially more promising application for the molecular knowledge about metastasis is towards the development of effective new prognostic tools in order to improve patient quality of life and enhance clinical decision making. Currently, the use of well-established prognostic indicators (such tumor size or grade) to predict outcome is helpful but imperfect, owing mainly to tumor plasticity and the reliance on subjective assessment criteria. Similarly, although some specific molecules are currently in clinical use as predictive markers of patient outcome and response to therapy (i.e. HER2/*neu* for breast cancer) (Bieche and Lidereau 1995), these too are imperfect as a result of tumor heterogeneity. This uncertainty in predicting disease outcome often results in under-treatment or over-treatment: some patients who need systemic therapy to treat metastatic disease may be missed, and other patients who have been successfully treated by local surgery and radiation and do not require systemic therapy may be unnecessarily exposed to toxic side-effects. There is therefore a clear need for improved tools that can be used to accurately and reliably predict disease outcome.

Several recent studies (Ramaswamy et al. 2003; van de Vijver et al. 2002; van't Veer et al. 2002) suggest that microarray-mediated gene expression profiling of primary tumors has the potential to benefit the patient and the clinician in many different ways. For instance, prognostic profiling might be able to distinguish between patients who need systemic treatment and those who do not. Furthermore, gene expression profiles could be used to look for one or more specific molecular changes which could then predict a patient's response to a particular treatment, and this could in turn lead the way to "tailored" treatments specific to each patient's prognostic profile (Kallioniemi 2002; van't Veer and De Jong 2002). Notwithstanding these exciting possibilities, there are several issues which need to be addressed before microarrays can be put into routine use in the clinic. First of all, the existing studies need to be independently validated by examining multiple large patient groups with various tumor types. Secondly, much stricter guidelines need to be established to govern various technical aspects of microarray use, such as array-to-array reproducibility and standardization of data analysis methods, in order to permit comparisons between studies. Finally, factors involving cost and tissue sample collection must also be taken into consideration, since microarrays are presently cost-prohibitive and work best with fresh tumor specimens rather than the standard formalin-fixed tumor specimens which are presently collected (Kallioniemi 2002; Sauter and Simon 2002; Schmidt and Begley 2003).

In conclusion, it is clear that a coordinated approach involving molecular biologists, pathologists, clinicians, and bioinformatic scientists will be required in order to translate experimental observations about molecular aspects of metastasis from the bench to the bedside. Such a multidisciplinary approach will hopefully lead to the development of novel prognostic and/or therapeutic strategies and the improved ability to manage, treat, and cure metastatic disease.

Acknowledgements We thank members of our laboratory and our collaborators for their research work and helpful discussions. We apologize to the authors whose work we could not cite because of space restrictions. Metastasis research in our laboratories is supported by grants from the Canadian Institutes of Health Research (#42511); the Canadian Breast Cancer Research Alliance, with special funding from the Canadian Breast Cancer Foundation and The Cancer Research Society (#016506); the Ontario Cancer Research Network (#04MAY00089); and an award from the Lloyd Carr-Harris Foundation. ALA is supported by the Imperial Oil Foundation. AFC is a Canada Research Chair in Oncology, supported by the Canada Research Chairs Program.

References

Abdel-Ghany M, Cheng HC, Elble RC, Pauli BU (2001) The breast cancer beta 4 integrin and endothelial human CLCA2 mediate lung metastasis. J Biol Chem 276:25438–25446

Agrawal D, Chen T, Irby R, Quackenbush J, Chambers AF, Szabo M, et al. (2002) Osteopontin identified as lead marker of colon cancer progression, using pooled sample expression profiling. J Nat Cancer Inst 94:513–521

Albelda SM, Mette SA, Elder DE, Stewart R, Damjanovich L, Herlyn M, et al. (1990) Integrin distribution in malignant melanoma: association of the beta 3 subunit with tumor progression. Cancer Res 50:6757–6764

Albini A, Melchiori A, Santi L, Liotta LA, Brown PD, Stetler-Stevenson WG (1991) Tumor cell invasion inhibited by TIMP-2. J Nat Cancer Inst 83:775–779

Al-Mehdi AB, Tozawa K, Fisher AB, Shientag L, Lee A, Muschel RJ (2000) Intravascular origin of metastasis from the proliferation of endothelium-attached tumor cells: a new model for metastasis. Nat Med 6:100–102

Arai J, Yasukawa M, Yakushijin Y, Miyazaki T, Fujita S (2000) Stromal cells in lymph nodes attract B-lymphoma cells

via production of stromal cell-derived factor-1. Eur J Haematol 64:323–332

Beer DG, Kardia SL, Huang CC, Giordano TJ, Levin AM, Misek DE, et al. (2002) Gene-expression profiles predict survival of patients with lung adenocarcinoma. Nat Med 8:816–824

Behrend EI, Craig AM, Wilson SM, Denhardt DT, Chambers AF (1994) Reduced malignancy of ras-transformed NIH 3T3 cells expressing antisense osteopontin RNA. Cancer Res 54:832–837

Bernards R, Weinberg RA (2002) A progression puzzle. Nature 418:82

Bernhard EJ, Gruber SB, Muschel RJ (1994) Direct evidence linking expression of matrix metalloproteinase 9 (92-kDa gelatinase/collagenase) to the metastatic phenotype in transformed rat embryo cells. Proc Natl Acad Sci USA 91:4293–4297

Berse B, Brown LF, Van de Water L, Dvorak HF, Senger DR (1992) Vascular permeability factor (vascular endothelial growth factor) gene is expressed differentially in normal tissues, macrophages, and tumors. Mol Biol Cell 3:211–220

Bieche I, Lidereau R (1995) Genetic alterations in breast cancer. Genes Chromosomes Cancer 14:227–251

Bienstock RJ, Barrett JC (2001) KAI1, a prostate metastasis suppressor: prediction of solvated structure and interactions with binding partners; integrins, cadherins, and cell-surface receptor proteins. Mol Carcinog 32:139–153

Blair DG, Cooper CS, Oskarsson MK, Eader LA, Vande Woude GF (1982) New method for detecting cellular transforming genes. Science 218:1122–1125

Blasi F, Carmeliet P (2002) uPAR: a versatile signalling orchestrator. Nat Rev Mol Cell Biol 3:932–943

Bondy GP, Wilson S, Chambers AF (1985) Experimental metastatic ability of H-ras-transformed NIH3T3 cells. Cancer Res 45:6005–6009

Bos JL (1989) ras oncogenes in human cancer: a review. Cancer Res 49:4682–4689

Brown LF, Papadopoulos-Sergiou A, Berse B, Manseau EJ, Tognazzi K, Perruzzi CA, et al. (1994) Osteopontin expression and distribution in human carcinomas. Am J Pathol 145:610–623

Brinckerhoff CE, Matrisian LM (2002) Matrix metalloproteinases: a tail of a frog that became a prince. Nat Rev Mol Cell Biol 3:207–214

Brunner N, Nielsen HJ, Hamers M, Christensen IJ, Thorlacius-Ussing O, Stephens RW (1999) The urokinase plasminogen activator receptor in blood from healthy individuals and patients with cancer. APMIS 107:160–167

Butler TP, Gullino PM (1975) Quantitation of cell shedding into efferent blood of mammary adenocarcinoma. Cancer Res 35:512–516

Cameron MD, Schmidt EE, Kerkvliet N, Nadkarni KV, Morris VL, Groom AC, et al. (2000) Temporal progression of metastasis in lung: cell survival, dormancy, and location dependence of metastatic inefficiency. Cancer Res 60:2541–2546

Campbell SL, Khosravi-Far R, Rossman KL, Clark GJ, Der CJ (1998) Increasing complexity of Ras signaling. Oncogene 17:1395–1413

Chakrabarti R, Robles LD, Gibson J, Muroski M (2002) Profiling of differential expression of messenger RNA in

normal, benign, and metastatic prostate cell lines. Cancer Genet Cytogenet 139:115–125

Chakravarti A, Chakladar A, Delaney MA, Latham DE, Loeffler JS (2002) The epidermal growth factor receptor pathway mediates resistance to sequential administration of radiation and chemotherapy in primary human glioblastoma cells in a RAS-dependent manner. Cancer Res 62:4307–4315

Chambers AF, Tuck AB (1993) Ras-responsive genes and tumor metastasis. Cri Rev Oncog 4:95–114

Chambers AF, MacDonald IC, Schmidt EE, Koop S, Morris VL, Khokha R, et al. (1995) Steps in tumor metastasis: new concepts from intravital videomicroscopy. Cancer Metastasis Rev 14:279–301

Chambers AF, Matrisian LM (1997) Changing views of the role of matrix metalloproteinases in metastasis. J Natl Canc Inst 89:1260–1270

Chambers AF (1999) The metastatic process: basic research and clinical implications. Oncol Res 11:161–168

Chambers AF, MacDonald IC, Schmidt EE, Morris VL, Groom AC (2000) Clinical targets for anti-metastasis therapy. Adv Cancer Res 79:91–121

Chambers AF, Naumov GN, Varghese HJ, Nadkarni KV, MacDonald IC, Groom AC (2001) Critical steps in hematogenous metastasis: an overview. Surg Oncol Clin N Am 10:243–255, vii

Chambers AF, Varghese HJ (2002) Oncogenes as therapeutic targets to prevent metastasis. In: Oncogene-directed therapies. Humana, NJ, pp 219–228

Chambers AF, Groom AC, MacDonald IC (2002) Dissemination and growth of cancer cells in metastatic sites. Nat Rev Cancer 2:563–572

Chan KC, Knox WF, Gee JM, Morris J, Nicholson RI, Potten CS, et al. (2002) Effect of epidermal growth factor receptor tyrosine kinase inhibition on epithelial proliferation in normal and premalignant breast. Cancer Res 62:122–128

Chang EH, Furth ME, Scolnick EM, Lowy DR (1982) Tumorigenic transformation of mammalian cells induced by a normal human gene homologous to the oncogene of Harvey murine sarcoma virus. Nature 297:479–483

Chekmareva MA, Kadkhodaian MM, Hollowell CM, Kim H, Yoshida BA, Luu HH, et al. (1998) Chromosome 17-mediated dormancy of AT6.1 prostate cancer micrometastases. Cancer Res 58:4963–4969

Chung YS, Maeda K, Sowa M (1996) Prognostic value of angiogenesis in gastrointestinal tumours. Eur J Cancer 32A:2501–2505

Ciardiello F, Tortora G (2001) A novel approach in the treatment of cancer: targeting the epidermal growth factor receptor. Clin Cancer Res 7:2958–2970

Ciardiello F, Caputo R, Troiani T, Borriello G, Kandimalla ER, Agrawal S, et al. (2001) Antisense oligonucleotides targeting the epidermal growth factor receptor inhibit proliferation, induce apoptosis, and cooperate with cytotoxic drugs in human cancer cell lines. Int J Cancer 93:172–178

Claffey KP, Brown LF, del Aguila LF, Tognazzi K, Yeo KT, Manseau EJ, et al. (1996) Expression of vascular permeability factor/vascular endothelial growth factor by melanoma cells increases tumor growth, angiogenesis, and experimental metastasis. Cancer Res 56:172–181

Clark GJ, Der CJ (1995) Aberrant function of the Ras signal transduction pathway in human breast cancer. Breast Cancer Res Treat 35:133–144

Clarke MR, Landreneau RJ, Finkelstein SD, Wu TT, Ohori P, Yousem SA (1997) Extracellular matrix expression in metastasizing and nonmetastasizing adenocarcinomas of the lung. Hum Pathol 28:54–59

Collen D (1999) The plasminogen (fibrinolytic) system. Thromb Haemost 82:259–270

Cooper CR, Chay CH, Gendernalik JD, Lee HL, Bhatia J, Taichman RS, et al. (2003) Stromal factors involved in prostate carcinoma metastasis to bone. Cancer 97:739–747

Coussens LM, Fingleton B, Matrisian LM (2002) Matrix metalloproteinase inhibitors and cancer: trials and tribulations. Science 295:2387–2392

Christofori G, Semb H (1999) The role of the cell-adhesion molecule E-cadherin as a tumour-suppressor gene. Trends Biochem Sci 24:73–76

Crooke ST (2001) Basic principles of antisense technology. In: Antisense drug technology: principles, strategies and applications. Marcel Dekker, New York, pp 1–28

Crowley CW, Cohen RL, Lucas BK, Liu G, Shuman MA, Levinson AD (1993) Prevention of metastasis by inhibition of the urokinase receptor. Proc Natl Acad Sci USA 90:5021–5025

Cuenda A (2000) Mitogen-activated protein kinase kinase 4 (MKK4). Int J Biochem Cell Biol 32:581–587

Debies MT, Welch DR (2001) Genetic basis of human breast cancer metastasis. J Mammary Gland Biol Neoplasia 6:441–451

De Marzo AM, Knudsen B, Chan-Tack K, Epstein JI (1999) E-cadherin expression as a marker of tumor aggressiveness in routinely processed radical prostatectomy specimens. Urology 53:707–713

Denhardt DT, Chambers AF (1994) Overcoming obstacles to metastasis–defenses against host defenses: osteopontin (OPN) as a shield against attack by cytotoxic host cells. J Cell Biochem 56:48–51

Dong G, Loukinova E, Chen Z, Gangi L, Chanturita TI, Liu ET, et al. (2001) Molecular profiling of transformed and metastatic murine squamous carcinoma cells by differential display and cDNA microarray reveals altered expression of multiple genes related to growth, apoptosis, angiogenesis, and the NF-kappaB signal pathway. Cancer Res 61:4797–4808

Dong JT, Lamb PW, Rinker-Schaeffer CW, Vukanovic J, Ichikawa T, Isaacs JT, et al. (1995) KAI1, a metastasis suppressor gene for prostate cancer on human chromosome 11p11.2. Science 268:884–886

Duffy MJ, Reilly D, O'Sullivan C, O'Higgins N, Fennelly JJ, Andreasen P (1990) Urokinase-plasminogen activator, a new and independent prognostic marker in breast cancer. Cancer Res 50:6827–6829

Edwards PA (2002) Metastasis: the role of chance in malignancy. Nature 419:559–560

Egeblad M, Werb Z (2002) New functions for the matrix metalloproteinases in cancer progression. Nat Rev Cancer 2:161–174

Eliceiri BP, Cheresh DA (1999) The role of alphav integrins during angiogenesis: insights into potential mechanisms of action and clinical development. J Clin Invest 103:1227–1230

Euer N, Schwirzke M, Evtimova V, Burtscher H, Jarsch M, Tarin D, et al. (2002) Identification of genes associated with metastasis of mammary carcinoma in metastatic versus non-metastatic cell lines. Anticancer Res 22:733–740

Ewing J (1928) Neoplastic diseases. A treatise on tumors. W.B. Saunders, London, pp 77–89

Fasano O, Birnbaum D, Edlund L, Fogh J, Wigler M (1984) New human transforming genes detected by a tumorigenicity assay. Mol Cell Biol 4:1695–1705

Fausto N (1991) Growth factors in liver development, regeneration and carcinogenesis. Prog Growth Factor Res 3:219–234

Fedarko NS, Jain A, Karadag A, Van Eman MR, Fisher LW (2001) Elevated serum bone sialoprotein and osteopontin in colon, breast, prostate, and lung cancer. Clin Cancer Res 7:4060–4066

Felding-Habermann B, O'Toole TE, Smith JW, Fransvea E, Ruggeri ZM, Ginsberg MH, et al. (2001) Integrin activation controls metastasis in human breast cancer. Proc Natl Acad Sci USA 98:1853–1858

Fidler IJ (1991) The biology of human cancer metastasis. Acta Oncol 30:668–675

Fidler IJ (1999) Critical determinants of cancer metastasis: rationale for therapy. Cancer Chemother Pharmacol 43: Suppl:S3–S10

Fidler IJ (2001) Seed and soil revisited: contribution of the organ microenvironment to cancer metastasis. Surg Oncol Clin N Am 10:257–269, vii–viii

Fidler IJ, Yano S, Zhang RD, Fujimaki T, Bucana CD (2002) The seed and soil hypothesis: vascularisation and brain metastases. Lancet Oncol 3:53–57

Firon M, Shaharabany M, Altstock RT, Horev J, Abramovici A, Resau JH, et al. (2000) Dominant negative Met reduces tumorigenicity-metastasis and increases tubule formation in mammary cells. Oncogene 19:2386–2397

Folkman J (1986) How is blood vessel growth regulated in normal and neoplastic tissue? Cancer Res 46:467–473

Folkman J (1992) The role of angiogenesis in tumor growth. Semin Cancer Biol 3:65–71

Folkman J (2002) Role of angiogenesis in tumor growth and metastasis. Semin Oncol 29:15–18

Frame S, Balmain A (2000) Integration of positive and negative growth signals during ras pathway activation in vivo. Curr Opin Genet Dev 10:106–113

Freije JM, MacDonald NJ, Steeg PS (1998) Nm23 and tumour metastasis: basic and translational advances. Biochem Soc Symp 63:261–271

Furger KA, Menon RK, Tuck AB, Bramwell VH, Chambers AF (2001) The functional and clinical roles of osteopontin in cancer and metastasis. Curr Mol Med 1:621–632

Gabrilovich DI, Chen HL, Girgis KR, Cunningham HT, Meny GM, Nadaf S, et al. (1996) Production of vascular endothelial growth factor by human tumors inhibits the functional maturation of dendritic cells. Nat Med 2:1096–1103

Gasparini G (2000) Prognostic value of vascular endothelial growth factor in breast cancer. Oncologist, 5 (Suppl 1) S37–S44

Gatenby RA, Maini P (2002) Modelling a new angle on understanding cancer. Nature 420:462

Gao AC, Lou W, Dong JT, Isaacs JT (1997) CD44 is a metastasis suppressor gene for prostatic cancer located on human chromosome 11p13. Cancer Res 57:846–849

Geissinger E, Weisser C, Fischer P, Schartl M, Wellbrock C (2002) Autocrine stimulation by osteopontin contributes to antiapoptotic signalling of melanocytes in dermal collagen. Cancer Res 62:4820–4828

Giancotti FG, Ruoslahti E (1999) Integrin signaling. Science 285:1028–1032

Gildea JJ, Seraj MJ, Oxford G, Harding MA, Hampton GM, Moskaluk CA, et al. (2002) RhoGDI2 is an invasion and metastasis suppressor gene in human cancer. Cancer Res 62:6418–6423

Glaves D (1983) Correlation between circulating cancer cells and incidence of metastases. Br J Cancer 48:665–673

Goldberg SF, Harms JF, Quon K, Welch DR (1999) Metastasis-suppressed C8161 melanoma cells arrest in lung but fail to proliferate. Clin Exp Metastasis 17:601–607

Goldberg SF, Miele ME, Hatta N, Takata M, Paquette-Straub C, Freedman LP, et al. (2003) Melanoma Metastasis Suppression by Chromosome 6: Evidence for a Pathway Regulated by CRSP3 and TXNIP. Cancer Res 63: 432–440

Grunstein J, Masbad JJ, Hickey R, Giordano F, Johnson RS (2000) Isoforms of vascular endothelial growth factor act in a coordinate fashion to recruit and expand tumor vasculature. Mol Cell Biol 20:7282–7291

Guo X, Friess H, Graber HU, Kashiwagi M, Zimmermann A, Korc M, et al. (1996) KAI1 expression is up-regulated in early pancreatic cancer and decreased in the presence of metastases. Cancer Res 56:4876–4880

Guo XZ, Friess H, Di Mola FF, Heinicke JM, Abou-Shady M, Graber HU, et al. (1998) KAI1, a new metastasis suppressor gene, is reduced in metastatic hepatocellular carcinoma. Hepatology 28:1481–1488

Hanahan D, Weinberg RA (2000) The hallmarks of cancer. Cell 100:57–70

Hart IR (1982) "Seed and soil" revisited: mechanisms of site-specific metastasis. Cancer Metastasis Rev 1:5–16

Hartsough MT, Morrison DK, Salerno M, Palmieri D, Ouatas T, Mair M, et al. (2002) Nm23-H1 metastasis suppressor phosphorylation of kinase suppressor of Ras via a histidine protein kinase pathway. J Biol Chem 277:32389–32399

Hasui Y, Marutsuka K, Asada Y, Osada Y (1996) Prognostic value of urokinase-type plasminogen activator in patients with superficial bladder cancer. Urology 47:34–37

Herold-Mende C, Steiner HH, Andl T, Riede D, Buttler A, Reisser C, et al. (1999) Expression and functional significance of vascular endothelial growth factor receptors in human tumor cells. Lab Invest 79:1573–1582

Holmgren L, O'Reilly MS, Folkman J (1995) Dormancy of micrometastases: balanced proliferation and apoptosis in the presence of angiogenesis suppression. Nat Med 1:149–153

Hotte SJ, Winquist EW, Stitt L, Wilson SM, Chambers AF (2002) Plasma osteopontin: associations with survival and metastasis to bone in men with hormone-refractory prostate carcinoma. Cancer 95:506–512

Hunter KW, Broman KW, Voyer TL, Lukes L, Cozma D, Debies MT, Rouse J, Welch DR (2001) Predisposition to efficient mammary tumor metastatic progression is linked to the breast cancer metastasis suppressor gene Brms1. Cancer Res 61:8866–8872

Itoh T, Tanioka M, Yoshida H, Yoshioka T, Nishimoto H, Itohara S (1998) Reduced angiogenesis and tumor progression in gelatinase A-deficient mice. Cancer Res 58:1048–1051

Janicke F, Prechtl A, Thomssen C, Harbeck N, Meisner C, Untch M, et al. (2001) Randomized adjuvant chemotherapy trial in high-risk, lymph node-negative breast cancer patients identified by urokinase-type plasminogen activator and plasminogen activator inhibitor type 1. J Natl Cancer Inst 93:913–920

Jiang WG, Hiscox S (1997) Hepatocyte growth factor/scatter factor, a cytokine playing multiple and converse roles. Histol Histopathol 12:537–555

Jiang WG, Grimshaw D, Lane J, Martin TA, Abounder R, Laterra J, et al. (2001) A hammerhead ribozyme suppresses expression of hepatocyte growth factor/scatter factor receptor c-MET and reduces migration and invasiveness of breast cancer cells. Clin Cancer Res 7:2555–2562

Junker K (2001) Prognostic factors in stage I/II non-small cell lung cancer. Lung Cancer, 33 (Suppl 1) S17–S24

Kallioniemi A (2002) Molecular signatures of breast cancer–predicting the future. N Engl J Med 347:2067–2068

Kauffman EC, Robinson VL, Stadler WM, Sokoloff MH, Rinker-Schaeffer CW (2003) Metastasis suppression: the evolving role of metastasis suppressor genes for regulating cancer cell growth at the secondary site. J Urol 169:1122–1133

Kawamata H, Kameyama S, Kawai K, Tanaka Y, Nan L, Barch DH, et al. (1995) Marked acceleration of the metastatic phenotype of a rat bladder carcinoma cell line by the expression of human gelatinase A. Int J Cancer 63:568–575

Khokha R, Zimmer MJ, Graham CH, Lala PK, Waterhouse P (1992) Suppression of invasion by inducible expression of tissue inhibitor of metalloproteinase-1 (TIMP-1) in B16-F10 melanoma cells. J Natl Cancer Inst 84: 1017–1022

Kim HL, Vander Griend DJ, Yang X, Benson DA, Dubauskas Z, Yoshida BA, et al. (2001) Mitogen-activated protein kinase kinase 4 metastasis suppressor gene expression is inversely related to histological pattern in advancing human prostatic cancers. Cancer Res 61:2833–2837

Kohn EC, Liotta LA (1995) Molecular insights into cancer invasion: strategies for prevention and intervention. Cancer Res 55:1856–1862

Kondo Y, Arii S, Mori A, Furutani M, Chiba T, Imamura M (2000) Enhancement of angiogenesis, tumor growth, and metastasis by transfection of vascular endothelial growth factor into LoVo human colon cancer cell line. Clin Cancer Res 6:622–230

Koop S, Khokha R, Schmidt EE, MacDonald IC, Morris VL, Chambers AF, et al. (1994) Overexpression of metalloproteinase inhibitor in B16F10 cells does not affect extravasation but reduces tumor growth. Cancer Res 54:4791–4797

Koop S, Schmidt EE, MacDonald IC, Morris VL, Khokha R, Grattan M, et al. (1996) Independence of metastatic ability and extravasation: metastatic ras-transformed and control fibroblasts extravasate equally well. Proc Natl AcadSci USA 93:11080–11084

Koshiba T, Hosotani R, Miyamoto Y, Ida J, Tsuji S, Nakajima S, et al. (2000) Expression of stromal cell-derived factor 1 and CXCR4 ligand receptor system in pancreatic cancer: a possible role for tumor progression. Clin Cancer Res 6:3530–3535

Kotani M, Detheux M, Vandenbogaerde A, Communi D, Vanderwinden JM, Le Poul E, et al. (2001) The metastasis suppressor gene KiSS-1 encodes kisspeptins, the natural ligands of the orphan G protein-coupled receptor GPR54. J Biol Chem 276:34631–34636

Lakka SS, Rajagopal R, Rajan MK, Mohan PM, Adachi Y, Dinh DH, et al. (2001) Adenovirus-mediated antisense urokinase-type plasminogen activator receptor gene transfer reduces tumor cell invasion and metastasis in non-small cell lung cancer cell lines. Clin Cancer Res 7:1087–1093

Lamszus K, Jin L, Fuchs A, Shi E, Chowdhury S, Yao Y, et al. (1997) Scatter factor stimulates tumor growth and tumor angiogenesis in human breast cancers in the mammary fat pads of nude mice. Lab Invest 76:339–353

Lee JH, Miele ME, Hicks DJ, Phillips KK, Trent JM, Weissman BE, et al. (1996) KiSS-1, a novel human malignant melanoma metastasis-suppressor gene. J Natl Cancer Inst 88:1731–1737

Leone A, Flatow U, King CR, Sandeen MA, Margulies IM, Liotta LA, et al. (1991) Reduced tumor incidence, metastatic potential, and cytokine responsiveness of nm23-transfected melanoma cells. Cell 65:25–35

Leone A, Flatow U, VanHoutte K, Steeg PS (1993) Transfection of human nm23-H1 into the human MDA-MB-435 breast carcinoma cell line: effects on tumor metastatic potential, colonization and enzymatic activity. Oncogene 8:2325–2333

Lin S, Rusciano D, Lorenzoni P, Hartmann G, Birchmeier W, Giordano S, et al. (1998) C-met activation is necessary but not sufficient for liver colonization by B16 murine melanoma cells. Clin Exp Metastasis 16:253–265

Liotta LA, Steeg PS, Stetler-Stevenson WG (1991) Cancer metastasis and angiogenesis: an imbalance of positive and negative regulation. Cell 64:327–336

Liotta LA, Kohn EC (2003) Cancer's deadly signature. Nat Genet 33:10–11

Look MP, van Putten WL, Duffy MJ, Harbeck N, Christensen IJ, Thomssen C, et al. (2002) Pooled analysis of prognostic impact of urokinase-type plasminogen activator and its inhibitor PAI-1 in 8377 breast cancer patients. J Natl Cancer Inst 94:116–128

Luzzi KJ, MacDonald IC, Schmidt EE, Kerkvliet N, Morris VL, Chambers AF, et al. (1998) Multistep nature of metastatic inefficiency: dormancy of solitary cells after successful extravasation and limited survival of early micrometastases. Am J Pathol 153:865–873

MacDonald IC, Groom AC, Chambers AF (2002) Cancer spread and micrometastasis development: quantitative approaches for in vivo models. Bioessays 24:885–893

Maecker HT, Todd SC, Levy S (1997) The tetraspanin superfamily: molecular facilitators. FASEB J 11:428–442

Masood R, Cai J, Zheng T, Smith DL, Hinton DR, Gill PS (2001) Vascular endothelial growth factor (VEGF) is an autocrine growth factor for VEGF receptor-positive human tumors. Blood 98:1904–1913

McCawley LJ, Matrisian LM (2000) Matrix metalloproteinases: multifunctional contributors to tumor progression. Mol Med Today 6:149–156

McGuire WL (1991) Breast cancer prognostic factors: evaluation guidelines. J Natl Cancer Inst 83:154–155

Melnyk O, Shuman MA, Kim KJ (1996) Vascular endothelial growth factor promotes tumor dissemination by a mechanism distinct from its effect on primary tumor growth. Cancer Res 56:921–924

Meng Q, Mason JM, Porti D, Goldberg ID, Rosen EM, Fan S (2000) Hepatocyte growth factor decreases sensitivity to chemotherapeutic agents and stimulates cell adhesion, invasion, and migration. Biochem Biophys Res Commun 274:772–779

Min HY, Doyle LV, Vitt CR, Zandonella CL, Stratton-Thomas JR, Shuman MA, et al. (1996) Urokinase receptor antagonists inhibit angiogenesis and primary tumor growth in syngeneic mice. Cancer Res 56:2428–2433

Miyazaki H, Fukuda M, Ishijima Y, Takagi Y, Iimura T, Negishi A, et al. (1999) Overexpression of nm23-H2/NDP kinase B in a human oral squamous cell carcinoma cell line results in reduced metastasis, differentiated phenotype in the metastasis site, and growth factor-independent proliferative activity in culture. Clin Cancer Res 5:4301–4307

Miyazaki T, Kato H, Shitara Y, Yoshikawa M, Tajima K, Masuda N, et al. (2000) Mutation and expression of the metastasis suppressor gene KAI1 in esophageal squamous cell carcinoma. Cancer 89:955–962

Mizejewski GJ (1999) Role of integrins in cancer: survey of expression patterns. Proc Soc Exp Biol Med 222:124–138

Moore MA (2001) The role of chemoattraction in cancer metastases. Bioessays 23:674–676

Morris VL, MacDonald IC, Koop S, Schmidt EE, Chambers AF, Groom AC (1993) Early interactions of cancer cells with the microvasculature in mouse liver and muscle during hematogenous metastasis: videomicroscopic analysis. Clin Exp Metastasis 11:377–390

Morrison DK (2001) KSR: a MAPK scaffold of the Ras pathway? J Cell Sci 114:1609–1612

Muller A, Homey B, Soto H, Ge N, Catron D, Buchanan ME, et al. (2001) Involvement of chemokine receptors in breast cancer metastasis. Nature 410:50–56

Murphy PM (2001) Chemokines and the molecular basis of cancer metastasis. N Engl J Med 345:833–835

Muschel RJ, Williams JE, Lowy DR, Liotta LA (1985) Harvey ras induction of metastatic potential depends upon oncogene activation and the type of recipient cell. Am J Pathol 121:1–8

Nagase H, Woessner JF Jr. (1999) Matrix metalloproteinases. J Biol Chem 274:21491–21494

Nakamura H, Nakamura K, Yodoi J (1997) Redox regulation of cellular activation. Ann Rev Immunol 15:351–369

Naramura M, Gillies SD, Mendelsohn J, Reisfeld RA, Mueller BM (1993) Therapeutic potential of chimeric and murine anti-(epidermal growth factor receptor) antibodies in a metastasis model for human melanoma. Cancer Immunol Immunother 37:343–349

Naumov GN, MacDonald IC, Chambers AF, Groom AC (2001) Solitary cancer cells as a possible source of tumour dormancy? Semin Cancer Biol 11:271–276

Nemoto H, Rittling SR, Yoshitake H, Furuya K, Amagasa T, Tsuji K, et al. (2001) Osteopontin deficiency reduces experimental tumor cell metastasis to bone and soft tissues. J Bone Miner Res 16:652–659

Newham P, Humphries MJ (1996) Integrin adhesion receptors: structure, function and implications for biomedicine. Mol Med Today 2:304–313

Nicolson GL, Custead SE (1982) Tumor metastasis is not due to adaptation of cells to a new organ environment. Science 215:176–178

Nicolson GL (1991) Gene expression, cellular diversification and tumor progression to the metastatic phenotype. Bioessays 13:337–342

Noel AC, Lefebvre O, Maquoi E, VanHoorde L, Chenard MP, Mareel M, et al. (1996) Stromelysin-3 expression promotes tumor take in nude mice. J Clin Invest 97:1924–1930

Odintsova E, Sugiura T, Berditchevski F (2000) Attenuation of EGF receptor signaling by a metastasis suppressor, the tetraspanin CD82/KAI-1. Curr Biol 10:1009–1012

Oka T, Ishida T, Nishino T, Sugimachi K (1991) Immunohistochemical evidence of urokinase-type plasminogen activator in primary and metastatic tumors of pulmonary adenocarcinoma. Cancer Res 51:3522–3525

Orr FW, Wang HH (2001) Tumor cell interactions with the microvasculature: a rate-limiting step in metastasis. Surg Oncol Clin N Am 10:357–381, ix–x

Otsuka M, Kato M, Yoshikawa T, Chen H, Brown EJ, Masuho Y, et al. (2001) Differential expression of the L-plastin gene in human colorectal cancer progression and metastasis. Biochem Biophys Res Commun 289:876–881

Overall CM, Lopez-Otin C (2002) Strategies for MMP inhibition in cancer: innovations for the post-trial era. Nat Rev Cancer 2:657–672

Paget S (1889) The distribution of secondary growths in cancer of the breast. Lancet 1:99–101

Paget S (1989) The distribution of secondary growths in cancer of the breast. 1889. Cancer Metastasis Rev 8:98–101 (Republication of the original 1889 Lancet article)

Parada LF, Tabin CJ, Shih C, Weinberg RA (1982) Human EJ bladder carcinoma oncogene is homologue of Harvey sarcoma virus ras gene. Nature 297:474–478

Paul CP, Good PD, Winer I, Engelke DR (2002) Effective expression of small interfering RNA in human cells. Nat Biotech 20:505–508

Philip S, Bulbule A, Kundu GC (2001) Osteopontin stimulates tumor growth and activation of promatrix metalloproteinase-2 through nuclear factor-kappa B-mediated induction of membrane type 1 matrix metalloproteinase in murine melanoma cells. J Biol Chem 276:44926–44935

Picksley SM, Lane DP (1994) p53 and Rb: their cellular roles. Curr Opin Cell Biol 6:853–858

Pomeroy SL, Tamayo P, Gaasenbeek M, Sturla LM, Angelo M, McLaughlin ME, et al. (2002) Prediction of central nervous system embryonal tumour outcome based on gene expression. Nature 415:436–442

Poste G, Fidler IJ (1980) The pathogenesis of cancer metastasis. Nature 283:139–146

Price JT, Wilson HM, Haites NE (1996) Epidermal growth factor (EGF) increases the in vitro invasion, motility and adhesion interactions of the primary renal carcinoma cell line, A704. Eur J Cancer 32A:1977–1982

Price JT, Bonovich MT, Kohn EC (1997) The biochemistry of cancer dissemination. Crit Rev Biochem Mol Biol 32:175–253

Ramaswamy S, Ross KN, Lander ES, Golub TR (2003) A molecular signature of metastasis in primary solid tumors. Nat Genet 33:49–54

Ree AH, Florenes VA, Berg JP, Maelandsmo GM, Nesland JM, Fodstad O (1997) High levels of messenger RNAs for tissue inhibitors of metalloproteinases (TIMP-1 and TIMP-2) in primary breast carcinomas are associated with development of distant metastases. Clin Cancer Res 3:1623–1628

Riedel F, Gotte K, Li M, Hormann K, Grandis JR (2002) EGFR antisense treatment of human HNSCC cell lines downregulates VEGF expression and endothelial cell migration. Int J Oncol 21:11–16

Robinson CJ, Stringer SE (2001) The splice variants of vascular endothelial growth factor (VEGF) and their receptors. J Cell Sci 114:853–865

Rosen EM, Goldberg ID (1995) Scatter factor and angiogenesis. Adv Cancer Res 67:257–279

Sakakura C, Hagiwara A, Nakanishi M, Shimomura K, Takagi T, Yasuoka R, et al. (2002) Differential gene expression profiles of gastric cancer cells established from primary tumour and malignant ascites. Br J Cancer 87:1153–1161

Salomon DS, Brandt R, Ciardiello F, Normanno N (1995) Epidermal growth factor-related peptides and their receptors in human malignancies. Crit Rev Oncol Hemat 19:183–232

Saunders MM, Seraj MJ, Li Z, Zhou Z, Winter CR, Welch DR, et al. (2001) Breast cancer metastatic potential correlates with a breakdown in homospecific and heterospecific gap junctional intercellular communication. Cancer Res 61:1765–1767

Sauter G, Simon R (2002) Predictive molecular pathology. N Engl J Med 34:1995–1996

Scherbarth S, Orr FW (1997) Intravital videomicroscopic evidence for regulation of metastasis by the hepatic microvasculature: effects of interleukin-1alpha on metastasis and the location of B16F1 melanoma cell arrest. Cancer Res 57:4105–4110

Schmidt U, Begley CG (2003) Cancer diagnosis and microarrays. Int J Biochem Cell Biol 35:119–124

Schmitz AA, Govek EE, Bottner B, Van Aelst L (2000) Rho GTPases: signaling, migration, and invasion. Exp Cell Res 261:1–12

Scholl FA, Betts DR, Niggli FK, Schafer BW (2000) Molecular features of a human rhabdomyosarcoma cell line with spontaneous metastatic progression. Br J Cancer 82:1239–1245

Segall JE, Tyerech S, Boselli L, Masseling S, Helft J, Chan A, et al. (1996) EGF stimulates lamellipod extension in metastatic mammary adenocarcinoma cells by an actin-dependent mechanism. Clin Exp Metastasis 14:61–72

Seraj MJ, Samant RS, Verderame MF, Welch DR (2000) Functional evidence for a novel human breast carcinoma metastasis suppressor, BRMS1, encoded at chromosome 11q13. Cancer Res 60:2764–2769

Shaw LM, Chao C, Wewer UM, Mercurio AM (1996) Function of the integrin alpha 6 beta 1 in metastatic breast carcinoma cells assessed by expression of a dominant-negative receptor. Cancer Res 56:959–963

Shaw LM, Rabinovitz I, Wang HH, Toker A, Mercurio AM (1997) Activation of phosphoinositide 3-OH kinase by the alpha6beta4 integrin promotes carcinoma invasion. A signaling adapter function for alpha6beta4 integrin in the control of HGF-dependent invasive growth. Cell 91:949–960

Shen GH, Ghazizadeh M, Kawanami O, Shimizu H, Jin E, Araki T, et al. (2000) Prognostic significance of vascular endothelial growth factor expression in human ovarian carcinoma. Br J Cancer 83:196–203

Sherley JL (2002) Metastasis: objections to the same-gene model. Nature 419:560

Shevde LA, Samant RS, Goldberg SF, Sikaneta T, Alessandrini A, Donahue HJ, et al. (2002) Suppression of human melanoma metastasis by the metastasis suppressor gene, BRMS1. Exp Cell Res 273:229–239

Shibata T, Kawano T, Nagayasu H, Okumura K, Arisue M, Hamada J, et al. (1996) Enhancing effects of epidermal growth factor on human squamous cell carcinoma motility and matrix degradation but not growth. Tumour Biol 17:168–175

Shih C, Shilo BZ, Goldfarb MP, Dannenberg A, Weinberg RA (1979) Passage of phenotypes of chemically transformed cells via transfection of DNA and chromatin. Proc Natl Acad Sci USA 76:5714–5718

Shih C, Weinberg RA (1982) Isolation of a transforming sequence from a human bladder carcinoma cell line. Cell 29:161–169

Shijubo N, Uede T, Kon S, Nagata M, Abe S (2000) Vascular endothelial growth factor and osteopontin in tumor biology. Crit Rev Oncog 11:135–146

Shipp MA, Ross KN, Tamayo P, Weng AP, Kutok JL, Aguiar RC, et al. (2002) Diffuse large B-cell lymphoma outcome prediction by gene-expression profiling and supervised machine learning. Nat Med 8:68–74

Singhal H, Bautista DS, Tonkin KS, O'Malley FP, Tuck AB, Chambers AF, et al. (1997) Elevated plasma osteopontin in metastatic breast cancer associated with increased tumor burden and decreased survival. Clin Cancer Res 3:605–611

Sodek J, Ganss B, McKee MD (2000) Osteopontin. Crit Rev Oral Biol Med 11:279–303

Stackpole CW (1981) Distinct lung-colonizing and lung-metastasizing cell populations in B16 mouse melanoma. Nature 289:798–800

Steeg PS (2003) Metastasis suppressors alter the signal transduction of cancer cells. Nat Rev Cancer 3:55–63

Stella MC, Comoglio PM (1999) HGF: a multifunctional growth factor controlling cell scattering. Int J Biochem Cell Biol 31:1357–1362

Stephens RW, Nielsen HJ, Christensen IJ, Thorlacius-Ussing O, Sorensen S, Dano K, et al. (1999) Plasma urokinase receptor levels in patients with colorectal cancer: relationship to prognosis. J Natl Cancer Inst 91:869–874

Stetler-Stevenson WG, Hewitt R, Corcoran M (1996) Matrix metalloproteinases and tumor invasion: from correlation and causality to the clinic. Semin Cancer Biol 7:147–154

Swartz MA, Skobe M (2001) Lymphatic function, lymphangiogenesis, and cancer metastasis. Microsc Res Tech 55:92–99

Tabin CJ, Bradley SM, Bargmann CI, Weinberg RA, Papageorge AG, Scolnick EM, et al. (1982) Mechanism of activation of a human oncogene. Nature 300:143–149

Tagashira H, Hamazaki K, Tanaka N, Gao C, Namba M (1998) Reduced metastatic potential and c-myc overexpression of colon adenocarcinoma cells (Colon 26 line) transfected with nm23-R2/rat nucleoside diphosphate kinase alpha isoform. Int J Mol Med 2:65–68

Taichman RS, Cooper C, Keller ET, Pienta KJ, Taichman NS, McCauley LK (2002) Use of the stromal cell-derived factor-1/CXCR4 pathway in prostate cancer metastasis to bone. Cancer Res 62:1832–1837

Taraboletti G, Garofalo A, Belotti D, Drudis T, Borsotti P, Scanziani E, et al. (1995) Inhibition of angiogenesis and murine hemangioma growth by batimastat, a synthetic inhibitor of matrix metalloproteinases. J Natl Cancer Inst 87:293–298

Thorgeirsson UP, Turpeenniemi-Hujanen T, Williams JE, Westin EH, Heilman CA, Talmadge JE, et al. (1985) NIH/3T3 cells transfected with human tumor DNA containing activated ras oncogenes express the metastatic phenotype in nude mice. Mol Cell Biol 5:259–262

Toi M, Ishigaki S, Tominaga T (1998) Metalloproteinases and tissue inhibitors of metalloproteinases. Breast Cancer Res Treat 52:113–124

To CT, Tsao MS (1998) The roles of hepatocyte growth factor/scatter factor and met receptor in human cancers. Oncol Rep 5:1013–1024

Trusolino L, Bertotti A, Comoglio PM (2001) A signaling adapter function for alpha6beta4 integrin in the control of HGF-dependent invasive growth. Cell 107:643–654

Tuck AB, Park M, Sterns EE, Boag A, Elliott BE (1996) Coexpression of hepatocyte growth factor and receptor (Met) in human breast carcinoma. Am J Pathol 148:225–232

Tuck AB, O'Malley FP, Singhal H, Tonkin KS, Harris JF, Bautista D, et al. (1997) Osteopontin and p53 expression are associated with tumor progression in a case of synchronous, bilateral, invasive mammary carcinomas. Arch Pathol Lab Med 121:578–584

Tuck AB, O'Malley FP, Singhal H, Harris JF, Tonkin KS, Kerkvliet N, et al. (1998) Osteopontin expression in a group of lymph node negative breast cancer patients. Int J Cancer 79:502–508

Tuck AB, Arsenault DM, O'Malley FP, Hota C, Ling MC, Wilson SM, et al. (1999) Osteopontin induces increased invasiveness and plasminogen activator expression of human mammary epithelial cells. Oncogene 18:4237–4246

Tuck AB, Elliott BE, Hota C, Tremblay E, Chambers AF (2000) Osteopontin-induced, integrin-dependent migration of human mammary epithelial cells involves activation of the hepatocyte growth factor receptor (Met). J Cell Biochem 78:465–475

Tuck AB, Hota C, Wilson SM, Chambers AF (2003) Osteopontin-induced migration of human mammary epithelial cells involves activation of EGF receptor and multiple signal transduction pathways. Oncogene 22:1198–1205

Tucker GC (2002) Inhibitors of integrins. Curr Opin Pharmacol 2:394–402

Turner T, Chen P, Goodly LJ, Wells A (1996) EGF receptor signaling enhances in vivo invasiveness of DU-145 human prostate carcinoma cells. Clin Exp Metastasis 14:409–418

Ullrich A, Schlessinger J (1990) Signal transduction by receptors with tyrosine kinase activity. Cell 61:203–212

van de Vijver MJ, He YD, van't Veer LJ, Dai H, Hart AA, Voskuil DW, et al. (2002) A gene-expression signature as a predictor of survival in breast cancer. N Engl J Med 347:1999–2009

van Diest PJ, Michalides RJ, Jannink L, van der Valk P, Peterse HL, de Jong JS, et al. (1997) Cyclin D1 expression in invasive breast cancer. Correlations and prognostic value. Am J Pathol 150:705–711

van't Veer LJ, De Jong D (2002) The microarray way to tailored cancer treatment. Nat Med 8:13–14

van't Veer LJ, Dai H, van de Vijver MJ, He YD, Hart AA, Mao M, et al. (2002) Gene expression profiling predicts clinical outcome of breast cancer. Nature 415:530–536

Varghese HJ, Davidson MT, MacDonald IC, Wilson SM, Nadkarni KV, Groom AC, et al. (2002) Activated ras regulates the proliferation/apoptosis balance and early survival of developing micrometastases. Cancer Res 62:887–891

Varner JA, Cheresh DA (1996) Integrins and cancer. Curr Opin Cell Biol 8:724–730

Vogelmann R, Kreuser ED, Adler G, Lutz MP (1999) Integrin alpha6beta1 role in metastatic behavior of human pancreatic carcinoma cells. Int J Cancer 80:791–795

Vogelstein B, Kinzler KW (1993) The multistep nature of cancer. Trends Genet 9:138–141

Vonlaufen A, Wiedle G, Borisch B, Birrer S, Luder P, Imhof BA (2001) Integrin alpha(v)beta(3) expression in colon carcinoma correlates with survival. Mod Pathol 14:1126–1132

Wang JM, Deng X, Gong W, Su S (1998) Chemokines and their role in tumor growth and metastasis. J Immunol Methods 220:1–17

Wang W, Wyckoff JB, Frohlich VC, Oleynikov Y, Huttelmaier S, Zavadil J, et al. (2002) Single cell behavior in metastatic primary mammary tumors correlated with gene expression patterns revealed by molecular profiling. Cancer Res 62:6278–6288

Warren RS, Yuan H, Matli MR, Gillett NA, Ferrara N (1995) Regulation by vascular endothelial growth factor of human colon cancer tumorigenesis in a mouse model of experimental liver metastasis. J Clin Invest 95:1789–1797

Watson JD, Gilman M, Witkowski J, Zoller M (1992) Recombinant DNA, 2nd edn. Scientific American Books, New York, pp 213–234

Weber GF, Ashkar S, Cantor H (1997) Interaction between CD44 and osteopontin as a potential basis for metastasis formation. Proc Assoc Am Physicians 109:1–9

Weiss L (1990) Metastatic inefficiency. Adv Cancer Res 54:159–211

Weiss L (1992) Comments on hematogenous metastatic patterns in humans as revealed by autopsy. Clin Exp Metastasis 10:191–199

Welch DR, Chen P, Miele ME, McGary CT, Bower JM, Stanbridge EJ, et al. (1994) Microcell-mediated transfer of chromosome 6 into metastatic human C8161 melanoma cells suppresses metastasis but does not inhibit tumorigenicity. Oncogene 9:255–262

Welch DR (1997) Technical considerations for studying cancer metastasis in vivo. Clin Exp Metastasis 15:272–306

Wells A (1999) EGF receptor. Int J Biochem Cell Biol 31:637–643

Witty JP, McDonnell S, Newell KJ, Cannon P, Navre M, Tressler RJ, et al. (1994) Modulation of matrilysin levels in colon carcinoma cell lines affects tumorigenicity in vivo. Cancer Res 54:4805–4812

Woodhouse EC, Chuaqui RF, Liotta LA (1997) General mechanisms of metastasis. Cancer 80:1529–1537

Xuan JW, Hota C, Shigeyama Y, D'Errico JA, Somerman MJ, Chambers AF (1995) Site-directed mutagenesis of the arginine-glycine-aspartic acid sequence. J Cell Biochem 57:680–690

Yamada SD, Hickson JA, Hrobowski Y, Vander Griend DJ, Benson D, Montag A, et al. (2002) Mitogen-activated protein kinase kinase 4 (MKK4) acts as a metastasis suppressor gene in human ovarian carcinoma. Cancer Res 62:6717–6723

Yamasaki H, Mesnil M, Omori Y, Mironov N, Krutovskikh V (1995) Intercellular communication and carcinogenesis. Mutat Res 333:181–188

Yamashita J, Ogawa M, Yamashita S, Nomura K, Kuramoto M, Saishoji T, et al. (1994) Immunoreactive hepatocyte growth factor is a strong and independent predictor of recurrence and survival in human breast cancer. Cancer Res 54:1630–1633

Yan C, Wang H, Boyd DD (2001) KiSS-1 represses 92-kDa type IV collagenase expression by down-regulating NF-kappa B binding to the promoter as a consequence of Ikappa Balpha -induced block of p65/p50 nuclear translocation. J Biol Chem 276:1164–1172

Yang X, Welch DR, Phillips KK, Weissman BE, Wei LL (1997) KAI1, a putative marker for metastatic potential in human breast cancer. Cancer Lett 119:149–155

Yang X, Wei LL, Tang C, Slack R, Mueller S, Lippman ME (2001) Overexpression of KAI1 suppresses in vitro invasiveness and in vivo metastasis in breast cancer cells. Cancer Res 61:5284–5288

Yano S, Shinohara H, Herbst RS, Kuniyasu H, Bucana CD, Ellis LM, et al. (2000) Expression of vascular endothelial growth factor is necessary but not sufficient for production and growth of brain metastasis. Cancer Res 60:4959–4967

Ye QH, Qin LX, Forgues M, He P, Kim JW, Peng AC, et al. (2003) Predicting hepatitis B virus-positive metastatic hepatocellular carcinomas using gene expression profiling and supervised machine learning. Nat Med 9:416–423

Yoshida BA, Dubauskas Z, Chekmareva MA, Christiano TR, Stadler WM, Rinker-Schaeffer CW (1999) Mitogen-activated protein kinase kinase 4/stress-activated protein/Erk kinase 1 (MKK4/SEK1), a prostate cancer metastasis suppressor gene encoded by human chromosome 17. Cancer Res 59:5483–5487

Yoshida BA, Sokoloff MM, Welch DR, Rinker-Schaeffer CW (2000) Metastasis-suppressor genes: a review and perspective on an emerging field. J Natl Cancer Inst 92:1717–1730

Yu W, Kim J, Ossowski L (1997) Reduction in surface urokinase receptor forces malignant cells into a protracted state of dormancy. J Cell Biol 137:767–777

Yu Y, Yang JL, Markovic B, Jackson P, Yardley G, Barrett J, et al. (1997) Loss of KAI1 messenger RNA expression in both high-grade and invasive human bladder cancers. Clin Cancer Res 3:1045–1049

Zhang K, Kim JP, Woodley DT, Waleh NS, Chen YQ, Kramer RH (1996) Restricted expression and function of laminin 1-binding integrins in normal and malignant oral mucosal keratinocytes. Cell Adhes Commun 4:159–174

Zhang L, Yang N, Garcia JR, Mohamed A, Benencia F, Rubin SC, et al. (2002) Generation of a syngeneic mouse model to study the effects of vascular endothelial growth factor in ovarian carcinoma. Am J Pathol 161:2295–2309

Zhang M, Shi Y, Magit D, Furth PA, Sager R (2000) Reduced mammary tumor progression in WAP-TAg/WAP-maspin bitransgenic mice. Oncogene 19:6053–6058

Zhu Z, Bohlen P, Witte L (2002) Clinical development of angiogenesis inhibitors to vascular endothelial growth factor and its receptors as cancer therapeutics. Curr Cancer Drug Targets 2:135–156

Zlotnik A, Yoshie O (2000) Chemokines: a new classification system and their role in immunity. Immunity 12:121–127

H.E. Kaiser and A. Nasir (eds.), Selected Aspects of Cancer Progression:
Metastasis, Apoptosis and Immune Response, 59–90.
© *Springer Science + Business Media B.V.* 2008

CHAPTER FIVE

The evolution of diversity within tumors and metastases

Rakesh K. Singh and James E. Talmadge

Abstract: The metastatic process is a series of steps – each of which must be completed for the development of a metastatic focus. Some steps consist of stochastic elements; however, overall the process selects for cellular phenotype(s) with metastatic properties from the heterogeneous cellular populations that develop within the primary tumor. Thus, although individual metastases are of clonal origin, different metastatic foci can originate from cellular variants. The development of cellular diversity is not a process unique to tumor cells. Cellular diversity rapidly evolves due to the genetic instability of metastatic cells, the selective pressures of the metastatic process and clinical interventions, and interactions with the microenvironment, including cellular immunity and tissue and organ microenvironments that can facilitate or suppress tumor growth, metastasis, and tumor cell survival. It is the cellular heterogeneity within the primary tumor, between metastatic foci, and within individual metastatic foci that provides a significant challenge for oncologists for successful clinical intervention.

Keywords: Metastasis, Macrophages, Selection, Heterogeneity

Laboratory of Transplantation Immunology, Department of
Pathology and Microbiology, University of Nebraska Medical
Center, Omaha, NE, USA

Introduction

Despite advances in the use of aggressive adjuvant chemotherapy and radiotherapy, which in combination with surgery are often successful in the eradication of the primary tumor, most deaths in cancer patients result from metastasis. This chapter reviews the process of metastasis on a cellular basis and is approached using as a goal the improvement of therapeutic protocols. Excellent reviews on the mechanism of metastasis and the characteristics of metastatic cells are provided by others elsewhere in this volume. A question important to our understanding of the pathogenesis of metastasis and to the improvement of cancer therapy is whether tumor cells that give rise to metastatic foci are random survivors of the cells within the primary tumor or represent a select subpopulation of tumor cells that pre-exists within the primary tumor population. If the metastatic process is selective and not random, then the cells within a metastatic focus represent an enlarged pool of tumor cells endowed with specialized characteristics and it may be possible to develop therapeutic modalities directed against the unique phenotype. The development of novel therapeutic modalities is important since tumor cells within primary tumors are heterogeneous with regard to their metastatic potential and their response to most therapeutic modalities, including chemotherapy, radiotherapy, and specific immunotherapy. This phenotypic heterogeneity is not unique to the primary tumor, but

is also observed among metastases (interlesional) as well as within metastases (intralesional). Clearly the only successful treatment of disseminated cancer will be one capable of overcoming the problems associated with the heterogeneity of malignant tumor cells. The development of screening protocols to identify novel anticancer agents must not only monitor the response of the primary tumor to therapy, but also examine the efficacy of such agents or protocols against the metastatic subpopulations within the primary tumor. Recent studies of metastasis have increased our understanding of the metastatic process as influenced by both tumor cell properties and host–tumor cell interactions. Many of these studies have challenged established paradigms, resulting in the modification of experimental techniques and models that are used to study metastasis, as well as, alter our outlook on therapeutic protocols designed to treat established secondary tumor foci. Therefore, the goal of this chapter is to provide an overview of these studies.

The heterogeneous nature of neoplasms

There is considerable evidence to suggest that human tumors are composed of heterogeneous cell subpopulations. The presence of heterogeneous cellular populations is observed in primary tumors based on numerous phenotypes (Nicolson 1984a). Early histological studies demonstrated morphologic differences among cells within the same tumor. For this reason, pathologists routinely examine several sections of a tumor to determine whether a tumor is benign or contains nests of invasive and malignant cells. Dunn (1959) examined the histology of numerous primary murine mammary tumors and concluded that cancer does not represent a single alteration of one cell that reproduces itself without change. Foulds (1956a, b, c, d) also noted that murine mammary tumors are composed of zones of tumor cells with different morphologies and that within each zone the cells appear homogeneous. To study this zonal heterogeneity, Henderson and Rous (1962) fragmented tumors of mixed morphology, which after transplantation as individual fragments, tended to develop into tumors with a uniform morphology. Other studies have described differences in cellular morphology (Dexter et al. 1978) and tumor histopathology (Hager et al.

1981; Kobori and Oota 1979; Mathieson et al. 1982; Pierce 1974; Woodruff 1983) within primary tumors of various histotypes. Heterogeneous histological patterns have been observed in multiple samples of breast carcinoma (Geier et al. 1979; Parbhoo 1981) and oat cell carcinoma (Ewing et al. 1980), as well as, from a histological and ultrastructural study of the tumor cells in a bronchial carcinoid (McDowell 1981). The coexistence of multiple subpopulations of tumor cells within a single neoplasm has been repeatedly demonstrated in animal tumors of diverse etiology and histological type. These include melanomas (Fidler et al. 1981; Fidler and Kripke 1977; Gray and Pierce 1964; Natali et al. 1983), lymphomas-leukemias (Mathieson et al. 1982; Brunson et al. 1978; Olsson and Ebbesen 1979), sarcomas (Kripke et al. 1978; Mantovani et al. 1981; Mitelman 1971; Raz et al. 1981; Suzuki et al. 1978; Vaupel et al. 1981; Wang et al. 1982), and carcinomas of different organs (Henderson and Rous 1962; Dexter et al. 1978; Hager et al. 1981; Pierce 1974; Danielson et al. 1980; Macinnes et al. 1981; Michalides et al. 1982; Miller and Heppner 1979; Dominguez and Huseby 1968; Talmadge et al. 1979; Trope 1975; Tsuruo and Fidler 1981; Zupi et al. 1980; Symmans et al. 1995; Naito et al. 1991; van Lamsweerde et al. 1983; Ware and Maygarden 1989; Harris and Best 1988). Heterogeneity in tumors induced by chemical agents (Wang et al. 1982; Talmadge et al. 1979), physical agents (Kripke et al. 1978), steroids (Dominguez and Huseby 1968), or viruses (Dexter et al. 1978; Mathieson et al. 1982; Danielson et al. 1980; Macinnes et al. 1981; Michalides et al. 1982; Colcher et al. 1981) has been widely described. Long-term passaged tumors (Fidler and Kripke 1977; Nicolin et al. 1981), tumors of recent origin (Fidler et al. 1981; Kripke et al. 1978), as well as, autochthonous tumors (Dexter et al. 1978; Olsson and Ebbesen 1979) have also been found to be comprised of multiple subpopulations.

Recent studies have demonstrated that at the time of diagnosis, most neoplasms are populated by cellular subpopulations with diverse phenotypes. This includes phenotypic heterogeneity with regard to antigenicity (Brunson et al. 1978; Miller and Heppner 1979; Colcher et al. 1981; Albino et al. 1981; Hager and Heppner 1982; Kerbel 1979; Killion and Kollmorgen 1976; Pimm and Baldwin 1980; Pimm et al. 1980; Prehn 1970; Schirrmacher and Bosslet

1982; Strzadala et al. 1981) and immunogenicity (Natali et al. 1983; Miller and Heppner 1979; Nicolin et al. 1981; Bosslet and Schirrmacher 1981, 1982; Fogel et al. 1979; Fuji and Mihich 1975; Fuji et al. 1977; Gorelik et al. 1979; Killion 1978; McCune et al. 1981; Miller 1982; Olsson et al. 1981; Schirrmacher et al. 1979; Thistlethwaithe et al. 1983). These variations in antigenicity and immunogenicity are important since they can profoundly influence the success of antigen specific immunotherapy. In a study using a number of AKR mouse lymphomas, Olsson and Ebbesen (1979) found that vaccination procedures against polyclonal tumors failed to prevent tumor growth following challenge since only the dominant subclone was restricted in growth. The minor subpopulations, which did not constitute a sufficient mass in the vaccine to stimulate the immune response, were able to proliferate following vaccination and eventually become the dominant population. In certain tumor systems, a successful host immune response to tumor cells bearing strong antigens may result in the emergence of tumor cell variants lacking the antigen. For example, Reading et al. (1980a) analyzed a number of in vivo and in vitro selected murine RAW117 lymphosarcoma cell lines (and clones derived from these cell lines) for their metastatic properties and cell surface antigen expression. They found that the ability to metastasize to the liver was inversely correlated with the expression of the antigenic RNA tumor virus envelope glycoprotein gp70. In this system, successful metastasis apparently requires escape from host immune surveillance via antigen deletion on the highly metastatic lymphosarcoma cells. However, in other metastatic systems such as the B16 melanoma, there is no relationship between metastasis and viral antigens as gp70 (Fidler and Nicolson 1981). In contrast, some tumors may express antigens that are increased on metastatic cells. For example, Shearman and Longenecker (1981) reported an increase in cellular antigen content that correlated with the ability of Marek's disease virus-transformed chick lymphoma cells to metastasize to the liver. In this system, the level of cell surface antigen detectable with monoclonal antibody increased concomitantly with the ability to colonize the liver. Thus, there appears to be no simple relationship between the display of cell surface antigens, immunogenicity, and metastasis.

Tumor cell populations residing within a parent neoplasm can also be heterogeneous with regard to drug sensitivity (Talmadge et al. 1984; Dexter and Leith 1986). Cells isolated from rat hepatomas (Barranco et al. 1978), methylcholanthrene-induced mouse sarcomas (Hakansson and Trope 1974a, b), murine lung cancers (Sacchi et al. 1981; Trope 1982), a murine melanoma (Tsuruo and Fidler 1981; Lotan and Nicolson 1979), and a mouse mammary tumor (Heppner et al. 1978) have different in vitro and in vivo sensitivities to various cytotoxic agents and radiation therapy (Trope 1975, 1982; Dexter and Leith 1986; Hakansson and Trope 1974a, b; Heppner et al. 1978; Biorklund et al. 1980; Hill et al. 1979; Leith et al. 1981; Leith et al. 1982a, b; Stephens and Peacock 1982; Trope et al. 1979; Trope et al. 1975). During an extensive study, Tsuruo and Fidler (1981) examined the in vitro sensitivity to various chemotherapeutic agents of tumor cells from parent tumors (rodent and human), their in vitro cloned populations, and spontaneous metastases from these tumor lines. Their findings demonstrated that differences in drug responsiveness exist among cells populating parent tumors (in vitro clones), as well as, between the parent line and its metastatic subpopulations. The differences observed in drug sensitivity between the primary and secondary tumors obviously have profound implications for the treatment of metastases with cytotoxic drugs.

Cells within individual tumors have also been shown to differ with regard to their growth rate both in vitro and in vivo (Dexter et al. 1978; Gray and Pierce 1964; Danielson et al. 1980; Zupi et al. 1980; Brock et al. 1982; Cifone et al. 1979; DeWys 1972; Miller et al. 1980; Soule et al. 1981; Talmadge et al. 1981). Tumor subpopulations can differ in the expression or production of "markers" of differentiation, including appropriate pigments (Fidler et al. 1981; Gray and Pierce 1964; Fidler and Hart 1981; Niles and Makarski 1978), receptors (Sluyser et al. 1976), cell products (Mathieson et al. 1982), and specialized biosynthetic enzymes (Dominguez and Huseby 1968). The subpopulations also differ on the basis of DNA content (Bohm and Sandritter 1975; Starace et al. 1982), karyotype (Dexter et al. 1978; Mitelman 1971; Becker et al. 1973; Ishidate et al. 1974; Ito and Moore 1967; Kusyk et al. 1981; Makino 1956; Nowell 1976; Ohno 1971; Rabotti 1959; Semple et al. 1982; Shapiro et al. 1981; Straus 1977; Vindelov 1977; Vindelov et al. 1980; Vindelov et al. 1982), as well as, the presence or absence of marker chromosomes in various tumor subpopulations (Shapiro et al. 1981;

Pathak 1990; Wolman 1986). Tumor cells also express a variety of cell surface receptors for lectins (Brunson KW, Nicolson 1978; Raz et al. 1980; Reading et al. 1980b; Talmadge et al. 1980; Tao and Burger 1977), hormone receptors (Sluyser et al. 1976; Brennan et al. 1979; Franks 1960; Isaacs and Coffey 1981; Isaacs et al. 1982), and metabolic characteristics (Semple et al. 1982; Angello et al. 1982; Baylin et al. 1975; Baylin et al. 1978; Kiricuta et al. 1965; Larner and Rutherford 1982). Using murine mammary tumor virus (MuMTV) DNA as a probe, cellular heterogeneity in the location and copy number of a specific gene has been demonstrated in the GR mouse mammary tumors (Macinnes et al. 1981; Michalides et al. 1982). This is in accordance with the heterogeneity observed in the expression of MuMTV-coded antigens within individual mammary tumors (Colcher et al. 1981). Studies on the differential response of BALB/c and C3H mammary tumor subpopulations to inducers of MuMTV gene expression suggest that the differences in regulation of MuMTV genes also correlate with tumor subpopulation heterogeneity (Hager and Heppner 1982). As discussed earlier, human tumors also exhibit marked intralesional heterogeneity in antigenicity and immunogenicity (Albino et al. 1981; McCune et al. 1981; Byers and Johnston 1977; MacLean et al. 1982; Tan et al. 1981). It was reported (McGee et al. 1982) that cancer antigens (Ashall et al. 1982) may be detected in some areas of a carcinoma, but not in others, even in those cells that were obviously malignant by morphological criteria. This suggests the possibility that cells from a tumor differ in the expression of the cancer antigen, either on the cell membrane or in the cytoplasm.

Intratumoral heterogeneity in tumor cell DNA content or ploidy levels has been reported for small cell carcinoma of the lung (Vindelov et al. 1980; Stich et al. 1960) and colon carcinoma (Vindelov et al. 1982; Wagner and Schulze 1978). Tumors have also been shown to be heterogeneous for markers that may be associated with the degree of differentiation, including β2-microglobulin (Weiss et al. 1981), estrogen receptors in breast cancer (Brennan et al. 1979; Lee 1978; Pertschuk et al. 1978), steroid receptors in prostatic cancer (Wagner and Schulze 1978; Ekman et al. 1979), and calcitonin levels in small cell carcinoma of the lung (Baylin et al. 1978). Tumor cell heterogeneity for calcitonin has also been described in virulent

medullary carcinoma (Lippman et al. 1982). This is especially interesting in that the heterogeneity for calcitonin staining was seen in medullary carcinomas with a high likelihood of metastatic spread, whereas uniform staining was observed in tumors with a small chance of recurrence.

Histological examination of tumor samples generally reveals differences in the morphology of tumor cancer cells within the same lesion. In addition, host infiltrating and connective tissues are not evenly distributed in tumors, and areas of necrosis may be present. Depending upon tumor size, marked disturbances in vasculature can occur, leading to focal differences in oxygen tension, pH, substrate supply, and waste drainage (Vaupel et al. 1981). Recent studies demonstrate individual tumor's reliance on vasculature in heterogeneous (Yu et al. 2001). Yu et al. (2001) demonstrated that hypoxia inducing factor-1 alpha (HIF-1α) expressing cells were highly dependent on proximity to blood vessels, whereas cells that had lost HIF-1α expression were much less reliant, suggesting that a selection for less vascular-dependent tumor cell variants occurs throughout the course of disease progression. The cells within a tumor may be cycling or noncycling, quiescent or reproductively dead (Dethlefsen 1980). Cells may be at any stage of the cell cycle, which may influence cellular properties such membrane biochemistry (Bosman and Winston 1970; Pasternak et al. 1971), antigen expression (Cikes and Klein 1972; Everson et al. 1974; Panem and Schauf 1974), sensitivity to immune killing (Lerner et al. 1971; Shipley 1971), drug cytotoxicity (Valeriote and van Putten 1975), and ability to metastasize (Sweeney et al. 1982; Weiss 1980a), resulting in the appearance of tumors that are heterogeneous with regard to all these properties. Therefore, *in situ* demonstration of tumor heterogeneity cannot constitute proof of a stable phenotypic heterogeneity. However, formal evidence vis-à-vis isolation and characterization of cells has been presented for a number of human tumors.

Tumor lines that differ in drug sensitivity (Barranco et al. 1973; Barranco et al. 1972), antigenicity (Albino et al. 1981), or tumorigenicity in nude mice (Aubert et al. 1980; Kozlowski et al. 1984) have been isolated from individual melanomas, both from primary lesions (Barranco et al. 1973; Barranco et al. 1972; Aubert et al. 1980) or multiple metastases of the

same patient (Albino et al. 1981; Aubert et al. 1980) as well as, from tumor subpopulations isolated from primary human colon carcinomas (Brattain et al. 1981; Dexter et al. 1981; Fidler 1990). Several of these subpopulations differ in karyotype (Brattain et al. 1981), *in vitro* growth properties (Brattain et al. 1981; Dexter et al. 1981), tumorigenicity (Brattain et al. 1981), and tumor histology in nude mice (Bohm and Sandritter 1975; Straus 1977). Similar isolations of tumor subpopulations have been reported for lung (Chu et al. 1979), ovarian (Mackintosh et al. 1981), and bladder (Hastings and Franks 1983) cancer. Other human neoplasms, including melanoma (Barranco et al. 1973; Barranco et al. 1972; Lotan 1979), colon adenocarcinoma and gastric carcinoma (Trope 1982; Trope et al. 1975), ovarian carcinoma (Trope et al. 1979), breast carcinoma (Lotan 1979; Baylin 1982; Siracky 1979a), lymphoma-leukemias (Biorklund et al. 1980; Siracky 1979b), and lung cancer (Brennan et al. 1979), also contain subpopulations of cells with different drug sensitivities.

Shirpo et al. (1981) studied the karyotypic heterogeneity within human tumors by karyotyping tumor cells from fresh samples of human gliomas within 6–72h after surgery. An array of unique karyotypes was found in each tumor (Pathak 1990; Wolman 1986). Simultaneously, dissociated tumor cells were cloned by dilution plating and the clones karyotyped. By matching karyotypes of the clones with those in the fresh sample, it was possible to show that the clones were present at the time of resection. Each of the eight gliomas was found in this way to have from 3 to 21 subpopulations – a minimal estimate since different subpopulations can have similar karyotypes. Different clones from the same tumor also varied in morphology and growth kinetics. A report from our laboratory also demonstrated the heterogeneity of the cloned subpopulations to chemotherapeutic agents (Talmadge et al. 1984).

Tumor heterogeneity for invasion and metastasis

The possibility that cells with differing metastatic capabilities might coexist within the same tumor was first suggested in 1939 by Koch (1939), who isolated a highly metastatic subline from the Ehrlich carcinoma tumor by serially transplanting lymph node metastases.

In 1955, Klein (Klein 1954, 1955) demonstrated that the gradual conversion of solid murine neoplasms into ascites variants was due to the selective overgrowth of a small number of cells that differed from the parent population in their ability to proliferate in the peritoneal cavity and metastasize to the lungs. Since the change was stable and heritable, Klein concluded that the gradual conversion of the solid tumor to the ascites form involved mutation and selection and was not attributable to adaptation.

Cells with different metastatic properties have been isolated from the same tumor, suggesting the hypothesis that all the cells in a primary tumor can successfully disseminate (Fidler 2002). Three different experimental approaches can be used to isolate tumor cell subpopulations with differing invasive and/or metastatic abilities. The first involves enrichment of the fraction of invasive–metastatic subpopulations in heterogeneous tumor cell populations. This uses the technique of repeated cycling to gradually enrich for a subpopulation with the desired metastatic phenotype. Spontaneous metastasis in a syngenic model is allowed to occur and the metastatic population recovered and recycled. The biological behavior of the selected cells is then compared with that of the parent tumor cells to determine whether there is an enhanced metastatic capacity. This procedure was used to obtain the B16-F10 line from the parent B16 melanoma (Fidler 1973). In these investigations, tumor cells were injected intravenously, lung tumor nodules excised 3 weeks later, and tumor cells from lung metastases established in culture, and then reinjected into new mice. After ten such cycles, a tumor line emerged that showed a marked increase in its ability to produce pulmonary tumor colonies (Fidler 1973). Studies from several laboratories, using similar strategies with animal and human tumors of diverse histologic origin, have also revealed significant variations in the metastatic capabilities of cells isolated from the same tumor (Dexter et al. 1978; Danielson et al. 1980; Zupi et al. 1980; Symmans et al. 1995; Naito et al. 1991; Ware and Maygarden 1989; Harris and Best 1988; Soule et al. 1981; Brunson and Nicolson 1978; Poste et al. 1980; Talmadge and Fidler 1982a; Tarin and Price 1979; Chambers et al. 1981; Kerbel et al. 1988b). Nonetheless, these selected tumor lines may still be heterogeneous and contain multiple subpopulations with differing metastatic potentials, as well as, other

phenotypes. The selection pressures serve only to enrich for a general tumor population that also contains subpopulations of tumor cells that may express differing invasive and metastatic properties.

The second and related method to isolate sublines with differing metastatic properties from a common parent tumor line is to select for (or against) properties considered important for successful metastasis. As in the enrichment method, variants displaying (or lacking) the property of interest are isolated and tested to determine whether their metastatic behavior is altered. This method has been used to examine whether properties as diverse as adhesive characteristics (Briles and Kornfeld 1978)[178], lectin resistance (Reading et al. 1980b; Tao and Burger 1977), invasive capacity (Poste et al. 1980; Hart 1979), resistance to cytotoxic T lymphocytes (Fidler and Bucana 1977; Frost and Kerbel 1981; Sandberg 1977), and resistance or sensitivity to natural killer (NK) cells (Gorelik et al. 1982; Hanna and Fidler 1981) influence the ability of tumor cells to metastasize. The third approach to demonstrate that malignant primary tumors contain subpopulations of cells with differing metastatic capabilities involves *in vitro* cloning tumor cell populations. The clones can be compared within the same population to determine if a defined phenotype affects the process of metastasis. The phenotypic analysis of tumor cell clones is the most direct and satisfactory of the three techniques to demonstrate cellular heterogeneity, as well as, to determine if a unique phenotype is important to metastasis. It must be remarked that in order to reach a generalized conclusion regarding the association of a particular attribute with the metastatic phenotype, one must examine multiple variants from multiple tumors. Thus, one might examine 10–20 clones from each of five breast cancer cell lines. However, even this effort would provide results that would only be attributable to breast cancer cell lines and would not support a more generalized conclusion, including other tumor histotypes. Thus, multiple variants from multiple cell lines of multiple histotypes are necessary in order to achieve a general conclusion such as an association of collogenase type IV activity with malignancy. Therefore, the level of effort to associate a specific phenotype with the metastatic phenotype/genotype is significant.

The demonstration of metastatic heterogeneity within a primary tumor was first reported in 1977 by Fidler and Kripke (1977) using the B16 melanoma.

To investigate whether primary tumors contained cells of differing or uniform metastatic potential, they prepared a cell suspension from a subcutaneous primary tumor and divided it into two parts. One part was immediately assayed for its ability to form experimental pulmonary metastases after intravenous injection into mice. From the other part of the original suspension, 17 clones were isolated (each one was established from an individual cell). After incubation for the same period of time, equal numbers of tumor cells in suspension from each of the cloned lines and from the parent tumor were injected into syngeneic mice. It was reasoned that if the tumor contained cells of uniform metastatic potential, then the cloned sublines should each produce the same number of pulmonary colonies as the uncloned parent population. This was not the case. The original uncloned parent tumor cell population produced similar numbers of metastases when injected into different animals, but the cloned sublines markedly differed in their metastatic potential. Control subcloning experiments showed that this variability was not introduced by the process of cloning, since groups of animals injected with a cloned parent tumor or subcloned lines all had a similar range and distribution of metastases (Fidler and Kripke 1977). Therefore, they concluded that tumors contained cellular subpopulations that were heterogeneous in their metastatic potential.

The B16 melanoma is an established tumor line that has been maintained by repeated passage in animals or cell culture for many times the life span of its natural host. Thus, the metastatic diversity in this tumor line could have been an artefact caused by its longevity; however, comparable data were observed with another murine melanoma of recent origin (Fidler et al. 1981). This tumor arose in a C3H mouse that had been subjected to ten 1-h exposures of UV radiation followed by the application of 2.5% croton oil in acetone to the skin of the scapular region for 2 years. The primary tumor that developed was removed and its fragments were transplanted into immunodeficient animals to circumvent the possibility of immune selection. Several weeks later, a tissue culture line was established and, during the fifth passage, the cell lines were cloned. Analysis of the metastatic capacity of the parent tumor (K-1735) and its cloned lines was performed in a manner similar to the original study. The clones differed dramatically from each other and

from the parent line in their production of lung tumor colonies with regard to their size, number, and pigmentation. Within each injected clone, however, these three characteristics were expressed uniformly. Thus, lung colonies produced by one clone could be distinguished readily from those produced by another on the basis of size and pigmentation. Statistical analysis of the results indicated that only two of 22 K-1735 clones were indistinguishable from the parent tumor (Fidler et al. 1981). With these figures as an indication of the degree of metastatic heterogeneity for metastasis, the K-1735 melanoma was shown to be no less heterogeneous than the B16 melanoma. These results indicate that the longevity of neoplasms is not the sole arbiter of metastatic heterogeneity. Further, clonal variation in metastatic properties is not unique to melanomas. Comparable extensive heterogeneity with regard to malignant-metastatic properties has been described in clones isolated from tumors of diverse histologic origin from the mouse (Dexter et al.1978; Suzuki et al. 1978; Kripke and Fidler 1980; Schmitt and Daynes 1981; Stackpole 1981), rat (Talmadge et al. 1979), chicken (Shearman and Longenecker 1980), hamster (Enders and Diamandopoulos 1969) and human (Nakajima et al. 1990; Morikawa et al. 1988; Saiki et al. 1991). In those studies, tumor cell subpopulations with differing metastatic phenotypes isolated from primary tumors or cultured tumor cell lines were as heterogeneous as those reported previously.

The value of tumor cell clones in analyzing any aspect of tumor cell behavior requires that the phenotypic characteristic(s) of interest remain stable during serial passage of the clones whether *in vivo* or *in vitro*. Poorly metastatic and highly metastatic clones were isolated from the UV-2237 fibrosarcoma (Kripke and Fidler 1980) and cultivated *in vitro* for 72 or 60 days, respectively (Cifone and Fidler 1981). Simultaneously, both clones were also grown subcutaneously in syngeneic mice. Then, cell cultures were established from these solid tumors and 1 week later, subclones were isolated from both the *in vitro* and the *in vivo* passaged tumor lines. The ability of the subclones to form experimental metastases was compared between the subclones derived from clones grown in culture or *in vivo* and subclones isolated and frozen when the parent clones were initially established. The patterns of behavior of all the subclones derived from the poorly metastatic clone were remarkably similar

to that of the parent clone, regardless of whether the subclones were derived at the time of isolation or after 72 days of continuous growth *in vitro* or *in vivo*. In contrast, after 60 days of cultivation *in vitro* or *in vivo*, the metastatic behavior of the subclones derived from the highly metastatic clone differed considerably from that of the parent clone, suggesting that the metastatic phenotype of the highly metastatic clone is unstable (Cifone and Fidler 1981). It was suggested that this rapid generation of diversity may have been caused in part by increased genetic instability.

Further studies using the B16 melanoma have also demonstrated that the invasive and metastatic properties of clones from this tumor are highly unstable during serial passage and that subclones with different invasive and metastatic properties are generated rapidly on serial passage either *in vitro* or *in vivo* (Bosslet and Schirrmacher 1982; Isaacs et al. 1982; Harris et al. 1982; Poste et al. 1981; Raz 1982; Stackpole 1983). Studies of individual B16 clones expressing a variety of stable biochemical markers has revealed this finding whereas the metastatic phenotype is unstable when clones are grown singly, mixing and cocultivation of clones eliminate this phenotypic instability, and the formation of variant subclones with altered metastatic properties is reduced dramatically (Poste et al. 1981). This suggests that some form of "interaction" occurs between the various cellular subpopulations in polyclonal populations, which "stabilizes" not only their invasive-metastatic properties, but also their relative proportions within the total populations (Poste et al. 1981). This type of interaction would conserve clonal diversity within a tumor cell population and prevent domination of the population by a few subpopulations or even a single subpopulation. This "stabilizing" effect produced by mixing clones is, however, specific for cells from the same tumor. Single clones of B16 melanoma cocultivated with clones from the Lewis lung carcinoma or UV-2237 fibrosarcoma show marked phenotypic instability and rapidly generate subclones with widely differing metastatic properties (Poste et al. 1981). The role of clonal interactions in regulating the stability of the metastatic phenotype is not unique to the B16 melanoma. Similar instability of the metastatic phenotype has been found in the mouse UV2237 fibrosarcoma (Cifone and Fidler 1981), mouse RAW117 lymphosarcoma (Brunson

and Nicolson 1978), mouse KHT sarcoma (Harris et al. 1982), rat lAR6 hepatocarcinoma (Talmadge et al. 1979), and spontaneously arising mammary tumors (Miller and Heppner 1979).

Metastasis as a selective and inefficient process

The pre-existence of metastatic variation within the parent tumor does not answer the question of whether the cells that ultimately form metastatic foci possess a greater metastatic potential than the cells that populate the parent neoplasm. To address this question, we examined the issue of whether metastases result from the random survival of cells released from the primary tumor or from the selective growth of cells with specialized properties that allow them to complete the metastatic process. If the process of metastasis is selective as suggested by previous studies, direct evidence to support this hypothesis would be the demonstration that cells populating spontaneous metastases should be demonstrably more metastatic than the cells within the heterogeneous parent neoplasm as long as (1) the starting tumor population was unse1ected and heterogeneous with respect to metastatic potential and (2) the metastatic process per se exerted selective pressures upon the tumor cells.

The initial studies from our laboratory to address this question used three metastatic variants of the B16 melanoma, thus minimizing variables that would be introduced by using tumor models with varying biological characteristics. The malignant melanoma, B16-F1 (Fidler 1973), is an unse1ected, tumor cell line that metastasizes poorly after intravenous injection (experimental metastasis) or footpad injection (spontaneous metastasis). The B16-F10 tumor was selected ten times for its ability to colonize the lungs after intravenous injection (Fidler 1973) and its high rate of experimental metastasis. The B16-BL6 tumor line was selected *in vitro* for its invasive ability and demonstrates a high incidence of spontaneous metastasis following intrafootpad implantation (Hart 1979).

The experimental design was as follows: the tumor lines were implanted into the footpads of syngeneic mice (Fig. 1). When the tumors reached a diameter of 1–1.2 cm, the tumor-bearing leg, including the popliteal lymph node, was resected. Several weeks later, when a few mice in each group appeared listless, the

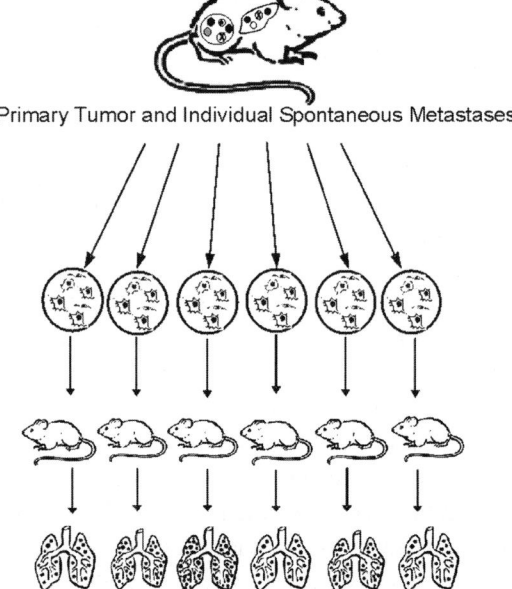

Primary Tumor and Individual Spontaneous Metastases

Fig. 1 Scheme outlining approaches to assess the heterogeneity between tumor metastases and the metastatic phenotypes of cells from metastatic lesions as compared to cells from the primary tumor

entire group was killed. From each group, well-isolated pulmonary metastases were surgically excised and established in culture as individual cell lines. The metastatic potential of cells from the parent tumor and their respective spontaneous metastases was then examined using assays of experimental and spontaneous metastasis.

Tumor cells harvested from spontaneous metastases of the poorly metastatic B16-F1 tumor line produced significantly more lung tumor colonies after intravenous injection than an equal number of cells from the parent line. Tumor cell lines from spontaneous metastases of the B16-BL6 tumor variant also produced significantly more lung colonies than the parent (B16-BL6) tumor line. However, the increase in metastatic potential of cell lines from spontaneous metastases of the B16-BL6 tumors was less than that observed with tumor cells from spontaneous metastases of B16-F1. Tumor cells from spontaneous metastases of the B16-F10 variant line, which was previously selected for the ability to form lung colonies, did not exhibit any increased lung-colonizing potential compared with the parent tumor. This latter observation was not attributable to the number of lung colonies involved since the injection of five-fold fewer tumor cells gave similar

results, i.e. the lung-colonizing potential of the tumor cells from spontaneous metastases was comparable to that of tumor cells from the parent tumor, although the total numbers of tumor foci were lower.

The spontaneous metastatic potential of these same tumor variants and their respective spontaneous metastases were studied, with results that were different overall from the results of the studies of lung-colonizing potential. As was found in the lung colony assays, tumor cells from spontaneous metastases of the B16-F1 tumor were significantly more spontaneously metastatic than the tumor cells from the parent tumor (t = 0.007), but tumor cells harvested from spontaneous metastases of the B16-F10 tumor also exhibited an increased propensity for spontaneous metastasis, whereas they did not express an increased lung-colonizing potential. Similar studies with metastases of the B16-BL6 tumor did not demonstrate a statistical difference in spontaneous metastases ability compared with that seen with the parent B16-BL6 tumor. The B16-BL6 tumor variant had been previously selected for an invasive phenotype and therefore was already highly spontaneously metastatic. Thus, the tumor cells from spontaneous metastases did not differ from the parent tumor with regard to their metastatic properties. We concluded that metastases obtained from an unselected, heterogeneous, and poorly metastatic cell line (Bl6-Fl) contained tumor cells with enhanced metastatic potential. In contrast, when the parent population was previously selected and was already near a peak metastatic phenotype (B16-F10 for lung colonization or 16-BL6 for spontaneous metastasis); further selection for the particular metastatic phenotype could not be achieved.

To rule out the possibility that these results were unique to the B16 melanoma tumor system, we repeated this study with four other tumor cell populations (Talmadge et al. 1981) in a manner similar to the original study. These tumors included a malignant melanoma of recent origin (K-1735); an ultraviolet-radiation-induced fibrosarcoma (UV-2237) and a cloned subline (clone 40); the Lewis lung carcinoma (3LL); and the reticulum cell sarcoma of histiocytic origin (M5076). We found in each of these tumor models that cells from spontaneous metastases of heterogeneous parent tumors always expressed a greater propensity to metastasize than tumor cells from the parent tumor. However, cells from spontaneous metastases of a cloned parent

tumor (UV-2237, clone 40), presumably a homogeneous tumor population, did not differ significantly in potential for experimental metastasis from the parent (cloned) tumor population. The metastatic homogeneity of clone 40 was not unique to this tumor since spontaneous metastases from the heterogeneous parent tumor population (UV-2237), from which the clone was obtained, exhibited a greater metastatic potential than the heterogeneous parent tumor line. Studies using the histiocytic lymphoma, M5076 (Talmadge et al. 1981), demonstrated that the increased metastatic potential of tumor cells from spontaneous metastases was not affected by the anatomical location of the metastases. The M5076 tumor consistently metastasizes to the liver and only rarely to the lungs (and then late in the course of tumor growth) (Hart et al. 1981). In these studies, we found that spontaneous hepatic metastases from the M5076 tumor exhibited an increased propensity to metastasize to the liver compared with the parent M5076 line (Talmadge and Fidler 1982b). This increase in metastatic potential occurred not only during experimental metastasis, but also in mice bearing unresected primary tumors.

We concluded from these studies that tumor cells from spontaneous metastases of unselected tumors, which are heterogeneous with regard to a metastatic phenotype, have an increased metastatic potential compared with the present tumor. In contrast, metastases from homogeneous (clonal) tumors or tumors previously selected for a metastatic phenotype vary only slightly from the parent tumor in their ability to metastasize. Therefore, although the process of metastasis is generally selective, it may give the appearance of being random if artifactually homogeneous tumors are examined.

These studies have been extended to other tumor models. Pollack and Fidler (1982) utilized young nude mice to investigate whether animals, which lack functional T lymphocytes and express low levels of NK cells, could provide a model to select for metastatic subpopulations from heterogeneous allogeneic melanoma. Three-week-old nude mice (T cell deficient) received a single cell suspension of either B16 or K-1735 melanoma tumor cells injected intravenously. Individual pulmonary metastases were harvested 3 weeks later and their metastatic potential was assessed in both nude mice and normal syngeneic mice. In all cases, the cells from the metastases

colonized the lungs with significantly higher efficiency than did cells from the parent tumors. Using a similar stratagey, Kozlowski et al. (1984) found that lung colonies from the human malignant melanoma A-375 in nude mice contained a select subpopulation of tumor cells of high metastatic potential compared with the parent tumor cell populations.

Raz et al. (1981) have shown that mouse UV-2237 fibrosarcoma cells recovered from lung metastases produced by a cloned parental cell line with low lung colonization potential do not express an increased capacity to colonize the lung when injected intravenously. This indicates that growth in the lung is not sufficient to augment the ability of cells from this tumor to localize and grow in the lung. Additional evidence indicating that growth in the lung is not required for tumor cells to selectively localize in the lung was presented by the same authors who found that an uncloned UV-2237 cell line grown in the peritoneal cavity contained multiple clonal subpopulations, including clones with a high capacity to colonize the lung. This observation is in agreement with that previously described by Klein (1954, 1955) i.e., adaptation to peritoneal growth resulted in cell lines with a high lung-colonizing potential.

Comparable findings investigating tumor adaptation have been obtained by Nicolson and Custead (1982). Mice were given intravenous injections of 120–180 ∝m diameter plastic microbeads coated with B16-F1 melanoma cells, which have a low lung colonization capacity. Entrapment of the beads within the pulmonary microcirculation serves to artificially enhance the localization of these cells within the lung. When visible lung colonies had formed, they were excised and dispersed to yield tumor cells that were grown *in vitro* for a brief period, reattached to new beads, and reinjected intravenously. After nine such cycles of *in vivo–in vitro* transfer employing carrier beads, cells recovered from lung colonies were reinjected into mice without beads and their metastatic ability was compared with the B16-F10 subline. The latter was isolated after an identical number of *in vivo–in vitro* transfers, but cells were injected as single cell suspensions rather than attached to beads. Nicolson and Custead found that even after ten cycles of growth in the lung, the B16-F1 cells attached to beads (B16-F1A10) were no more metastatic than parental B16-F1 cells. In contrast, the B16-F10 cell line was highly efficient in colonizing the lung; producing between 5 and 13 times more lung metastases than the parental B16-F1 cell line. In a series of studies, Poste et al. (1982a, b) examined the metastatic properties of tumor cell clones isolated from individual metastatic lesions obtained following the intravenous injection of various B16 melanoma variants. The individual metastatic lesions were examined at different stages in the evolution of metastasis (i.e. larger or smaller pulmonary nodules). They found that the progressive growth of metastatic lesions was accompanied by the emergence of variant tumor cellular subpopulations with altered metastatic properties within the lesion. They investigated the experimental metastatic potential of the clones obtained from the different metastases and determined that the cells populating the individual metastases were all metastatic. In contrast, the tumor cells within the parent tumor included nonmetastatic variants. They concluded that subpopulations isolated from different metastases in the same host differ markedly in their metastatic ability and that those cells with both high and low, but not nonmetastatic, metastatic capacities can be recovered from metastases.

In a study similar to ours (Talmadge and Fidler 1982a, b), Neri et al. (1982) examined the selective or random nature of metastases using the rat 13,762 mammary adenocarcinoma. They reported that following subcutaneous implantation into the mammary footpad of syngeneic rats, the unresected parental mammary adenocarcinoma metastasizes at a low frequency to the lymph nodes and lungs. Cell lines adapted to tissue culture from individual secondary sites were, in contrast to the parent tumor, inevitably metastatic (without resection) from a subcutaneous site within 23 days of transplantation; confirming the selective nature of metastasis in a rat mammary tumor model. Another study using a *Herpesvirus hominis* type 2 induced tumor line from a Syrian hamster (Walker et al. 1982) also demonstrated the selective nature of metastasis. The parent tumor exhibited a low level of spontaneous metastases from the primary subcutaneous tumor site. However, tumor lines established from lung foci showed an elevated metastatic potential compared with the parent cell line such that all animals injected with cells from the metastases developed secondary foci within 40 days following resection of the primary tumor.

Despite the experimental evidence supporting the selective process of metastasis, the conclusion that

metastatic cells arise by a selective process has not been universal. Mantovani et al. (1981) and Giaviazzi et al. (1980) reported that cells with high and low metastatic potential gave rise to metastases and that the metastases were not always formed by cells with the highest metastatic potential. The selection hypothesis as outlined in this chapter states that in order for tumor cells to form metastases, they must express a phenotype that allows them to complete all the steps in the metastatic process, as well as, avoid destruction during metastasis. Tumor cells that lack any of these attributes will be unable to produce metastases; an observation supported by our studies (Talmadge and Fidler 1982a, b) as in every case, the tumor cells within metastases were metastatic. However, metastases are not necessarily composed of cellular subpopulations with the greatest metastatic potential. Because of the heterogeneous nature of tumors (discussed earlier) and because metastases from one tumor may derive from multiple progenitor cells (Talmadge et al. 1982), one would expect that tumor cells within different individual metastases would vary in their metastatic potentials, which is indeed the case. (See also the review by Fidler (2002), Poste (1982), Nicolson (1984b).) This heterogeneity would occur if the primary tumor were initially heterogeneous. If a tumor had been previously selected for a metastatic phenotype (for example, B16-F10 for lung colonization or B16-BL6 for spontaneous metastases), then the metastatic process would be unlikely to select for cellular subpopulations with an increased metastatic potential (Talmadge and Fidler 1982a) and the process of metastasis could not select for a metastatic variant from a homogeneous (clonal) tumor population (Raz et al. 1981; Raz 1982; Talmadge and Fidler 1982b). If the parent tumor population was passaged by trocar transplantation, a technique that tends to limit tumor heterogeneity (Trope 1982; Fidler and Hart 1981), then metastases from such a population exhibit increased metastatic potential (Mantovani et al. 1981; Giavazzi et al. 1980).

The conclusion that metastasis is a selective process does not rule out the occurrence of random (chance) events. Subpopulations of tumor cells with metastatic capabilities may be killed during the metastatic process and thus, not complete the metastatic "decathalon" to form metastatic foci. During metastasis, tumor cells are exposed to vascular turbulence, the mononuclear phagocytic system, Natural Killer cells, and other detrimental conditions that could prevent completion of the metastatic process. There is, therefore, some element of chance during metastasis and certainly not all cells with a metastatic phenotype survive to form metastatic foci. However, a tumor cell probably will not form a metastatic focus if it does not express all the attributes needed to metastasize given the constraints imposed by the host.

Stackpole (1983) and Weiss et al. (1983) have argued against the concept of a selective process during metastasis. Stackpole examined the metastatic properties, in both experimental and spontaneous metastases, of a large number (>150) of clones and subclones from B16 melanoma cell lines. In this study, Stackpole reported that the phenotypic diversity with regard to three distinct dissemination-related biological parameters (metastasis, colonization, and cell proliferation rate) was generated so rapidly and with such regularity within B16 melanoma clones that tumor longevity may be inconsequential to the issue of heterogeneity. He suggested that significant fluctuation in tumor cell subpopulations occurred periodically, affecting both the metastatic and colonizing predilection of a tumor cell line. He also suggested that this fluctuation in metastatic potential was sufficiently extensive that one should question the reproducibility of any one observation.

Based on a variation of the Luria-Delbruck fluctuation analysis, the early heterogeneity studies by Fidler and Kripke (1977) and Kripke et al. (1978) required a control demonstrating that the metastatic phenotype of a tumor line was reproducible. As they stated, the analysis of metastatic phenotype by cloning experiments requires the demonstration that subclones from a recent clone were statistically similar in metastatic potential compared with the metastatic potential of the parent/clone, thereby demonstrating that the heterogeneity observed with the parental tumor was not an artefact of the cloning process. Therefore, the fluctuation observed by Stackpole (1983), unsupported by statistical analysis, needed to demonstrated assay reproducible for metastatic predilection.

The concept that metastasis is a nonrandom process involving the selective survival of cellular subpopulations with the phenotypic properties required to complete each step in the metastatic process does not infer that random events do not occur. Tumor cell

subpopulations, even those with metastatic properties, are at constant risk during the metastatic process due to random events. Similarly, metastatic competence to complete any specific step in the metastatic process may be subject to a reversible impairment imposed by its microenvironment. Weiss et al. (1983) has suggested that alterations in nutrition or metabolism secondary to cellular proliferation or degenerative events within solid tumors may modulate a number of cellular properties, i.e. site dependent, reversible changes. These studies suggest that phenotypic modulation could then induce a situation in which metastatic subpopulations may be subject to transitory phenotypic alterations that enhance or restrict metastatic competence (Weiss 1980b, 1990, 1996, 2000). It is proposed that cells entering the metastatic process do so from "transient metastatic compartments" and that, after allowance is made for pathophysiologic differences between primary and metastatic lesions, metastases are no more likely to metastasize than their parent primary tumor. Weiss et al. (1983) report that, in the case of KHT and B16 tumors, when selected primary and secondary groups of animals are compared, cancer cells from the latter gave rise to significantly more metastases than the former. In contrast, cells from selected primary lesions of 3LL and T-24 tumors gave rise to significantly more metastases than cells derived from pulmonary metastases. Therefore, the process of metastasis appeared selective for two tumors, while the process appeared random with two other tumors. However, all the tumor variants were metastatic and no tumor line, established from either primary or metastatic lesions, was nonmetastatic. Thus, the cells within the metastatic lesions did express the phenotypic characteristics required for the metastatic process, suggesting that the primary tumor was composed of tumor populations, the majority of which were metastatic. These results are in agreement with the hypothesis that spontaneous metastases arise from preexistent metastatic populations within the primary tumor, since this does not infer that the metastatic cell populations compose a minor subpopulation within the primary tumor. The studies with homogeneous tumors (clone 40 from UV-2237) or previously selected tumors, B16-F10 or B16-BL6, demonstrated that metastases from homogeneous or highly metastatic tumors were similar in metastatic ability to the parental tumor.

Whereas large number of tumor cells from primary tumor may gain access to the circulation, few of them result in metastases (Schirrmacher et al. 1982). The mechanisms responsible for the elimination of these tumor cells, termed metastatic inefficiency, are poorly understood (Weiss 1980c). Experimental evidence suggest that the metastatic process is inefficient due to phenotypic differences in heterogeneous cell population with respect to genotypes associated with different metastatic process (Weiss 1986, 1990). The majority of cells entering the metastatic process do not form metastasis. A recent report demonstrated that cellular apoptosis *in vivo* may be associated with a decreased incidence of metastasis *in vivo* (Wong et al. 2001). Thus, it is the attributes of the surviving minority of cancer cells that result in metastases. We have discussed earlier that a tumor contains subpopulations of cancer cells capable of producing either more or fewer metastases than other subpopulations. The overall metastatic behavior of a cancer cell subpopulation could, therefore, depend on the number of cells with a metastatic phenotype and this could be either a transient or stable property of these cells. The expression of a metastatic phenotype is further complicated by synergism between tumor and host cells (Talmadge et al. 1984; Weiss 1983, 1990; Singh et al. 1997; Singh and Fidler 1996) Recent studies indicate that genetic mutations can result in either the acquisition or increased metastatic potential of a tumor cell population. Many of the mechanisms involved in the enhancement of the metastatic phenotype appear to be initiated by individual malignant cells and may include an interaction with the host microenvironment. An important consequence of metastatic inefficiency is the organ-specific patterns of metastasis. It has been suggested that a temporal and stepwise progression occurs in lung metastasis and that it is dependent on cell survival, dormancy and location dependence of metastatic inefficiency (Cameron et al. 2000). In this study, B16F10 cells were injected in a manner to target the lung, and at sequential times two parameters were quantified: (1) overall cell survival and metastatic development; and (2) local cell survival and growth with respect to lung surface and specific interior structure. An initial high rate of cellular survival was observed for cells trapped in the lung circulation, including extravasation into lung tissue and subsequent survival of extravasated solitary cells before metastasis formation (Cameron et al. 2000). However at the time

of cell division, a major loss occurred, such that only a small portion of tumor cells started to form metastases, but most of these developed into and formed macroscopic foci. In contrast, solitary cells found at a later time were dormant (Cameron et al. 2000). These results suggest that metastatic inefficiency is largely due to post-extravasation events affecting solitary cells. They also demonstrated that trapping and early growth of injected cells is unaffected by location within the lung, and that subsequent metastatic growth can be enhanced by the location within lung (Cameron et al. 2000), suggesting that continued growth of metastases occurs preferentially in a specific tissue microenvironment.

Dissimilarities among metastases from a primary tumor

The observation that a tumor population is heterogeneous, be it either a primary tumor or a metastatic focus, does not suggest that such individual lesions do not, as a population, differ phenotypically from one another. Thus, metastases within one host can exhibit heterogeneity with regard to many phenotypes and genotypes besides metastatic capacity such as hormone receptors (Brennan et al. 1979; Sluyser and Van Nie 1974); marker enzymes (Baylin et al. 1978; Fialkow 1976); antigenicity or immunogenicity (Olsson and Ebbesen 1979; Albino et al. 1981; Kerbel 1979; Pimm and Baldwin 1980; Fogel et al. 1979; McCune et al. 1981; Thistlethwaithe et al. 1983; Schirrmacher et al. 1982); macrophage content (Key et al. 1982); metabolic characteristics (Baylin et al. 1978); androgen response (Brennan et al. 1979); karyotypic expression (Kusyk et al. 1981; Sandberg 1977); DNA content (Starace et al. 1982; Reeve and Twentyman 1982)[98, 227]; growth factor preference and dependence (Nicolson 1984a; Fidler 2002; Lu and Kerbel 1994; Filmus and Kerbel 1993; Kerbel 1990, 1992); and response to various chemotherapeutic agents (Tsuruo and Fidler 1981; Dexter and Leith 1986; Heppner et al. 1978; Fugmann et al. 1977; Schlag and Schreml 1982).

Evidence that human cancers contain tumor cell subpopulations with divergent phenotypes comes from the comparison of tumor cells from both primary tumors and metastases. Here again, one may see divergence in histological type (Parbhoo 1981). The problem of cellular diversity inherent in some

neoplasms is clearly demonstrated by the work of Baylin et al. (1978). The elevation serum histaminase, L-dopa decarboxylase, and calcitonin are used as clinical markers for the presence of small-cell lung cancer in humans. However, the simultaneous sampling of primary and metastatic lesions of small-cell lung cancer by Baylin and colleagues demonstrated significant differences in the levels of these markers between the primary tumor and metastases. All primary (chest) lesions produced high levels of these markers. In contrast, either low levels or none of these three products could be detected in four of seven metastases isolated from the livers of several patients. Moreover, immunohistochemical tests for histaminase demonstrated that cells within the primary tumor were heterogeneous in their enzyme content. Therefore, the level of these three markers in a patient's serum may not accurately represent the tumor burden in patients with small-cell lung tumors. Another study reported differences in sensitivity to antineoplastic drugs *in vitro* between cells from primary ovarian carcinomas and their metastases (Trope et al. 1979; Kusyk et al. 1981). A variable, estrogen receptor content between primary breast cancers and their metastases, as well as, among multiple metastases in the same patient has been reported (Brennan et al. 1979).

Striking evidence that metastases of primary human tumors may not be uniformly susceptible to control by immunologic manipulation comes from a study performed by McCune et al. (1981). These investigators used active specific immunotherapy directed against advanced renal carcinoma and its metastases by giving weekly injections of autologous irradiated tumor cells obtained from the primary neoplasm admixed with *Corynebacterium parvum*. They found that not only did the degree of therapeutic efficacy vary from patient to patient, but also some metastatic lesions regressed, while others simultaneously progressed in patients whose overall response was favourable. This variable responsiveness of metastatic lesions even within the same patient was attributed to the antigenic diversity of the metastatic subpopulations. These observations suggest that antitumor immune responses can be evoked in patients with renal carcinoma; however, equally importantly, they show that the heterogeneity of the metastatic cells with respect to antigenicity is a problem that must be overcome if specific immunotherapy is to be truly effective in eradicating metastatic disease.

In another study, Albino et al. 1981 examined the antigenic characteristics of six individual metastases from a single patient with melanoma. These lines showed marked phenotypic diversity as indicated by characteristic differences in growth rate, morphology, pigmentation, and the expression of surface antigens and glycoproteins. Some lines expressed HLA-DR products, whereas others lacked HLA-DR expression. These lines could also be distinguished on a basis of their glycoprotein profile. Additional quantitative differences in the membrane antigenic phenotype of the three cell lines were revealed by serological tests with a battery of monoclonal and conventional antibodies defining melanoma differentiation antigens. Another study of melanoma-bearing patients by Natali et al. (1983) examined the antigenic heterogeneity of primary and metastatic lesions surgically removed from nine patients with nodular melanoma. The antigenic expression was investigated using monoclonal antibodies to HLA-A and B antigens to β2-microglobulin and to melanoma-associated antigens (MAA). The latter included three types of membrane-bound MAA as well as, a cytoplasmic MAA. In spite of a homogeneous morphologic appearance, multiple lesions removed from the same patient differed significantly in their reactivity with the panel of monoclonal antibodies when studied using an indirect immunofluorescence test. The extent of antigenic heterogeneity did not correlate with melanin synthesis, site of origin of the primary tumor, site of metastatic foci, or treatment and appeared to have decreased heterogeneity in patients carrying the primary tumor.

Easty et al. 1981 isolated from a single patient five tumor lines, including the primary squamous carcinoma of the tongue, two subsequent local recurrences, and two lymph node metastases. The cell lines from the recurrences demonstrated the greatest morphological divergence from the primary tumor line and also demonstrated the greatest differences at the ultrastructural level, in increased production of plasminogen activator, and in the composition of cell-surface glycoproteins. In another study by Feder and Gilbert (1983), cell lines were established from a primary neuroblastoma and, 11 months later, from four different individual metastases. Cells from the primary tumor demonstrated considerable heterogeneity in terms of chromosome number, while the cells from four subsequent metastases were all nearly diploid.

However, all of the tumor samples contained the same marker chromosome rearrangements, indicating their origin from a common precursor. In addition, each of the cell lines analyzed (including those from the metastases and those from the primary tumor) also contained unique distinguishing chromosomal abnormalities. Feder and Gilbert (1983) concluded that the differences in karyotype among these tumor samples and cell lines reflected the different selection pressures at work in each instance. These differences also represent karyotypic variations between the individual metastases.

In a rat mammary adenocarcinoma model, Neri et al. (1982) studied the phenotypic characteristics of several spontaneous metastases, as well as, the primary tumor. They examined their cell culture morphologies, histologic structures at the primary site and secondary metastatic sites, and growth characteristics *in vivo* and *in vitro*. In agreement with other studies that we have discussed, there was considerable variation in these phenotypes among the individual metastases (Neri et al. 1982). A subsequent study by Welch et al. (1983) examined the sensitivity of clonal populations from mammary adenocarcinoma metastases and the primary tumor to γ-radiation *in vitro*. This study demonstrated considerable clonal heterogeneity within the tumor and among its metastases with respect to the response to γ-radiation. They found that the inherent sensitivity to radiation could change with time, thereby altering the radiation survival responses.

As we discussed earlier, the development of cells with a metastatic phenotype in tumors is, in part, due to the selection of cells with a metastasis favouring genotype (Liotta and Stetler-Stevenson 1991; Liotta et al. 1991; Fidler 1995; Fidler and Radinsky 1996). Gene expression profiling using a high density microarray has contributed to our analysis of differences between metastatic and nonmetastatic cells. Clark et al. (2000) described the use of this approach to identify genes selectively upregulated in metastatic mouse and human melanoma cells compared to their nonmetastatic counterparts. They found a remarkable overexpression of RhoC, which they reported could stimulate metastasis. Similarly, Bittner et al. (2000) compared different subgroups of human melanoma by microarrays and also found a distinct patterns of gene expression. They demonstrate that the gene expression profile of human melanoma varied greatly among

different subgroups. What is interesting is that three of the genes identified by Clarke et al. (2000) (fibronectin, thymosin β4 and RhoC) showed increased expression in all of human and mouse melanoma derived metastasis. One drawback to such studies is that microarray analysis cannot identify post-transcriptional modifications of proteins. Additionally approaches used in microarray studies do not provide an unbiased method to pinpoint important and potentially new contributors of cancer. However, these studies will provide a basis to identify one or more key genetic or epigenetic events that induce metastasis. Another limitation is the heterogeneous nature of neoplasms; however, the approaches used in these studies do provide an unbiased method to identify a potentially new protein important to the metastatic process (Ridley 2000). Furthermore, a study examining the molecular signature of metastasis in primary solid tumors by Ramaswamy et al. (2003) has provided additional support for the non randomness of the metastatic process and development of a metastatic phenotype. In this report, the authors identified differences in gene expression patterns between primary tumors and metastases. However, they also found that the expression profile associated with metastasis was found in some primary tumors. Ramaswamy et al. examined the primary tumors with metastases-associated gene expression signatures and found that this was associated with a poor prognosis. Thus, his results are in accord with the idea that the tumor cells in a metastatic site have a stronger metastatic fitness, challenging the idea that metastasis arises from rare metastatic cells. In a subsequent discussion, however, Fidler and Kripke (2003) argued that the study by Ramaswamy and colleagues provided no evidence to contradict a selection process by metastasis. The study by Ramaswamy et al. looked at primary tumors in aggregate; therefore, could not rule out a selective process. The significance of the study by Ramaswamy et al. is not that it runs contrary to popular dogma, but that it enables the identification of the subset of tumors, designated as early stage by pathological criteria that nonetheless have already released metastatic cells (Fidler and Kripke 2003). Thus, this study constitutes an important step in the search to predict the behaviour of tumors detected at an early stage, even though it does not address the prevalence of metastatic cells (Fidler and Kripke 2003).

Clonal origin of metastases

Metastases, in general, do not result from the random survival of cells released from the primary tumor, but rather from the selected growth of specialized subpopulations of malignant cells that exist as subpopulations within the parent tumor as previously discussed in this chapter. However, this observation does not address the nature of the emboli released from the primary tumor. The embolus, which ultimately develops into a metastatic focus, may originate from a single cell or a cellular aggregate comprising of both tumor cells and host cells. Such a cell aggregate may be composed of tumor cells from a clonal population due to the zonal nature of neoplasms (Trope 1982; Fidler and Hart 1981) or a heterogeneous population of tumor cells. In addition, all the secondary tumor foci in a host may originate from a single progenitor cell or from multiple progenitor cells. We performed a series of experiments (Fig. 2) to determine (1) whether individual metastases are clonal or multicellular in origin and (2) whether metastases produced by one tumor originate from a single or different progenitor cell (Talmadge et al. 1982). These studies were patterned after the classical study by Becerk et al. (1963), which demonstrated the pluripotent nature of bone marrow stem cells.

In our experiments, cells from a metastatic variant of the K-1735 melanoma (K-1735 met-2) were exposed to x-radiation, which randomly induces chromosomal breaks and rearrangements. A certain number of these breaks and rearrangements (if not lethal) result in the formation of centric fusion chromosomes, and these provided the marker chromosomes for this study. We reasoned that if all the tumor cells populating a single spontaneous metastasis that arose from a primary tumor of x-irradiated tumor cells exhibited the same chromosomal arrangements, then the metastasis would have been derived from one cell. However, if the tumor cells populating individual, spontaneous metastases exhibited multiple chromosomal arrangements, then the metastasis would have arisen from more than one progenitor cell. The demonstration of a multiple tumor cell origin of a metastasis would be predicated on the stable expression of the various marker chromosomes. We chose to induce and use marker chromosomes rather than drug resistance since a larger number of markers would be available. In addition, the frequency

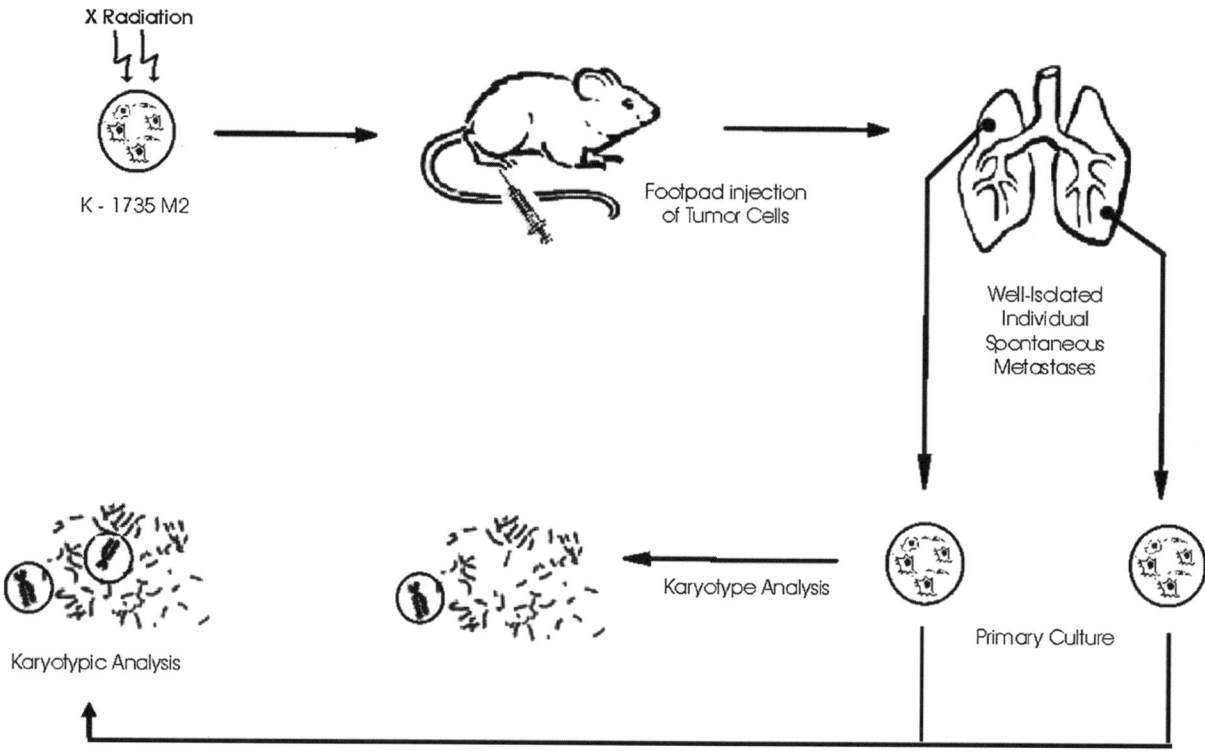

Fig. 2 Outline of the strategy used to assess the clonal origin of spontaneous metastases. In this approach, highly metastatic melanoma cells were irradiated to induce random chromosomal markers and injected to form a primary tumor, which were resected a month later. Once the mice were moribund, they were autopsied and individual metastatic lesions established in culture and karyotyped. Karotypic analysis was used to assess the cellular origin of the metastases, revealing a clonal origin

of reversion to drug resistance is a constant concern in the interpretation of results when drug markers are used. If a metastasis did prove to be clonal in origin, the demonstration of clonality based on two or three parent populations would be far weaker than that associated with a study utilizing a large number of cells with different marker chromosomes. In initial experiments, cells from the metastatic line of the K-1735 melanoma (K-1735, M2) were exposed to 650 rad of x-radiation and cells from the x-irradiated line were implanted into the footpad of syngeneic C3H/HeN mice. When the tumors reached a diameter of 1 cm, the tumor-bearing leg, including popliteal lymph node, was resected at mid-femur. Eight to ten weeks after tumor inoculation, spontaneous lung metastases were isolated, grown in culture as individual lines, and karyotyped. In ten of 21 lines, all the chromosomes were telocentric; therefore, these metastases were noninformative. In the other 11 lines, single or multiple marker chromosomes (submetacentric, metacentric, minute) were observed. In eight of these lines, unique patterns of chromosome(s) were

found in most spreads, suggesting that each metastasis originated from a single progenitor cell. In the remaining three lines, the pattern of markers varied, suggesting a bimodal or multimodal cell origin. However, G-band analysis indicated that these variations probably represented evolution within the individual metastasis. These data showed that, although metastasis is a highly selective process, different metastases can originate from different progenitor cells and that most metastases appear to be clonal in origin. This finding of multiple progenitors could account for the biological heterogeneity that exists among various metastases (Fidler and Hart 1981). The results suggested that metastases result from either the proliferation of a single viable cell or a single viable cell within a heterogeneous embolus or that a circulating tumor embolus is likely to be homogeneous because it originated from a clonal zone of a primary neoplasm (Fidler and Hart 1981).

Previous studies (Nicolson et al. 1978) have shown that tumor emboli composed of cellular aggregates, either homotypic (tumor cells) or heterotypic (host

cells, generally leukocytes and platelets), are arrested more rapidly in the first capillary bed encountered and have a better rate of survival. Both incidences resulted in a higher frequency of metastasis compared with the circulation of a similar number of single cells. The more rapid rate of arrest is a physical phenomenon associated with the larger embolus size, while the prolonged survival of tumor cell emboli is believed to be due to the "protection" afforded against host effector cells, NK cells, and monocytes, as well as, the turbulence within the circulation. If larger emboli are better able to metastasize, it would appear logical that metastasis could develop and, in part, be composed of a progenitor cell population that by itself was unable to complete the metastatic cascade, but was able to when associated in an embolus with a metastatic competent cell. Therefore, we designed experiments to address the following questions: (1) do tumor emboli that survive the many steps of metastasis consist of single cells or cell aggregates of monoclonal origin and (2) can metastatic competent cells provide a suitable environment within cellular aggregates whereby metastatic compromised or metastatic incompetent cells could survive and proliferate within the metastatic focus?

In a collaborative study with Fidler, we formed heterotypic cellular aggregates of either a single population of metastatic tumor cells (with a marker chromosome); a mixture of metastatic tumor cell populations, each with a different characteristic marker karyotypes or heterogeneous cellular aggregates composed of metastatic competent tumor cells; and benign tumor cells each with different marker karyotypes. The size and extent of cellular aggregation was monitored by autoradiographic studies using admixed radiolabeled and unlabeled cell populations. In this study, the metastatic variant K-1735 M2-X21 (chromosome mode of 42 and a 3:4 arm length ratio submetacentric chromosome) was admixed with the metastatic parent tumor population, K-1735, M2 (chromosome mode of 44, without any marker chromosomes), or K-1735-cl 2~, a nonmetastatic tumor variant (chromosome mode of 54, without any marker chromosomes). These heterotypic (or homotypic) cellular aggregates were injected intravenously into the lateral tail vein of syngeneic mice and, 21–24 days later, well-isolated metastatic tumor foci were established individually in tissue culture and karyotyped. The karyotypes of the metastases obtained from mice injected with K-1735, M2, or K-1735 M2-X21 (11/11 and 10/10, respectfully) expressed the appropriate characteristics. The mice injected with K-1735 cl 26, a nonmetastatic variant, did not develop metastatic foci. In contrast, all the metastases examined (9/9) from mice injected with cellular aggregates of the metastatic variant K-1735 M2-X21 and the nonmetastatic tumor cell line K-1735 cl 26 expressed the karyotype characteristics of K-1735 M2-X2 in every spread examined (at least 70 per metastasis). The karyotypes of the cells from the metastases of mice injected with cellular aggregates of K-1735, M2, and K-1735 M2-X21 (both metastatic variants) were composed of a characteristic spread of one (K-1735 M2 [4/12] and K-1735 M2-X21 [8/12]) or the other metastatic line. We concluded from this study that, although cellular aggregates are arrested more rapidly and result in an increased metastatic frequency, only a single cell from within the embolus normally survives to form the metastatic focus. Therefore, the metastatic process is clonal in origin and the metastatic foci appear to arise from a single cell that could have been only one of several within the tumor embolus.

In a previous study, Poste et al. (1982b) used tumor cells bearing stable biochemical markers. Syngeneic mice were given intravenous injections of aliquots of wild type B16-F10 cells admixed with equal aliquots of TFTr and Ouar B16-F10 variants, and individual lung metastases established in culture. Clones (Barranco et al. 1973) from each of the 22 metastases obtained from three animals were isolated and tested for resistance to TFT or Oua. Nineteen of 22 lesions were populated by cells with the same drug sensitivity. In addition, two metastases that contained cells with different drug sensitivities were identified, suggesting to the authors a polyclonal origin. One of the metastases expressed cells with either a wild type or an Oua phenotype, while the other was composed of both TFTr and Ouar cells. This latter metastasis, discounting the possibility of metastases coalescence, is very probably of polyclonal origin. Nonetheless, their study also suggests that a polyclonal event is a very rare occurrence and that most metastases are both clonal in origin and of single cell origin.

Reeve and Twentyman (1982), by the flow cytometric analysis of the x-radiation-induced sarcoma RIF-1, have shown that the parent tumor is composed of both diploid and tetraploid subpopulations of cells.

They examined both experimental and spontaneous metastases from mice bearing the RIF-1 tumor and observed that, unlike the parent tumor population, they exhibited a single level of ploidy, which was a stable characteristic of cell lines established from metastases. Based on this observation, they suggested that metastasis is a clonal event at least in this tumor system. However, a clonal origin denotes a population arising from a single cell and, since only a single ploidly level was observed in the metastases, their observation need not imply a clonal origin, but is rather an assessment of the relative distribution of each ploidy level in spontaneous metastases. Current studies of human tumors based on different cellular markers, including an ovarian carcinoma (Kusyk et al. 1981), neuroblastoma (Feder and Gilbert 1983), neurofibrosarcoma Friedman (1982), and breast and renal cells cancer (Symmans et al. 1995; Naito et al. 1991), have addressed the clonal versus multicellular origin of metastatic foci. More recent studies using a variety of genetic and phenotypic markers such as chromosome/cytogenetic markers, enzyme polymorphism, immunoglobulin markers, genetic instability, and drug resistance markers, have successfully established the clonal nature of metastases (Wolman 1986; Kerbel 1990; Nicolson 1984c; Kerbel et al. 1988a; Clark 1991). Further, a novel method using random integration of a foreign gene, which was used to examine cell lineage in development, was successfully used to demonstrate the clonal origin of metastasis (Kerbel 1990; Kerbel et al. 1988a; Talmadge and Zbar 1987; Singh et al. 1997). This method exploits the random integration of foreign DNA followed by restriction digestion and analysis using Southern hybridization. Assuming that single copy of foreign DNA is inserted, each clone contains a DNA unique marker (restriction fragment of variable length) detectable by Southern blotting using appropriate probe. The evidence in all these studies, while not definitive, supports the observation of a clonal origin for metastases.

The rapid generation of biological diversity within clonal metastases

The demonstration that spontaneous metastases result from the clonal expansion of highly specialized cells and that cells of metastases demonstrate a high rate of spontaneous mutation (Cifone and Fidler 1981) compared with nonmetastatic tumorigenic cells suggests that clonal metastases may rapidly become heterogeneous. We were interested, therefore, in investigating whether rapid tumor evolution and progression could occur within a clonal metastasis. If so, this finding would provide a rational explanation for the observed heterogeneity within and among metastases. The demonstration of clonal origin metastases (Talmadge et al. 1982; Poste et al. 1982b) provided the experimental basis for this study (Fig. 3). The experimental metastatic potential was examined for the cells from the original, demonstrable, clonal origin, X-met-21 line (parent); ten *in vitro* isolated clones; the X-met-21 after growth *in vivo* for 60 days; and seven individual spontaneous lung metastases from the X-rnet-21 primary tumors. Six of ten clones differed significantly from the parent tumor in their capacity to produce lung tumor colonies, while the metastatic potential of the X-met-21 population did not change after subcutaneous growth for 60 days when compared with X-met-21 cultured line. However, cells from six of seven spontaneous metastases from mice bearing the X-met-21 tumor differed significantly in their metastatic potential from the X-met-21 line growing at a primary site. The cells from the cloned lines and from spontaneous metastases differed greatly in their ability to produce experimental metastases (t = 0.0001, Kruskal-Wallis test, Chi square approximation).

In another set of experiments, we used an *in vitro* colony-forming inhibition assay to determine the relative sensitivity of tumor cells from the parent line, tumor cells from the cloned lines, and spontaneous metastases to various chemotherapeutic drugs. The study used the chemotherapeutic drugs amsacrine (AMSA), adriamycin (ADR), bleomycin (BLEO), and vincristine (VCR). Statistical analysis of the differences in drug sensitivity revealed that the following numbers of clones and metastases differed significantly from the parent tumor line: for AMSA, five of ten clones and two of seven metastases; for VCR, one of ten clones and three of seven metastases; for ADR, seven of ten clones and four of seven metastases; and for BLEO, six of ten clones and three of seven metastases. This variability was reproducible and was not caused by artifacts associated with the cloning or selection procedures. This conclusion is based on a study in which five subclones isolated from a benign

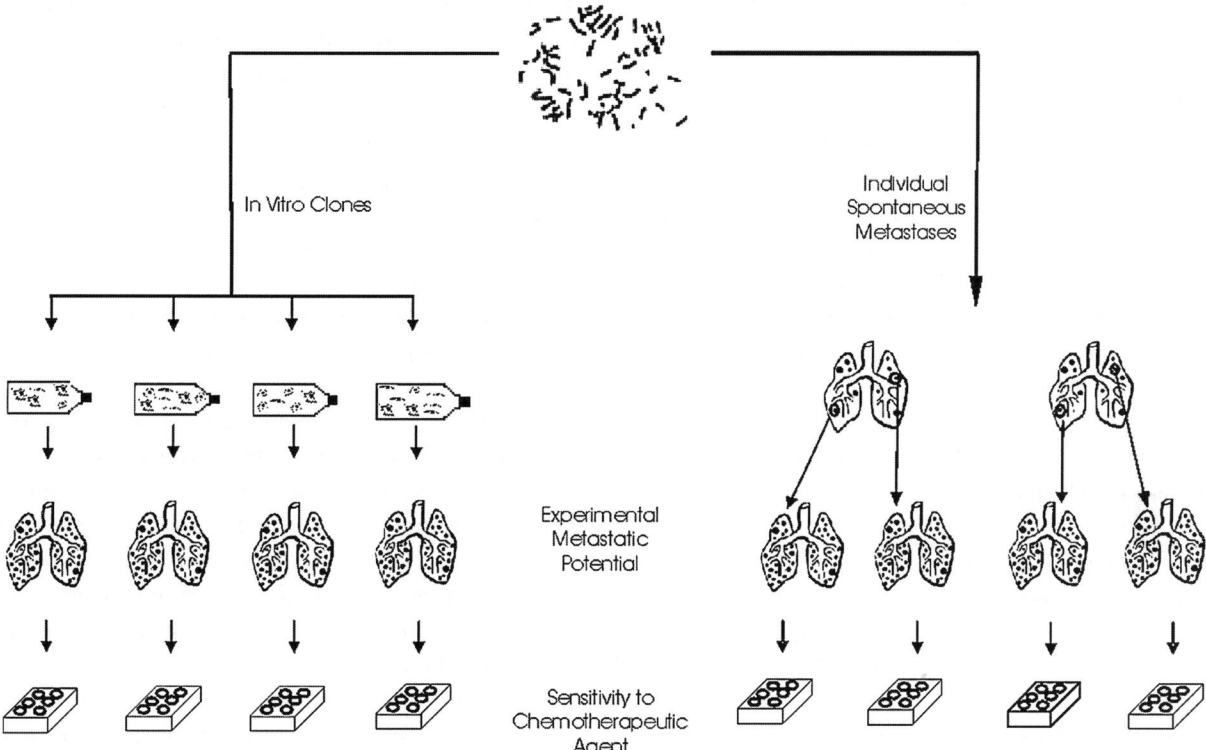

Fig. 3 Scheme outlining an experimental design to assess how rapidly metastases of clonal origin develop intralesional heterogeneity with respect to metastatic potential and sensitivity to cytostatic agents

clone of the K-1735 melanoma were not distinguishable from the parent clone in their response to cytotoxic drugs. In contrast to this diversity in metastatic potential and drug sensitivity, all cells examined expressed the unique submetacentric chromosome marker, suggesting that its expression was very stable.

In the study of clonal origin by Poste et al. (1982b), the metastatic properties of several tumor cell clones isolated from individual B16 melanoma pulmonary metastases at different stages during the evolution of metastasis were investigated. They found that during the early stages of metastatic growth following intravenous injection of tumor cells, the majority of metastatic lesions contain cells with indistinguishable metastatic phenotypes (intralesional clonal homogeneity). In contrast, the progressive growth of metastatic lesions was accompanied by the emergence of variant tumor cells with altered metastatic properties within clonal homogeneous lesions (intralesional heterogeneity).

In summary, distinct differences in metastatic properties and drug sensitivity were found in most of the *in vitro* and *in vivo* isolated clones. In contrast to the rapid evolution in sensitivity to chemotherapeutic drugs and metastatic potential in our studies, as well as, the three clinical studies (Kusyk et al. 1981; Feder and Gilbert 1983; Friedman et al. 1982), we observed that marker chromosomes appear to be expressed in an extremely stable manner both *in vitro* and *in vivo*. These intralesional differences could not be attributed to *in vivo* fusion of tumor cells with each other or normal host cells since tetraploid karyotypes were only rarely observed. Recent studies provide another explanation for the inherently metastatic and heterogeneous

nature of cancers – their derivation from distinct stem cells [254]. The type of stem cell from which a neoplasm arises determines both the metastatic potential and the phenotypic diversity of that neoplasm (Owens and Watt 2003; Perez-Losada and Balmain 2003; Medina 2002). Hence, tumours originating from an early stem cell or its progenitor cells metastasise readily and have a more heterogeneous phenotype, whereas tumours originating from a later stem cell or its progenitor cells have limited metastatic potential and a more homogeneous phenotype (Owens and Watt 2003; Perez-Losada and Balmain 2003; Medina 2002). It appears, therefore, that the progressive growth of metastases results in the rapid development of biological intralesional heterogeneity (Tu et al. 2002).

Role of host–tumor interaction

Our contemporary understanding of the events involved in human tumor progression centers rests largely on genetic changes, which include mutations, translocations and amplifications of oncogenes, and loss of heterozygocity of suppressor genes, all of which confers a growth advantage on cells resulting in clonal dominance. Most events in tumor progression are genetic, conferring clonal dominance and metastatic competence. However, other events, which regulate the metastatic process, are epigenetic in nature and are not inherited by the tumor cells, but are dependent on the host response. Therefore, the process of tumor growth and metastasis is complex, involving highly orchestrated interactions between tumor and stromal cells. Once metastatic cells reach a specific organ parenchyma, they must proliferate to give rise to clinically detectable lesions (Fidler 1990). This growth depends upon the interactions of the metastatic cells with the organ microenvironment which itself is mediated by autocrine and paracrine growth factors (Fidler 1990; Singh et al. 1997; Fidler et al. 2000; Hart and Fidler 1980). Many host derived factors can actually enhance the survival and growth of tumor cells in specific organs (Singh et al. 1997; Fidler et al. 2000). Moreover, organ-specific factors can affect the capacity of tumor cells to produce angiogenic molecule (Singh et al. 1994; Singh and Fidler 1997; Singh and Varney 2000), invade host stroma (Morikawa et al. 1988), and even respond to chemotherapeutic drugs

(Wilmanns et al. 1992). Studies from our laboratory and others support the importance of the epigenetic regulation of metastatic processes (Singh et al. 1997; Talmadge 1983). These data demonstrate that spontaneously metastasizing human cell lines metastasize in athymic nude mice only when the cells are injected orthotopically (natural site) (Fidler et al. 2000).

Metastasis appears to follow a predefined pattern of spread to particular organs, depending on the primary tumor site Rusciano and Burger 1992). The lodgment of viable tumor cells in a given organ is a necessary, but by itself, insufficient condition for metastases to develop (Fidler 1988; Nicolson 1991a, b, c). However, determination of the mechanism(s) regulating the organ preference of metastatic tumor cells for lodgement is a first and crucial step in the development of organ-specific metastases. The organ arrest of circulating malignant cells could be due to either mechanical trapping in the capillary bed of an organ's microcirculation (in which case the vascular connections of the primary tumor might have an effect on the patterns of organ metastases) or to specific interactions between the malignant cells and the target organ. The non-specific arrest of blood-borne tumor cells may be facilitated by their interaction with platelets. If the resulting embolus is formed while cells are circulating, then it may enhance the arrest of the tumor cells in the vasculature of the first organ downstream of the primary tumor site. If this organ represents a favorable milieu for tumor growth, then the interaction with platelets will enhance tumor metastases at that site; if not, then it may prevent tumor cells from reaching their preferred organ and, thus, cause a reduction in metastatic potential (Gasic 1984). In addition, platelets shield the tumor cells from physical damage and provide additional adhesion mechanism(s) and serve as a potential source of growth factors (Karpatkin et al. 1988; Nierodzik et al. 1994).

Among the organ-specific interactions that lead tumor cells to lodge in particular organs, several properties have been widely recognised as having a special relevance. The first close interaction between circulating malignant cells and their target occurs at the level of the microvasculature, where tumor cells contact the capillary endothelium. Capillary endothelial cells derived from different organs are not alike, and organ specific patterns of cell surface glycoproteins and ligands

have been identified on various organ endothelial cells (Nicolson 1991a; Rajotte et al. 1998; Pauli and Lee 1988; Pauli et al. 1990; Pauli and Knudson 1988; Johnson et al. 1991). Experimental evidence suggests an active role for vascular endothelial cells in directing the organ preference of metastasis (Pauli and Lee 1988; Pauli et al. 1990; Johnson et al. 1991; Auerbach et al. 1987; Auerbach 1988; Sher et al. 1988). Auerbach et al. (1987) and Auerbach (1988) showed that different malignant cell lines preferentially attach to endothelial cells of their *in vivo* target organs. Similar findings have been reported with different murine cell lines selected for their organ-specific homing to lung, liver and brain (Nicolson 1991a, b, c). Several mechanisms for the adhesion of malignant cells to organ-specific endothelial cells and eventual metastases have been proposed, including a role for local components of extracellular matrix (Pauli and Lee CL 1988; Pauli et al. 1990; Johnson et al. 1991); ICAM-1 (Auerbach et al. 1987; Auerbach 1988) and the expression of variant forms of membrane glycoprotein CD44 (Gunthert et al. 1996; Gunthert et al. 1991). In addition to well-characterized anatomical diversity *in situ*, specific differences are increasingly being recognized between surface antigens on endothelial cells from different tissues, including an absence of the classic endothelial marker factor VIII-related antigen (von Willebrand factor) from many endothelial cells (Zetter 1990, 1993, 1997; McCarthy et al. 1991). Microvascular heterogeneity extends to the properties of endothelial cells that are thought to be involved in tumor angiogenesis and metastasis, such as growth factor responsiveness and expression of cell adhesion molecules (Zetter 1990, 1993, 1997). These findings are not only of relevance to the unambiguous identification and characterization of endothelial cells, but also may explain the phenomenon of organ preferential tumor metastasis and provide novel opportunities for therapy.

Once malignant cells have adhered to the endothelium, they can cause endothelial cells to retract, with consequent exposure to the basement membrane (Nicolson 1991a, b, c). The exposed sub-endothelial matrix is a better adhesive substrate for tumors cells than the endothelial cell membrane and serve as determinants in organ-specific metastasis (Lichtner et al. 1989, 1995). Several basement membrane components, including laminin, fibronectin, collagens,

heparan sulfate proteoglycans and vitronectin, have all been identified as tumor cell adhesion molecules. The interaction of malignant cells with these glycoproteins that may bind to cellular receptors and matrix proteins may also influence the metastatic behaviour.

After arrest of malignant cells in a given organ, following either specific or random events, a further component of the successful establishment of an organ-specific metastatic foci is the presence of organ-specific factors that selectively direct the malignant cell migration into an organ itself (Hart and Fidler 1980; Hert 1982). Hujanen and Terranova (1985) and Nicolson and Moustafa (1998) have demonstrated the preferential migration of melanoma, sarcoma and breast carcinoma cells to extracts from the organs consistent with metastatic predilection. To invade, tumor cells need to possess enzymatic machinery able to degrade the local cell-extracellular matrix (ECM), which could cause organ-specific factors to be released from the degraded matrix and, in turn, induce specific cell migration (Cerra and Nathanson 1989; Orr et al. 1995). Using the indium-labeling technique to monitor the homing kinetic of both cell lines, Aoudjit et al. (1998a, b) showed that the critical step for the successful metastasis of the lymphoma cell was determined in the final steps of the disseminating process, namely after homing. Their results indicate that, whereas binding of tumor cells to vascular endothelium through specific adhesion mechanisms is a prerequisite for dissemination of tumor cells, the resistance of a tumor cell to the antagonist action of the host and/or its ability to grow tumors occurs only after homing to the target organ Aoudjit et al. 1998a, b).

Specific organ colonization may also be influenced by the ability of malignant cells to survive the local, non-immune tumoricidal activity. Li et al. (1991) and Nolibe and Poupon (1986) showed that lung microvascular endothelial cells might be activated to lyse tumor target cells and, thereby, to influence organ-specific metastasis. In addition, the involvement of natural killer cells in controlling incidence of metastasis to specific organs have been documented (Nolibe and Poupon 1986; Hao and Joshi 1990; Phillips 1989, 1990).

Organ-specific localization of tumor cells has a rate-limiting influence at various stages within the metastatic process, such as tumor cell arrest and

extravasation. In addition, the ability of the immune system to recognize and successfully eradicate tumors is highly dependent on the adhesion of activated lymphocytes to tumor cells. Despite the rapid accumulation of information on the molecular basis of cell adhesion, our understanding of organ-specific metastasis *per se* is incomplete. Nevertheless, progress has been made both in understanding the molecular basis of tumor cell adhesion and its relationship to tumor metastasis. In addition, understanding the mechanism and regulation of phenotypes associated with metastasis in different organ-microenvironment will help in understanding the role of host–tumor interaction and metastasis. We have demonstrated that the expression of growth and angiogenic factors is modulated by epigenetic mechanism(s), and may be due to the presence of regulatory factors, both stimulatory and inhibitory in nature, and which are expressed by site-specific host stroma (Singh et al. 1997; Fidler et al. 2000). These observations also suggest that one of the mechanisms for the maintenance of tumor growth is the dysregulation of the balance between growth stimulators and inhibitors, both of which may be expressed by host cells, tumor cells or both. This dysregulation allows tumor growth and eventual metastasis. Based on a growing body of evidence, it is clear that numerous cytokines/growth factors have a role in the regulation of site-specific tumor growth and metastasis. However, individual cytokines may not express this full potential, but rather participate within a cytokine network. Thus, a better understanding of the relative expression of these epigenetic regulatory mechanism(s) will increase our understanding of organ-specific regulation of tumor growth and metastasis.

Clinical implications

The most feared and devastating aspect of cancer is the propensity of tumor cells to disseminate from their primary site to distant organs and once there, develop into metastatic tumor foci. Despite impressive advances in the surgical resection of primary neoplasms and aggressive adjuvant therapies, most cancer deaths are attributable to metastases. There appear to be several reasons for this lack of success in controlling metastatic foci. First, by the time cancer is diagnosed, metastases may already be present in several organs of patient's body, which makes surgical resection or destruction by radiotherapy or chemotherapy unlikely. Moreover, the metastases may be located in organs that are difficult to treat with effective therapeutic doses without undue toxic effects. Second, in spite of the development of promising anticancer drugs and regimes, their efficiency is hindered by the occurrence of drug-resistant variants within tumors. This tumor cell resistance to conventional therapy is probably the single most important factor responsible for the refractory response of tumor therapy. The emergence of a resistant tumor subpopulation is a serious problem and limits the effectiveness of chemotherapy, hormonal therapy, radiotherapy, hyperthermia, or antigen specific immunotherapy. This phenotypic diversity, which allows selected variants to develop within the primary tumor, means not only that primary tumors and metastases can differ in their responses to treatment, but also that individual tumor cells within a metastasis differ from one another (diagrammatically represented in Fig. 4). This diversity can be generated rapidly even when the tumors originate from a single transformed cell. This complication appears to arise because metastatic lesions are fairly large by the time they are diagnosed. For example, a tumor mass at the lower limit of radiographic detection, i.e. 1 cubic centimeter, may contain as many as 10^9 cells. Eradication of 99.9% of these cells, a remarkable therapeutic achievement, still leaves 10^6 cells to proliferate, thus providing a large base for the further generation of biological heterogeneity.

The problem of selecting effective therapy for heterogeneous tumors may be further compounded by the existence of interactions between tumor subpopulations, as well as, between tumor cells and normal cells, which affect the measured sensitivity of the whole tumor, a concept that has received extensive study by Heppner and Miller (1983). She and her co-workers have shown that subpopulations of a mouse mammary tumor, which differ in sensitivity to chemotherapeutic agents, can interact in such a way that the apparent sensitivity of one subpopulation is changed in the presence of the other (Miller et al. 1981). Interactions between cells appear to act through metabolic processes affecting drug metabolism, through factors affecting cell growth, or through immune mechanisms. For example, the well-known

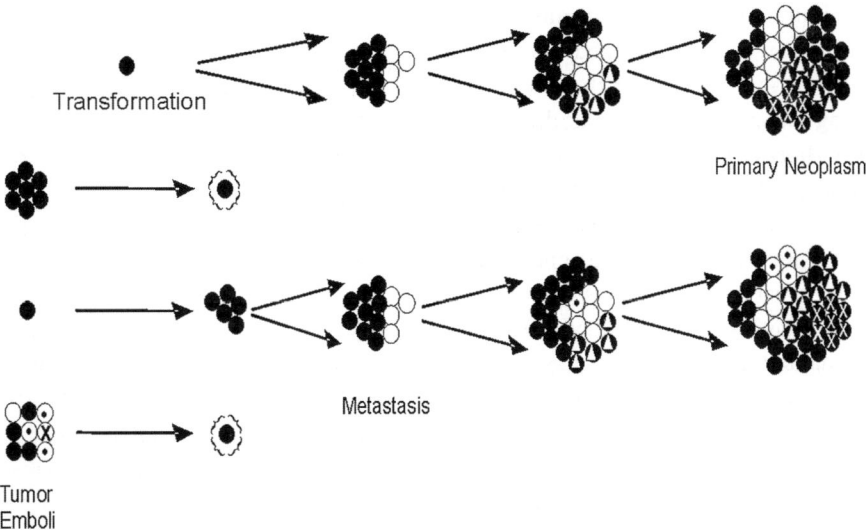

Fig. 4 In the clonal origin hypothesis of tumor development, a single cell undergoes one or more transforming events and with growth, it becomes heterogenous, including heterogeneity for metastatic phenotype(s). The cells within the primary tumor with a metastatic phenotype, as well as cells with a partial metastatic phenotype, can from homogeneous emboli, from which a single cell survives to form a metastatic focus. Alternatively, a metastatic focus can arise from a single cell tumor embolus or from a heterogeneous cellular embolus from which a single cell survives to establish a metastasis. Regardless of how the clonality arises, heterogeneity develops in association with tumor progressions from which therapeutic resistant variants can be selected and overgrown to dominate the metastatic lesion

phenomenon of "metabolic cooperation," a process by which small molecules pass between cells in contact, presumably through gap junctions (Loewenstein 1979; Subak-Sharpe et al. 1969), is one way in which interactions between cells can affect response to therapy. Metabolic cooperation could also result in the rescue of a sensitive cell by molecules from a resistant cell or in the death of a resistant cell due to passage of molecules from a sensitive cell.

Taken together, the growing body of experimental and clinical evidence suggests that cells within the same cancer exhibit different susceptibilities to the broad range of conventional treatments; therefore, the successful therapy of malignant tumors will require the development of new approaches capable of overcoming this variation and against which resistance is unlikely to develop. Given the extraordinary level of cell diversity evidently present in many tumors, the probability is small that a single anticancer drug, or any other treatment used alone, will be capable of killing all of the cancer cells in a malignant tumor and

its metastases. New treatment strategies may be possible that slow the potential of tumor cells that survive initial waves of treatment(s) to generate new variants. To achieve this with existing therapeutic modalities, one has to reduce the time interval between successive administrations of different anticancer agents in an attempt to destroy the cancer cell subpopulations that survive each successive treatment before the generation of large numbers of new variants could occur. This conclusion is reflected in the growing trend in cancer medicine to use a combination of anticancer treatments for the patients with malignant tumors.

However it originates, the heterogeneity of tumors has important implications for the treatment of metastases. For example, cells obtained from a primary tumor are not necessarily representative of cancer cells populating metastases or even cells in different regions of the same primary tumor. Thus, experimental efforts must be concentrated on identification of the features that permit malignant cells to metastasize. In addition, the test systems for new therapeutic agents

or modalities must address the problem of tumor cell heterogeneity. More effort must be devoted to testing the efficacy of combination therapies with the objective of circumventing the problem of cellular diversity within tumors. The short-term therapeutic goal must be to choose the combination of antitumor agents and to determine the sequence of administration that will be most effective against a particular tumor. The mechanisms by which cancer cells diversify are not fully understood. By limiting the number of different subpopulations of cancer cells within a tumor, we may be able to improve the odds that a combination of anticancer drugs, perhaps in conjunction with other adjunct therapeutic modalities, will destroy all of the subpopulations of tumor cells. Clearly, the only successful treatment of metastatic disease will be one that circumvents the different phenotypes of tumor cells within individual metastases of a patient and will probably require multiple therapeutic agents and multiple therapeutic modalities.

Acknowledgments This research was sponsored by the National Institute of Health grant, Avon-NCI Progress for Patients Award Program (J.E.T.); the Nebraska Research Initiative grant (J.E.T.); and grant CA72781 (R.K.S.), from National Cancer Institute, National Institutes of Health.

References

Albino AP, Lloyd KO, Houghton AN, Oettgen HF, Old LJ (1981) Heterogeneity in surface antigen and glycoprotein expression of cell lines derived from different melanoma metastases of the same patient. Implications for the study of tumor antigens. J Exp Med 154:1764–1778

Angello JC, Danielson KG, Anderson LW, Hosick HL (1982) Glycosaminoglycan synthesis by subpopulations of epithelial cells from a mammary adenocarcinoma. Cancer Res 42:2207–2210

Aoudjit F, Potworowski EF, Springer TA, St Pierre Y (1998a) Protection from lymphoma cell metastasis in ICAM-1 mutant mice: a posthoming event. J Immunol 161:2333–2338

Aoudjit F, Potworowski EF, St Pierre Y (1998b) The metastatic characteristics of murine lymphoma cell lines in vivo are manifested after target organ invasion. Blood 91:623–629

Ashall F, Bramwell ME, Harris H (1982) A new marker for human cancer cells. 1 The Ca antigen and the Ca1 antibody. Lancet 2:1–6

Aubert C, Rouge F, Galindo JR (1980) Tumorigenicity of human malignant melanocytes in nude mice in relation to their differentiation in vitro. J Natl Cancer Inst 64:1029–1040

Auerbach R (1988) Patterns of tumor metastasis: organ selectivity in the spread of cancer cells. Lab Invest 58:361–364

Auerbach R, Lu WC, Pardon E, et al. (1987) Specificity of adhesion between murine tumor cells and capillary endothelium: an in vitro correlate of preferential metastasis in vivo. Cancer Res 47:1492–1496

Barranco S, Luce J, Romsdahl M, Humphrey R (1973) Bleomycin as a possible synchronizing agent for human tumor cells in vivo. Cancer Research 33(4):882–887

Barranco SC, Haenelt BR, Gee EL (1978) Differential sensitivities of five rat hepatoma cell lines to anticancer drugs. Cancer Res 38:656–660

Barranco SC, Ho DH, Drewinko B, Romsdahl MM, Humphrey RM (1972) Differential sensitivites of human melanoma cells grown in vitro to arabinosylcytosine. Cancer Res 32:2733–2736

Baylin SB (1982) Clonal selection and heterogeneity of human solid neoplams. In: Fidler IJ, White RJ (eds) Design of models for testing cancer therapeutics agents, pp 50–63

Baylin SB, Abeloff MD, Wieman KC, Tomford JW, Ettinger DS (1975) Elevated histaminase (diamine oxidase) activity in small-cell carcinoma of the lung. N Engl J Med 293:1286–1290

Baylin SB, Weisburger WR, Eggleston JC, et al. (1978) Variable content of histaminase, L-dopa decarboxylase and calcitonin in small-cell carcinoma of the lung. Biologic and clinical implications. N Engl J Med 299:105–110

Becerk AJ, McCulloch EA, Till JE (1963) Cytological demonstration of the clonal nature of spleen colonies derived from transplanted mouse marrow cells. Nature 197:452–454

Becker FF, Klein KM, Wolman SR, Asofsky R, Sell S (1973) Characterization of primary hepatocellular carcinomas and initial transplant generations. Cancer Res 33:3330–3338

Biorklund A, Hakansson L, Stenstam B, Trope C, Akerman M (1980) On heterogeneity of non-Hodgkin's lymphomas as regards sensitivity to cytostatic drugs. An in vitro study. Eur J Cancer 16:647–654

Bittner M, Meltzer P, Chen Y, et al. (2000) Molecular classification of cutaneous malignant melanoma by gene expression profiling. Nature 406:536–540

Bohm N, Sandritter W (1975) DNA in human tumors: a cytophotometric study. Curr Top Pathol 60:151–219

Bosman HB, Winston RA (1970) Synthesis of glycoprotein, glycolipid, protein, and lipid in sychronized L5178Y cells. J Cell Biol 45:23–33

Bosslet K, Schirrmacher V (1981) Escape of metastasizing clonal tumor cell variants from tumor-specific cytolytic T lymphocytes. J Exp Med 154:557–562

Bosslet K, Schirrmacher V (1982) High-frequency generation of new immunoresistant tumor variants during metastasis of a cloned murine tumor line (esb). Int J Cancer 29:195–202

Brattain MG, Fine WD, Khaled FM, Thompson J, Brattain DE (1981) Heterogeneity of malignant cells from a human colonic carcinoma. Cancer Res 41:1751–1756

Brennan MJ, Donegan WL, Appleby DE (1979) The variability of estrogen receptors in metastatic breast cancer. Am J Surg 137:260–262

Briles EB, Kornfeld S (1978) Isolation and metastatic properties of detachment variants of B16 melanoma cells. J Natl Cancer Inst 60:1217–1222

Brock WA, Swartzendruber DE, Grdina DJ (1982) Kinetic heterogeneity in density-separated murine fibrosarcoma subpopulations. Cancer Res 42:4999–5003

Brunson KW, Beattie G, Nicolsin GL (1978) Selection and altered properties of brain-colonising metastatic melanoma. Nature 272:543–545

Brunson KW, Nicolson GL (1978) Selection and biologic properties of malignant variants of a murine lymphosarcoma. J Natl Cancer Inst 61:1499–1503

Byers VS, Johnston JO (1977) Antigenic differences among osteogenic sarcoma tumor cells taken from different locations in human tumors. Cancer Res 37:3173–3183

Cameron MD, Schmidt EE, Kerkvliet N, et al. (2000) Temporal progression of metastasis in lung: cell survival, dormancy, and location dependence of metastatic inefficiency. Cancer Res 60:2541–2546

Cerra RF, Nathanson SD (1989) Organ-specific chemotactic factors present in lung extracellular matrix. J Surg Res 46:422–426

Chambers AF, Hill RP, Ling V (1981) Tumor heterogeneity and stability of the metastatic phenotype of mouse KHT sarcoma cells. Cancer Res 41:1368–1372

Chu MY, Takeuchi T, Yeskey KS, Bogaars H, and Calabresi P (1979) Tumor cell heterogeneity in human lung carcinoma LX-1 [abstract]. Proc Am Assoc Cancer Res 20:151–152

Cifone MA, Fidler IJ (1981) Increasing metastatic potential is associated with increasing genetic instability of clones isolated from murine neoplasms. Proc Natl Acad Sci USA 78:6949–6952

Cifone MA, Kripke ML, Fidler IJ (1979) Growth rate and chromosome number of tumor cell lines with different metastatic potential. J Supramol Struct 11:467–476

Cikes M, Klein G (1972) Quantitative studies of antigen expression in cultured murine lymphoma cells. I. Cell-surface antigens in "Asynchronous" cultures. J Natl Cancer Inst 49:1599–1606

Clark EA, Golub TR, Lander ES, Hynes RO (2000) Genomic analysis of metastasis reveals an essential role for rhoc. Nature 406:532–535

Clark WH (1991) Tumour progression and the nature of cancer. Br J Cancer 64:631–644

Colcher D, Horan HP, Teramoto YA, Wunderlich D, Schlom J (1981) Use of monoclonal antibodies to define the diversity of mammary tumor viral gene products in virions and mammary tumors of the genus Mus. Cancer Res 41:1451–1459

Danielson KG, Anderson LW, Hosick HL (1980) Selection and characterization in culture of mammary tumor cells with distinctive growth properties in vivo. Cancer Res 40:1812–1819

Dethlefsen L (1980) The growth dynamics of murine mammary tumor cells in situ. In: McGrath CM, Brennan MJ, Rich MA (eds) Cell biology of breast cancer. Academic, New York, pp 145–160

DeWys WD (1972) Studies correlating the growth rate of a tumor and its metastases and providing evidence for tumor-related systemic growth-retarding factors. Cancer Res 32:374–379

Dexter DL, Kowalski HM, Blazar BA, et al. (1978) Heterogeneity of tumor cells from a single mouse mammary tumor. Cancer Res 38:3174–3181

Dexter DL, Leith JT (1986) Tumor heterogeneity and drug resistance. J Clin Oncol 4:244–257

Dexter DL, Spremulli EN, Fligiel Z, et al. (1981) Heterogeneity of cancer cells from a single human colon carcinoma. Am J Med 71:949–956

Dominguez OV, Huseby RA (1968) Heterogeneity of induced testicular interstitial cell tumors of mice as evidenced by steroid biosynthetic enzyme activities. Cancer Res 28:348–353

Dunn TB (1959) Morphology of mammary tumors in mice. In: Homburger F, Risheman NH (eds) Physiopathology of cancer. Karger, New York, pp 38–84

Easty DM, Easty GC, Carter RL, et al. (1981) Five human tumour cell lines derived from a primary squamous carcinoma of the tongue, two subsequent local recurrences and two nodal metastases. Br J Cancer 44:363–370

Ekman P, Snochowski M, Zetterberg A, Hogberg B, Gustafsson JA (1979) Steroid receptor content in human prostatic carcinoma and response to endocrine therapy. Cancer 44:1173–1181

Enders JF, Diamandopoulos GT (1969) A study of variation and progression in oncongenicity in an SV 40-transformed hamster heart cell line and its clones. Proc R Soc Lond B Biol Sci 171:431–443

Everson LK, Plocinik BA, Rogentine GN, Jr (1974) HL-A expression on the G1, S, and G2 cell-cycle stages of human lymphoid cells. J Natl Cancer Inst 53:913–920

Ewing SL, Sumner HW, Ophoven JJ, Mayer JE, Humphrey EW (1980) Small cell anaplastic carcinoma with differentiation: a report of 14 cases. [abstract]. Lab Invest 42:115–116

Feder MK, Gilbert F (1983) Clonal evolution in a human neuroblastoma. J Natl Cancer Inst 70:1051–1056

Fialkow PJ (1976) Clonal origin of human tumors. Biochim Biophys Acta 458:283–321

Fidler IJ (1973) Selection of successive tumour lines for metastasis. Nat New Biol 242:148–149

Fidler IJ (1988) Origin of cancer metastases and its implications for therapy. Isr J Med Sci 24:456–463

Fidler IJ (1990) Critical factors in the biology of human cancer metastasis: twenty-eighth G.H.A. Clowes memorial award lecture. Cancer Res 50:6130–6138

Fidler IJ (1995) Cancer biology: invasion and metastasis. In: Abeloff MD, Armitage JO, Lichter AS, Niederhuber JE (eds) Clinical oncology. Churchill Livingstone, New York, pp 55–76

Fidler IJ (2002) Critical determinants of metastasis. Semin Cancer Biol 12:89–96

Fidler IJ, Bucana C (1977) Mechanism of tumor cell resistance to lysis by syngeneic lymphocytes. Cancer Res 37:3945–3956

Fidler IJ, Gruys E, Cifone MA, Barnes Z, Bucana C (1981) Demonstration of multiple phenotypic diversity in a murine melanoma of recent origin. J Natl Cancer Inst 67:947–956

Fidler IJ, Hart IR (1981) Biological and experimental consequences of the zonal composition of solid tumors. Cancer Res 41:3266–3267

Fidler IJ, Hart IR (1981) Biological and experimental consequences of the zonal composition of solid tumors. Cancer Res 41:3266–3267

Fidler IJ, Kripke ML (1977) Metastasis results from preexisting variant cells within a malignant tumor. Science 197:893–895

Fidler IJ, Kripke ML (2003) Genomic analysis of primary tumors does not address the prevalence of metastatic cells in the population. Nat Genet 34:23

Fidler IJ, Nicolson GL (1981) The immunobiology of experimental metastatic melanoma. Cancer Biol Rev 2:171–234

Fidler IJ, Radinsky R (1996) Search for genes that suppress cancer metastasis. J Natl Cancer Inst 88:1700–1703

Fidler IJ, Singh RK, Yoneda J, et al. (2000) Critical determinants of neoplastic angiogenesis. Cancer J Sci Am 6(Suppl 3):S225–S236

Filmus J, Kerbel RS (1993) Development of resistance mechanisms to the growth-inhibitory effects of transforming growth factor-beta during tumor progression. Curr Opin Oncol 5:123–129

Fogel M, Gorelik E, Segal S, Feldman M (1979) Differences in cell surface antigens of tumor metastases and those of the local tumor. J Nat Cancer Inst 62:585–588

Foulds L (1956a) The hisologic analysis of mammary tumors of mice. I. Scope of investigations and general principles of anlysis. J Natl Cancer Inst 17:701–712

Foulds L (1956b) The hisologic analysis of mammary tumors of mice. II. The hisotology of responsiveness and progression. The origins of tumors. J Nat Cancer Inst 17:713–754

Foulds L (1956c) The hisologic analysis of mammary tumors of mice. III. Organoid tumors. J Nat Cancer Inst 17:755–782

Foulds L (1956d) The hisologic analysis of mammary tumors of mice. IV. Secretion. J Nat Cancer Inst 17:783–801

Franks LM (1960) Estrogen-treated prostatic cancer. Cancer 13:490–501

Friedman JM, Fialkow PJ, Greene CL, Weinberg MN (1982) Probable clonal origin of neurofibrosarcoma in a patient with hereditary neurofibromatosis. J Natl Cancer Inst 69:1289–1292

Frost PH, Kerbel RS (1981) Immunoselection in vitro of a nonmetastatic variant from a highly metastatic tumor. J Natl Cancer Inst 69:1289–1292

Fugmann RA, Anderson JC, Stolfi RL, Martin DS (1977) Comparison of adjuvant chemotherapeutic activity against primary and metastatic spontaneous murine tumors. Cancer Res 37:496–500

Fuji H, Mihich E (1975) Selection for high immunogenicity in drug-resistant sublines of murine lymphomas demonstrated by plaque assay. Cancer Res 35:946–952

Fuji H, Mihich E, Pressman D (1977) Differential tumor immunogenicity of L1210 and its sublines. I. Effect of an increased antigen density on tumor cell surfaces on primary B cell responses in vitro. J Immunol 119:983–986

Gasic GJ (1984) Role of plasma, platelets, and endothelial cells in tumor metastasis. Cancer Metastasis Rev 3:99–114

Geier GR, Schwarz JA, Schlag P (1979) Cytologic uniformity of breast cancer from different localizations: a pattern analysis study. Exp Cell Biol 47:241–249

Giavazzi R, Alessandri G, Spreafico F, Garattini S, Mantovani A (1980) Metastasizing capacity of tumour cells from spontaneous metastases of transplanted murine tumours. Br J Cancer 42:462–472

Gorelik E, Feldman M, Segal S (1982) Selection of 3LL tumor subline resistant to natural effector cells concomitantly selected for increased metastatic potency. H supramol Struct 12:385–402

Gorelik E, Fogel M, Segal S, Feldman M (1979) Tumor-associated antigenic differences between the primary and the descendant metastatic tumor cell populations. J Supramol Struct 12:385–402

Gray JM, Pierce GB (1964) Relationship between growth rate and differentiation of melanoma in vivo. J Nat Cancer Inst 32:1201–1211

Gunthert U, Birchmeier W, Schlag PM (1996) Attempts to understand metastasis formation. II. Regulatory factors. Introduction. Curr Top Microbiol Immunol 213(Pt 2): V–VII

Gunthert U, Hofmann M, Rudy W, et al. (1991) A new variant of glycoprotein CD44 confers metastatic potential to rat carcinoma cells. Cell 65:13–24

Hager JC, Fligiel S, Stanley W, Richardson AM, Heppner GH (1981) Characterization of a variant-producing tumor cell line from a heterogeneous strain BALB/cfc3h mouse mammary tumor. Cancer Res 41:1293–1300

Hager JC, Heppner GH (1982) Heterogeneity of expression and induction of mouse mammary tumor virus antigens in mouse mammary tumors. Cancer Res 42:4325–4329

Hakansson L, Trope C (1974a) Cell clones with different sensitivity to cytostatic drugs in methylcholanthrene-induced mouse sarcomas. Acta Pathol Microbiol Scand [A] 82:41–47

Hakansson L, Trope C (1974b) On the presence within tumours of clones that differ in sensitivity to cytostatic drugs. Acta Pathol Microbiol Scand [A] 82:35–40

Hanna N, Fidler IJ (1981) Relationship between metastatic potential and resistance to KN cell-mediated cytoxicity in three murine tumor systems. J Natl Cancer Inst 66:1183–1190

Hao WM, Joshi SS (1990) Differential susceptibility of metastatic lymphoma cells to natural immunity. Oncology 47:483–487

Harris JF, Best MW (1988) Dynamic heterogeneity: metastatic variants to liver are generated spontaneously in mouse embryonal carcinoma cells. Clin Exp Metastasis 6:451–462

Harris JF, Chambers AF, Hill RP, Ling V (1982) Metastatic variants are generated spontaneously at a high rate in mouse KHT tumor. Proc Natl Acad Sci USA 79:5547–5551

Hart IR (1979) The selection and characterization of an invasive variant of the B16 melanoma. Am J Pathol 97:587–600

Hart IR (1982) 'Seed and soil' revisited: mechanisms of site-specific metastasis. Cancer Metastasis Rev 1:5–16

Hart IR, Fidler IJ (1980) Role of organ selectivity in the determination of metastatic patterns of B16 melanoma. Cancer Res 40:2281–2287

Hart IR, Talmadge JE, Fidler IJ (1981) Metastatic behavior of a murine reticulum cell sarcoma exhibiting organ-specific growth. Cancer Res 41:1281–1287

Hastings RJ, Franks LM (1983) Cellular heterogeneity in a tissue culture cell line derived from a human bladder carcinoma. Br J Cancer 47:233–244

Henderson JS, Rous P (1962) The plating of tumor components on the subsutaneous expanses of young mice. J Exp Med 115:1211–1230

Heppner GH, Dexter DL, DeNucci T, Miller FR, Calabresi P (1978) Heterogeneity in drug sensitivity among tumor cell subpopulations of a single mammary tumor. Cancer Res 38:3758–3763

Heppner GH, Miller BE (1983) Tumor heterogeneity: biological implications and therapeutic consequences. Cancer Metastasis Rev 2:5–23

Hill HZ, Hill GJ, Miller CF, Kwong F, Purdy J (1979) Radiation and melanoma: response of B16 mouse tumor cells and clonal lines to in vitro irradiation. Radiat Res 80:259–276

Hujanen ES, Terranova VP (1985) Migration of tumor cells to organ-derived chemoattractants. Cancer Res 45:3517–3521

Isaacs JT, Coffey DS (1981) Adaptation versus selection as the mechanism responsible for the relapse of prostatic cancer to androgen ablation therapy as studied in the Dunning R-3327-H adenocarcinoma. Cancer Res 41:5070–5075

Isaacs JT, Wake N, Coffey DS, Sandberg AA (1982) Genetic instability coupled to clonal selection as a mechanism for tumor progression in the Dunning R-3327 rat prostatic adenocarcinoma system. Cancer Res 42:2353–2371

Ishidate M, Jr., Aoshima M, Sakurai Y (1974) Population changes of a rat leukemia by different routes of transplantation. J Natl Cancer Inst 53:773–781

Ito E, Moore GE (1967) Characteristic differences in clones isolated from an S37 ascites tumor in vitro. Exp Cell Res 48:440–447

Johnson RC, Augustin-Voss HG, Zhu DZ, Pauli BU (1991) Endothelial cell membrane vesicles in the study of organ preference of metastasis. Cancer Res 51:394–399

Karpatkin S, Pearlstein E, Ambrogio C, Coller BS (1988) Role of adhesive proteins in platelet tumor interaction in vitro and metastasis formation in vivo. J Clin Invest 81:1012–1019

Kerbel RS (1979) Implications of immunological heterogeneity of tumours. Nature 280:358–360

Kerbel RS (1990) Growth dominance of the metastatic cancer cell: cellular and molecular aspects. Adv Cancer Res 55:87–132

Kerbel RS (1992) Expression of multi-cytokine resistance and multi-growth factor independence in advanced stage metastatic cancer. Malignant melanoma as a paradigm. Am J Pathol 141:519–524

Kerbel RS, Waghorne C, Korczak B, Breitman ML (1988a) Clonal changes in tumours during growth and progression evaluated by southern gel analysis of random integrations of foreign DNA. Ciba Found Symp 141:123–148

Kerbel RS, Waghorne C, Korczak B, Lagarde A, Breitman ML (1988a) Clonal dominance of primary tumours by metastatic cells: genetic analysis and biological implications. Cancer Surv 7:597–629

Key M, Talmadge JE, Fidler IJ (1982) Lack of correlation between the progressive growth of spontaneous metastases and their content of infiltrating macrophages. J Reticuloendothel Soc 32:387–396

Killion JJ (1978) Immunotherapy with tumor subpopulations. I. Active, specific immunotherapy of L1210 Leukemia. Cancer Immunol Immunother 4:115–119

Killion JJ, Kollmorgen GM (1976) Isolation of immunogenic tumor cells by cell-affinity chromatography. Nature 259:674–676

Kiricuta I, Mustea I, Rogozan I, Simu G (1965) Relations between tumor metastasis. I. Aspects of the crabtree effect. Cancer 18:978–984

Klein E (1954) Gradual transformation of solid into ascites tumors: Permanent differences between the original and the transformed sublines. Cancer Res 14:482–485

Klein E (1955) Gradual transformation of solid into ascites tumors: evidence favoring the mutation-selection theory. Exp Cell Res 23:188–212

Kobori O, Oota K (1979) Neuroendocrine cells in serially passaged rat stomach cancers induced by MNNG. Int J Cancer 23:536–541

Kosh F (1939) Zur frage der metastazenbildung bei Impftumoren. Z Krebsforsch 48:495–505

Kozlowski JM, Fidler IJ, Campbell D, et al. (1984) Metastatic behavior of human tumor cell lines grown in the nude mouse. Cancer Res 44:3522–3529

Kripke ML, Fidler IJ (1980) Enhanced experimental metastasis of ultraviolet light-induced fibrosarcomas in ultraviolet light-irradiated syngeneic mice. Cancer Res 40:625–629

Kripke ML, Gruys E, Fidler IJ (1978) Metastatic heterogeneity of cells from an ultraviolet light-induced murine fibrosarcoma of recent origin. Cancer Res 38:2962–2967

Kusyk CJ, Seski JC, Medlin WV, Edwards CL (1981) Progressive chromosome changes associated with different sites of one ovarian carcinoma. J Natl Cancer Inst 66:1021–1025

Larner EH, Rutherford CL (1982) Implementation of micromethods to resolve problems of human breast tumor heterogeneity in analysis of cyclic 3:5 nucleotide phosphodiesterase. Cancer Res 42:1661–1668

Lee SH (1978) Cytochemical study of estrogen receptor in human mammary cancer. Am J Clin Pathol 70:197–203

Leith JT, Brenner HJ, DeWyngaert JK, et al. (1981) Selective modification of the X ray survival response of two mouse mammary adenocarcinoma sublines by N,N-dimethylformamide. Int J Radiat Oncol Biol Phys 7:943–947

Leith JT, Dexter DL, DeWyngaert JK, et al. (1982a) Differential responses to x-irradiation of subpopulations of two heterogeneous human carcinomas in vitro. Cancer Res 42:2556–2561

Leith JT, Gaskins LA, Dexter DL, Calabresi P, Glicksman AS (1982b) Alteration of the survival response of two human colon carcinoma subpopulations to x-irradiation by N,N-dimethylformamide. Cancer Res 42:30–34

Lerner RA, Oldstone MB, Cooper NR (1971) Cell cycle-dependent immune lysis of Moloney virus-transformed lymphocytes: presence of viral antigen, accessibility to antibody, and complement activation. Proc Natl Acad Sci USA 68:2584–2588

Li LM, Nicolson GL, Fidler IJ (1991) Direct in vitro lysis of metastatic tumor cells by cytokine-activated murine vascular endothelial cells. Cancer Res 51:245–254

Lichtner RB, Belloni PN, Nicolson GL (1989) Differential adhesion of metastatic rat mammary carcinoma cells to organ-derived microvessel endothelial cells and subendothelial matrix. Exp Cell Biol 57:146–152

Lichtner RB, Kaufmann AM, Kittmann A, et al. (1995) Ligand mediated activation of ectopic EGF receptor promotes matrix protein adhesion and lung colonization of rat mammary adenocarcinoma cells. Oncogene 10:1823–1832

Liotta LA, Steeg PS, Stetler-Stevenson WG (1991) Cancer metastasis and angiogenesis: an imbalance of positive and negative regulation. Cell 64:327–336

Liotta LA, Stetler-Stevenson WG (1991) Tumor invasion and metastasis: an imbalance of positive and negative regulation. Cancer Res 51:5054s–5059s

Lippman SM, Mendelsohn G, Trump DL, Wells SA, Jr., Baylin SB (1982) The prognostic and biological significance of cellular heterogeneity in medullary thyroid carcinoma: a study of calcitonin, L-dopa decarboxylase, and histaminase. J Clin Endocrinol Metab 54:233–240

Loewenstein WR (1979) Junctional intercellular communication and the control of growth. Biochim Biophys Acta 560:1–65

Lotan R (1979) Different susceptibilities of human melanoma and breast carcinoma cell lines to retinoic acid-induced growth inhibition. Cancer Res 39:1014–1019

Lotan R, Nicolson GL (1979) Heterogeneity in growth inhibition by beta-trans-retinoic acid of metastatic B16 melanoma clones and in vivo-selected cell variant lines. Cancer Res 39:4767–4771

Lu C, Kerbel RS (1994) Cytokines, growth factors and the loss of negative growth controls in the progression of human cutaneous malignant melanoma. Curr Opin Oncol 6:212–220

Macinnes JI, Chan EC, Percy DH, Morris VL (1981) Mammary tumors from GR mice contain more than one population of mouse mammary tumor virus-infected cells. Virology 113:119–129

Mackintosh FR, Louie AC, Evans TL, Amylon MD, Sikic BI (1981) Clonocal heterogeneity in a human ovarian adenocarcinoma. [abstract]. Proc AACR 22:379

MacLean GD, Seehafer J, Shaw AR, Kieran MW, Longenecker BM (1982) Antigenic heterogeneity of human colorectal cancer cell lines analyzed by a panel of monoclonal antibodies. I. Heterogeneous expression of Ia-like and HLA-like antigenic determinants. J Natl Cancer Inst 69:357–364

Makino S (1956) Further evidence favoring the concept of the stem cell in ascites tumors of rats. Ann NY Acad Sci 63:818–830

Mantovani A, Giavazzi R, Alessandri G, Spreafico F, Garattini S (1981) Characterization of tumor lines derived from spontaneous metastases of a transplanted murine sarcoma. Eur J Cancer 17:71–76

Mathieson BJ, Zatz MM, Sharrow SO, et al. (1982) Separation and characterization of two component tumor lines within the AKR lymphoma, AKTB-1, by fluorescence-activated cell sorting and flow microfluorometry analysis. II. Differential histopathology of sig+ and sig– sublines. J Immunol 128:1832–1838

McCarthy JB, Skubitz AP, Iida J, et al. (1991) Tumor cell adhesive mechanisms and their relationship to metastasis. Semin Cancer Biol 2:155–167

McCune CS, Schapira DV, Henshaw EC (1981) Specific immunotherapy of advanced renal carcinoma: evidence for the polyclonality of metastases. Cancer 47:1984–1987

McDowell EM, Sorokin SP, Hoyt RF, Jr., Trump BF (1981) An unusual bronchial carcinoid tumor: light and electron microscopy. Hum Pathol 12:338–348

McGee JO, Woods JC, Ashall F, Bramwell ME, Harris H (1982) A new marker for human cancer cells, 2 immunohistochemical detection of the Ca antigen in human tissues with the Ca1 antibody. Lancet 2:7–10

Medina D (2002) Biological and molecular characteristics of the premalignant mouse mammary gland. Biochim Biophys Acta 1603:1–9

Michalides R, Wagenaar E, Sluyser M (1982) Mammary tumor virus DNA as a marker for genotypic variance within hormone-responsive GR mouse mammary tumors. Cancer Res 42:1154–1158

Miller BE, Miller FR, Heppner GH (1981) Interactions between tumor subpopulations affecting their sensitivity to the antineoplastic agents cyclophosphamide and methotrexate. Cancer Res 41:4378–4381

Miller BE, Miller FR, Leith J, Heppner GH (1980) Growth interaction in vivo between tumor subpopulations derived from a single mouse mammary tumor. Cancer Res 40:3977–3981

Miller FR (1982) Intratumor immunologic heterogeneity. Cancer Metastasis Rev 1:319–334

Miller FR, Heppner GH (1979) Immunologic heterogeneity of tumor cell subpopulations from a single mouse mammary tumor. J Natl Cancer Inst 63:1457–1463

Mitelman F (1971) The chromosomes of fifty primary Rous rat sarcomas. Hereditas 69:155–186

Morikawa K, Walker SM, Nakajima M, et al. (1988) Influence of organ environment on the growth, selection, and metastasis of human colon carcinoma cells in nude mice. Cancer Res 48:6863–6871

Naito S, Kumazawa J, von Eschenbach AC, Fidler IJ (1991) Metastatic heterogeneity of human renal cell carcinoma. Urol Int 47(Suppl 1):90–95

Nakajima M, Morikawa K, Fabra A, Bucana CD, Fidler IJ (1990) Influence of organ environment on extracellular matrix degradative activity and metastasis of human colon carcinoma cells. J Natl Cancer Inst 82(24):1890–1898

Natali PG, Cavaliere R, Bigotti A, et al. (1983) Antigenic heterogeneity of surgically removed primary and autologous metastatic human melanoma lesions. J Immunol 130:1462–1466

Neri A, Welch D, Kawaguchi T, Nicolson GL (1982) Development and biologic properties of malignant cell sublines and clones of a spontaneously metastasizing rat mammary adenocarcinoma. J Natl Cancer Inst 68:507–517

Nicolin A, Canti G, Marelli O, Veronese F, Goldin A (1981) Chemotherapy and immunotherapy of L1210 leukemic mice with antigenic tumor sublines. Cancer Res 41:1358–1362

Nicolson GL (1984a) Cell surface molecules and tumor metastasis. Regulation of metastatic phenotypic diversity. Exp Cell Res 150:3–22

Nicolson GL (1984b) Generation of phenotypic diversity and progression in metastatic tumor cells. Cancer Metastasis Rev 3:25–42

Nicolson GL (1984c) Tumor progression, oncogenes and the evolution of metastatic phenotypic diversity. Clin Exp Metastasis 2:85–105

Nicolson GL (1991a) Gene expression, cellular diversification and tumor progression to the metastatic phenotype. Bioessays 13:337–342

Nicolson GL (1991b) Molecular mechanisms of cancer metastasis: tumor and host properties and the role of oncogenes and suppressor genes. Curr Opin Oncol 3:75–92

Nicolson GL (1991c) Tumor and host molecules important in the organ preference of metastasis. Semin Cancer Biol 2:143–154

Nicolson GL, Brunson KW, Fidler IJ (1978) Specificity of arrest, survival, and growth of selected metastatic variant cell lines. Cancer Res 38:4105–4111

Nicolson GL, Custead SE (1982) Tumor metastasis is not due to adaptation of cells to a new organ environment. Science 215:176–178

Nicolson GL, Moustafa AS (1998) Metastasis-associated genes and metastatic tumor progression. In Vivo 12:579–588

Nierodzik ML, Klepfish A, Karpatkin S (1994) Role of platelet integrin gpiib-gpiiia, fibronectin, von Willebrand factor, and thrombin in platelet-tumor interaction in vitro and metastasis in vivo. Semin Hematol 31:278–288

Niles RM, Makarski JS (1978) Hormonal activation of adenylate cyclase in mouse melanoma metastatic variants. J Cell Physiol 96:355–359

Nolibe D, Poupon MF (1986) Enhancement of pulmonary metastases induced by decreased lung natural killer cell activity. J Natl Cancer Inst 77:99–103

Nowell PC (1976) The clonal evolution of tumor cell populations. Science 194:23–28

Ohno S (1971) Genetic implication of karyological instability of malignant somatic cells. Physiol Rev 51:496–526

Olsson L, Ebbesen P (1979) Natural polyclonality of spontaneous AKR leukemia and its consequences for so-called specific immunotherapy. J Natl Cancer Inst 62:623–627

Olsson L, Kiger N, Kronstrom H (1981) Sensitivity of cloned high- and low-metastatic murine Lewis lung tumor cells to lysis by cytotoxic autoreactive cells. Cancer Res 41:4706–4709

Orr FW, Sanchez-Sweatman OH, Kostenuik P, Singh G (1995) Tumor-bone interactions in skeletal metastasis. Clin Orthop 19–33

Owens DM, Watt FM (2003) Contribution of stem cells and differentiated cells to epidermal tumours. Nat Rev Cancer 3:444–451

Panem S, Schauf V (1974) Cell-cycle dependent appearance of murine leukemia-sarcoma virus antigens. J Virol 13:1169–1175

Parbhoo SP (1981) Heterogeneity in human mammary cancer. In: Stoll BA (ed) Systemic control of breast cancer. William Heinemann, London, pp 55–77

Pasternak CA, Warmsley AM, Thomas DB (1971) Structural alterations in the surface membrane during the cell cycle. J Cell Biol 50:562–564

Pathak S (1990) Cytogenetic abnormalities in cancer: with special emphasis on tumor heterogeneity. Cancer Metastasis Rev 8:299–318

Pauli BU, Augustin-Voss HG, el Sabban ME, Johnson RC, Hammer DA (1990) Organ-preference of metastasis. The role of endothelial cell adhesion molecules. Cancer Metastasis Rev 9:175–189

Pauli BU, Knudson W (1988) Tumor invasion: a consequence of destructive and compositional matrix alterations. Hum Pathol 19:628–639

Pauli BU, Lee CL (1988) Organ preference of metastasis. The role of organ-specifically modulated endothelial cells. Lab Invest 58:379–387

Perez-Losada J, Balmain A (2003) Stem-cell hierarchy in skin cancer. Nat Rev Cancer 3:434–443

Pertschuk LP, Tobin EH, Brigati DJ, et al. (1978) Immunofluorescent detection of estrogen receptors in breast cancer. Comparison with dextran-coated charcoal and sucrose gradient assays. Cancer 41:907–911

Phillips NC (1989) Kupffer cells and liver metastasis. Optimization and limitation of activation of tumoricidal activity. Cancer Metastasis Rev 8:231–252

Phillips NC (1990) Macrophages, metastasis and immunity. Prog Clin Biol Res 354A:257–270

Pierce GB (1974) Cellular heterogeneity of cancers. In: Pop T, DiPaolo JA (eds) World symposium on model studies in chemical carcinogensis. Marcel Dekker, New York, pp 463–472

Pimm MV, Baldwin RW (1980) Antigenic heterogeneity of primary and metastatic tumors and its implications for immunotherapy. In: Grundmann E (ed) Metastatic tumor growth. Gustav Fischer, Stuttgart, p 305

Pimm MV, Embleton MJ, Baldwin RW (1980) Multiple antigenic specificities within primary 3-methylcholanthrene-induced rat sarcomas and metastases. Int J Cancer 25:621–629

Pollack VA, Fidler IJ (1982) Use of young nude mice for selection of subpopulations of cells with increased metastatic potential from nonsyngeneic neoplasms. J Natl Cancer Inst 69:137–141

Poste G (1982) Experimental systems for analysis of the malignant phenotype. Cancer Metastasis Rev 1:141–199

Poste G, Doll J, Fidler IJ (1981) Interactions among clonal subpopulations affect stability of the metastatic phenotype in polyclonal populations of B16 melanoma cells. Proc Natl Acad Sci USA 78:6226–6230

Poste G, Doll J, Hart IR, Fidler IJ (1980) In vitro selection of murine B16 melanoma variants with enhanced tissue-invasive properties. Cancer Res 40:1636–1644

Poste G, Doll J, Brown AE, Tzeng J, Zeidman I (1982a) Comparison of the metastatic properties of B16 melanoma clones isolated from cultured cell lines, subcutaneous tumors, and individual lung metastases. Cancer Res 42:2770–2778

Poste G, Tzeng J, Doll J, et al. (1982b) Evolution of tumor cell heterogeneity during progressive growth of individual lung metastases. Proc Natl Acad Sci USA 79:6574–6578

Prehn RT (1970) Analysis of antigenic heterogeneity within individual 3-mthylcholanthrene-induced mouse sarcomas. J Nat Cancer Inst 45:1039–1045

Rabotti G (1959) Ploidy of primary and metastatic human tumours. Nature 183:1276–1277

Rajotte D, Arap W, Hagedorn M, et al. (1998) Molecular heterogeneity of the vascular endothelium revealed by in vivo phage display. J Clin Invest 102:430–437

Ramaswamy S, Ross KN, Lander ES, Golub TR (2003) A molecular signature of metastasis in primary solid tumors. Nat Genet 33:49–54

Raz A (1982) Regional emergence of metastatic heterogeneity in a growing tumor. Cancer Lett 17:153–160

Raz A, Hanna N, Fidler IJ (1981) In vivo isolation of a metastatic tumor cell variant involving selective and nonadaptive processes. J Natl Cancer Inst 66:183–189

Raz A, McLellan WL, Hart IR, et al. (1980) Cell surface properties of B16 melanoma variants with differing metastatic potential. Cancer Res 40:1645–1651

Reading CL, Belloni PN, Nicolson GL (1980a) Selection and in vivo properties of lectin-attachment variants of malignant murine lymphosarcoma cell lines. J Natl Cancer Inst 64:1241–1249

Reading CL, Brunson KW, Torrianni M, Nicolson GL (1980b) Malignancies of metastatic murine lymphosarcoma cell lines and clones correlate with decreased cell surface display of RNA tumor virus envelope glycoprotein gp70. Proc Natl Acad Sci USA 77:5943–5947

Reeve JG, Twentyman PR (1982) Ploidy distribution of tumour cells derived from induced and spontaneously arising metastases of a murine radiation-induced sarcoma, RIF-1. Eur J Cancer Clin Oncol 18:1001–1006

Ridley A (2000) Molecular switches in metastasis. Nature 406:466–467

Rusciano D, Burger MM (1992) Why do cancer cells metastasize into particular organs? Bioessays 14:185–194

Sacchi A, Calabresi F, Greco C, Zupi G (1981) Different metastatic potential of in vitro and in vivo lines selected from Lewis lung carcinoma: correlation with response to different bleomycin schedulings. Invasion Metastasis 1:227–238

Saiki I, Naito S, Yoneda J, et al. (1991) Characterization of the invasive and metastatic phenotype in human renal cell carcinoma. Clin Exp Metastasis 9:551–566

Sandberg AA (1977) Chromosome markers and progression in bladder cancer. Cancer Res 37:2950–2956

Schirrmacher V, Bosslet K (1982) Clonal analysis of expression of tumor-associated transplantation antigens and of metastatic capacity. Cancer Immunol Immunother 13:62–68

Schirrmacher V, Bosslet K, Shantz G, Clauer K, Hubsch D (1979) Tumor metastases and cell-mediated immunity in a model system in DBA/2 mice. IV. Antigenic differences between a metastasizing variant and the parental tumor line revealed by cytotoxic T lymphocytes. Int J Cancer 23:245–252

Schirrmacher V, Fogel M, Russmann E, et al. (1982) Antigenic variation in cancer metastasis: immune escape versus immune control. Cancer Metastasis Rev 1:241–274

Schlag P, Schreml W (1982) Heterogeneity in growth pattern and drug sensitivity of primary tumor and metastases in the human tumor colony-forming assay. Cancer Res 42:4086–4089

Schmitt M, Daynes RA (1981) Heterogeneity of tumorigenicity phenotype in murine tumors. I. Characterization of regressor and progressor clones isolated from a nonmutagenized ultraviolet regressor tumor. J Exp Med 153:1344–1359

Semple TU, Moore GE, Morgan RT, Woods LK, Quinn LA (1982) Multiple cell lines from patients with malignant melanoma: morphology, karyology, and biochemical analysis. J Natl Cancer Inst 68:365–380

Shapiro JR, Yung WK, Shapiro WR (1981) Isolation, karyotype, and clonal growth of heterogeneous subpopulations of human malignant gliomas. Cancer Res 41:2349–2359

Shearman PJ, Longenecker BM (1980) Selection for virulence and organ-specific metastasis of herpesvirus-transformed lymphoma cells. Int J Cancer 25:363–369

Shearman PJ, Longenecker BM (1981) Clonal variation and functional correlation of organ-specific metastasis and an organ-specific metastasis-associated antigen. Int J Cancer 27:387–395

Sher BT, Bargatze R, Holzmann B, et al. (1988) Homing receptors and metastasis. Adv Cancer Res 51:361–390

Shipley WU (1971) Immune cytolysis in relation to the growth cycle of Chinese hamster cells. Cancer Res 31:925–929

Singh RK, Bucana CD, Gutman M, et al. (1994) Organ site-dependent expression of basic fibroblast growth factor in human renal cell carcinoma cells. Am J Pathol 145:365–374

Singh RK, Fidler IJ (1996) Regulation of tumor angiogenesis by organ-specific cytokines. In: Gunthert U, Birchmeier W (eds) Attempts to understand metastasis formation II. Springer, New York, pp 1–11

Singh RK, Fidler IJ (1997) Clinical applications of ifns "Future Directions". In: Penny R, Stuart-Harris R (eds) Clinical applications of the interferons. Chapman & Hall, London, pp 391–404

Singh RK, Tsan R, Radinsky R (1997) Influence of the host microenvironment on the clonal selection of human colon carcinoma cells during primary tumor growth and metastasis. Clin Exp Metastasis 15:140–150

Singh RK, Tsan R, Radinsky R (1997) Influence of the host microenvironment on the clonal selection of human colon carcinoma cells during primary tumor growth and metastasis. Clin Exp Metastasis 15:140–150

Singh RK, Varney ML (2000) IL-8 expression in malignant melanoma: implications in growth and metastasis. Histol Histopathol 15:843–849

Siracky J (1979a) Approach to the problem of heterogeneity of human tumor-cell populations. Br J Cancer 39:570–577

Siracky J (1979b) Origin of the resistance of leukemic cells to folic acid antagonists. Nature 169:628–629

Sluyser M, Evers SG, De Goeij CC (1976) Sex hormone receptors in mammary tumours of GR mice. Nature 263:386–389

Sluyser M, Van Nie R (1974) Estrogen receptor content and hormone-responsive growth of mouse mammary tumors. Cancer Res 34:3253–3257

Soule HD, Maloney T, McGrath CM (1981) Phenotypic variance among cells isolated from spontaneous mouse mammary tumors in primary suspension culture. Cancer Res 41:1154–1164

Stackpole CW (1981) Distinct lung-colonizing and lung-metastasizing cell populations in B16 mouse melanoma. Nature 289:798–800

Stackpole CW (1983) Generation of phenotypic diversity in the B16 mouse melanoma relative to spontaneous metastasis. Cancer Res 43:3057–3065

Starace G, Badaracco G, Greco C, Sacchi A, Zupi G (1982) DNA content distribution of in vivo and in vitro lines of Lewis lung carcinoma. Eur J Cancer Clin Oncol 18:973–978

Stephens TC, Peacock JH (1982) Clonal variation in the sensitivity of B16 melanoma to m-AMSA. Br J Cancer 45:821–829

Stich HF, Florian SF, Emson HE (1960) The DNA content of tumor cells. I. Polyps and adenocarcinomas of the large intestine of man. J Natl Cancer Inst 24:471–482

Straus MJ (1977) Growth characteristics of lung cancer. In: Straus MJ (ed) Lung cancer. Gune & Stratton, New York, pp 19–32

Strzadala L, Opolski A, Radzikowski C, Mihich E (1981) Differential expression of murine leukemia antigen on L1210 parental and drug-resistant sublines. Cancer Res 41:4934–4937

Subak-Sharpe H, Burk RR, Pitts JD (1969) Metabolic co-operation between biochemically marked mammalian cells in tissue culture. J Cell Sci 4:353–367

Suzuki N, Withers HR, Koehler MW (1978) Heterogeneity and variability of artificial lung colony-forming ability among clones from mouse fibrosarcoma. Cancer Res 38:3349–3351

Sweeney FL, Pot-Deprun J, Poupon MF, Chouroulinkov I (1982) Heterogeneity of the growth and metastatic behavior of cloned cell lines derived from a primary rhabdomyosarcoma. Cancer Res 42:3776–3782

Symmans WF, Liu J, Knowles DM, Inghirami G (1995) Breast cancer heterogeneity: evaluation of clonality in primary and metastatic lesions. Hum Pathol 26:210–216

Talmadge JE (1983) The selective nature of metastasis. Cancer Metastasis Rev 2:25–40

Talmadge JE, Benedict K, Madsen J, Fidler IJ (1984) Development of biological diversity and susceptibility to chemotherapy in murine cancer metastases. Cancer Res 44:3801–3805

Talmadge JE, Fidler IJ (1982a) Cancer metastasis is selective or random depending on the parent tumour population. Nature 297:593–594

Talmadge JE, Fidler IJ (1982b) Enhanced metastatic potential of tumor cells harvested from spontaneous metastases of heterogeneous murine tumors. J Natl Cancer Inst 69:975–980

Talmadge JE, Key ME, Hart IR (1981) Characterization of a murine ovarian reticulum cell sarcoma of histiocytic origin. Cancer Res 41:1271–1280

Talmadge JE, Starkey JR, Davis WC, Cohen AL (1979) Introduction of metastatic heterogeneity by short-term in vivo passage of a cloned transformed cell line. J Supramol Struct 12:227–243

Talmadge JE, Starkey JR, Stanford DR (1981) In vitro characteristics of metastatic variant subclones of restricted genetic origin. J Supramol Struct Cell Biochem 15:139–151

Talmadge JE, Wolman SR, Fidler IJ (1982) Evidence for the clonal origin of spontaneous metastases. Science 217:361–363

Talmadge JE, Zbar B (1987) Clonality of pulmonary metastases from the bladder 6 subline of the B16 melanoma studied by southern hybridization. J Natl Cancer Inst 78:315–320

Tan MH, Shimano T, Chu TM (1981) Differential localization of human pancreas cancer-associated antigen and carcinoembryonic antigen in homologous pancreatic tumoral xenograft. J Natl Cancer Inst 67:563–569

Tao TW, Burger MM (1977) Non-metastasising variants selected from metastasising melanoma cells. Nature 270:437–438

Tarin D, Price JE (1979) Metastatic colonization potential of primary tumour cells in mice. Br J Cancer 39:740–754

Thistlethwaithe P, Davidson DD, Fidler IJ, Roth JA (1983) Syngeneic humoral immune responses to tumor-associated antigens expressed by K-1735 UV-induced melanoma and its metastases. Cancer Immunol Immunother 15:11–16

Trope C (1975) Different sensitivity to cytostatic drugs of primary tumor and metastasis of the Lewis carcinoma. Neoplasma 22:171–180

Trope C (1982) Different susceptibilities of tumor cell subpopulations to sytotoxic agents. In: Fidler IJ, White RJ (eds) Design of models for testing cancer chemotherapeutic agents. D. Van Nostrand, New York, pp 64–79

Trope C, Aspegren K, Kullander S, Astedt B (1979) Heterogeneous response of disseminated human ovarian cancers to cytostatics in vitro. Acta Obstet Gynecol Scand 58:543–546

Trope C, Hakansson L, Dencker H (1975) Heterogeneity of human adenocarcinomas of the colon and the stomach as regards sensitivity to cytostatic drugs. Neoplasma 22:423–429

Tsuruo T, Fidler IJ (1981) Differences in drug sensitivity among tumor cells from parental tumors, selected variants, and spontaneous metastases. Cancer Res 41:3058–3064

Tu SM, Lin SH, Logothetis CJ (2002) Stem-cell origin of metastasis and heterogeneity in solid tumours. Lancet Oncol 3:508–513

Valeriote F, van Putten L (1975) Proliferation-dependent cytotoxicity of anticancer agents: a review. Cancer Res 35:2619–2630

van Lamsweerde AL, Henry N, Vaes G (1983) Metastatic heterogeneity of cells from Lewis lung carcinoma. Cancer Res 43:5314–5320

Vaupel PW, Frinak S, Bicher HI (1981) Heterogeneous oxygen partial pressure and ph distribution in C3H mouse mammary adenocarcinoma. Cancer Res 41:2008–2013

Vindelov LL (1977) Flow microfluorometric analysis of nuclear DNA in cells from solid tumors and cell suspensions. A new method for rapid isolation and straining of nuclei. Virchows Arch B Cell Pathol 24:227–242

Vindelov LL, Hansen HH, Christensen IJ, et al. (1980) Clonal heterogeneity of small-cell anaplastic carcinoma of the lung demonstrated by flow-cytometric DNA analysis. Cancer Res 40:4295–4300

Vindelov LL, Spang-Thomsen M, Visfeldt J, et al. (1982) Clonal evolution demonstrated by flow cytometric DNA analysis of a human colonic carcinoma grown in nude mice. Exp Cell Biol 50:216–221

Wagner RK, Schulze RA (1978) Clinical relevance of androgen receptor in human prostate carcinoma. Acra Endocrinol Suppl 215:139–140

Walker JR, Rees RC, Teale D, Potter CW (1982) Properties of a Herpes virus-transformed hamster cell line – I. Growth and culture characteristics of sublines of high and low metastatic potential. Eur J Cancer Clin Oncol 18:1017–1026

Wang N, Yu SH, Liener IE, et al. (1982) Characterization of high- and low-metastatic clones derived from a methylcholanthrene-induced murine fibrosarcoma. Cancer Res 42:1046–1051

Ware JL, Maygarden SJ (1989) Metastatic diversity in human prostatic carcinoma: implications of growth factors and growth factor receptors for the metastatic phenotype. Pathol Immunopathol Res 8:231–249

Weiss L (1980a) Brain Metastasis, Metastasis a monographic series. The Hague, Martinus Nijhoff 2, 30–49. 1980.

Weiss L (1980b) Differences between cancer cells in primary and secondary tumors. In: Ioachim HL (ed) Pathobiology annual 1980 V 10. Raven, New York, pp 51–81

Weiss L (1980c) Cancer cell traffic from the lungs to the liver: an example of metastatic inefficiency. Int J Cancer 25:385–392

Weiss L (1983) Random and nonrandom processes in metastasis, and metastatic inefficiency. Invasion Metastasis 3:193–207

Weiss L (1986) Metastatic inefficiency: causes and consequences. Cancer Rev 3:1–24

Weiss L (1990) Metastatic inefficiency. Adv Cancer Res 54:159–211.

Weiss L (1996) Metastatic inefficiency: intravascular and intraperitoneal implantation of cancer cells. Cancer Treat Res 82:1–11

Weiss L (2000) Metastasis of cancer: a conceptual history from antiquity to the 1990s. Cancer Metastasis Rev 19:I–XI

Weiss L, Holmes JC, Ward PM (1983) Do metastases arise from pre-existing subpopulations of cancer cells? Br J Cancer 47:81–89

Weiss MA, Michael JG, Pesce AJ, DiPersio L (1981) Heterogeneity of beta 2-microglobulin in human breast carcinoma. Lab Invest 45:46–57

Welch DR, Milas L, Tomasovic SP, Nicolson GL (1983) Heterogeneous response and clonal drift of sensitivities of metastatic 13762NF mammary adenocarcinoma clones to gamma-radiation in vitro. Cancer Res 43:6–10

Wilmanns C, Fan D, O'Brian CA, Bucana CD, Fidler IJ (1992) Orthotopic and ectopic organ environments differentially influence the sensitivity of murine colon carcinoma cells to doxorubicin and 5- fluorouracil. Int J Cancer 52:98–104

Wolman SR (1986) Cytogenetic heterogeneity: its role in tumor evolution. Cancer Genet Cytogenet 19:129–140

Wong CW, Lee A, Shientag L, et al. (2001) Apoptosis: an early event in metastatic inefficiency. Cancer Res 61:333–338

Woodruff MF (1983) Cellular heterogeneity in tumours. Br J Cancer 47:589–594

Yu JL, Rak JW, Carmeliet P, et al. (2001) Heterogeneous vascular dependence of tumor cell populations. Am J Pathol 158:1325–1334

Zetter BR (1990) The cellular basis of site-specific tumor metastasis. N Engl J Med 322:605–612

Zetter BR (1993) Adhesion molecules in tumor metastasis. Semin Cancer Biol 4:219–229

Zetter BR (1997) On target with tumor blood vessel markers. Nat Biotechnol 15:1243–1244

Zupi G, Mauro F, Sacchi A (1980) Cloning in vitro and in vivo of Lewis lung carcinoma: properties and characteristics. Br J Cancer Suppl 41:309–310

H.E. Kaiser and A. Nasir (eds.), Selected Aspects of Cancer Progression:
Metastasis, Apoptosis and Immune Response, 91–102.
© *Springer Science + Business Media B.V. 2008*

CHAPTER SIX

Tumor angiogenesis, antiangiogenic therapy and anti-antiangiogenesis response

Mengfeng Li

Abstract: Tumor growth and metastasis require new blood vessel formation, i.e. angiogenesis. Antiangiogenic therapy is emerging as a novel and potentially promising anti-cancer approach. A variety of strategies targeting tumor angiogenesis have been developed and tested in preclinical and clinical settings. The overall efficacy of antiangiogenic therapy, however, is yet to be improved. In this regard, how tumors respond to antiangiogenic therapy needs to be further investigated. Recent research suggests that tumor cells could counteract the effects of anti-endothelial inhibitors, i.e. they are not only ang-iogenic but also anti-antiangiogenic. The anti-antiang-iogenic phenotype of tumor cells might be genetically driven. Specifically, oncogenic alterations acquired in tumor cells could provide surviving signals to endothe-lial cells and confer a resistance to endothelial targeting therapies. When using an anti-endothelial treatment, simultaneously suppressing the anti-antiangiogenic property of tumor cells might be essential. Based on this consideration, combination of direct and indirect antian-giogenic therapies might represent a useful strategy to achieve effective inhibition of tumor angiogenesis.

Keywords: Angiogenesis, Antiangiogenic therapy, Head and neck cancer, Epidermal growth factor receptor

University of Pittsburgh Cancer Institute, Pittsburgh, PA 15213 USA

Tumor angiogenesis

Angiogenesis refers to the biological process via which new blood vessels are formed from existing ones. New blood vessel formation is required in a wide variety of physiological conditions, such as embryonic development, menstrual cycle and wound healing, as well as in various pathological entities. It is now widely acknowledged that tumor growth and metastasis require effective blood supply and thus, angiogenesis. In 1970s, Folkman first demonstrated that without new blood vessel formation, a tumor is unable to grow beyond a microscopic size (Folkman 1971). Furthermore, newly formed blood vessels also serve as a gateway for the dissemination of can-cer cells, and hence are important for tumor metas-tasis. During the past three decades, great advances have been made in recognizing the essential role of angiogenesis in tumor progression, in characterizing and detailing the angiogenic processes in various tumor types, in uncovering the molecular and cellular mechanisms that regulate tumor angiogenesis, and in developing antiangiogenic approaches as an anticancer modality.

Tumor angiogenesis is a complex and a dynamic process involving functioning of several vascular compartments, including endothelial cells that con-stitute the vascular endothelium, vascular pericytes surrounding the endothelium and the extracellular matrix. The initiation and progression of angiogenesis require endothelial cells to survive, proliferate and migrate in response to angiogenic stimuli. This process is promoted by a network of angiogenic factors. Up to date, there have been over a hundred angiogenesis regulators identified. The first class group of pro-angiogenic factors is composed of molecules that promote endothelial survival and proliferation, including vascular endothelial growth factors (VEGFs), angiopoietins, epithelial growth factor (EGF), fibroblast growth factors (FGFs), platelet-derived growth factor (PDGF), placental growth factor (PlGF), insulin-like growth factor IGF), angiogenic cytokines such as interleukin-8 (IL-8), and their endothelial receptors. Matrix metalloproteinases (MMP) constitute a group of proteolytic enzymes that break down the extracellular matrix (ECM) so that endothelial cells can migrate to form new vessels. A third group of angiogenic regulators have to do with the cell–cell and cell–matrix interactions, which in turn provide signals for endothelial survival, adhesion and vascular integrity. These involve integrins and other adherent molecules. The chemical structures and biological functions of these pro-angiogenic factors are summarized elsewhere (Tonini et al. 2003). In addition, recent evidence suggests that endothelial progenitor cells (EPC), mainly derived from bone marrow, also contribute to constituting vascular endothelium, and that VEGF plays an important role in the differentiation and homing of EPC to an angiogenic site (Lyden et al. 2001).

The initiation and progression of blood vessel formation are under sophisticated regulation by molecular machinery that comprises not only angiogenesis promoters, but also angiogenesis inhibitors. Negative regulators of angiogenesis include a large number of endogenous molecules that inhibit endothelial survival proliferation and migration, and ECM breakdown (Table 1). This list of angiogenesis inhibitors is still growing, and new inhibitory molecules are continuously being identified from various experimental systems. In adult, the process of vascularization largely remains quiescent, with a few exceptions such as wound healing and the menstrual cycle. It is believed that such a state of quiescence is maintained by a balance between angiogenesis promoters and inhibitors.

A widely recognized theory describes that under non-angiogenic conditions, an excess of antiangiogenic factors tilt the balance toward the inhibitory side, and as a result, the presence of inhibitory signals in endothelial cells is sustained. A direct deduction from this theory is, naturally, that angiogenesis occurs when the proangiogenic signal becomes dominant, either due to increase of angiogenic promoters or decrease of suppressors, or both. This theory provides a straightforward explanation how, when extra blood vessels are needed, the quiescent state of vascularization can be ended.

A classic illustration of this hypothesis is the scenario of tumor angiogenesis. It has been extensively shown that the production of proangiogenic factors is increased in a wide variety of experimental and human tumors. Upregulation of VEGF and/or other angiogenesis promoters has been found to be associated with tumor progression and in some cases, with poor prognosis (Linderholm et al. 2001; Blann et al. 2001; Salven et al. 2000; Ishigami et al. 1998; Tokunaga et al. 1998). In many experimental systems, ecotopic overexpression of VEGF gene promotes tumor growth and metastasis. On the other hand, many tumors display an angiogenic profile with endogenous antiangiogenic factors downregulated, such as pigment epithelium-derived factor (PEDF) and thrombospondin-1 (TSP-1) (Guan et al. 2003; Watnick et al. 2003). Thus, by increasing proangiogenic factors and decreasing antiangiogenic factors, tumor switches the non-angiogenic phenotype to an angiogenic phenotype that favors the initiation and progression of neovessel formation. This phenotype switch is believed to occur at early stages during tumor development.

A fundamental question is, then, how a tumor switches on its angiogenic phenotype prior to or at the point when new blood vessels are needed for its further growth. Hypoxia in the microenvironment is an obvious and important driving force. Necrosis at the center of a tumor, largely due to a lack of blood supply when the tumor grows beyond a thorough reach of oxygen, nutrients and growth factors from pre-existing blood vessels via diffusion, triggers expression of angiogenic factors such as VEGF. It is well established that hypoxia inducible factor 1-α (HIF-1α) is upregulated and stabilized under hypoxia and serves as a positive regulator of VEGF transcription (Shweiki et al. 1992; Tsuzuki et al. 2000; Levy et al. 1995; Forsythe et al. 1996). On the other hand, fundamentally tumor is a consequence of sequential

Table 1 Endogenous Angiogenesis Inhibitors (EAI)

Factor	Structure/biochemistry	References
Avascular tissue-derived EAI		
Pigment epithelium-derived factor	50 kD protein	Dawson et al. (1999)
Troponin I	21 kD subunit of troponin complex	Moses et al. (1999)
Angiogenic factor antagonists		
sFlt-1	Extracellular domains of Flt-1 (soluble VEGFR-1)	Kendall and Thomas (1993)
sFGF receptor	60–85 kD protein	Hanneken et al. 1994
Angiopoietin-2		Maisonpierre et al. (1997)
Antiangiogenic cytokines		
Interferon-alpha	Cytokine	Singh et al. (1995) and von Marschall et al. (2003)
Interferon-gamma	Cytokine	Friesel et al. (1987)
IP-10	C-X-C chemokine lacking ELR motif induced by IFN-γ	Cao et al. (1995) and Strieteret al. (1995)
IL-12	Heterodimer of 35 kD and 40 kD subunits	Voest et al. (1995) and Sgadari et al. (1996)
IL-1	Cytokine	Cozzolino et al. (1990)
IL-4	Cytokine	Volpert et al. (1998)
IL-10	Cytokine	Huang et al. (1996)
IL-18	Cytokine	Cao et al. (1999)
TNF-alpha	Cytokine	Leibovich et al. (1987) and Roberts et al. (1998)
PF-4 (platelet factor-4)	28 kD heparin-binding protein	Maione et al. (1990)
VEGI	174aa cytokine of TNF family	Zhai et al. (1999) and Yu et al. (2001)
Proteolytic fragment of proteins		
Endostatin	20 kD proteolytic fragment of collagen XVIII	O'Reilly et al. (1997)
Angiostatin and plasminogen kringles	Proteolytic fragment of plasminogen	O'Reilly et al. (1994); Joe et al. (1999); Cao et al. (1997)
Serpin Antithrombin	53–55 kD cleaved conformation	O'Reilly (1999)
Arrestin	26 kD C NC domain of collagen IV alpha1 chain	Pablo et al. (2000)
Canstatin	24 kD C NC domain of collagen IV alpha2 chain	Kamphaus et al. (2000)
Restin	22 kD C-fragment of collagen XV NC domain	Ramchandran et al. (1999)
Tumstatin	28 kD NC1 fragment of collagen IV	Maeshima et al. (2000)
Vasostatin	N-fragment of calreticulin	Pike et al. (1998)
Other protein EAI		
Endothelial monocyte activating polypeptide II	20 kD inflammatory cytokine	Schwarz et al. (1999); Berger et al. (2000)
Maspin	42 kD serpin	Zhang et al. (2000)
METH1, 2	110 and 98 kD proteins with MP and disintegrin-like Domains and TSP-1 repeats	Vazquez et al. (1999)
PEX	C-fragment of MMP-2	Brooks et al. (1998)
Plasminogen activator inhibitor-1, 2		Stefansson et al. (2001); McMahon et al. (2001); Mueller et al. (2005)
Prolactin (16 kD fragment)	Naturally occurring cleaved N-fragment of prolactin	Ferrara et al. (1991); Clapp et al. (1993)
TGF-β1	25 kD, 3 isoforms	Chaudhury and D'Amore (1991)
Thrombospondin 1, 2 (TSP-1, 2)	450 kD trimeric glycoprotein	Good et al. (1990); Volpert et al. (1995)
Tissue factor inhibitors	containing 3 Kunitz-type proteinase inhibitor domains	Hembrough et al. (2001)
TIMP-1, 2, 3	8.5, 21 kD	Takigawa et al. (1990); Anand-Apte et al. (1997)
Non-protein EAI		
1,23-(OH)2-Vit D3	417 Da	Oikawa et al. (1990)
2-methoxyestradiol	300 Da estrogen metabolite	Fotsis et al. (1994)
ApoE	Apolipoprotein E	Vogel et al. (1994)
Dopamine	Neurotransmitter	Basu et al. (2001)
Retinoic acid	300 Da	Lingen et al. (1996)

genetic and molecular alterations. These alterations include loss of tumor suppressor genes and activation of transforming oncogenes. In parallel with this process of malignant development is the changed balance of angiogenesis promoters and inhibitors. For example, functional loss of tumor suppressor genes (e.g. p53, PTEN, VHL) has been linked to upregulation of VEGF and downregulation of thrombospondin-1 (Dameron et al. 1994; Stratmann et al. 1997; Wen et al. 2001). Similar changes in the angiogenic profile can also result from activation of oncogenes such as Ras and epidermal growth factor receptor (EGFR) (Watnick et al. 2003; Rak et al. 1995; Arbiser et al. 1997; Perrotte et al. 1999). In this context, tumor angiogenesis is genetically controlled. Thus, genetic aberrations not only trigger surviving and proliferating mechanisms within a cancer cell, but also contribute to establishing an external angiogenic environment required for tumor growth. Such a contribution of oncogenes and tumor suppressor genes to the angiogenic phenotype may have important implications when considering targeting tumor angiogenesis. One implication is, as the genetic alterations evolve over the course of tumor progression, that the angiogenic profile may change in parallel. Consequently, an antiangiogenic strategy might be effective at certain stages of tumor development but ineffective at other stages. Another implication is that when an antiangiogenic approach is designed to only target the endothelial compartment directly (e.g. inducing endothelial apoptosis), potentially it could be compromised by the surviving or proliferating signal derived from the tumor cells. In both cases, a combination of multiple angiostatic approaches, as well as the temporal sequence of administrations, might be essential.

Antiangiogenic cancer therapy

Antiangiogenesis is probably one of the most visible areas in the field of cancer therapy over the past few years. The requirement of blood vessel formation for tumor progression makes angiogenesis a reasonable therapeutic target. The underlying hypotheses for this new modality of cancer therapy are that without angiogenesis, a tumor remains in a state of dormancy, and that such a dormant state can be established and maintained by pharmacological intervention with angiogenesis inhibitors. Initially, the primary target of antiangiogenic strategies was the tumor associated endothelial cells. Because endothelial cells do not mutate like malignant cells, in theory antiangiogenic therapy does not develop drug resistance when administered over time, providing an advantage other anti-cancer cell modalities. Indeed, this perspective was later evidenced in some experimental systems (Boehm et al. 1997).

Many antiangiogenic agents have been developed in laboratories and pharmaceutical industry thus far. The race was preluded by the discovery of the antiangiogenic function of several cytokines and agents, which had been originally identified as bioactive molecules in non-angiogenesis systems. Interferons, interleukin-12, tumor necrosis factor-α and thalidomide belong to this category. While all these "circumstantial" antiangiogenic agents were found to be angiostatic to various degrees by in vitro or in vivo angiogenesis assays, usually it is hard to differentiate their antiangiogenic efficacy from other activities. Efforts were then shifted to developing agents that specifically inhibit angiogenic endothelial cells. Although it took a long journey to identify such agents, in 1994 and 1997, respectively, the entire field was excited by the discovery of angiostatin and endostatin, two endogenous angiogenesis inhibitors found in tumor systems (O'Reilly et al. 1994, 1997). Numerous other endogenous angiogenesis inhibitors have also been identified, and in parallel, pharmacologically designed antiangiogenic agents are under active development. One interesting example is a neutralizing antibody of VEGF, Avastin. Recent human trials have provided evidence that Avastin is effective in slowing down the progression of late-stage cancers and prolonging patient survival (Willett et al. 2004 74,).

Production of endogenous angiogenesis inhibitors by tumor cells is an interesting phenomenon. This phenomenon was initially derived from early observations that removal of a primary tumor could promote the growth of a secondary tumor or metastases, indicating an inhibitory effect exerted by the primary tumor. In 1980s, Gorelik described the observation using the Lewis lung carcinoma as a model and initially attributed it to "concomitant immunity". Later on, the same group noted that the same observation could be made in immune-comprised mice, suggesting non-immune factors might be involved, with the exact mechanism

unclarified (Gorelik 1983; Gorelik et al. 1981). The puzzle was then solved, at least partly, by the identification of angiostatin by O'Reilly and Folkman in the same tumor model. Another endogenous angiogenesis inhibitor, a cleaved form of Serpin antithrombin, was also identified in a similar system (O'Reilly et al. 1994). It is yet to be understood, however, why tumor cells generate proteins that restrict their own growth. Is this an active process, or instead, these inhibitors are by-products of the malignant phenotype? Many of these endogenous inhibitors are indeed proteolytic products of proteinases that are synthesized in tumor cells to break down the ECM and thus to promote invasion of tumor cells and endothelial cells. There is no clear answer to these questions yet. Nonetheless, the aforementioned experimental systems might represent interesting tools with which potent antiangiogenic factors can be developed.

Since tumor angiogenesis is a multi-step process involving numerous molecular pathways, a wide variety of antiangiogenic strategies have been proposed and tested to target almost every single pathway involved. Examples of these strategies include: (1) blockers of endothelial survival and growth factors/receptors (e.g. VEGF antibody and VEGF receptor tyrosine kinase inhibitors); (2) molecules interrupting endothelial adhesion or cell–cell interactions (e.g. antibodies of integrins); (3) inhibitors of matrix proteases and ECM breakdown (e.g. MMP inhibitors); (4) a class of endogenous angiogenesis inhibitors found in various screening system, either circumstantially or rational-based, whose molecular mode of actions have not yet been completely clarified (e.g. endostatin and angiostatin); and (5) "non-specific" angiogenesis inhibitors, referring to those cytokines or molecules originally identified in biological systems other than angiogenesis (e.g. interferons, IL12). Many of these agents have been shown to suppress angiogenesis in preclinical studies, either by experimental angiogenesis assays such as the in vivo Matrigel assay and cornea vascularization assay, or by testing them in tumor animal models. It is of note that approaches combining different antiangiogenic agents have also been tested, and enhancement in the antiangiogenic efficacies has been seen in many of these studies. These data might indicate either that targeting only one angiogenic pathway could be leaky, or that alternative pathways are usually available to support angiogenesis. A related

issue, then, is whether comprehensive, yet efficient, approaches can be established to identify synergistic interaction among different angiogenesis inhibitors. Advances in technology may provide new insights in this area. High-throughput drug screening systems may be useful in this aspect. New tools of genomics and proteomics are also available to facilitate the optimization of combined antiangiogenic therapy. A recent report proposed to employ genomic profiling in identifying potential synergism between two angiogenesis inhibitors (Cline et al. 2002). This approach was designed based on the assumption that angiogenesis inhibitors with complementary functions should modify the expression of different gene subsets in endothelial cells. By profiling changes in gene expression using microarray and clustering the changed genes, one might be able to identify functionally complementary antiangiogenic agents. Although many concerns, some being fundamental, are to be addressed, it provides a prototype of using genomic approach to improve the efficacy of antiangiogenic therapy.

One other important advance in the field of antiangiogenic cancer therapy is the development of metronomic chemotherapy, which refers to using conventional chemotherapy drugs in a modified administration schedule to shift the therapeutic target from cancer cells to endothelial cells (Browder et al. 2000; Klement et al. 2000). Metronomic schedule is usually characterized by low-doses, more frequent administrations and longer overall therapeutic courses. Under such antiangiogenic chemotherapy, endothelial death precedes the apoptosis of tumor cell, no matter whether the tumor cells are resistant to the chemotherapy drug or not. These studies are exciting because by redirecting chemotherapy against endothelial cells, drug resistance acquired by the malignant cancer cells can be overcome.

One notion to be kept in mind is that tumor angiogenesis involves not only the endothelial cell compartment (and pericytes), but also the tumor cells, which supply angiogenic signals. This notion gives rise to the development of strategies of suppressing the angiogenic phenotype of tumor cells. In a broader sense, approaches designed to interfere with molecular pathways (oncogenes, tumor suppressor genes, growth factors/receptors) might also be antiangiogenic because as discussed above, many of these pathways are in close association with the angiogenic phenotype of tumor cells. One typical example in this context is the anti-EGFR agents, including antagonizing antibodies, small-molecule inhibitors of EGFR tyrosine kinase and antisense constructs. Suppression of EGFR activity by these inhibitors has been found to lead to angiogenesis suppression, accompanied by downregulation of VEGF, IL-8 and MMPs (Perrotte et al. 1999; Ciardiello et al. 2001; Huang et al. 2002; Li et al. 2002; Kedar et al. 2002). This mode of action is designated "*indirect*" antiangiogenesis, as it inhibits the capability of tumor cells to stimulate endothelial cells, rather than directly causing endothelial inhibition or death. Indirect antiangiogenic modality, in contrast to the "*direct*" antiangiogenic modality that directly causes endothelial inhibition or death, might carry the tendency to develop drug resistance, as its immediate target is the tumor cells that mutate at high frequencies. On the other hand, however, indirect antiangiogenesis, when combined with a direct antiangiogenic regimen, might enhance the overall therapeutic efficacy by cutting down the proangiogenic stimuli.

Angioprevention

One area of particular interest is using antiangiogenic modality for cancer prevention purposes. Treatment of advanced cancers has encountered challenges, largely due to a relatively large number of genetic and molecular abnormalities required for the development and maintenance of malignancies. In contrast, fewer molecular aberrations are present at the early phase of cancer. Recently, chemoprevention, which uses natural or synthetic agents with low or no toxicity to prevent, delay or reverse cancer progression, is increasingly gaining attention. With the improvement of early detection, the chronic feature of cancer might offer a window for early intervention. Because the initiation of angiogenesis is an early event during tumorigenesis, antiangiogenic strategies might be a promising approach to cancer chemoprevention. In this regard, the concept of "angioprevention" has begun to emerge (Bisacchi et al. 2003). By appropriately using angiogenesis inhibitors at the early stage of tumor development, one might be able to prevent the angiogenic switch. Key to the success of angioprevention is whether non-toxic and yet effective antiangiogenic strategies can be identified and developed

for early intervention. Notably, most of the antiangiogenic agents identified so far demonstrate relatively less unwanted toxicity compared with conventional anticancer chemotherapy. Furthermore, many chemopreventive agents originated from natural sources, which are carry less toxic than chemotherapy drugs, have been found to be antiangiogenic. These features appear to be promising.

Response to antiangiogenic therapy: is tumor anti-antiangiogenic?

The emergence of antiangiogenic therapy adds a new field to the "war against cancer". While this therapeutic modality has begun to show promises, as demonstrated by a large number of preclinical studies and some clinical trials, use of antiangiogenic agents to treat tumors has proved challenging. Despite the initial high expectation, significant inhibition of tumor vascularization or progression in cancer patients has rarely been evidenced. Moreover, even when tested in animal tumor models, the antiangiogenic efficacy of angiogenesis inhibitors is not always parallel with their in vitro anti-endothelial activity. Thus, like all other emerging anti-cancer modalities, more in-depth research is urgently needed for enhancing the therapeutic efficacy.

Some problems associated with current clinical tests are relatively obvious. Lack of appropriate surrogate markers makes it difficult to monitor the dynamics of, and to evaluate the in vivo effects, antiangiogenic treatment. Advances in imaging technology will help ease the issue. Recent attempts at using circulating endothelial cells as a surrogate marker has shed light in this area, but further validation work has yet to be done (Mancuso et al. 2003). A second issue related to appropriately designing human trials is the limited knowledge on the pharmacokinetics of antiangiogenic agents. As such, optimal timing and schedule of antiangiogenic intervention are to be identified. Animal studies have suggested that a low-dose, long-lasting administration schedule might be more efficacious than one-bolus use of angiogenesis inhibitors (Kisker et al. 2001). It is also likely that angiogenesis inhibitors might be more effective in early stages of cancer than in terminal diseases when a large and complex vasculature has been established.

Indeed, patients with terminal cancers have been the major target population for almost all completed or ongoing clinical trials. Furthermore, the proved phenotypic heterogeneity of endothelial cells in different organs or tissues and in different tumor types adds another dimension of complexity to the selection and use of antiangiogenic agents (Achilles et al. 2001; Fidler 2001). In this regard, animal models used to screen angiostatic drugs are an influential factor for the experimental outcome. A subcutaneous xenograft model might respond differently to such drugs than an orthotopic model where even the same tumor cell line was implanted.

While the above issues are currently under intensive investigation, one subject to which sufficient attention has not been paid is whether tumor cells per se can act as an impediment to antiangiogenic therapy. It is of note that many potent anti-endothelial agents fail to replicate their antiangiogenic potency in tumor models. When tested in vitro, these agents may cause tremendous endothelial inhibition or death. They may also display profound inhibitory effects on experimental angiogenesis induced by exogenous VEGF/FGF in vivo, as demonstrated in the Matrigel assay or the cornea assay. However, when used to suppress angiogenesis in tumors, they produce modest to little antiangiogenic effect. Complete suppression of angiogenesis or tumor shrinkage has been uncommon. These observations then led to a speculation that the intratumoral endothelial cells might be less sensitive to the action of anti-endothelial agents. This notion is consistent with the observation that even residual tumor cells surviving in a dormant state during antiangiogenic therapy could re-initiate profound neovascularization (Kendall and Thomas 1993). Thus, it is possible that tumor cells not only stimulate

angiogenesis (i.e. they are angiogenic), but also may impede the efficacy of an angiostatic treatment (i.e. they are also "anti-antiangiogenic").

There is one notable difference between tumor-induced angiogenesis and non-tumor angiogenesis (e.g. experimental angiogenesis induced by exogenous VEGF/FGF in Matrigel assay). In the latter case, tumor cells may continuously produce large amount of proangiogenic factors that support endothelia survival and proliferation. This notion was well illustrated by our study in which endostatin was used to treat squamous cell carcinoma (SCC) xenograft. We found that during the entire course of endostatin treatment, SCC cells continuously produced a large amount VEGF even though tumor vessel formation was partially inhibited by endostatin (Li et al. 2002). Further study showed that the systemic level of VEGF produced by SCC cells in vivo remained high during endostatin monotherapy (Table 2, second column). This is not surprising because endostatin only targets endothelial cells without affecting the angiogenic phenotype of SCC cells. It raises the question whether the molecular basis for such an "anti-antiangiogenic" phenotype can be identified.

As discussed in the first section of this chapter, the angiogenic capacity of tumor cells is largely attributable to the intrinsic genetic alterations that also contribute to the establishment of the malignant phenotype. In this context, activated oncogenes and growth factors/receptors are usually the genetic basis for the angiogenic phenotype. In our SCC model, the 1,483 SCC cells of head and neck (SCCHN) were used to establish the experimental tumor. Like most of SCCHN cell lines, 1,483 cells overexpress EGFR on cell surface. Overexpression or overactivation of EGFR is a major genetic change found in the vast majority of SCCHN

Table 2 EGFR-signalling in SCCHN cells sustains a high serum level of VEGF during endostatin monotherapy

Days of treatment	Level of HuVEGF* in mouse serum (pg/ml) during endostatin, EGFR-AS, or combined treatment		
	Endostatin treatment	EGFR-antisense treatment	Combined treatments
Day 1	195.14 ± 14.65	207.33 ± 15.62 ($p > 0.05$)	187.34 ± 19.34 ($p > 0.05$)
Day 10	218.85 ± 16.76 ($p > 0.05$)	153.06 ± 13.76 ($p < 0.05$)	140.02 ± 11.43 ($p < 0.01$)
Day 20	233.12 ± 14.53 ($p > 0.05$)	70.74 ± 10.3 ($p < 0.01$)	65.21 ± 9.72 ($p < 0.01$)

*An ELISA kit that only detects human VEGF was used to measure the VEGF in mouse sera, which should specifically determine the VEGF secreted by the *Human* 1,483 HNSCC cells implanted in the mice. p value was calculated by comparing each VEGF concentration with that of the "Endostatin treatment" of "Day 1" (195.14 ± 14.65 pg/ml), using 2-tail t test.

and is believed to contribute to SCCHN progression. Inhibition of EGFR expression or activity has been demonstrated to suppress the growth of SCCHN in vitro as well as in vivo (Li et al. 2002). On the other hand, experimental evidence has linked EGFR-mediated signalling to upregulation of proangiogenic factors in various types of tumor cells, including SCCHN (Ciardiello et al. 2001; Huang et al. 2002; Li et al. 2002; Kedar et al. 2002; Huang et al. 1999). Indeed, as shown in Table 2 (third column), the systemic level of VEGF production in 1483 SCCHN tumor was significantly reduced by anti-EGFR treatment. The intratumoral level of VEGF was also decreased (Li et al. 2002). Moreover, in addition to VEGF, several other proangiogenic factors, such as IL-8, MMP2 and MMP-9, appear to be downstream to EGFR signalling (Perrotte et al. 1999; Huang et al. 1999; Bruns et al. 2000). The effects of EGFR inhibition on the expression of these factors were also confirmed in one of our in vitro studies (Table 3). Based on these findings, it is possible that EGFR signalling, via stimulating and sustaining an upregulation of multiple endothelial survival factors, might create an intratumoral environment in which endothelial cells can survive an anti-endothelial treatment.

To demonstrate the endothelial survival supported by EGFR-signalling in tumor cells, an experimental system was designed to mimic the paracrine interaction between SCCHN cells and endothelial cells. In this system, endostatin and camptothecin, a general apoptosis inducer, were used to induce apoptosis of human umbilical endothelial cells (HUVEC), as shown in Table 4 (left column). It is of note that the conditioned medium (CM) of 1,483 SCCHN cells

can protect HUVEC from the apoptosis induced by camptothecin or endostatin, suggesting a presence of endothelial survival factors in CM of 1,483 cells. When 1,483 cells were pre-treated with Iressa, an EGFR inhibitor that inhibits the phosphorylation and activation of EGFR, the endothelial protective function was diminished, indicating that EGFR-signalling in 1,483 cells is responsible for the pro-survival activity in the CM. These data support the hypothesis that by inhibiting EGFR signalling, the anti-endothelial efficacies of angiogenesis inhibitors against SCCHN can be enhanced. This anti-antiangiogenesis model is summarized in Fig. 1.

Whether this model can be generalized to other tumor systems, where other oncogenic pathways than EGFR might be dominating, is yet to be tested. In this regard, however, other oncogenes have also been shown to stimulate the expression of endothelial survival/growth factors in tumor cells of various tissue or organ origins. It would not be surprising if similar observations are made with respect to these oncogenes. One important implication suggested by this model is that to achieve effective angiogenesis inhibition, a combination of anti-endothelial strategies (direct antiangiogenesis) and approaches to suppressing the anti-antiangiogenic phenotype of tumor cells (indirect antiangiogenesis) might be necessary (O'Reilly 2002). In vivo studies, indeed, have shown enhanced antiangiogenic efficacies by this type of combined therapies (Ciardiello et al. 2001; Huang et al. 2002; Li et al. 2002; Kedar et al. 2002).

The proposed model, as illustrated in Fig. 1, has underscored the importance of the endothelial

Table 3 Modulation of angiogenic factors by EGFR inhibition in 1483 HNSCC cells[a]

	VEGF		IL-8		MMP-2		MMP-9	
	Iressa	Antisense	Iressa	Antisense	Iressa	Antisense	Iressa	Antisense
RNA Change[b]	−54.5%	−62.2%	−62.8%	−57.3%	−48.3%	−52.4%	−73.9%	−82.1%
Protein Change[c]	−82.3%	−78.2%	−38.4%	−23.5%	−79.3%	−68.3%	−84.2%	−78.9%

[a]1,483 cells were treated with EGFR tyrosine kinase Iressa (0.5 μM), EGFR-antisense oligonucleotide (1 μg/ml) or vehicle DMSO, followed by EGF stimulation (50 ng/ml) for 5 h. RNA was extracted for GEArray assay. Conditioned medium was used for ELISA assay. Decreases of EGFR phosphorylation were confirmed by immunoprecipitation.
[b]RNA levels were examined by GeaArray assay. Effects of EGFR inhibition (Iressa or EGFR-antisense) on the RNA levels of angiogenic factors are presented as percentages decreased relative to the DMSO control, normalized with β-actin mRNA signal as determined by UVP Imaging Quantification software.
[c]Levels of angiogenic factors in conditioned medium were determined by ELISA method. Effects of EGFR inhibition (Iressa or EGFR-antisense) on soluble angiogenic factors are presented as percentages decreased relative to the DMSO control.

Table 4 Endothelial protection activity mediated by EGFR-signaling in SCCHN cells

Anti-endothelial agent	Regular medium		CM from 1483 cells		CM from Iressa-treated 1483 cells	
	Camp	Endo	Camp	Endo	Camp	Endo
Annexin binding	22.4%	58.3%	3.4%	2.3%	19.3%	55.3%
Caspase activation	24.3%	43.1%	3.8%	6.1%	23.1%	34.4%

Camp = camptothecin (2 μM), Endo = endostatin (300 ng/ml). Numbers show the percentages of Annexin+/PI– or activated caspase+ HUVEC assessed by flow cytometry

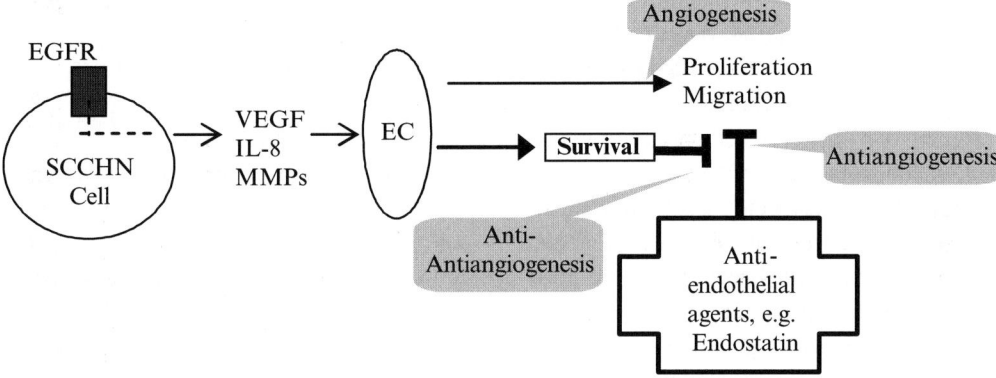

Fig. 1 Schematic illustration of the interaction between SCCHN cells and endothelial cells (EC)

survival/apoptosis cascade in the response of tumor to antiangiogenic therapy. At the first look, this appears to have overlooked other biological processes such as endothelial migration, proliferation and differentiation. Indeed, all these biological processes are required for the development of angiogenesis and have become targets of antiangiogenic strategies. An interesting notion related to this issue is that many, if not all, antiangiogenic agents developed thus far seem to be pleiotropic modifiers of angiogenesis. Very often is an antiangiogenic agent found to simultaneously inhibit more than one angiogenic step. Thus, it has been puzzling how one particular agent could effect on multiple cellular processes. It remains unclear whether one such agent simultaneously acts on different molecular targets, or alternatively, it interferes with one cellular mechanism that is critical for all other endothelial functions. Apparently, there is no clear-cut answer to this question at this moment. However, a model suggested by Jimenez and Volpert might have provided some interesting insight in this regard

(Jimenez and Volpert 2001). This model hypothesizes that the survival/apoptosis state of endothelial cells is a determinant in endothelial functions in the responses to angiogenic stimuli. It proposes that once an apoptotic state is induced, conflicting signals are generated "inside the endothelial cell that are incompatible with the processes needed to create a complex angiogenic response to the angiogenic stimuli." In this context, antiangiogenic agents probably exert their pleiotropic activities by triggering the apoptotic cascade in endothelial cells, and as a result, the endothelial cells fail to go through the angiogenic process, displaying a spectrum of angiostatic phenotype: inability to migrate, irresponsiveness to mitogenic signals, and so forth. According to this model, antiangiogenic agents are in essence apoptosis inducers, and the establishment of an apoptotic state is a primary response of endothelial cells to antiangiogenic agents. If this holds true, then the survival signal provided by the tumor cells, should be able to counteract the effect of angiogenesis inhibitors and thus serve as an impediment to

antiangiogenic therapy. Such an ability of tumor cells to impede antiangiogenic therapy appears to overlap with the anti-antiangiogenic property highlighted in our model.

Summary

Research in the last 30 years has greatly advanced our understanding of the mechanisms via which tumor angiogenesis is developed. Consequently, antiangiogenic therapy is emerging as a novel and potentially promising anti-cancer approach. A variety of strategies targeting tumor angiogenesis have been developed and tested in preclinical and clinical settings. It is entirely possible that the use of antiangiogenic modality will extend to the area of cancer prevention in the near future. Efforts are being made to address issues key to the enhancement of antiangiogenic and antitumor efficacies. In this regard, how tumors respond to antiangiogenic therapy needs to be further investigated. Recent research suggests that tumor cells could counteract the effects of anti-endothelial inhibitors, i.e. they are not only *angiogenic* but also *anti-antiangiogenic*. The anti-antiangiogenic phenotype of tumor cells might be genetically driven. Specifically, oncogenic alterations acquired in tumor cells could provide surviving signals to endothelial cells and confer a resistance to endothelial targeting therapies. When using an anti-endothelial treatment, simultaneously suppressing the anti-antiangiogenic property of tumor cells might be needed. Based on this consideration, combination of direct and indirect antiangiogenic therapies might represent a useful strategy to achieve effective inhibition of tumor angiogenesis.

References

Anand-Apte B, Pepper MS, Voest E, Montesano R, Olsen B, Murphy G, Apte SS, Zetter B (1997) Inhibition of angiogenesis by tissue inhibitor of metalloproteinase-3. Invest Ophthalmol Vis Sci 38:817–823

Basu S, Nagy JA, Pal S, Vasile E, Eckelhoefer IA, Bliss VS, Manseau EJ, Dasgupta PS, Dvorak HF, Mukhopadhyay D (2001) The neurotransmitter dopamine inhibits angiogenesis induced by vascular permeability factor/vascular endothelial growth factor. Nat Med 7:569–574

Berger AC, Alexander HR, Tang G, Wu PS, Hewitt SM, Turner E, Kruger E, Figg WD, Grove A, Kohn E, Stern D, Libutti

SK (2000) Endothelial monocyte activating polypeptide II induces endothelial cell apoptosis and may inhibit tumor angiogenesis. Microvasc Res 60:70–80

Blann AD, Li JL, Li C, Kumar S (2001) Increased serum VEGF in 13 children with Wilms' tumour falls after surgery but rising levels predict poor prognosis. Cancer Lett 173:183–186

Brooks PC, Silletti S, von Schalscha TL, Friedlander M, Cheresh DA (1998) Disruption of angiogenesis by PEX, a noncatalytic metalloproteinase fragment with integrin binding activity. Cell 92:391–40

Cao R, Farnebo J, Kurimoto M, Cao Y (1999) Interleukin-18 acts as an angiogenesis and tumor suppressor. FASEB J 13:2195–2202

Cao Y, Chen A, An SS, Ji RW, Davidson D, Llinas M (1997) Kringle 5 of plasminogen is a novel inhibitor of endothelial cell growth. J Biol Chem 272:22924–22928

Cao Y, Chen C, Weatherbee JA, Tsang M, Folkman J (1995) gro-beta, a -C-X-C- chemokine, is an angiogenesis inhibitor that suppresses the growth of Lewis lung carcinoma in mice. J Exp Med 182:2069–2077

Chaudhury RA, D'Amore PA (1991) Endothelial cell regulation by transforming growth factor-beta. J Cell Biochem 47:224–229

Clapp C, Martial JA, Guzman RC, Rentier-Delure F Weiner RI (1993) The 16-kilodalton N-terminal fragment of human prolactin is a potent inhibitor of angiogenesis. Endocrinol 133:1292–1299

Cozzolino F, Torcia F, Aldinucci D (1990) Interleukin-1 is an autocrine regulator of human endothelial cell growth. Proc Natl Aca Sci 87:6487–6491

Dameron KM, Volpert OV, Tainsky MA, Bouck N (1994) The p53 tumor suppressor gene inhibits angiogenesis by stimulating the production of thrombospondin. Cold Spring Harb Symp Quant Biol 59:483–489

Dawson DW, Volpert OV, Gillis P, Crawford SE, Xu H, Benedict W, Bouck NP (1999) Pigment epithelium-derived factor: a potent inhibitor of angiogenesis. Science 285:245–248

Ferrara N, Clapp C Weiner R (1991) The 16K fragment of prolactin specifically inhibits basal or fibroblast growth factor stimulated growth of capillary endothelial cells. Endocrinology 129:896–900

Folkman J (1971) Tumor angiogenesis: therapeutic implications. Engl J Med 285:1182–1186

Forsythe JA, Jiang BH, Iyer NV, Agani F, Leung SW, Koos RD, Semenza GL (1996) Activation of vascular endothelial growth factor gene transcription by hypoxia-inducible factor 1. Mol Cell Biol 16:4604–4613

Fotsis T, Zhang Y, Pepper MS, Adlercreutz H, Montesano R, Nawroth PP, Schweigerer L (1994) The endogenous oestrogen metabolite 2-methoxyoestradiol inhibits angiogenesis and suppresses tumour growth. Nature 368:237–239

Friesel R, Komoriya A, Maciag T (1987) Inhibition of endothelial cell proliferation by gamma-interferon. J Cell Biol 104:689–696

Good DJ, Polverini PJ, Rastinejad F, Beau MML, Lemons RS, Frazier WA Bouck NP (1990) A tumor suppressor-dependent inhibitor of angiogenesis is immunologically and functionally indistinguishable from a fragment of thrombospondin. Proc Natl Acad Sci USA 87:6624–6628

Guan M, Yam HF, Su B, Chan KP, Pang CP, Liu WW, Zhang WZ, Lu Y (2003) Loss of pigment epithelium derived factor expression in glioma progression. J Clin Pathol 56:277–282

Hanneken A, Ying W, Ling N Baird A (1994) Identification of soluble forms of the fibroblast growth factor receptor in blood. Proc Natl Acad Sci USA 91:9170–9174

Hembrough TA, Ruiz JF, Papathanassiu AE, Green SJ, Strickland DK (2001) Tissue factor pathway inhibitor inhibits endothelial cell proliferation via association with the very low density lipoprotein receptor. J Biol Chem 276:12241–12248

Huang S, Xie K, Bucana CD, Ullrich SE, Bar-Eli M (1996) Interleukin 10 suppresses tumor growth and metastasis of human melanoma cells: potential inhibition of angiogenesis. Clin Cancer Res 2:1969–1979

Ishigami SI, Arii S, Furutani M, Niwano M, Harada T, Mizumoto M, Mori A, Onodera H, Imamura M (1998) Predictive value of vascular endothelial growth factor (VEGF) in metastasis and prognosis of human colorectal cancer. Br J Cancer 78:1379–1384

Joe YA, Hong YK, Chung DS, Yang YJ, Kang JK, Lee YS, Chang SI, You WK, Lee H, Chung SI (1999) Inhibition of human malignant glioma growth in vivo by human recombinant plasminogen kringles 1–3. Int J Cancer 82:694–699

Kamphaus GD, Colorado PC, Panka DJ, Hopfer H, Ramchandran R, Torre A, Maeshima Y, Mier JW, Sukhatme VP, Kalluri R (2000) Canstatin, a novel matrix-derived inhibitor of angiogenesis and tumor growth. J Biol Chem 275:1209–1215

Kendall RL, Thomas KA (1993) Inhibition of vascular endothelial cell growth factor activity by an endogenously encoded soluble receptor. Proc Natl Acad Sci USA 90:10705–10709

Leibovich SJ, Polverini PJ, Shepard HM, Wiseman DM, Shively V, Nuseir N (1987) Macrophage-induced angiogenesis is mediated by tumour necrosis factor-alpha. Nature 329:630–632

Levy AP, Levy NS, Wegner S, Goldberg MA (1995) Transcriptional regulation of the rat vascular endothelial growth factor gene by hypoxia. J Biol Chem 270:13333–13340

Linderholm BK, Lindahl T, Holmberg L, Klaar S, Lennerstrand J, Henriksson R, Bergh J (2001) The expression of vascular endothelial growth factor correlates with mutant p53 and poor prognosis in human breast cancer. Cancer Res 61:2256–2260

Lingen MW, Polverini PJ, Bouck NP (1996) Inhibition of squamous cell carcinoma angiogenesis by direct interaction of retinoic acid with endothelial cells. Lab Invest 74:476–483

Lyden D, Hattori K, Dias S, Costa C, Blaikie P, Butros L, Chadburn A, Heissig B, Marks W, Witte L, Wu Y, Hicklin D, Zhu Z, Hackett NR, Crystal RG, Moore MA, Hajjar KA, Manova K, Benezra R, Rafii S (2001) Impaired recruitment of bone-marrow-derived endothelial and hematopoietic precursor cells blocks tumor angiogenesis and growth. Nat Med 7:1194–1201

Maeshima Y, Colorado PC, Torre A, Holthaus KA, Grunkemeyer JA, Ericksen MB, Hopfer H, Xiao Y, Stillman IE, Kalluri R (2000) Distinct antitumor properties of a type IV collagen domain derived from basement membrane. J Biol Chem 275:21340–21348

Maione TE, Gray GS, Petro J, Donner AJ, Bauer SI, Carson HF, Sharpe RJ (1990) Inhibition of angiogenesis by recombinant human platelet 4 and related peptides. Science 247:2077–2083

Maisonpierre PC, Suri C, Jones PF, Bartunkova S, Wiegand SJ, Radziejewski C, Compton D, McClain J, Aldrich TH, Papadopoulos N, Daly TJ, Davis S, Sato TN, Yancopoulos GD (1997) Angiopoietin-2, a natural antagonist for Tie2 that disrupts in vivo angiogenesis. Science 277:55–60

McMahon GA, Petitclerc E, Stefansson S, Smith E, Wong MK, Westrick RJ, Ginsburg D, Brooks PC, Lawrence DA (2001) Plasminogen activator inhibitor-1 regulates tumor growth and angiogenesis. J Biol Chem 276:33964–33968

Moses MA, Wiederschain D, Wu I, Fernandez CA, Ghazizadeh V, Lane WS, Flynn E, Sytkowski A, Tao T, Langer R (1999) Troponin I is present in human cartilage and inhibits angiogenesis. Proc Natl Acad Sci USA 96:2645–2650

Mueller BM, Yu YB Laug WE (1995) Overexpression of plasminogen activator inhibitor 2 in human melanoma cells inhibits spontaneous metastasis in scid/scid Mice. Proc Natl Acad Sci USA 92:205–209

O'Reilly MS, Boehm T, Shing Y, Fukai N, Vasios G, Lane WS, Flynn E, Birkhead JR, Olsen BR, Folkman J (1997) Endostatin: an endogenous inhibitor of angiogenesis and tumor growth. Cell 88:277–285

O'Reilly MS, Holmgren L, Shing Y, Chen C, Rosenthal RA, Moses M, Lane WS, Cao Y,

Sage EH, Folkman J (1994) Angiostatin: a novel angiogenesis inhibitor that mediates the suppression of metastases by a Lewis lung carcinoma. Cell 79: 315–328

O'Reilly MS, Pirie-Shepherd S, Lane WS, Folkman J (1999) Antiangiogenic activity of the cleaved conformation of the serpin antithrombin. Science 285:1926–1928

Oikawa T, Hirotani K, Ogasawara H, Katayama T, Nakamura O, Iwaguchi T, Hiragun A (1990) Inhibition of angiogenesis by vitamin D3 analogues. Eur J Pharmacol 178:247–16

Pablo CC, Adriana T, George K, Yohei M, Helmut H, Keiko T, Ruediger V, Eric DZ, Seth H, Pradip KS, Mark BE, Mohanraj D, Michael S, Mark P, Donald WK, Ralph RW, Vikas PS, Raghu K (2000) Anti-angiogenic Cues from Vascular Basement Membrane Collagen. Cancer Res 60:2520–2526

Pike SE, Yao L, Jones KD, Cherney B, Appella E, Sakaguchi K, Nakhasi H, Teruya-Feldstein J, Wirth P, Gupta G, Tosato G (1998) Vasostatin, a calreticulin fragment, inhibits angiogenesis and suppresses tumor growth. J Exp Med 188:2349–2356

Rak J, Mitsuhashi Y, Bayko L, Filmus J, Shirasawa S, Sasazuki T, Kerbel RS (1995) Mutant ras oncogenes upregulate VEGF/VPF expression: implications for induction and inhibition of tumor angiogenesis. Cancer Res 55:4575–4580

Ramchandran R, Dhanabal M, Volk R, Waterman MJ, Segal M, Lu H, Knebelmann B, Sukhatme VP (1999) Antiangiogenic activity of restin, NC10 domain of human collagen XV: comparison to endostatin. Biochem Biophys Res Commun 255:735–739

Roberts AB, Sporn MB, Assoian RK, Smith JM, Roche NS, Wakefield LM, Heine UI, Liotta LA, Falanga V, Kehrl JH, et al. (1998) Transforming growth factor type beta: rapid induction of fibrosis and angiogenesis in vivo and stimulation of collagen formation in vitro. Proc Natl Acad Sci USA 83:4167–4171

Salven P, Orpana A, Teerenhovi L, Joensuu H (2000) Simultaneous elevation in the serum concentrations of

the angiogenic growth factors VEGF and bFGF is an independent predictor of poor prognosis in non-Hodgkin lymphoma: a single-institution study of 200 patients. Blood 96:3712–3718

Schwarz MA, Kandel J, Brett J, Li J, Hayward J, Schwarz RE, Chappey O, Wautier JL, Chabot J, Lo Gerfo P, Stern D (1999) Endothelial-monocyte activating polypeptide II, a novel antitumor cytokine that suppresses primary and metastatic tumor growth and induces apoptosis in growing endothelial cells. J Exp Med 190:341–354

Sgadari C, Angiolillo AL, Tosato G (1996) Inhibition of angiogenesis by interleukin-12 is mediated by the interferon-inducible protein 10. Blood 87:3877–3882

Shweiki D, Itin A, Soffer D, Keshet E (1992) Vascular endothelial growth factor induced by hypoxia may mediate hypoxia-initiated angiogenesis. Nature 359:843–845

Singh RK, Gutman M, Bucana CD, Sanchez R, Llansa N Fidler IJ (1995) Interferons α and ß down-regulate the expression of basic fibroblast growth factor in human carcinomas. Proc Natl Acad Sci USA 92:4562–4566

Stefansson S, Petitclerc E, Wong MK, McMahon GA, Brooks PC, Lawrence DA (2001) Inhibition of angiogenesis in vivo by plasminogen activator inhibitor-1. J Biol Chem 276:8135–8141

Stratmann R, Krieg M, Haas R, Plate KH (1997) Putative control of angiogenesis in hemangioblastomas by the von Hippel-Lindau tumor suppressor gene. J Neuropathol Exp Neurol 56:1242–1252

Strieter RM, Polverini PJ, Kunkel SL, Arenberg DA, Burdick MD, Kasper J, Dzuiba J, Van Damme J, Walz A, Marriott D, et al (1995) The functional role of the ELR motif in CXC chemokine-mediated angiogenesis. J Biol Chem 270:27348–27357

Takigawa M, Nishida Y, Suzuki F, Kishi J, Yamashita K, Hayakawa T (1990) Induction of angiogenesis in chick yolk-sac membrane by polyamines and its inhibition by tissue inhibitors of metalloproteinases (TIMP and TIMP-2). Biochem Biophys Res Commun 171(3):1264–1271

Tokunaga T, Oshika Y, Abe Y, Ozeki Y, Sadahiro S, Kijima H, Tsuchida T, Yamazaki H, Ueyama Y, Tamaoki N, Nakamura M (1998) Vascular endothelial growth factor (VEGF) mRNA isoform expression pattern is correlated with liver metastasis and poor prognosis in colon cancer. Br J Cancer 77:998–1002

Tonini T, Rossi F, Claudio PP (2003) Molecular basis of angiogenesis and cancer. Oncogene 22:6549–6556

Tsuzuki Y, Fukumura D, Oosthuyse B, Koike C, Carmeliet P, Jain RK (2000) Vascular endothelial growth factor

(VEGF) modulation by targeting hypoxia-inducible factor-1alpha – hypoxia response element – VEGF cascade differentially regulates vascular response and growth rate in tumors. Cancer Res 60:6248–6252

Vazquez F, Hastings G, Ortega MA, Lane TF, Oikemus S, Lombardo M, Iruela-Arispe ML (1999) METH-1, a human ortholog of ADAMTS-1, and METH-2 are members of a new family of proteins with angio-inhibitory activity. J Biol Chem 274:23349–23357

Voest EE, Kenyon BM, O'Reilly MS, Truitt G, D'Amato RJ, Folkman J (1995) Inhibition of angiogenesis in vivo by interleukin 12. J Natl Cancer Inst 87:581–586

Vogel T, Guo N, Drezlich N, Krutzsch HC, Blake DA, Panet A, Roberts DD (1994) Apolipoprotein E: A potent inhibitor of endothelial and tumor cell proliferation. J Cell Biochem 54:299–307

Volpert OV, Fong T, Koch AE, Peterson JD, Waltenbaugh C, Tepper RI, Bouck NP (1998) Inhibition of angiogenesis by interleukin 4. J Exp Med 188:1039–1046

Volpert OV, Tolsma SS, Pellerin S, Feige JJ, Chen H, Mosher DF, Bouck N (1995) Inhibition of angiogenesis by thrombospondin-2. Biophys Biochem Res Commun 217:326–332

von Marschall Z, Scholz A, Cramer T, Schafer G, Schirner M, Oberg K, Wiedenmann B, Hocker M, Rosewicz S (2003) Effects of interferon alpha on vascular endothelial growth factor gene transcription and tumor angiogenesis. J Natl Cancer Inst 95:437–448

Watnick RS, Cheng YN, Rangarajan A, Ince TA, Weinberg RA (2003) Ras modulates Myc activity to repress thrombospondin-1 expression and increase tumor angiogenesis. Cancer Cell 3:219–231

Wen S, Stolarov J, Myers MP, Su JD, Wigler MH, Tonks NK, Durden DL (2001) PTEN controls tumor-induced angiogenesis. Proc Natl Acad Sci USA 98:4622–4627

Yu J, Tian S, Metheny-Barlow L, Chew LJ, Hayes AJ, Pan H, Yu GL, Li LY (2001) Modulation of endothelial cell growth arrest and apoptosis by vascular endothelial growth inhibitor. Cir Res 89:1161–1167

Zhai Y, Nia J, Jianga G, Lua L, Xinga L, Cartera KC, Janata F, Kozaka D, Xua S, Aggarwalb BB, Rubena S, Li L, Gentza R Yu G-1 (1999) VEGI, a novel cytokine of the tumor necrosis factor family, is an angiogenesis inhibitor that suppresses the growth of colon carcinomas in vivo. FASEB J 13:181–189

Zhang M, Volpert O, Shi YH, Bouck N (2000) Maspin is an angiogenesis inhibitor. Nat Med 6:196–199

H.E. Kaiser and A. Nasir (eds.), Selected Aspects of Cancer Progression:
Metastasis, Apoptosis and Immune Response, 103–113.
© Springer Science + Business Media B.V. 2008

CHAPTER SEVEN

Apoptosis and cancer

László Kopper and István Peták

Abstract: Impaired molecular regulation of active (programmed) cell death is one of the most important hallmark of cancer. Apoptosis is the best characterized biochemical and morphological form of cell death, although other forms also exist. Positive regulators of apoptosis are often lost or inactivated, while inhibitor proteins of apoptosis are often upregulated in cancer. Individual variabilities in these molecular strategies contribute to the differences in sensitivity to current anti-cancer therapies and also provide new existing molecular targets for future therapeutic approaches.

Keywords: Apoptosis, programmed cell death, necrosis, survival pathways, targetted therapy

Cell birth – via proliferation – and cell death are the two endpoints in the control of tissue homeostasis in multicellular organisms. Disturbances of either process have pathological consequences and can lead, e.g. to malformations, neurodegenerative diseases, autoimmune diseases or cancer. Till now cell death was considered as a dichotomy of apoptosis and necrosis. Whereas apopto-sis is an inherent, controlled cell death programme, the counterpart, necrosis, is a more chaotic and accidental way of dying (Danial and Korsmeyer 2004). However, in the past few years different forms of non-apoptotic cell death also appeared and these should be considered when explaining cell killing process (Okada and Mak 2004).

The regulated (programmed) cell death (PCD) has an important function: to save the organism against unwanted or potentially harmful cells. According to the current paradigm the inactivation of PCD is central both in the development of cancer and its response to therapy (Okada and Mak 2004; Brown and Attardi 2005). The resistance to cell death – particularly apoptotic cell death – will lead to one of the most critical events in carcinogenesis: the accumulation of genetic errors in proliferating cells due to the quantitative or qualitative defects of DNA repair leading to genomic instability. Cancers with relative resistance to apoptosis can withstand significant DNA damage, unfavourable environments (as encountered by metastasizing cells) and the action of cytotoxic therapy. Research on new cancer therapies has therefore focused to reverse this resistance and trigger apoptosis in tumor cells.

This short review outlines the main features of different types of cell death in normal circumstances and in certain aspects of tumor growth and therapy, including certain attempts to induce cell death by modulating the activity of those components that are considered as key elements in the cell death programs.

Department of Pathology and Experimental Cancer Research, Semmelweis University, Budapest, Hungary
Correspondence to: Professor László Kopper, Ist Department of Pathology and Experimental Cancer Research, Semmelweis University, Üllöi út 26, 1085 Budapest, Hungary.
Tel/fax: 36-1-3170891; e-mail: kopper@korb1.sote.hu

Programs for cell death

It is difficult to give an exclusive definition for PCD, because the different death programs have many overlaps in their signaling pathways. In addition, a cell may switch back and forth between different death pathways. It is suggested, that the dominant cells death phenotype is decided by the relative speed of the available programs, and the fastest and most effective pathway is usually the prominent.

Apoptotic pathways (caspase-dependent cell death)

Apoptosis is characterized by disintegration of the cell into small fragments ("apoptotic bodies") and that can be removed by phagocytosis without inflammatory reaction. The apoptotic cascade is initiated by two main pathways, involving either the activation of cell death receptors to respond to death ligands, or the release of cytochrome c from mitochondria. Both pathways will trigger a specific family of cystein proteases, the caspases, to execute the self-killing process (Hengartner 2000; Degterev et al. 2003).

Cell death receptor (extrinsic) pathway

Cell death receptors appear at the highest level of evolution, in mammalian cells. Death ligands are cytokines and able to bind to the death receptors and start the apoptosis program. Death receptors, -ligands and their signal transmitting pathway have the potential to initiate "self-killing" or to order another cell to kill itself ("instructive" apoptosis).

This pathway is regulated by the "death receptors" of the TNF-receptor family: FAS (CD95/APO-1), DR4 (TNF-related apoptosis-inducing ligand receptor 1, TRAIL-R1) and DR5 (TRAIL-R2). Death receptors are transmembrane glycoproteins, with a death domain in the intracytoplasmic part, to transmit the signal. Death ligands are also transmembrane proteins liberated from the surface mainly by matrix metalloproteases. The distribution of the receptors and ligands are different: Receptors are present in most tissues, while their ligand are expressed only in certain cell types or upon a stress stimulus and have systemic toxicity. On the contrary, only few normal tissue is sensitive to TRAIL.

When FAS is activated by FASL, caspase-8 will be recruited via an adapter molecule FADD (FAS-associated

death domain) to make a death-inducing signalling complex (DISC). Here, caspase-8 is activated and can subsequently activate caspase cascade leading to apoptosis. In certain cells the DISC-mediated apoptotic signal, i.e. caspase-8 activity is sufficient to fully fulfill the cascade (type I cells), while in others the contribution of mitochondrial pathway by caspase-8 cleaved BID is required (type II cells) to amplify the apoptotic process. FADD has a unique domain (death effector domain, DAD, belonging to the caspase recruiting protein family, CARD). The signal complex can contain other proteins as well, supporting or inhibiting death receptors (e.g. silencer of death domain, SODD; c-FLIP (c: cellular, FADD-like interleukin-1-β-converting enzyme-inhibitory protein).

All death receptors triggers apoptosis with cytosolic death domain-containing adapter protein FADD and the FADD interacting caspase-8. The death receptor-FADD-caspase-8 pathway is called extrinsic pathway. FAS and TRAIL receptors form a membrane-associated complex also with FADD and caspase-8 named death inducing signaling complex (DISC). TNFR1 makes a complex with RIP and TRAF2 at the plasma membrane and send signals via the anti-apoptotic NF-kB pathway, whereas TNFR1 should be internalized to form pro-apoptotic complex containing TRADD, FADD and caspase-8. In these signaling complexes FADD mediates the assembly of procaspase-8 dimers, which are maturated by autoproteolysis. Mature active caspase-8 can cleave and activate effector caspases, as caspase-3 and -7. Furthermore, caspase-8 can cleave proapoptotic BCL-2 family member BID producing a truncated form (tBID). tBID is translocated to mitochondria and activate the intrinsic apoptotic pathway through the conformational change of BAX and BAK. In the TNFR1 initiated complex TRADD can also mediate recruitment of TRAF2, which interferes with TNFR1 induced apoptosis partly by inhibiting caspase-8, partly by the activation of NF-kB. TNFR2, a non-death domain containing member of the family, has a proapoptotic function, recruiting TRAF2/IAP complex, triggering its proteasomal degradation, and so liberating TNFR1 from the inhibitory action of this complex.

Death receptor-mediated apoptosis can be inhibited at different level. Proteolitic cleavage or alternative mRNA can produce soluble forms of death receptors, which can bind to the death ligands outside the cell as a protection against membrane-bound receptor activation by consuming the available specific ligands. Furthermore FASL and TRAIL can bind to decoy receptors (DcR3 for FASL and DcR1 and 2 for TRAIL), however, these receptors have no intracellular domains necessary for DISC formation, therefore they can bind the ligand, but are unable to transmit the signal. Another, remarkable inhibitor is c-FLIP, which binds to the DISC and inhibits caspase-8 activation.

Mitochondrial (intrinsic) pathway

The mitochondrial pathway can be induced, e.g. by extra- or intracellular stress (hypoxia, DNA-damage, insufficient amount of growth factors, etc.). Inducing factor could be the p53 activated by DNA damage, ceramid, reactive oxygen species, increased intracellular Ca++ concentration, caused by the damage of different cell membranes due to metabolic failures. Some agents, as nitrogen-monoxid (NO), may have pro- and anti-apoptotic effect as well.

In most cases of PCD the "point of no return" is permeabilization of the outer mitochondrial membrane leading to the release of toxic proteins. The membrane permeability is controlled by pro-apoptotic (e.g. BAX, BAK, BAD, BID, BIM, BMF, NOXA) and anti-apoptotic (e.g. BCL-2, BCL-XL) members of the BCL-2 family, performing or preventing heterodimerization of proapoptotic members. BCL-2 family members are located into membranes of different cell organelles (mitochondrium, endoplasmic reticulum, even in the nucleus). Some of them (e.g. BID, BAX, BAD) sit in the cytoplasm and signal is required to translocate them into a membrane, usually into the outer membrane of the mitochondrium. All family members contain at least one of the BCL-2-homologous domains (BH1-BH4). BH3 is responsible for the anti- or proapoptotic behaviour, and certain pro-apoptotic members contain only BH3 domain. Functionally VDAC (voltage dependent anionic channel) and ANT (adeninnucleotide translocator) are active participants of the permeability control.

Permeabilization can also be resulted by the opening of a permeability transition pore in the inner mitochondrial membrane, which allows the accumulation of water and small molecules (up to 1.5 kDa), leading to the swelling of the intermembrane space and the rupture of the outer membrane. Among the released proteins cytochrome c, an important component of

the respiratory chain, form a complex ("apoptosome") with ATP, APAF-1 (apoptotic protease activating factor 1) and caspase-9. Two other mitochondrial proteins, Smac/DIABLO and OMI/HtrA2 can inhibit the catalytic function of the complex. OMI/HtrA2 in this way takes part in caspase-dependent apoptosis, but using its protease activity can act as an effector in the necrosis-like PCD. Another released protein, endonuclease G, can also contribute to both caspase-dependent and -independent PCD. Apoptosis inducing factor (AIF) is normally performs an oxidoreductase function in the intermembrane space, but becomes an active cell killer when it is released to the cytosol. It is translocated to the nucleus, and helps, probably with endonuclease G, chromatin condensation and high molecular weight (50 kb) DNA fragmentation. This activity is regulated by the antiapoptotic heat shock protein 70 (HSP70). Several observation support that AIF activity can be triggered [e.g. with overactivation of poly(ADP-ribose)polymerase I by excessive calcium influx, or in pneumococcus induced apoptosis] independently from caspases. It is thought that AIF could be a safeguard death promoter in cancer cells with faulty caspase activation (Joseph et al. 2002).

Caspases

Apoptosis in mammalian cells is mediated by a family of cystein proteases, called caspases. As part of the apoptosis control, caspases are initially expressed as inactive precursors, procaspases. When initiator caspases, as caspase-8 and caspase-9, are activated by oligomerization, they cleave the precursor forms of effector caspases, such as caspase-3, caspase-6, caspase-7. Activated effector caspases further cleave various cellular substrates (e.g. DNase from the ICAD/CAD complex, or even more procaspases), resulting in the well-known biochemical and morphological changes (e.g. chromatin condensation, nuclear fragmentation, DNA laddering) that are associated with apoptosis.

Family members of IAP (inhibitors of apoptosis) (c-IAP1, c-IAP2, X-IAP, survivin) can bind directly to caspases (Degterev et al. 2003; Wang et al. 2004; Leist and Jaattela 2001) and inhibit their activity. Anti-apoptotic action requires at least one BIR (baculovirus IAP domain). IAPs are negatively regulated by proteins from the mitochondrium, Smac/DIABLO and

OMI/HtrA2 using ubiquitinization and proteasome degradation. Heat-shock proteins can also interfere with apoptosis: HSP27 blocks DAXX or cytochrome c, HSP 70 the binding of APAF-1 and procaspase-9, HSP90 inhibits the oligomerization of APAF-1, and the latter as a chaperon protects the phosphorylated form of anti-apoptotic AKT.

It should be mentioned that granzym B, a powerful protease, the product of the cytotoxic cells and injected into the target cell through a pore-forming unit (porin), also can switch on the apoptotic machinery.

Non-apoptotic pathways (caspase-independent cell death)

In recent years, it has become evident that although caspases are key players in the apoptotic process, the caspase activation is not the only determinant of decisions in programmed cell death. These alternative caspase-independent models include autophagy, paraptosis, mitotic catastrophe, and apoptosis-like or necrosis-like PCD, as well as senesence (Bröker et al. 2005). These potentials protect the organism against unwanted cells when caspase-mediated pathways fail, but can also be triggered by various stimuli.

In a descriptive model cells are divided into four subclasses, based on their nuclear morphology (Leist and Jaattela 2001). In apoptosis the chromatin is condensated in often globular or crescent shaped compact figures (stage II chromatin condensation), while in apoptosis-like cell death the condensation is less compact (stage I). In necrosis-like cell death there is no chromatin condensation, but at best, clustering to loose speckles, whereas necrosis shows cell membrane rupture with cytoplasmic swelling.

Autophagy (also called type II cell death to distinguish it from apoptosis, the type I cell death) eliminate long-lived proteins and organelles by sequestering them into multimembrane autophagic vesicles to a subsequent degradation by the lysosomal system. There is evidence that lysosomal degradation of organelles is required for cellular remodelling due to differentiation, stress or damage by cytotoxins. In experimental systems the breakdown of the autophagy process, together with the heterogenous disruption of the autophagy gene (Beclin 1) may support carcinogenesis (Qu et al. 2003). AKT can also inhibit the autophagy induced proteolysis. On the contrary, tumor cells may need autophagy to survive hostile microenvironment and cytotoxic therapies. Today, the exact role of autophagy in mammalian cell death is still only partially understood.

Paraptosis is best described by cytoplasmic vacuolation due to the swelling of mitochondria and the endoplasmic reticulum. It is mediated by mitogen-activated protein kinases and switched on by TAJ/TROY (member of TNF receptor family) and insulin-like growth factor I receptor. Paraptosis but not apoptosis can be inhibited by AIP1/ALIX, which can interact with ALG-2 (calcium-binding death-related protein) (Wang et al. 2004; Sperandio et al. 2004). This also suggests that paraptosis is different from apoptosis. It is still unanswered how far autophagy and paraptosis represent independent types of programmed cell death.

Mitotic catastrophe is caused by mitotic failure due to the faulty checkpoints with the threatening possibility of the appearance aneuploid cells. Morphologically mitotic catastrophe is associated with the formation of multinucleate, giant cells that contain uncondensed chromosomes, and is distinct from apoptosis, necrosis or autophagy. It can be mainly triggered by DNA damage as well as agents stabilizing or destabilizing the microtubules (Castedo et al. 2004). However, it is still debated whether mitotic catastrophe is a fully caspase-independent type of programmed cell death. Many proteins take part in the regulation of G2 and mitotic checkpoints, e.g. CDK1 (inhibited by WEE1, MYT, and stimulated by CDC25C). If the checkpoint is not working properly, the cell can start a premature mitosis (aberrant mitosis), before DNA-synthesis is completed or DNA-damage is repaired. The same is the consequence of the damage of microtubules or mitotic spindle, caused, e.g. by cytotoxic agents (paclitaxel induces an abnormal metaphase in which the sister chromatids fail to segregate properly). All mitotic regulators, e.g. PLK (polo-like kinase), NIMA (never in mitosis, gene A9, Aurora family members, BUB1, can participate in mitotic catastrophe. It seems that survivin maintains the spindle checkpoint and prevent of accumulation of stressed cells that would otherwise undergo aberrant mitosis (Carvalho et al. 2003).

Premature senesence (senesence: type of "living cell death") could also be a way to dye, and in that sense inhibits tumor development. Cells loosing their

proliferative activity enter a form of permanent cell cycle arrest (replicative senesence). Senescent cells are metabolically active but non-dividing and show an increase in size. These cells express senesence-associated β-galactosidase and this process is generally p53-dependent. Other cell cycle inhibitors are also activated, as CDKNA1 gene (coding p21waf), CDKNA2 gene (p16), and retinoblastoma gene. One of the main duties of senesence program is to suppress tumorigenesis. Therefore, to turn the tumor cells from proliferative to senescent phase could be a strategy for therapy.

Necrosis is usually the result of extensive cellular trauma caused by pathological conditions, as trauma, infection, ischemia, etc. The normal physiological pathways that essential to maintain cellular homeostasis (regulation of ion transport, pH balance, energy production) are severely damaged. The rupture of the cell membrane and the release of intracellular components into the microenvironment trigger inflammatory response (which is absent in apoptotic cell death). The molecular program for necrosis is still a mystery.

Organelles in PCD

There is an attempt to classify caspase-independent cell death according to the organelles (mitochondria, lysosomes, endoplasmic reticulum, plasma membrane) involved. The signals from the different cellular organelles are linked and may effect both downstream and upstream of each other.

Mitochondria

See above as apoptotic (intrinsic) pathway.

Lysosomes

Recently, it has become evident that lysosomal proteins (e.g. cathepsin B) has an active role in cell death induced by several stimuli, including oxidative stress, TNF-α, bile salts or chemotherapeutic drugs. It seems that partial, selective permeabilization of the lysosomal membrane triggers apoptotic-like PCD, massive damage of the lysosomes results in unregulated necrosis. Several mechanisms are described to achieve this selective permeabilization or to control it (e.g. reactive oxygen species, proapoptotic members

of BCL-2 family, HSP70). Cystein protease cathepsin B and L and the asparatic protease cathepsin D are the most abundant lysosomal proteases, and cathepsin B and D seem to have the most prominent role in apoptotic- and necrotic-like PCD (Guicciardi et al. 2004). In cell lines cathepsin B can act as effector protease downstream of caspases and execute cell death independently from apoptotic machinery. Another set of data showed that lysosomal proteases promote cell death indirectly by triggering mitochondrial dysfunction and release of mitochondrial proteins, moreover they directly cleave and activate caspases. It seems that lysosomal proteases induce PCD via multiple pathways that may overlap with the traditional mediators of apoptosis. These mediators may vary depending on the cell type and the death stimulus.

Endoplasmic reticulum

ER in case of cellular stress can maintain homeostasis by withholding protein synthesis and metabolism. If the damage to the ER is too severe unfolded protein response or release of calcium into the cytoplasm can initiate PCD (Breckenridge et al. 2003). This, or BCL-2 family member BiM, leads to the activation of caspase-12, which in its inactive state is localized at the cytosolic face of the ER. Activated caspase-12 triggers downstream caspases. However, ER stress can also induce permeabilization of the mitochondrial membrane and activate mitochondrial death proteins. BCL-2 family members and the shifts in cytoplasmic calcium ensure cross talks between ER and mitochondria (Annis et al. 2004). ER stress via intracellular calcium influx can activate calpains (calcium-activated neutral proteases), which act downstream of caspases. Calpains are controlled by calpastatin, what in turn inactivated by calpain- or caspase-mediated cleavage. Different experiments showed the cooperation of calpains and caspases, as well as calpains and cathepsins.

Cell death in tumor development and progression

Cancer cells are among the main enemies in a multicellular organism, therefore they must be eliminated. If the cell death programs fail to fulfill this task, the neoplastic cells will proliferate and accumulate with further geno- and phenotypic changes allowing invasion

and metastatization. Essentially, the loss of PCD provides a survival and growth advantage.

Antiapoptotic effect of virus proteins

It is trivial that viruses can contribute to tumorigenesis. Besides integration into the host genome, certain viral proteins have anti-apoptotic effect. It is the interest of the virus to keep the infected cell alive, at least for a while, and this function can support the survival of cells with DNA-damage. Here only few examples are mentioned. KSHV can inactivate p53 by LANA (latency associated nuclear antigen of KSHV), and caspase-8 by producing v-FLIP. EBV maintain the latent infection by the inhibition of a protein kinase (by EBER) and inducing BCL-2 (by LMP1). HHV8 release a BCL-2 homologue protein, HPV E6 blocks p53, HTLV Tax protein activates NF-kB and the FASL inactivating decoy receptor (DcR3).

Arrest in the cell cycle versus apoptosis

Along the carcinogenesis or after cytotoxic therapy there is a critical balance for the damaged cells between cell cycle arrest (for DNA repair and survival) and cell death. DNA damage activates kinases as ATM (ataxia-telangiectasia mutated), ATR (ATM and Rad-3 related) and DNA-PK (DNA-dependent protein kinase), which can directly or indirectly phosphorylate p53. The negative regulator of this step is MDM-2, which holds p53 for proteasome degradation. P53 has many functions as transcription factor: one of the most important is either to stop the cell is cycle (up-regulating p21waf1, GADD45) or promote its apoptosis (up-regulating, e.g. BAX, NOXA, TRAIL-R2, FAS) (Schuler and Green 2001; Petak et al. 2001). There are several models to explain how cells choose between p53-mediated cell cycle arrest and apoptosis, emphasizing the role of the extent of DNA damage, the cell types with different sets of genes available for p53, or the presence of transcriptional co-factors. The gene encoding p53, TP53, is the most frequently mutated gene in human cancers (~50%). Furthermore, the activity of wild-type p53 may be compromised by the loss of positive regulators (as p14arf), overactivation of negative regulators (as AKT), or by the mutant p53, MDM2, or viral proteins. Activity of p53 is further modulated by different regulatory proteins, e.g. ASSP, (apoptosis stimulating protein of p53). (The lack of ASSP is responsible for the inactivity of wild-type p53 in about 70% of breast cancers.) Huge amount of in vitro data support that cytotoxic agents are more effective in tumors with wild-type p53, however, the clinical relevance of p53 status in drug sensitivity (or resistance) remains controversial, therefore, need to be determined.

Besides p53, BRCA1 has also been implicated in the regulation of apoptosis. BRCA1 may function as a sensor of cell stress (DNA-damage). A recent study found that BRCA1 acts differently in breast cancer cells depending on the cytotoxic agent: increased the sensitivity to anti-microtubule agents (paclitaxel, vinorelbine), but inhibited apoptosis when etoposide or cisplatin were given (Quinn et al. 2003). This suggests BRCA1 could be a useful predictive marker to therapeutic response.

Overexpression of anti-apoptotic proteins

In many human tumors the anti-apoptotic regulators are frequently overexpressed serving the survival of tumor cells (Longley and Johnston 2005). The classical example is the constitutively active BCL-2 in follicular lymphomas as a consequence of (Sjostrom et al. 2002; Nicholson and Anderson 2002) translocation or amplification of BCL-2. BCL-2 overexpression has also ben shown to induce MYC-dependent lymphomagenesis. In this case the proliferative potential of MYC is preserved, indicating that BCL-2 counteract MYC-induced apoptosis. The clinical significance of the overexpression of decoy receptors (e.g. in colorectal cancers) is still not known, but it may influence TRAIL effectiveness. In vitro studies found that FLIP is overexpressed of in various cell lines (melanoma, or Sternberg–Reed cells in Hodgkin-lymphoma). Survivin overexpression has been found in a wide range of human tumors in vivo and was identified as being among the most common transcripts up-regulated in cancer compared with normal tissues. The finding, that clinically the low levels of survivin correlated with better response to therapy, suggested that it may be a useful clinical marker. A firm answer is still missing. In MALT lymphoma (MALT: mucosa associated lymphoid tissue) the translocation involves cIAP2 and the MLT/MALT1 genes, producing the caspase-inhibitor cIAP2 and a paracaspase (Guicciardi et al. 2004; Sjostrom et al. 2002). (It is interesting that lymphomas carrying this

translocation are resistant to anti-Helicobacter pylori therapy.) Melanomas overexpressing ML-IAP proved to be more resistant against chemotherapy-induced apoptosis, than the non-producers.

Defects in autophagic pathway of protein degradation might also be connected to cancer. For example, autophagy is partly controlled by the PI3K pathway, and constitutive activation of PI3K signalling is common in human cancers. Such activation could inhibit both apoptotic and autophagic cell death. The best evidence that impaired autophagy is connected to tumorigenesis comes from studies of BECN1 gene, encoding beclin-1. In mammalian cells beclin-1 interacts with PI3K and take part in the induction of autophagy in response to starvation. The gene (17q21) is monoallelically deleted in many ovarian, breast and prostate cancers.

Decreased production of pro-apoptotic molecules

One of the most prominent pro-apoptotic proteins is BAX. In many human tumors BAX gene has loss-of-function mutations or shifts in the reading frame. In metastatic melanoma the tumor cells can escape apoptosis by inactivate APAF-1 gene, partly by deletion or by promoter methylation. The strategy is similar in neuroblastoma, where the caspase-8 gene could be deleted or hypermethylated. In about 80% of small-cell lung cancer the procaspase-8 activity is very low. FAS gene mutation in the death domain region or deletion resulting in truncated receptor were observed in multiple myeloma and T-cell leukemia. FAS expression was decreased or absent in different tumor types (e.g. hepatocellular cc, colon cc, melanoma) compared to the relevant normal counterpart. Deletions and mutations can occur in TRAIL receptors as well. In head and neck tumors and in NSCLC the deletion in the 8p21–22 region involved TRAILR2. Mutations changed the death domain (similarly to FAS). In certain tumors the decreased production of a pro-apoptotic protein (XAF1, XIAP-associated factor 1) led to the failure antagonizing anti-apoptotic XIAP.

Pro-survival signalling (apoptosis is blocked by the overproduction of pro-survival signals)

Tyrosine kinase receptors (e.g. EGFR family, including EGFR/ERBB1/HER1, HER2/ERBB2/NEU, HER3/ ERBB3, HER4/ERBB4) can influence efficacy

of cytotoxic agents by regulating anti-apoptotic signalling. Receptor dimerization results in cross phosphorylation of the key tyrosine residues in the cytoplasmic domain, which offers docking sites for downstream signal transducers. Such signals are generated along the phosphatydilinositol-3-kinase (PI3K)/AKT (protein kinase B, PKB) pathway and the STAT (signal transducers and activators of transcription) pathway.

Activation of PI3K can lead to phosphatidylinositol 3,4,5-triphosphate, which translocates AKT to the plasma membrane, where it is phosphorylated by 3-phosphoinositide-dependent kinase I (PDK-1). AKT further activates and regulates the function of many cellular proteins, including key regulators of apoptosis, e.g. BAD (Longley and Johnston 2005). Phosphorylation of BAD stops its inhibition by anti-apoptotic BCL-xL, furthermore, can inhibit caspase-9 and activate transcription factor –kB (NF-kB). As a whole, activated BAD will work against apoptosis. AKT also effect p53 activity as it promotes phosphorylation and translocation of MDM-2 to the nucleus, where it down-regulates p53 expression. AKT is frequently activated in human cancers due to mutations or amplifications of upstream regulators. In vitro studies demonstrated that inhibiting the PI3K/AKT pathway increases the cytotoxic effects of different chemotherapeutic agents (Nguyen et al. 2004). Recently, in NSCLC patients the phospho-AKT-positive tumors showed better clinical response following gefitinib therapy, than the negative tumors (Cappuzzo et al. 2004).

STAT proteins carry cytoplasmic signals from growth factor and cytokine receptors to the nucleus and activate transcription of various target genes. Recruitment of STATs to the activated receptors and their phosphorylation is usually mediated by a receptor-associated tyrosine kinase of the Janus kinase (JAK) family (Yu and Jove 2004). Persistent activation of STATs, especially STAT3 and STAT5 is a frequent finding in human cancers due to the constitutive activation of upstream tyrosine kinase (similarly to AKT activation). Among others STAT3 and STAT5 can regulate the expression of different anti-apoptotic proteins (e.g. BCL-xL, Mcl-1, BCL-2, survivin).

NF-kB is a key player in oncogenesis by promoting proliferation and inhibiting apoptosis. NF-kB is not a single gene but a family of closely related transcription factor genes: NF-kB1 (p50/p105), NF-kB2 (p52/p100), RELA (p65), c-REL, RELB, which produce

seven of proteins with REL homology domain mediating their dimerization, interaction with specific inhibitors and DNA binding. NF-kB dimers are mainly cytoplasmic kept in a transcriptionally inactive form by IkBs. Upon phsphorylation by IkB kinases (IKKs), IkB undergo proteasome-dependent degradation and so the NF-kB is activated, translocated into the nucleus (Lin and Karin 2003; Dolcet et al. 2005). In most instances NF-kB has an anti-apoptotic effect by up-regulating the expression of various anti-apoptotic proteins: e.g. IAPs, TNF-receptor associated factors (TRAFs), c-FLIP, BCL-2, BFL-1 (A1), BCL-X_L or down-regulate pro-apoptotics, e.g. PTEN. (PTEN suppresses the pro-survival PI3K/AKT pathway.) Whether inducibly or constitutively activated, NF-kB seems to be a critical factor in drug resistance. In vitro studies demonstrated that inhibition of NF-kB can sensitize cancer cells to chemotherapy-induced apoptosis. NF-kB is probably a major target for proteasome inhibitors, as proteasome inhibition prevents degradation of IkB, blocking NF-kB nuclear translocation (Cusack 2003). In contrast, recent evidences support that certain dimers of NF-kB could have pro-apoptotic effect, therefore it is possible, that NF-kB exerts dual function, either activator or inhibitor of apoptotic cell death, depending on the levels of RELA and c-REL.

Therapeutic induction of tumor-cell death

Development of cancer is partly due to the failure to eliminate damaged cells by the apoptotic program, either because the program is faulty, or the overproduction of survival factors inhibit the function of otherwise existing program. Since the cancer cell's susceptibility to apoptosis is severely compromised, other forms of cell killing become more important in a response to DNA-damaging (cytotoxic) agent. Success of chemotherapy is largely dependent on the ability of cytotoxic agents to induce cell death. Majority of agents trigger mitochondria pathway, but the death receptors are also involved. Recent evidences suggest that certain forms of chemotherapy induced cell death are more apoptosis-like/necrosis-like PCD than apoptosis or necrosis. It is a question, what will determine the form of cell death induced by a particular chemotherapeutic agent or radiotherapy? Presumably it depends on the context, including cell type, the type of DNA damage to which

the cell is exposed or the dose of the agent (Bröker et al. 2005; Abend 2003).

There are normal tissues (e.g. thymocytes, spermatogonia, hair-follicle cells, stem cells of the small intestine and bone marrow and tissues of the developing embryos) which are sensitive to the induction of apoptosis by DNA-damaging agents. Tumors originated from these tissues (e.g. lymphomas, some hematological tumors) are also sensitive to DNA-damaging cytotoxic agents. The induced apoptosis is p53-dependent, therefore if this pathway is inhibited (e.g. p53 mutation, BCL-2 overexpression), the sensitivity to the treatment will decrease (Gudkov and Komarova 2003). In solid tumors, however, the main reason for cell death is not the induction of apoptosis. Although it can happen, but mitotic catastrophe or senesence-like irreversible growth arrest are more frequent. It has been shown in vitro and in vivo that changing the sensitivity to apoptosis will not influence the overall sensitivity of these tumor cells to cytotoxic treatment. One can conclude that the view which made apoptosis synonymous to "cell killing" and any manipulation that altered the level of apoptosis was considered a way to reach a similar change in the overall cell killing, is probably wrong. It would be a consequence that other types of cell death should be involved and also that the resistance to therapy is not explained by inadequate apoptosis program. Nevertheless, considering the importance of this issue, more preclinical and clinical data are needed to make firm, clinically useful statements.

Induction of pro-apoptotic effect

Therapy by targeting or inducing death receptors

Targeting and activation of death receptors to induce apoptosis as a therapeutic goal has received enormous interest over the past decade, when it was proved that death receptor ligands and cytotoxic agents operate via the same mechanism, by inducing PCD, often in a cooperative manner. The problems for protein-based and virus-based gene therapy are similar: improving delivery and tumor-directed actions, preventing off-target action, and preventing or reducing immunogenicity of the used therapeutics (Wajant et al. 2005).

FAS/FASL The loss of the death-receptor expression have been reversed by chemotherapy and the FAS-negative tumor cells expressed FAS. Such induction

is usually p53-dependent, but in p53-mutated tumors interferon-γ can make the stimulation of FAS expression. This explains how interferon-γ increases the therapeutic effect of 5-fluorouracil (Schwartzberg et al. 2002).

It would be a good idea to give death-ligand, if the tumor cells have death-receptor. Recombinant death ligands and agonistic antibodies worked effectively in vitro inducing and/or supporting chemotherapy-induced apoptosis. However, activation of FAS as a therapeutic action was challenged due to the acute hepatotoxicity of FAS-agonistic antibodies. Therefore, the principal aim is to avoid the unwanted side-effects. The soluble, trimeric FASL has no bioactivity per se, but becomes activated, when immobilized by binding to the extracellular matrix. Immobilization of FASL was achieved with a trimeric fusion protein, consisting an antibody domain-recognizing tumor stroma marker fibroblast activation protein (FAP) and soluble FASL. The anti-FAP-FASL fusion protein showed no hepatotoxicity or systemic toxic effect in mice. It is a question how active is this fusion protein in the local activation of FAS. Another possibility to circumvent systemic side-effects in gene therapy is the use adenovirus vector for targeting with inducible and/or tissue-specific promoters. The usefulness of this method in vivo has not been shown yet. Limitations of the broader clinical use are the inefficient delivery to the target, strong immunogenicity and inflammatory response. Intratumoral application can reduce these obstacles (Moon et al. 2003). Further possibility is to use FAS-specific antibodies, that, for poorly understood reasons, have tissue-restricted agonistic properties. Future studies will show whether such antibodies can make clinical use.

TRAIL Nowadays TRAIL is probably is the best candidate of a death ligand for systemic administration. In normal (non-transformed) cells apoptosis induction by soluble TRAIL is prevented by an unknown mechanism what is missing or less active in tumor cells. It seems that aggregated (cross-linked) TRAIL variants are more efficient than the non-aggregated TRAIL, but serious side effects can be expected. The task is to ensure strictly tumor localized action of such reagents. Trimeric single chain antibody-TRAIL fusion proteins (immobilized on the cell surface as FASL, see above) activate TRAILR2 more efficiently target antigen expressing than non-expressing cells, although they are equally TRAIL sensitive. The potential systemic toxicity of trimeric TRAIL variants when combined

with cytotoxic agents is only begun to study in preclinical models. For example, proteasome inhibitors, which are often used to sensitize tumor cells for apoptotic action of TRAIL, also makes non-transformed cells sensitive to TRAIL-induced apoptosis (Leverkus et al. 2003). To fully activate the TRAIL receptors on tumor cells may require to stimulate with transmembrane TRAIL and/or to sensitize the cells for death receptor-induced apoptosis, e.g. by cytotoxic drugs. It is important to avoid hepatotoxicity and systemic side effects. In animal model the local administration of transmembrane TRAIL encoding adenovirus produced strong anti-tumoral effect. Studies are at early phase with TRAILR-specific antibodies.

In preclinical studies TRAIL proved to be effective, e.g. in glioma and colon cc, while in breast cancer the result was dependent on the chemotherapeutic drug used in the combination: effect of doxorubicin and 5-fluorouracil was enhanced by TRAIL, but not with melphalan, methotrexat or paclitaxel. In certain tumor cell type the contribution of the mitochondrial apoptotic pathway was necessary, which predict that the TRAIL will not be effective in all tumor types.

Other attempts

A peculiar way to increase the probability of apoptosis induction is the artificial enhancement of mitochondrial membrane permeability (e.g. by lodinamine, arsenit, betulinic acid, CD437, amphipatic cationic a-helical peptide), which would promote the escape the pro-apoptotic molecules, if the conventional chemotherapy fails.

Replacement of missing or inactive pro-apoptotic molecule is still a challenge. In experimental systems the introduction of BAX gene using an Ad-DF3-BAX vector-gene complex destroyed almost all implanted tumor cells. Another option is to decrease the methylation (e.g. by 5-aza-deoxy-cytidine), when the promoter region of a pro-apoptotic gene (e.g. caspase-8) is hypermethylated.

Inhibition of antiapoptotic effect

Antisense therapy

With the revolutionary development of high-throughput genomic, transcriptomic and proteomic technologies,

hundreds of potential therapeutic targets have been identified. Many of these gene products are not easily reached by small molecules or antibodies, and so other strategies to influence gene-expression are attractive. Known nucleotide sequences of a given gene offer the possibility to rapidly design antisense oligonucleotides (ASO) or short interfering RNA (siRNA) duplexes. The better chemical modifications of ASO increase resistance to nucleases, prolong tissue half-lives and improve scheduling. Recent clinical trials support the activity of these drugs to effectively suppress target-gene expression (Gleave and Monia 2005).

Among the most promising targets for antisense therapy are anti-apoptotic regulators overexpressed during tumor development (e.g. BCL-2, BCL-X$_L$, survivin, XIAP, MCL-1). G3139 (oblimersen, Genasense, Genta) is a first generation 18-mer phosphorothioate ASO, complementer to the first six codons of the initiating sequence of human BCL-2 mRNA. Promising preclinical results allowed to start clinical studies and they still continue in Phase I-III trials in different tumor types, in combination with different cytotoxic agents (Rai and Moore 2004; Chanan-Khan 2005; Chi et al. 2003). BCL-X$_L$ is another anti-apoptotic member of BCL-2 family and in certain tumors both BCL-2 and BCL-X$_L$ are overexpressed. A bispecific ASO targeting both molecules was recently tested as a second-generation 2 -MOIE modified ASO, and found to be a powerful inhibitor of BCL-2, BCL-X$_L$ and MCL-1. Similarly, an ASO was developed against MCL-1, which acts as an anti-apoptotic agent, heterodimerizing with proteins known to promote apoptosis. Survivin, a member of IAP family, is expressed at a high level in various human tumor types (e.g. lung colon, pancreas, breast, prostate). Survivin antisense (LY2181308) showed pro-apoptotic activity in tumor cell line, including the sensitization of tumor cells to chemotherapy-induced apoptosis (e.g. in non-Hodgkin lymphoma) (Ansell et al. 2004). Further apoptosis-related targets for ASOs are, e.g. XIAP (AEG35156/GEM640, Aegera Therapeutics), clusterin (OGX-011, OncoGeneX Technologies), STAT3 (ISIS 345794), HSP27 (OGX-427, OncoGeneX Technologies), MDM2 (GEM240, Hybridon). As is true for clinical development of all targeted therapies, important issues include optimal biological dose, relevance in the patient population being studied, and

rationales for combination strategies to yield unambiguous end points.

RNAi targeting c-FLIP dramatically sensitizes colon cancer cell lines to 5-FU, oxaliplatin and CPT-11. FLIP can inhibit the perforin–granzym B pathway. Overexpression of survivin has been shown to challenge chemotherapy-induced apoptosis in vitro, which was overruled by RNAi targeting (Nakamura et al. 2004).

Inhibition of the survival pathways, e.g. inhibition of the lipid-kinase route (PI3K-AKT-TOR) is probably an efficient indirect way to help cell death induction. It is highly possible, that this mechanism is at least partly responsible for the therapeutic effectiveness of tyrosine-kinase inhibitors, as imatinib (Gleevec) targeting BCR-ABL fusion product.

Conclusion

At the advent of targeted tumor therapy we have to learn more about those mechanisms which are mainly responsible for tumor growth and progression. Cell death is a critical cellular phenomenon, and today the monopolium of apoptosis seems to be shaken. Certain technologies of a therapy based on the stimulation of death are already known, and many of these are over the preclinical stage and entered into the clinical trials.

References

Abend M (2003) Reasons to reconsider the significance of apoptosis for cancer therapy. J Radiat Biol 79:927–941

Annis MG, Yethon JA, Leber B, Andrews DW (2004) There is more to life and death than mitochondria: BCL-2 proteins at the endoplasmic reticulum. Biochem Biophys Acta 1644:115–123

Ansell SM, Arendt BK, Grote DM, Jelinek DF, Novak AJ, Wellik LE et al. (2004) Inhibition of survivin expression suppresses the growth of aggressive non-Hodgkin's lymphoma. Leukemia 18:616–623

Breckenridge DG, Germain M, Mathai JP, Nguyen M, Shore GC (2003) Regulation of apoptosis by endoplasmic reticulum pathways. Oncogene 22:8608–8618

Bröker LE, Kruyt FAE, Giaccone G (2005) Cell death independent of caspases: a review. Clin Cancer Res 11:3155–3162

Brown JM, Attardi LD (2005) The role of apoptosis in cancer development and treatment response. Nature Rev Cancer 5:231–237

Cappuzzo F, Magrini E, Ceresoli GL, Bartolini S, Rossi E, Ludovini V et al. (2004) AKT phosphorylation and

gefitinib efficacy in patients with advanced non-small-cell lung cancer. J Natl Cancer Inst 96:1133–1141

Carvalho A, Carmena M, Sambade C, Earnshaw WC, Wheatley SP (2003) Survivin is required for stable checkpoint activation on taxol-treated HeLa cells. J Cell Sci 116:2987–2998

Castedo M, Perfettini JL, Roumier T, Andreau K, Medema R, Kroemer G (2004) Cell death by mitotic catastrophe: a molecular definition. Oncogene 23:2825–2837

Chanan-Khan A (2005) BCL-2 antisense therapy in B-cell malignancies. Blood Rev 19:213–221

Chi KN, Murray RN, Gleave ME (2003) A phase II study of oblimersen sodium (G3139) and docetaxel (D) in patients with metastatic hormone-refractory prostate cancer. Proc Am Soc Clin Oncol 22:393

Cusack JC (2003) Rationale for the treatment of solid tumors with proteasome inhibitor bortezomib. Cancer Treat Rev 29(Suppl 1):21–31

Danial NN, Korsmeyer SJ (2004) Cell death: critical control points. Cell 116:205–219

Degterev A, Boyce M, Yuan J (2003) A decade of caspases. Oncogene 22:8543–8567

Dolcet X, Llobet D, Pallares J, Matias-Guiu X (2005) NF-kB in development and progression of human cancer. Virchows Arch 446:475–482

Gleave ME, Monia BP (2005) Antisense therapy for cancer. Nature Rev Cancer 5:468–479

Gudkov AV, Komarova EA (2003) The role of p53 in determining sensitivity to radiotherapy. Nature Rev Cancer 3:117–129

Guicciardi ME, Leist M, Gores GJ (2004) Lysosomes in cell death. Oncogene 23:2881–2890

Hengartner MO (2000) The biochemistry of apoptosis. Nature 407:770–776

Joseph B, Marchetti P, Formstecher P, Kroemer G, Lewensohn R, Zhivotovsky B (2002) Mitochondrial dysfunction is an essential step for killing of non-small cell lung carcinomas resistant to conventional treatment. Oncogene 21:65–77

Leist M, Jaattela M (2001) Four deaths and a funeral: from caspases to alternative mechanisms. Nat Rev Mol Cell Biol 2:589–598

Leverkus M, Sprick MR, Wachter T, Mangling T, Baumann B, Serfling E et al. (2003) Proteasome inhibition results in TRAIL sensitization of primary keratinocytes by removing the resistance-mediating block of effector caspase maturation. Mol Cell Biol 23:777–790

Lin A, Karin M (2003) NF-kappaB in cancer: a marked target. Semin Cancer Biol 13:107–114

Longley DB, Johnston PG (2005) Molecular mechanisms of drug resistance. J Pathol 205:275–292

Moon C, Oh Y, Roth JA (2003) Current status of gene therapy for lung cancer and head and neck cancer. Clin Cancer Res 9:5055–5067

Nakamura M, Tsuji N, Asanuma K, Kobayashi D, Yagihashi A, Hirata K et al. (2004) Survivin as a predictor of cis-diammine dichloroplatinum sensitivity in gastric cancer patients. Cancer Sci 95:44–51

Nguyen DM, Chen GA, Reddy R, Tsai W, Schrump WD, Cole G Jr et al. (2004) Potentiation of paclitaxel cytotoxicity in lung and esophageal cancer cells by pharmacologic inhibition of the phosphoinositide 3-kinase/protein kinase B (AKT)-mediated signaling pathway. J Thorac Cardiovasc Surg 127:365–375

Nicholson KM, Anderson NG (2002) The protein kinase B/AKT signalling pathway in human malignancy. Cell Signal 14:381–395

Okada H, Mak TW (2004) Pathways of apoptotic and non-apoptotic death in tumour cells. Nature Rev Cancer 4:592–603

Petak I, Tillman DM, Houghton JA (2001) p53-dependence of FAS induction and acute apoptosis in response to 5-fluorouracil-leucovorin in human colon carcinoma cell lines. Clin Cancer Res 6:4432–4441

Qu X, Yu J, Bhagat G, Furuya N, Hibshoosh H, Troxel A et al. (2003) Promotion of tumorigenesis by heterozygous disruption of the beclin 1 autophagy gene. J Clin Invest 112:1809–1820

Quinn JE, Kennedy RD, Mullan PB, Gilmore PM, Carty M, Johnston PG, et al. (2003) BRCA1 functions as a differential modulator of chemotherapy-induced apoptosis. Cancer Res 63:6221–6228

Rai KR, Moore JO (2004) Phase 3 randomized trial of fludarabine/cyclophosphamide chemotherapy with or without oblimersen sodium (BCL-2 antisense; genasense; G3139) for patients with relapsed or refractory chronic lymphocytic lukemia (CLL). Blood 104:100a

Schuler M, Green DR (2001) Mechanisms of p53-dependent apoptosis. Biochem Soc Trans 29:14517–14522

Schwartzberg LS, Petak I, Stewart C, Turner PK, Ashley J, Tillman DM et al. (2002) Modulation of the FAS signaling pathway by IFN-gamma in therapy of colon cancer: phase I trial and correlative studies of IFN-gamma, 5-fluorouracil and leucovorin. Clin Cancer Res 8:2488–2498

Sjostrom J, Blomquist C, von Boguslawski K, Bengtsson NO, Mjaaland I, Malmstrom P et al. (2002) The predictive value of BCL-2, BAX, BCL-xL, bag-1, FAS and FASL for chemotherapy response in advanced breast cancer. Clin Cancer Res 8:811–816

Sperandio S, Poksay K, de Belle I, Lafuente MJ, Liu B, Nasir J et al. (2004) Paraptosis: mediation by MAP kinases and inhibition by AIP-1/Alix. Cell Death Differ 11:1066–1075

Wajant H, Gerspach J, Pfizenmaier K (2005) Tumor therapeutics by design: targeting and activation of death receptors. Cytokine Growth Factor Rev 16:55–76

Wang Y, Li X, Wang L, Ding P, Zhang Y, Han W et al. (2004) An alternative form of paraptosis-like cell death, triggered by TAJ/TROY and enhanced by PDC5 overexpression. J Cell Sci 117:1525–1532

Yu H, Jove R (2004) The STATs of cancer – new molecular targets come of age. Nature Rev Cancer 4:97–105

H.E. Kaiser and A. Nasir (eds.), Selected Aspects of Cancer Progression:
Metastasis, Apoptosis and Immune Response, 115–137.
© Springer Science + Business Media B.V. 2008

CHAPTER EIGHT

Macrophages in tumour development and metastasis

Alexandra Eichten[1,2], Karin E. de Visser[1,3], and Lisa M. Coussens[2,4,5,6]

Abstract: Activated stromal cells, e.g. inflammatory cells, fibroblasts and vascular cells, present in tumour microenvironments profoundly influence neoplastic development and progression to the tumour state. Macrophages are multifunctional immune cells that often constitute a major component of the inflammatory cell repertoire associated with premalignant and malignant tissues. Macrophages are recruited from the blood circulation by tumour-derived chemoattractants and preferentially localize to hypoxic tumour regions. Depending on their activation status and microenvironment, macrophages can impact tumour development and progression by either positive or negative mechanisms. Based upon this duality, the macrophage balance theory was proposed to emphasize complex relationships between tumour-associated macrophages and neoplastic cells. When appropriately activated, as during acute inflammatory responses, macrophages manifest an M1 phenotype and gain tumouricidal capacities; however, under adverse conditions present within tumour microenvironments, macrophages adopt an M2 phenotype and functionally contribute to neoplastic progression and overall tumour development.

Keywords: Angiogenesis, Cancer, Inflammation, Macrophage, Metastasis

Introduction

Macrophages are the major terminally differentiated cells of the mononuclear phagocyte system (Janeway et al. 2001). They are released from bone marrow as immature monocytes and circulate in the bloodstream before entering tissues (Gordon 2003). After extravasation from the blood vasculature, monocytes undergo final differentiation into resident organ specific macrophages (Mantovani et al. 2007) such as alveolar macrophages in lung, Küpffer cells in liver and osteoclasts in bones. Macrophage literally means "big eater". Initially, macrophages were described as cells specialized in uptake and digestion of foreign intruders and removal of dead cells; however, our knowledge of the vast repertoire of functions they regulate has expanded significantly.

Macrophages are versatile cells with capability to adapt their metabolism, phenotype and functional capacities to their microenvironment. As a consequence

[1]Both authors contributed equally to this work
[2]Department of Pathology,
[3]Department of Molecular Biology
[4]Cancer Research Institute,
[5]Comprehensive Cancer Center, University of California, San Francisco, 2340 Sutter St., San Francisco, CA 94143
[6]Address for correspondence: L.M. Coussens, Ph.D. University of California, San Francisco 2340 Sutter Street, N-221 San Francisco, CA 94143 Phone: 415-502-6378 Fax: 415-514-0878; e-mail: coussens@cc.ucsf.edu

of this flexibility, macrophages survive and function under adverse circumstances, including healing wounds and hypoxic areas within tumours (Lewis et al. 1999; Crowther et al. 2001). Macrophages require activation in order to exhibit their cytotoxic activity (Gordon 2003; Mantovani et al. 2002). Appropriately activated macrophages can exert direct cytotoxic effects on "abnormal" cells, phagocytose and process foreign material, debris and dead cells, and present antigens to lymphocytes, thus linking the innate with the adaptive immune system (Gordon 2003; Mantovani et al. 2002). In addition, macrophages produce various factors, including pro-and/or anti-inflammatory cytokines, chemokines, proteases, reactive oxygen species (ROS) and growth factors (Gordon 2003; Mantovani et al. 2002) thereby influencing a wide array of cells and processes in their local microenvironment; thus, macrophages are considered crucial players in diverse physiological and pathological processes including wound healing (Crowther et al. 2001), asthma (Poulter et al. 1994) and cancer (Mantovani 1994; Mantovani et al. 1992).

Cancer is a progressive disease typically requiring initial mutations in proliferating cells that are necessary but not sufficient for progression to the tumour state (Hanahan and Weinberg 2000). Additional genetic and epigenetic changes render initiated cells self-sufficient for growth, insensitive to growth-inhibitory signals and resistant to programs of terminal differentiation, senescence or apoptosis (Hanahan and Weinberg 2000) that together support unlimited self-renewal, activation of angiogenic and tissue remodeling programs and the ability to survive and invade into ectopic tissue environments (Hanahan and Weinberg 2000; Bissell and Radisky 2001). While tumours are composed of neoplastic cells, they also contain a diverse array of activated stromal cells, including endothelial cells forming the blood vasculature and lymphatics, fibroblasts and immune cells, all of which co-exist in a dynamic extracellular matrix (ECM) that together foster cancer development. Macrophages compose a large percentage of the total immune cell repertoire in many tumour types (Coussens and Werb 2001; Balkwill and Mantovani 2001; Ishigami et al. 2003; Noguchi et al. 2003; Li et al. 2002; Hamada et al. 2002; Funada et al. 2003; Lewis and Pollard 2006) and accumulating clinical data suggest that their presence influences

tumour development by both pro- and anti-tumour mechanisms (Mantovani 1994; Mantovani et al. 1992; Bingle et al. 2002; Ohno et al. 2002). In human gastric cancer, the degree of macrophage infiltration positively correlates with worse clinical outcome (Ishigami et al. 2003). Likewise, clinical prognosis of patients with renal cell carcinoma containing high numbers of tumour-associated macrophages (TAMs) is poor (Hamada et al. 2002), and in patients with oral cancer, abundance of TAMs is associated with local invasion and increased vessel density (Li et al. 2002). In contrast, other studies have described a beneficial effect of macrophage infiltration in human cancer. The overall survival rate of patients with colorectal cancer containing high numbers of tumour-infiltrating macrophages and CD8+ T cells is higher than those with low numbers of macrophages and CD8+ T cells (Funada et al. 2003). In addition, the number of intra-tumoural CD68+ macrophages is significantly greater in esophageal carcinoma patients without lymph node metastasis (Noguchi et al. 2003). The exact cellular and/or molecular mechanisms underlying these contradictory correlations between macrophage infiltration and tumour progression remain to be elucidated; however, it is likely that the dual role of macrophages during tumour development owes to their activation and differentiation status. In this chapter, we discuss the controversial relationship between tumours and macrophages, mechanisms by which macrophages are recruited towards neoplastic tissues and the influence of the tumour microenvironment on their differentiation and activation status, as well as how pro- and anti-tumour functions of macrophages either contribute to or retard neoplastic development.

Recruitment of macrophages to neoplastic lesions

Neoplastic cells and activated stromal cells present in tumour microenvironments secrete a diverse array of polypeptide growth factors, cytokines and chemokines that regulate recruitment of leukocytes (Mantovani 1994; Balkwill and Mantovani 2001; Baggiolini and Loetscher 2000; Zlotnik and Yoshie 2000; Payne and Cornelius 2002; Silzle et al. 2003) (Table 1), the presence of which can differentially regulate either pro- or anti-tumourigenic programs. Accumulation of macrophages is a common phenomenon in many human

Table 1 Regulation of differentiation, recruitment and activation of TAMs by mediators and conditions in vitro and in vivo

Mediators and conditions	Source	Macrophage response	References	
			in vitro	*in vivo*
CCL2 (MCP-1)	Neoplastic cells Endothelial cells Fibroblasts Inflammatory cells (including TAMs)	• Recruitment	Silzle et al. (2003)	Negus et al. (1995, 1997); Monti et al. (2003); Sato et al. (1995); Wong et al. (1998); Ueno et al. (2000); Riethdorf et al. (1996); Leung et al. (1997); Ohta et al. (2003); Nesbit et al. (2001)
CCL5 (RANTES)	Neoplastic cells	• Recruitment • Production of MMP-9	Azenshtein et al. (2002); Luboshits et al. (1999)	Negus et al. (1997); Luboshits et al. (1999); von Luettichau et al. (1996); Mrowietz et al. (1999)
CSF-1	Tumour	• Recruitment • Prolonged survival		Tang et al. (1992); Scholl et al. (1994); Lin et al. (2002)
CCL3 and/or CCL4	Neutrophils	• Recruitment		von Stebut et al. (2003)
Hypoxia	Low oxygen levels in center of tumours or experimentally induced hypoxia	• Accumulation due to recruitment and/or retention	Lewis et al. (1999); Rofstad et al. (2002); Maity and Solomon (2000); Turner et al. (1999); Negus et al. (1998); Butterick et al. (1981); Cazin et al. (1990); Ghezzi et al. (1991); Scannell et al. (1993); Yun et al. (1997); Tsukamoto et al. (1996); Guida and Stewart (1998); Leeper-Woodford and Mills (1992); te Koppele et al. (1991); Yu et al. (1998); Harmey et al. (1998)	Lewis et al. (1999); Negus et al. (1997); Leek et al. (1999); Lewis et al. (2000); Angermuller et al. (1995)
		• Morphological changes • Change in cell surface marker expression • Escape of apoptosis • Switch to anaerobic glycolytic pathway • Change of cytokine expression profile and phagocytotic activity • Induction of transcription factors		
VEGF	Xenografts and melanomas engineered to overexpress VEGF-C	• Recruitment	Kappel et al. (1999)	Duyndam et al. (2002); Skobe et al. (2001); Takahashi et al. (2003); Gabrilovich et al. (1996)

(Continued)

Table 1 Regulation of differentiation, recruitment and activation of TAMs by mediators and conditions in vitro and in vivo — Cont'd

Mediators and conditions	Source	Macrophage response	References	
			in vitro	*in vivo*
IL-6	Neoplastic cells	• Inhibition of maturation • Expression of TF • Inhibition maturation	Menetrier-Caux et al. (1998)	
M-CSF	Neoplastic cells	• Inhibition maturation • Blocking APC function	Menetrier-Caux et al. (1998)	
Decoy receptor 3	Tumours	• Skew T cell responses to Th2 • Impaired phagocytic activity • Impaired production free radicals and pro-inflammatory cytokines (in vitro)	Chang et al. (2003)	
Unknown mediators	Tumours	• Shift from M1 to M2 phenotype resulting in production of IL-10, TGF-β	Mantovani et al. (2002); Kambayashi et al. (1995); Loercher et al. (1999); Blot et al. (2003); Mytar et al. (2003); Chang et al. (2001); Lwaleed et al. (2001); Torisu et al. (2000); Ko et al. (2002); Sica et al. (2000)	Mantovani et al. (2002); Sica et al. (2000); Ibe et al. (2001); Maeda et al. (1995); Dinapoli et al. (1996); Andrade et al. (1992); Jenkins et al. (1995); Thomsen et al. (1995)
TGF-β and/or IL-10	Neoplastic cells	• Reduction in TNF-α production	Kambayashi et al. (1995); Sica et al. (2000)	Maeda et al. (1995); Sica et al. (2000)
MIF	Neoplastic cells Tumours Macrophages	• Stimulation of phagocytosis • Increase of cytolytic activity • Stimulation of TNF-α release	Onodera et al. (1997); Pozzi and Weiser (1992); Ren et al. (2003)	Onodera et al. (1997); Ren et al. (2003); Meyer-Siegler and Hudson (1996); Shimizu et al. (1999)

tumours, suggesting that signals from the tumour microenvironment trigger either their recruitment from ECM or blood circulation or, alternatively, regulate their retention, proliferation and/or differentiation at the neoplastic site.

Recruitment by chemokines and cytokines Chemokines are structurally related small polypeptide signalling molecules classified into four families based on amino acid sequence similarities: CXC, CC, C and CX$_3$C (Zlotnik and Yoshie 2000). They were originally named according to their function or expression profile, but recently a more systematic nomenclature was proposed (Zlotnik and Yoshie 2000). Chemokines

bind to and activate chemokine receptors, e.g. small GTP-binding protein coupled seven-transmembrane domain receptors (Murphy 1996). Chemokine receptors are expressed on diverse cell types common to neoplastic environments, including inflammatory cells, fibroblasts, endothelial and neoplastic cells (Homey et al. 2002).

CCL2 (monocyte chemoattractant protein-1, MCP-1) is a member of the CC chemokine family and a potent chemoattractant for monocytes (Yoshimura et al. 1989; Roth et al. 1995; Sozzani et al. 1995; Lu et al. 1998; Chae et al. 2002). CCL2 is expressed in a variety of human carcinomas, including ovarian (Negus et al. 1995, 1997), breast (Ueno et al. 2000)

and pancreatic (Monti et al. 2003), as well as gliomas (Leung et al. 1997), meningomas (Sato et al. 1995) and lymphoepithelioma-like carcinoma of the lung (Wong et al. 1998). For some tumour types, a positive correlation has been observed between CCL2 levels and abundance of tumour-associated macrophages (TAMs) which, depending on the tumour type, reflects either a positive or negative clinical outcome (Monti et al. 2003; Sato ct al. 1995; Ucno ct al. 2000; Leung et al. 1997; Ohta et al. 2003). In human gastric and breast carcinoma for example, high levels of CCL2 correlate with macrophage infiltration, high levels of vascular endothelial growth factor (VEGF) and increased microvessel density (Ueno et al. 2000; Ohta et al. 2003), that together direct a pro-tumour effect by positively regulating angiogenesis. In contrast, in pancreatic cancer high serum levels of CCL2 positively correlate with TAM infiltration and are associated with higher survival rate following surgical resection (Monti et al. 2003). Thus, CCL2 expression correlates with recruitment of macrophages towards neoplastic lesions, the biological outcome and significance of which, however, remains unclear.

Monocytes can elicit either proliferative or apoptotic regulatory effects on target cells dependent upon many factors including the repertoire of chemokine receptors expressed on the cell surface, as well as the local chemokine environment; thus, local chemokine concentration as well as chemokine diversity is important. Experimental mouse models of human cancer development have been utilized to address some of these ambiguities regarding how chemokines, specifically CCL2, regulate TAM infiltration by manipulating either chemokine or chemokine receptor expression. Using an experimental model of melanoma development, it was demonstrated that local concentration of CCL2 was important for regulating macrophage infiltration, the degree of which profoundly affected tumour formation capability (Nesbit et al. 2001); melanoma cells with low CCL2 expression induce a modest monocyte infiltration and tumour growth, while high CCL2 production results in extensive macrophage infiltration and subsequent tumour destruction (Nesbit et al. 2001), suggesting a concentration dependent balance for the biological function of CCL2. Although the exact molecular and/or cellular mechanisms regulated by variable CCL2 concentration are still largely unknown, it has been

speculated that macrophage-derived tumour necrosis factor alpha (TNFα) plays a central role in regulating the ambigious effects of CCL2 (Sato et al. 1986; Leibovich et al. 1987; Fajardo et al. 1992). Supporting data has emerged from studies demonstrating that low levels of macrophage-derived TNFα stimulate migration of endothelial cells, thereby activating angiogenesis, whereas high levels of TNFα are known to inhibit endothelial cell proliferation, thereby inhibiting angiogenesis (Sato et al. 1986; Leibovich et al. 1987; Fajardo et al. 1992). Low levels of tumour derived CCL2 triggered modest monocyte infiltration into the tumour, which subsequently resulted in modest TNFα (Nesbit et al. 2001). These modest TNFα levels resulted in activation of angiogenesis and thereby supported tumour growth (Nesbit et al. 2001). In contrast, high levels of CCL2 induced extensive monocyte infiltration resulting in high levels of TNFα, subsequently inhibiting angiogenesis and tumour growth (Nesbit et al. 2001). Taken together, these data suggest that the abundance of macrophages recruited towards neoplastic lesions is critically dependent on local expression levels of CCL2 that in turn affects the pro- and/or anti-tumour effects of macrophages dependent upon local levels of bioavailable TNFα.

Chemokines are not only expressed by neoplastic cells in developing tumours, but their expression can also be induced in activated fibroblasts and endothelial cells, as well as in other inflammatory cells and TAMs themselves thereby providing a positive amplification circuit for further macrophage recruitment (Silzle et al. 2003; Wong et al. 1998; Ueno et al. 2000; Riethdorf et al. 1996; Leung et al. 1997). The significance of chemokine expression emanating from activated stromal cells has been addressed experimentally. In a coculture system, it was observed that human breast carcinoma-associated fibroblasts mediate recruitment of macrophages via induced expression of CCL2, as neutralization with CCL2-specific antibodies blocked recruitment (Silzle et al. 2003), thus underscoring the participation of stromal cells, such as fibroblasts, in the recruitment of macrophages towards neoplastic lesions. Taken together, these experimental data and clinical correlates suggest that CCL2 is a key player regulating macrophage recruitment into neoplastic microenvironments. Once macrophages have entered the tumour microenvironment, downregulation

of cell surface expression of CCR2 might provide a possible mechanism for their retention in tumours (Sica et al. 2000). The fact that CCL2 also plays a key role in macrophage recruitment into tissue during acute inflammatory responses (Lu et al. 1998; Chae et al. 2002; Boring et al. 1997) suggests that CCL2 and TAM recruitment pathways are conserved in chronic inflammation associated with neoplastic progression.

CCL5 (RANTES) is another chemokine that has gained attention for its role in regulating macrophage recruitment (Negus et al. 1997; Azenshtein et al. 2002; Luboshits et al. 1999; von Luettichau et al. 1996; Mrowietz et al. 1999). Similar to CCL2, CCL5 expression is increased in several human cancers, including ovarian and breast carcinoma and melanoma (Negus et al. 1997; Luboshits et al. 1999; von Luettichau et al. 1996) where elevated expression levels correlate with worse clinical outcome (Luboshits et al. 1999). *In vitro* experiments have demonstrated that CCL5 produced by breast carcinoma cells induce monocyte recruitment (Azenshtein et al. 2002), while increased expression of CCL5 by human melanoma cells *in vivo* correlates with increased tumourgenicity upon injection into nude mice (Mrowietz et al. 1999).

Chemokines are not the only factors possessing strong chemoattractant capabilities for macrophages. Cytokines such as colony-stimulating factors, e.g. CSF-1, also regulate recruitment of macrophages into tissues as well as prolonging survival of TAMs (Tang et al. 1992; Scholl et al. 1994; Lin et al. 2002). CSF-1 is a macrophage lineage-specific dimeric polypeptide growth factor that acts through a cell surface tyrosine kinase receptor encoded by the *c-fms* proto-oncogene (CSF-1R) on effector cells (Kacinski 1995, 1997). Interestingly, elevated expression of *CSF-1* and *c-fms/ CSF-1R* has long been associated with poor prognosis in several types of human epithelial cancer, e.g. breast, uterine and ovarian (Kacinski 1995). Elevated expression of CSF-1 also correlates with abundant leukocyte infiltration during development and progression of human breast and ovarian cancer (Tang et al. 1992; Scholl et al. 1994; Lin et al. 2002; Tang et al. 1990), the significance of which has been addressed experimentally utilizing a transgenic mouse model of mammary carcinogenesis (Lin et al. 2001). In this study, the authors compared dynamics of mammary carcinoma development by comparing transgenic mice susceptible to de novo development of mammary

carcinomas (PyMT mice) (Cardiff 2001; Cardiff et al. 2000) with PyMT mice harboring a recessive null mutation in the CSF-1 gene (*Csf1$^{op/op}$*) (Lin et al. 2001). Whereas absence of CSF-1 during early neoplastic development was without apparent consequence, development of late-stage invasive carcinoma and its metastatic pulmonary derivatives were significantly attenuated in PyMT/*Csf1$^{op/op}$* mice. The key difference between PyMT mice and PyMT/*Csf1$^{op/op}$* mice was not in the proliferative capacity of neoplastic epithelial cells, but failure to recruit mature macrophages into neoplastic tissue in the absence of CSF-1 (Lin et al. 2001). Macrophage recruitment was restored by transgenic CSF-1 expression in mammary epithelium in PyMT/*Csf1$^{op/op}$* mice, as was characteristic primary and metastatic tumour development (Lin et al. 2001). Together, these genetic experiments provide a causal link between CSF-1-dependent infiltrating macrophages and malignant potential of epithelial cells.

In conclusion, neoplastic cells and activated stromal cells present in neoplastic microenvironments promote recruitment of macrophages via secretion of soluble mediators, such as CCL2, CCL5 and CSF-1, that regulate cell behavior by interaction and binding to respective receptors expressed on macrophages. Moreover, the levels of macrophage chemoattractants appears to directly reflect the number of macrophages recruited to tumour sites, thus significantly impacting biological outcome (Nesbit et al. 2001).

Recruitment by Hypoxia During neoplastic progression, inadequate perfusion of diseased tissue can result in formation of areas of low oxygen tension (hypoxia) (Harris 2002). Several clinical and experimental studies have suggested that TAMs preferentially accumulate in such hypoxic regions (Lewis et al. 1999; Sica and Bronte 2007). Macrophages are documented to accumulate in avascular and necrotic areas in human ovarian and breast carcinomas (Negus et al. 1997; Leek et al. 1999; Lewis et al. 2000). Under hypoxic condition, hypoxia-inducible factor 1 (HIF-1) is activated in TAMs, resulting in upregulation of CXCR4 (Sica and Bronte 2007; Schioppa et al. 2003). At the same time, HIF-1 activation results in induction of SDF-1 (a ligand for CXCR-4) in hypoxic tissue (Ceradini et al. 2004), thus guiding the localization and function of TAMs Besides influencing recruitment of macrophages, hypoxia can also

mediate accumulation of macrophages via retainment or entrapment. *In vitro* studies suggest that various cell types respond to hypoxia with increased expression of surface molecules involved in chemotaxis and migration, suggesting that migration of these cells may be increased under hypoxic conditions (Lewis et al. 1999). For instance, the urokinase plasminogen activator receptor (uPAR), a GPI-anchored cell surface receptor expressed on many cell types, mediates a wide variety of biological responses and is upregulated on neoplastic cells under hypoxic conditions (Rofstad et al. 2002; Maity and Solomon 2000). Since uPAR has been reported to be involved in macrophage migration (Gyetko et al. 1994), it is possible that recruitment of macrophages to hypoxic areas within tumours is mediated by upregulation of uPAR expression on their cell surface. In contrast, studies with ovarian carcinoma cells have found that CCL2 is downregulated under hypoxic conditions and thereby imposes inhibitory effects on migratory capacity of macrophages (Turner et al. 1999; Negus et al. 1998). In addition, hypoxia also inhibits macrophage expression of CCR2 and CCR5, thus further immobilizing TAMs (Sica et al. 2000; Bosco et al. 2004). These results in combination with the finding that the chemotactic response of monocytes to CCL2, CCL3, CCL5 and N-formyl-met-leu-phe (fMLP) is diminished under hypoxic conditions (Turner et al. 1999; Negus et al. 1998), suggest that hypoxia severely affects macrophage migration indirectly by downregulating chemokine levels, as well as directly by inhibiting the capacity of macrophages to migrate. While the underlying mechanisms of these hypoxia-induced changes regulating differential macrophage recruitment and/or migration are to date largely unknown, it is clear that hypoxic conditions influence monocyte migration in a rapid and reversible fashion (Turner et al. 1999).

How can the paradox of inhibited macrophage migration by hypoxia on one hand, and high macrophage density in hypoxic tumour regions on the other, be explained? One hypothesis is that macrophages migrate towards or into a tumour in response to chemotactic signals provided by cells present in normoxic tumour areas. Once macrophages enter hypoxic zones however, they become immobilized, resulting in accumulation. Although experimental evidence supporting this hypothesis is provided by evaluation of TAMs

with regards to VEGF expression in neoplastic tissues (Crowther et al. 2001), the causal relationship between presence of TAMs and VEGF expression levels is unclear. While significantly higher numbers of TAMs have been reported in hypoxic areas of VEGF-positive compared to VEGF-negative tumours (Lewis et al. 2000) and expression of macrophage-derived VEGF is upregulated in hypoxic areas of invasive breast carcinomas (Lewis et al. 2000) correlating with macrophage infiltration (Leek et al. 2000), it remains unclear whether the correlation between VEGF and abundance of TAMs is a consequence of paracrine regulation of VEGF inducing recruitment of TAMs or alternatively autocrine production of VEGF by recruited TAMs. A study by Duyndam et al. (2002) utilizing human ovarian carcinoma xenografts overexpressing VEGF revealed enhanced infiltration by macrophages indicating that VEGF possesses chemotactic activity towards macrophages. This finding suggests that the presence of VEGF is important for accumulation of macrophages in neoplastic environments, and since TAMs are a major source of VEGF, the chemoattracting effect of VEGF in hypoxic areas on macrophages directs a positive feedback loop. Altogether, there are many pathways via which macrophages recruitment is regulated, some may affect recruitment towards neoplastic cells, while others may affect their retention once present, and dependent upon the scenario, can either regulate potent anti-tumour or pro-tumour effects. These recruitment pathways offer tremendous therapeutic potential as they can be manipulated to effect the number of macrophages in tumours as well as to prevent M2 macrophages from exhibiting pro-tumour effects.

Macrophage differentiation and activation in tumour microenvironments

Terminal differentiation and activation of macrophages is largely dependent upon their local microenvironment (Mantovani et al. 2002). Since macrophages are versatile and respond rapidly to environmental cues, their morphology, viability and functional capacity within tumours greatly depends on tumour type, tumour stage and their localization within the tumour; thus, understanding these spatial and temporal restrictions is an important aspect of understanding macrophage regulation of cancer development.

Differentiation of macrophages and neoplastic progression. Recent experimental data suggest that tumour growth is associated with accumulation of immature myeloid cells (lmC) in peripheral blood (Almand et al. 2000; Kusmartsev and Gabrilovich 2003). ImC isolated from tumour-bearing mice display impaired differentiation capacity to mature myeloid cells, e.g. macrophages and dendritic cells (DCs) (Kusmartsev and Gabrilovich 2003). The impaired differentiation of ImC induced in the presence of tumours persists following transfer of ImC into tumour-free hosts or following culture in the absence of neoplastic cells (Kusmartsev and Gabrilovich 2003) suggesting they represent a terminal cell type. Mechanistically, it has been proposed that sustained ImC immaturity is due to increased endogenous levels of hydrogen peroxide and hyperproduction of reactive oxygen species (ROS) (Kusmartsev and Gabrilovich 2003). Several tumour-derived factors including VEGF (Takahashi et al. 2003; Gabrilovich et al. 1996), interleukin (IL)-6, CSF-1 (Menetrier-Caux et al. 1998) and decoy receptor 3 (DcR3), a soluble receptor of the tumour necrosis factor (TNF) receptor superfamily (Chang et al. 2003), can significantly affect differentiation of multiple hematopoietic cell lineages. Failure of ImC to differentiate into mature macrophages and/or DCs in the presence of tumours, together with the observation that ImC exert inhibitory effects on DC and T cell function via direct cell–cell contact, suggest their involvement in regulating overall immune-suppression often observed in cancer patients and escape of malignant cells from immune surveillance (Almand et al. 2000, 2001 ; Kusmartsev and Gabrilovich 2003; Serafini et al. 2004; Gabrilovich et al. 2001) suggesting that tumour derived factors profoundly impact maturation of macrophages and subsequently affect the ability of cancer patients to mount efficient anti-cancer immune responses.

Activation of macrophages and neoplastic progression. "Classical" macrophage activation occurs in response to microbial agents such as lipopolysaccharide (LPS) and pro-inflammatory cytokines such as interferon-gamma (IFNγ resulting in full activation of macrophages to the M1 phenotype (Mantovani et al. 2002). Other soluble factors also influence macrophage activation and differentiation, including various anti-inflammatory molecules (IL-4, -13, -10) and glucocorticoids (Mantovani et al. 2002); however, these factors trigger an antagonistic activation program classified as alternative macrophage activation resulting in M2 macrophage phenotypes (Mantovani et al. 2002). Since macrophages are exposed to a mixture of pro- and anti-inflammatory mediators, M1 and M2 phenotypes are extremes of a continuum of diverse functional states (Mantovani et al. 2002). Macrophages of the M1 phenotype are typically present during early phases of acute inflammation (Topoll et al. 1989) and have the potential to induce neoplastic cell death after activation by IL-12 and IFNγ (Tsung et al. 2002). Macrophages of the M2 phenotype on the other hand are often found during later phases of acute inflammation and in chronic inflammatory states including arthritis, psoriasis and cancer (Djemadji-Oudjiel et al. 1996; Sica et al. 2006). In these situations, M2 macrophages produce angiogenic growth factors, cytokines and proteases (Mantovani et al. 2002) thereby providing a diverse armament of pro-tumour factors that positively regulate cancer development.

M1 and M2 macrophages differ in terms of their cell surface receptor expression, cytokine and chemokine expression profiles, ability to regulate immune responses and cytotoxic activities (Mantovani et al. 2002, 2007). A major property of activated macrophages is to modulate cellular responses via secretion of cytokines and chemokines. In general, M1 macrophages produce pro-inflammatory cytokines while M2 macrophages produce anti-inflammatory mediators and cytokines. For example, M2 macrophages secrete CCL24 (eotaxin-2), a CC chemokine with potent chemotactic activity towards Th2 cells (Watanabe et al. 2002). M1 macrophages in contrast, produce TNFα and IL-12, pro-inflammatory cytokines favouring Th1 responses (Trinchieri 1995). In addition, administration of Th2 cytokines IL-4 and IL-13 to macrophage cultures promote production of CCL22 (Macrophage-Derived Chemokine, MDC) by macrophages, whereas the Th1 cytokine IFNγ inhibits the IL-4 and IL-13 mediated induction of CCL22 (Bonecchi et al. 1998). Since Th2 lymphocytes express the CCL22 receptor CCR4, CCL22 production by M2 macrophages preferentially attracts Th2 cells, thus creating an amplification loop of polarized Th2 responses (Bonecchi et al. 1998). Consistent with these *in vitro* studies, monocytes isolated from ascites

of patients with ovarian carcinoma demonstrate an M2 phenotype as characterized by production of IL-10 and transforming growth factor beta (TGFβ), thereby inhibiting autologous Th1 responses (Loercher et al. 1999). Thus, Th2-associated alternative macrophage activation results in a macrophage M2 phenotype characterized by reduced pro-inflammatory cytokine secretion, while Th1-associated classical macrophage activation leads to an M1 phenotype characterized by secretion of pro-inflammatory cytokines.

Consequences of inducing the M2 macrophage phenotype. It is generally accepted that tumour infiltrating macrophages are regulated by both tumour- and immune cell-derived factors favouring acquisition of polarized M2 phenotypes. Why is this phenotype advantageous to a developing tumour? TAMs that fail to elicit anti-tumour activity towards neoplastic cells, instead promote tumour growth and progression by providing a diverse assortment of proangiogenic and pro-growth soluble factors (Mantovani et al. 2002; Sica et al. 2006). For example, co-culture of macrophages with colon carcinoma cells results in increased production of anti-inflammatory cytokines (IL-10) and failure to produce pro-inflammatory cytokines (TNFα) upon LPS stimulation, suggesting a tumour-induced shift from M1 to the M2 phenotype (Kambayashi et al. 1995). Likewise, *in vitro* exposure of human blood monocytes to human tumour cell lines of epithelial and nonepithelial origin results in selective reduced production of pro-inflammatory cytokines (TNFα and IL-12) by monocytes (Mytar et al. 2003), whereas monocytes isolated from ascites of patients with ovarian carcinoma produced anti-inflammatory cytokines (IL-10 and TGFβ) (Loercher et al. 1999), suggesting that tumour cells modulate cytokine production by macrophages and regulate acquisition of the M2 phenotype. In an experimental plasmacytoma and fibrosarcoma transplantation model, it was reported that treatment of tumour-bearing mice with cyclophosphamide switched the cytokine production of TAMs from IL-10 to IFNγ that coincided with tumour destruction (Ibe et al. 2001). This study suggests the possibility for reversion of the M2 to the M1 phenotype *in vivo* – a change with profound consequences for tumour progression.

Besides regulating changes in cytokine production, other alterations in the functional capacity of M2 macrophages have been observed including reduced nitric oxide (NO) production as a consequence of reduced expression of the inducible nitric oxide sythase (iNOS) gene (Dinapoli et al. 1996; Mills et al. 1992; Sotomayor et al. 1995). NO is an important effector molecule possessing potent tumouricidal activity that is produced by macrophages (and other immune cells) in response to LPS and IFNγ (Xie et al. 1992; Lorsbach et al. 1993). Decreased NO production by TAMs correlates with severe reduction in tumouricidal activity (Dinapoli et al. 1996); thus, tumour-induced diminished NO production by TAMs is a mechanism by which tumours can escape cytolytic attack by macrophages.

Importantly, several studies have demonstrated that M2 macrophages in tumours are not terminal phenotypes (Kambayashi et al. 1995; Maeda et al. 1995; Sica et al. 2000). For example, co-culture of macrophages with colon carcinoma cells in the presence of neutralizing anti-IL-10 antibodies resulted in reversal of TNFα production, a cytokine that is produced by M1 macrophages but not by M2 macrophages (Kambayashi et al. 1995). Likewise, treatment of tumour bearing mice with anti- TGFβ and anti-IL-10 antibodies restored TNFα production by macrophages present in malignant ascites (Maeda et al. 1995), indicating that TGFβ and IL-10 levels in tumour-bearing mice affect TNFα production by macrophages, suggesting a shift from the M1 to M2 phenotype that is reversible. Together, these data suggest that in the presence of malignant cells, macrophages acquire an anti-inflammatory cytokine profile and defective tumouricidal activity that is consistent with an M2 phenotype and as a consequence, macrophages present in tumour microenvironments potentiate neoplastic progression. Importantly, the M2 phenotype is reversible, and understanding the mechanisms by which the M2 phenotype is reversible to M1 phenotypes may provide therapeutic opportunities for novel anti-cancer agents.

Differentiation and activation of macrophages under hypoxic conditions. Not only are macrophages specifically recruited towards areas of hypoxia, but also their activation and differentiation patterns are severely changed under these adverse conditions (Murdoch and Lewis 2005; Lewis and Pollard 2006). Many macrophage characteristics and activities including

morphology, expression of cell surface markers, survival, phagocytosis, metabolic activity and cytokine secretion are altered under hypoxic conditions, resulting in survival and functionality of macrophages (Lewis et al. 1999). For example, macrophages adapt their metabolic activity by producing adenosine triphosphate (ATP) via anaerobic glycolytic pathways favouring survival under low oxygen conditions (Butterick et al. 1981) and thereby undergo morphological changes such as loss of mitochondria, cell flattening and retraction of lamellipodia (Cazin et al. 1990; Angermuller et al. 1995). Likewise, expression levels of several surface markers and secretion of certain soluble molecules (e.g. TNFα) are increased in human monocytes by hypoxia (Lewis and Pollard 2006; Ghezzi et al. 1991; Scannell et al. 1993; Murdoch and Lewis 2005). Since TNFα is a potent regulator of inflammation, upregulation of TNFα and its receptor on macrophages in hypoxic areas regulates macrophage response during carcinogenesis. Besides changes in morphology, metabolism, expression of cell surface receptors and cytokine secretion, macrophage survival is also altered under low oxygen conditions. Although a significant number of macrophages undergo apoptosis under the adverse conditions of hypoxia, a selective population of macrophages adapts to hypoxia by becoming resistant to apoptosis (Yun et al. 1997; Tsukamoto et al. 1996; Guida and Stewart 1998). The exact mechanisms promoting survival of macrophages under low oxygen levels remain to be determined.

Hypoxic conditions induce expression of diverse transcription factors in macrophages that also potentially contribute to neoplastic progression. For example, hypoxic conditions can induce hypoxia-inducible factor (*HIF*)-*1* expression in alveolar macrophages (Yu et al. 1998). HIF-1 is a basic helix–loop–helix transcription factor that induces the expression of proteins involved in tissue responses to hypoxia such as the angiogenic factors VEGF and erythropoietin (Harris 2002). In agreement with this, macrophages release several proangiogenic cytokines including VEGF, bFGF and TNFα in response to hypoxia (Ghezzi et al. 1991; Scannell et al. 1993; Guida and Stewart 1998; Harmey et al. 1998; Lewis et al. 2000; Lewis and Pollard 2006; Murdoch and Lewis 2005). In combination with the observation that the number of macrophages positively correlates with angiogenesis in many human tumours, recruitment

and/or retention of macrophages in hypoxic areas may specifically stimulate the pro-angiogenic capacity of macrophages. Thus, while macrophages require oxygen for phagocytosis, efficient microbicidal and digestive activity (Leeper-Woodford and Mills 1992; te Koppele et al. 1991; Angermuller et al. 1995; Babior 1984) and demonstrate an M1 biased cytokine repertoire when exhibiting anti-tumour activities, under low oxygen levels, these versatile cells adapt M2 phenotypes characterized by changes in metabolism, morphology, cytokine production and cell surface expression, resulting in impaired tumour cell killing and promotion of tumour growth.

Pro- and Anti-Tumour TAM Function

Tumours consist of neoplastic cells as well as a diverse array of activated responding host cells, including immune cells, a major component of which are macrophages (Coussens and Werb 2001; Balkwill and Mantovani 2001; Ishigami et al. 2003; Noguchi et al. 2003; Li et al. 2002; Hamada et al. 2002; Funada et al. 2003). Accumulating clinical data suggest that their presence influences tumour development by both pro- and anti-tumour mechanisms (Mantovani 1994; Mantovani et al. 1992; Bingle et al. 2002; Ohno et al. 2002).

Macrophages and anti-tumour cytotoxicity. In vitro studies indicate that classically activated M1 macrophages kill tumour cells either via direct or indirect cytotoxicity (Fig. 1), suggesting that cytotoxic activity of M1 macrophages is not limited to states of acute inflammation. Direct cytotoxic activity of M1 macrophages can be mediated by either of two mechanisms: macrophage-mediated tumour cytotoxicity (MTC) or antibody-dependent cellular cytotoxicity (ADCC). MTC is a slow cell-to-cell contact-dependent process involving cell surface adhesion molecules, such as ICAM-1 (Webb et al. 1991) that depends on the cell cycle state of the target cell (Horn et al. 1991). ADCC on the other hand, depends on the presence of antibodies bound to foreign or aberrant proteins on tumour cell surfaces recognized by Fc receptors expressed on macrophages (Shaw et al. 1978; Kawase et al. 1985). Cell killing mechanisms of both pathways are similar and include release of cytotoxic factors such as TNFα (Urban et al. 1986), serine proteases and

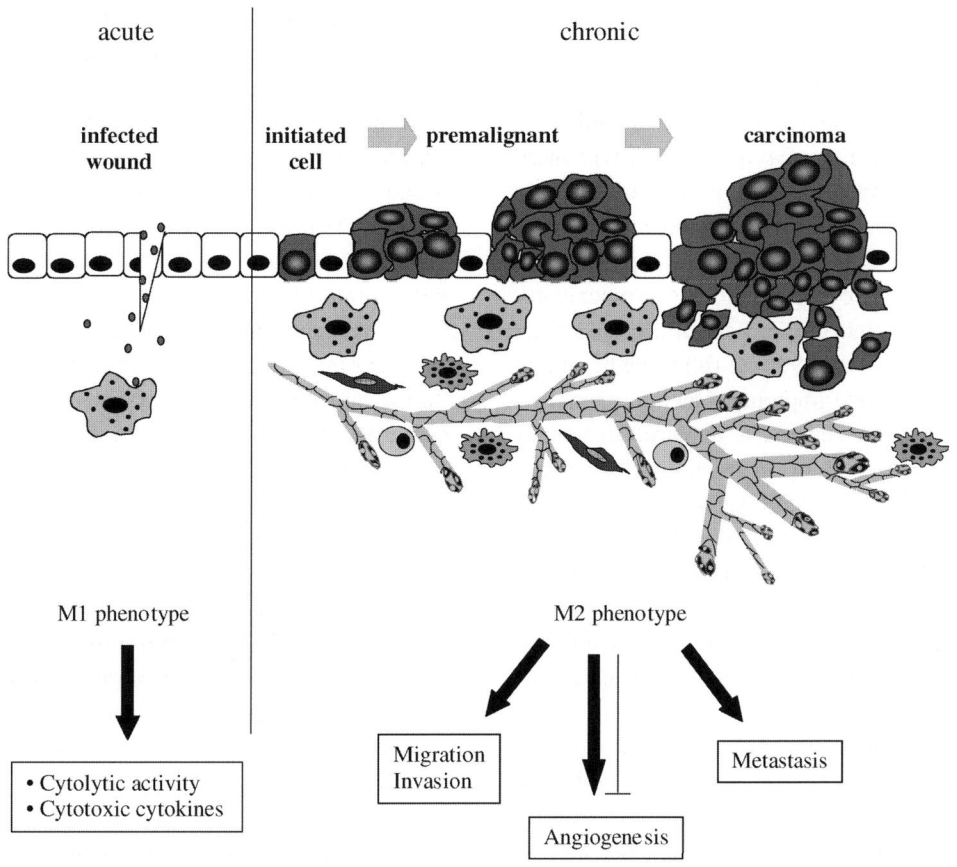

Fig. 1 Effects of M1 and M2 macrophage phenotypes. Macrophages of the M1 phenotype exhibit cytolytic activity and release cytotoxic cytokines upon acute inflammation. In neoplastic microenvironments, macrophages (TAMs) acquire M2 phenotypes that regulate induction of angiogenic programs, tissue remodeling via release of proteases, pro-migratory environments and scenarios favoring metastatic processes. M2 TAMs can also exhibit some anti-tumour functions; however, their pro-tumour effects typically dominate and result in an overall tumour promoting effect

ROS (Hibbs et al. 1988; Keller et al. 1990). In contrast to direct cytotoxicity, indirect cytotoxicity of activated M1 macrophages is mediated by secretion of different factors, e.g. NO, IL-12, TNFα and reactive oxygen intermediates (ROI) (Johnson et al. 1986; Klassen and Sagone 1980; Pullyblank et al. 1995). For example, IFNγ induces release of reactive oxygen and nitrogen species by macrophages (Young and Hardy 1995) thought to contribute to macrophage-mediated cell lysis (Hibbs et al. 1988; Keller et al. 1990). Macrophages express inducible nitric oxide synthase (iNOS) to generate NO, which, upon release, contributes to tumouricidal activity (Keller et al. 1990). In addition, M1 macrophages produce IL-12 (Trinchieri 1998), that has anti-tumourigenic potential by inducing production of other cytokines, particularly IFNγ

and IL-2, proliferation and cytotoxic activity in NK cells and T cells (Grohmann et al. 2001; Munder et al. 1998; Ryffel 1997). Importantly, macrophages present in tumour microenvironments often obtain the alternatively activated M2 phenotype, resulting in defective cytotoxic, and thus defective tumouricidal activity (Fig. 1).

TAMs, in contrast to classically activated M1 macrophages, are not able to kill other cell types via production of ROIs (Martin and Edwards 1993; Siegert et al. 1999). Co-culture of tumour cells with macrophages triggered arginase induction that resulted in reduction of NO production and suppressed macrophage-mediated cytotoxicity towards breast carcinoma cells (Chang et al. 2001). Furthermore, *in vitro* exposure of human blood monocytes to various human tumour cell lines

(pancreatic adenocarcinoma HPC-4, colorectal adeno-carcinoma DeTa and lung carcinoma A549) resulted in diminished cytotoxic activity towards tumour cells monocytes were previously exposed to, as well as to the other tumour cell lines (Mytar et al. 2003). These *in vitro* results are supported by *in vivo* observations in mammary carcinoma tumour-bearing mice where mac-rophages located within tumours exhibit a pronounced suppression of NO production and anti-tumour cyto-toxic activity compared to more distally located mac-rophages (Dinapoli et al. 1996). Taken together, these studies suggest that one important function of macro-phages, e.g. cytotoxic activity, is impaired in tumour microenvironments resulting in promotion of tumour development (Fig. 1). Thus, understanding the under-lying mechanisms involved in reactivating cytolytic activity of TAMs may represent a powerful therapeutic opportunity.

Cytokines have also been implicated in the tumour-icidal activity of TAMs. When macrophages obtain the M2 phenotype in a tumour microenvironment their cytokine production differs from the cytokine profile observed in M1 macrophages, resulting in promotion of tumour progression (Mantovani et al. 2002). The M2 phenotype is reversible, since admin-istration of proinflammatory cytokines IL-12 and IL-15 results in increased cytotoxicity and subsequent delay or complete inhibition of tumour growth (Lasek et al. 2003). These studies suggest that expression or administration of cytokines typically secreted by M1 macrophages results in anti-tumour effects on tumours implying that shifting cytokine patterns in TAMs from the M2 phenotype to the M1 phenotype could be beneficial for therapeutic purposes.

Macrophages and angiogenesis. The process of ang-iogenesis describes the development of new blood vessels to meet the metabolic needs of the growing tumour (Folkman and Shing 1992; Weidner et al. 1992; Folkman 1994, 1995; Folkman and D'Amore 1996). Experimental studies have revealed that one of the mechanisms by which TAMs contribute to cancer development is through activation of the angiogenic switch (Lin et al. 2006; Lin and Pollard 2007). M2 macrophages promote angiogenesis via secretion of pro-angiogenic factors (Lwaleed et al. 2001; Torisu et al. 2000; Sunderkotter et al. 1994; Barbera-Guillem et al. 2002; Vrana et al. 1996; Dupuy et al. 2003;

Zhang et al. 1994). One pro-angiogenic factor secreted by macrophages is VEGF. During carcinogenesis, VEGF is expressed and secreted by both neoplastic cells and TAMs (Leek et al. 1994). *In vitro* coculture experiments suggested that malignant cells promote VEGF secretion by macrophages (Barbera-Guillem et al. 2002) and *in vivo* studies have demonstrated that angiogenesis correlates with infiltration of macro-phages into growing tumour and that VEGF secreted by TAMs represents an essential support for tumour angiogenesis and growth (Barbera-Guillem et al. 2002). Several clinical and experimental studies have suggested that TAMs preferentially accumulate in hypoxic regions (Lewis et al. 1999) where VEGF expression and secretion are induced (Harris 2002). Altogether these studies suggest that TAMs represent essential mediators of tumour angiogenesis by modu-lating VEGF. Recruitment of TAMs to hypoxic areas due to high VEGF levels and subsequent expression of VEGF by TAMs results in further TAM recruitment, thereby promoting angiogenesis and tumour growth.

A recent report identified a subset of Tie-2 express-ing monocytes as important inducers of angiogenesis in both orthotopic and spontaneous tumour models. Since knock-out of the angiopoietin receptor Tie-2-expressing cells markedly reduced angiogenesis in xenograph tumours and prompted substantial tumour regression, these data suggest that Tie-2 expressing monocytes/macrophages might account for a large portion of the pro-angiogenic activity in tumours (De Palma et al. 2005). These data suggest that Tie2-expressing monocytes account for most of the proan-giogenic activity of myeloid cells in tumours.

Tissue factor (TF) is a membrane bound pro-tein expressed by endothelial cells and monocytes (Kappel et al. 1999) involved in blood coagulation following various immune responses under physi-ological conditions (Edwards and Rickles 1992). During carcinogenic progression however, TF has been implicated in tumour development and tumour angiogenesis (Lwaleed et al. 2001; Vrana et al. 1996; Dupuy et al. 2003; Zhang et al. 1994). TF potenti-ates angiogenesis by inducing thrombin formation and fibrin generation (Chen et al. 2001) and thus stimulates migration of endothelial cells to form new tubes and vessels (Senger et al. 1996; Dvorak et al. 1992). Coculture experiments have suggested that TF expression by macrophages increases upon

contact with breast carcinoma cells (Lwaleed et al. 2001) suggesting that TF expression by macrophages might be increased in the tumour microenvironment *in vivo*. Studies utilizing a hepatocellular carcinoma mouse model support these *in vitro* data and further reveal that tumoural angiogenesis is associated with TF expression by macrophages and endothelial cells (Dupuy et al. 2003). Furthermore, the findings that TF-positive tumours express more VEGF than TF-negative tumours (Zhang et al. 1994) suggest that TF might regulate VEGF expression. Macrophage-derived TF might thus play a role in impacting angiogenesis by regulating VEGF expression. Taken together, an important mechanism by which TAMs impact tumour progression is activation of angiogenesis (Fig. 1) by differentially regulating expression of proangiogenic factors such as VEGF and TF.

Macrophages and metastasis formation. Metastases form when malignant cells exit primary tumour sites and spread distally where they grow into secondary tumours (Woodhouse et al. 1997; Yokota 2000; Fidler 2001, 2002, 2003; Comoglio and Trusolino 2002; Jussila and Alitalo 2002; Engers and Gabbert 2000). The underlying mechanisms regulating tumour cell dissemination and metastasis formation are to date poorly understood; however, accumulating data suggested TAMs involvement in the process (Fig. 1). Perhaps most compelling has been the study by Lin et al. (2001) where CSF-1 was implicated in macrophage recruitment and metastasis formation as discussed previously. This study revealed that the absence of CSF-1 resulted in failure to recruit mature macrophages into neoplastic tissue and subsequently significantly attenuated development of late-stage invasive carcinoma and its metastatic pulmonary derivatives (Lin et al. 2001). The failure to recruit macrophages and the subsequent attenuated development of pulmonary metastases was restored by transgenic CSF-1 expression in mammary epithelium in PyMT/*Csf1*$^{op/op}$ mice (Lin et al. 2001). Furthermore, intravital imaging of murine mammary carcinoma models revealed that macrophages affect metastasis in at least two ways: first, the motility of tumour cells away from the primary tumour mass always occurs in juxtaposition of macrophages, suggesting that macrophages attract tumour cells and second, extravasation of tumour cells from the blood stream into the tissue often occurs

at clusters of macrophages on abluminal surfaces of angiogenic vessels. These cell movements critically require CSF-1 and EGF signaling in the macrophages and tumour cells respectively, with the ligand being produced by the reciprocal cell type (Goswami et al. 2005; Wyckoff et al. 2004). In accordance with these observations using mouse models, a correspondence between the number of macrophages in the tumour stroma and the metastatic potential of the tumour was observed in human endometrial cancer (Ohno et al. 2004). Oosterling et al. (2005) found a link between macrophages in metastatic sites and the growth of metastatic tumours. When macrophages were selectively depleted in the peritoneal cavity or liver, CC531 colon tumour cell lines injected into either the peritoneum or the liver portal vein formed tumours more slowly compared to control mice. Taken together these data demonstrate involvement of CSF-1 and EGF in macrophage-mediated effects on metastasis formation. However, it is likely that other macrophage-derived factors in addition to CSF-1 and EGF are also involved in metastasis formation. The underlying mechanisms of how macrophages contribute to tumour cell dissemination and outgrowth of metastases remain to be elucidated. A further identification of macrophage-derived mediators involved in regulating invasion and metastasis and their functional contribution and biological relevance may contribute to the development of new anti-cancer therapeutics.

Involvement of matrix metalloproteinases in pro-and anti-tumour effects of TAMs. Matrix metalloproteinases (MMPs) represent a large family of proteolytic enzymes that play key roles in cancer progression by promoting ECM remodelling, proliferation, angiogenesis and tumour formation (Stamenkovic 2000; Egeblad and Werb 2002; Lafleur et al. 2003). MMPs regulate tumour development by remodelling ECM components as well as non-ECM substrates such as cytokines, growth factors, cell–cell and cell–matrix adhesion molecules (Egeblad and Werb 2002). Various members of the MMP family are expressed by stromal cells as well as by malignant cells in a variety of tumour contexts (Egeblad and Werb 2002). *In vitro* studies have demonstrated that M1 macrophages secrete MMPs upon activation by various pro-inflammatory cytokines (Zhou et al. 2003). For example, monocytes that are exposed *in vitro* to either TNFα

128 Macrophages in tumour development and metastasis

or granulocyte/macrophage colony-stimulating factor (G/M-CSF) express higher MMP-9 levels, while a combinatorial treatment with both G/M-CSF and IFNγ results in upregulation of MMP-1 production (Zhou et al. 2003). These studies suggest that macrophages express and secrete a different repertoire of MMPs depending on the cytokines they are exposed to. Tumour associated M2 macrophages also express MMPs. For example, TAMs express MMP-12 that has the capacity to generate the potent angiogenesis inhibitor angiostatin by cleavage of plasminogen (O'Reilly et al. 1994). Angiostatin inhibits tumour angiogenesis by blocking proliferation of endothelial cells, thus affecting both primary and metastatic tumour growth (O'Reilly et al. 1994; Gorrin-Rivas et al. 2000; Dong et al. 1997). These finding suggest that M2 macrophage-derived MMP-12 might promote neoplastic progression by promoting angiogenesis. Similar to MMP-12, two other members of the MMP family, namely MMP-7 and MMP-9, also generate angiostatin (Patterson and Sang 1997). Complementary to *in vitro* studies, TAMs have been found to be major sources of MMP-9 in invasive tumours such as breast (Davies et al. 1993a), bladder (Davies et al. 1993b) skin (Coussens et al. 2000) and ovarian carcinomas (Naylor et al. 1994). The spatial and temporal expression of MMP-9 by inflammatory cells including macrophages has been investigated utilizing a transgenic mouse model of skin carcinogenesis in the absence of MMP-9 and in chimeric mice harbouring bone marrow derived cells from MMP-9 proficient donor mice (Coussens et al. 2000). Tumour incidence and tumour growth were reduced in the absence of MMP-9, but were restored upon bone marrow transplantation of MMP-9 proficient cells (Coussens et al. 2000) suggesting that MMP-9 is sufficiently expressed by inflammatory cells at different stages of tumour progression. Macrophages expressed MMP-9 at the tumour stage rather than the premalignant stages in this model (Coussens et al. 2000) suggesting that macrophage-derived MMP-9 plays a critical role in tumour growth and invasion rather than in premalignant progression. It has been shown that macrophage derived MMP-9 is crucial for angiogenesis in a mouse model of human cervical carcinogenesis, thus promoting tumour development (Giraudo et al. 2004).

Taken together, *in vitro* and *in vivo* studies suggest that TAM-derived MMPs play a dual role in tumour

progression: On one hand, macrophage-derived MMP-9 exerts pro-tumour effects by remodeling ECM components, thus facilitating cell survival, migration and invasion, while *in vitro* studies on the other hand suggest that macrophage-derived MMPs are involved in the generation of angiostatin, thereby exerting anti-tumour functions. Most likely, in a developing tumour the pro-tumour functions of macrophages are favoured as opposed to anti-tumour effects, resulting in an overall promotion of tumour development.

Macrophages as targets for anti-cancer therapy

In order to exploit macrophages for anti-cancer purposes, it will be crucial to develop diagnostic tools to assess whether macrophages exhibit pro- or anti-tumour properties in a specific cancer type. In human cancers that contain M2 macrophages with pro-tumour activities therapeutic strategies should focus on blocking recruitment of additional pro-tumour macrophages or selective removal of macrophages from the tumour microenvironment. In addition, it may be desirable to inhibit the pro-tumour activation status of macrophages resident in tumours or to convert their pro- into anti-tumour activation status.

Strategies to block recruitment of macrophages. Neoplastic and activated stromal cells present in tumour microenvironments secrete many growth factors and cytokines that induce recruitment of macrophages. Chemokines in particular have been reported to play a central role in recruiting macrophages towards neoplastic cells (Mantovani 1994; Balkwill and Mantovani 2001; Payne and Cornelius 2002; Silzle et al. 2003). Neutralizing the effects of macrophage chemoattractants has been proposed as an approach to inhibit influx of macrophages into tumours, thus preventing macrophages from exhibiting pro-tumour activity. One key chemokine involved in macrophage recruitment towards tumours is CCL2 (Silzle et al. 2003; Monti et al. 2003; Ueno et al. 2000; Ohta et al. 2003). The feasibility of blocking tumour growth by neutralization of CCL2 has been demonstrated in a study where mice bearing CCL2 expressing melanomas were treated with neutralizing antibodies against CCL2 that resulted in inhibition of macrophage recruitment and importantly, abrogated

 Springer

tumour growth (Nesbit et al. 2001). In contrast, another study reported that neutralization of CCL2 resulted in tumour promotion (Zhang et al. 1997). One possible explanation for this disparity is based on differences in CCL2 expression and subsequent differences in the number of infiltrating macrophages. It has been reported that experimental melanomas secreting high levels of CCL2 attracted extensive numbers of macrophages resulting in suppressed tumour growth, whereas low CCL2 expression levels result in modest macrophage infiltration and subsequently in promotion of tumour growth (Nesbit et al. 2001). Thus, depending on chemokine expression levels, approaches aimed at neutralizing chemokine activity may be efficacious and result in inhibited tumour-cell mediated macrophage recruitment and subsequently inhibition of tumour growth.

Another potential therapeutic opportunity would be to block macrophage recruitment towards or into the neoplastic microenvironment by inhibiting production of chemoattractant mediators produced by neoplastic cells. As neoplastic cells are an important source of macrophage-attracting chemokines, therapies aimed at blocking secretion of chemokines may inhibit macrophage recruitment and subsequently prevent macrophages from contributing to neoplastic progression. The feasibility of blocking secretion of macrophage chemoattractants by neoplastic cells has been reported *in vitro* where exposure of human head and neck squamous cell carcinoma cells with retinoic acid (RA) decreased their secretion of macrophage chemotactic factors, e.g. CCL2 and TGFβ. As a result, tumour cells were unable to chemotactically attract peripheral blood monocytes *in vitro* (Liss et al. 2002). Moreover, due to decreased TGFβ production by RA-treated tumour cells, macrophages failed to produce the pro-angiogenic factors VEGF and IL-8 (Liss et al. 2002). RA also directly inhibited migration capacity of macrophages and their ability to induce corneal neovascularization *in vivo* (Liss et al. 2002). Another strategy to target macrophage recruitment is via local modulation of HIF-1α expression levels in tumours, as HIF-1α has been reported as being a critical regulator of myeloid cell infiltration (Sica et al. 2006; Cramer et al. 2003). Thus, future anti-cancer therapies could modulate macrophage recruitment either by neutralization of macrophage-chemoattractants or by ablating chemokine production by neoplastic cells.

Strategies to block the pro-tumour activation status of macrophages Macrophages frequently obtain an M2 phenotype with pro-tumour activation status upon infiltration into the tumour microenvironment (Mantovani et al. 2002). Several studies have shown that this is a reversible process, and therefore might offer tremendous therapeutic advantage. The effectiveness of blocking the pro-tumour activation status of macrophages has been underscored by several experimental observations. One mechanism by which macrophages and other inflammatory cells contribute to tumourigenesis is in a paracrine cyclooxygenase-dependent manner. For example, macrophage-derived cyclooxygenase-2 (COX-2) present in human colorectal adenomas promote malignant behaviour (Ko et al. 2002; Sonoshita et al. 2001; Seno et al. 2002; Kamate et al. 2002). *In vitro* exposure to a selective COX-2 inhibitor suppresses macrophage-epithelial cell signalling during early stages of intestinal tumourigenesis (Ko et al. 2002). Likewise, when mice bearing P815 mastocytoma tumours containing massive macrophage infiltration were treated with COX-1 and -2 inhibitors, a reduced size and growth rate of tumours was observed (Kamate et al. 2002). These *in vitro* and *in vivo* studies underscore the feasibility of inhibiting macrophage-derived mediators as potential anti-cancer modalities. Moreover, long-term usage of anti-inflammatory drugs, including aspirin and non-steroidal anti-inflammatory drugs (NSAIDs) significantly reduces the risk of diverse human cancers, including colorectal, breast and gastric cancer (Williams et al. 1999; Garcia-Rodriguez and Huerta-Alvarez 2001; Meier et al. 2002; Sharpe et al. 2000; Cotterchio et al. 2001; Akre et al. 2001; Gonzalez-Perez et al. 2003) indicating that inhibition of macrophage-derived factors can modulate tumour outgrowth.

Once a tumour is established, it is, however, very likely that inhibition of macrophage-derived mediators alone is not sufficient to obtain tumour regression. In addition to inhibiting the pro-tumour functions of macrophages, it may be desirable to alter their pro-tumour into anti-tumour activities. One approach would be to modulate cytokine profiles present in tumour microenvironments such that macrophages exhibit anti-tumour functions. For example, human colon carcinoma cells expressing G/M-CSF display increased recruitment of tumouricidal macrophages, and were subsequently found to be highly

sensitive to lysis by TAMs *in vitro* (Shinohara et al. 2000). In addition, vaccination with irradiated autologous melanoma cells engineered to secrete GM-CSF resulted in increased anti-tumour immunity in patients with metastatic melanoma (Soiffer et al. 2003). These studies suggest that ectopic expression and subsequent secretion of G/M-CSF results in tumour cell killing by TAMs, whose M2 phenotype had been shifted favouring the M1 phenotype. An alternative approach would be to vary cytokine profiles in tumour microenvironments by transferring *in vitro* modified macrophages into tumour-bearing hosts. For instance, it has been reported that transduction of macrophages with IL-12 results in increased surface expression of major histocompatability complex (MHC) class I and II and F4/80 antigens (macrophage-specific antigens), indicating an M1 phenotype differentiation (Satoh et al. 2003). Injection of IL-12 transduced macrophages into prostate tumour-bearing mice resulted in increased migration of macrophages into draining lymph nodes, increased infiltration of CD4[+] and CD8[+] T lymphocytes and an increased survival (Satoh et al. 2003). Thus, induction of a pro-inflammatory profile in macrophages results in their ability to promote a successful anti-tumour immune response. However, adoptive transfer therapy requires long culture periods, is very labour intensive and brings the risk of introducing unwanted pathogens into the host. For clinical use, it will be desirable to develop approaches that induce conversion of pro-tumour M2 phenotypes into anti-tumour M1 phenotypes. Treatment of tumour-bearing mice with cyclophosphamide results in a switch in cytokine production of TAMs from IL-10 to IFNγ (Ibe et al. 2001). This switch from M2 to M1 cytokine profile coincides with destruction of tumour vasculature and tumour necrosis (Ibe et al. 2001). Likewise, combined administration of IL-15 injections and intratumoural injections with IL-12 overexpressing melanoma cells eradicated established experimental melanoma tumours (Lasek et al. 2003). Importantly, administration of IL-15 and IL-12 did not only result in local anti-tumour effects, but also induced systemic anti-tumour immunity, delaying or inhibiting outgrowth of metastases (Lasek et al. 2003). Subsequent *in vitro* studies revealed a shift from a Th2 to Th1 response, which resulted in an increase in IFNγ production as well as a general shift in the cytokine expression pattern in macrophages towards a M1 phenotype (Lasek et al. 2003), thus enhancing anti-tumour functions of macrophages.

An alternative approach is to selectively remove macrophages from the tumour microenvironment. Proof-of-principle of such a drastic strategy was recently demonstrated by Luo et al. (2006) who employed a DNA vaccination strategy to elicit a CD8[+] T cell response against legumain, a protease overexpressed in TAMs. Vaccination of mice was sufficient to reduce the density of TAMs in tumour tissue and resulted in suppression of tumour angiogenesis and growth.

Macrophages have the unique ability to infiltrate and accumulate into hypoxic regions of tumours (Lewis et al. 1999). Strategies aimed at using this functional behaviour of macrophages to deliver drugs to hypoxic tumour areas are currently under investigation (Griffiths et al. 2000). The concept of using macrophages as a cell-based delivery system for hypoxia regulated genes into tumours has been studied (Carta et al. 2001; Burke et al. 2002). *In vitro* studies have shown that infiltration of macrophages expressing the cytochrome P450 enzyme into three-dimensional hypoxic multicellular aggregates (spheroids) of breast cancer cells resulted in conversion of the prodrug cyclophosphamide into its active metabolite inducing an enhanced cytotoxic effect of cyclophosphamide on the spheroids (Egeblad and Werb 2002), suggesting that the unique characteristic of macrophages to accumulate in hypoxic areas can be used to specifically target therapeutics to tumours without affecting normoxic tissues. The possibility of specifically targeting drugs to hypoxic regions would not only be a potential therapy for cancer, but also for inflammatory and cardiovascular disease in which hypoxia is a commonly observed phenomenon. In conclusion, multiple anti-cancer approaches focussed on inhibiting recruitment of M2 macrophages and/or in blocking their anti-tumour activity, in combination with approaches focussed on converting M2 into M1 macrophages or drug-delivery macrophages are under investigation and might in the future complement already existing anti-cancer therapies.

Concluding remarks

Macrophages are versatile cells that play dual roles during tumour progression. Recruitment to and/ or accumulation of macrophages in tumour micro-

environments are induced by various macrophage-chemoattracting mediators secreted by neoplastic cells and activated stromal cells present in the neoplastic environment. However, differentiation of macrophages in close proximity to neoplastic cells differs greatly from that observed in tissues undergoing acute inflammation. In the tumour microenvironment, macrophages differentiate into alternative M2 phenotypes, where they, unlike classically activated M1 macrophages, exhibit pro-tumour functions. Pro-tumour M2 phenotypes can be converted into anti-tumour M1 phenotypes under certain conditions, a conversion that may offer possibilities for development of successful future clinical anti-cancer therapies.

Acknowledgements We thank Evelyn Galenski for administrative assistance. The authors acknowledge support from the Serono Foundation for the Advancement of Medical Science (AE) the Dutch Cancer Society (KEdV) and grants from the National Institutes of Health, Sandler Program in Basic Sciences, National Technology Center for Networks and Pathways and a Department of Defense Era of Hope Scholar Award (LMC).

References

Akre K, Ekstrom AM, Signorello LB, Hansson LE, Nyren O (2001) Aspirin and risk for gastric cancer: a population-based case-control study in Sweden. Br J Cancer 84:965–968

Almand B, Clark JI, Nikitina E, van Beynen J, English NR, Knight SC, Carbone DP, Gabrilovich DI (2001) Increased production of immature myeloid cells in cancer patients: a mechanism of immunosuppression in cancer. J Immunol 166:678–689

Almand B, Resser JR, Lindman B, Nadaf S, Clark JI, Kwon ED, Carbone DP, Gabrilovich DI (2000) Clinical significance of defective dendritic cell differentiation in cancer. Clin Cancer Res 6:1755–1766

Andrade SP, Hart IR, Piper PJ (1992) Inhibitors of nitric oxide synthase selectively reduce flow in tumor-associated neovasculature. Br J Pharmacol 107:1092–1095

Angermuller S, Schunk M, Kusterer K (1995) Alteration of xanthine oxidase activity in sinusoidal endothelial cells and morphological changes of Kupffer cells in hypoxic and reoxygenated rat liver. Hepatology 21:1594–1601

Azenshtein E, Luboshits G, Shina S, Neumark E, Shahbazian D, Weil M, Wigler N, Keydar I, Ben-Baruch A (2002) The CC chemokine RANTES in breast carcinoma progression: regulation of expression and potential mechanisms of promalignant activity. Cancer Res 62:1093–1102

Babior BM (1984) The respiratory burst of phagocytes. J Clin Invest 73:599–601

Baggiolini M, Loetscher P (2000) Chemokines in inflammation and immunity. Immunol Today 21:418–420

Balkwill F, Mantovani A (2001) Inflammation and cancer: back to Virchow? Lancet 357:539–545

Barbera-Guillem E, Nyhus JK, Wolford CC, Friece CR, Sampsel JW (2002) Vascular endothelial growth factor secretion by tumor-infiltrating macrophages essentially supports tumor angiogenesis, and igg immune complexes potentiate the process. Cancer Res 62:7042–7049

Bingle L, Brown NJ, Lewis CE (2002) The role of tumour-associated macrophages in tumour progression: implications for new anticancer therapies. J Pathol 196:254–265

Bissell MJ, Radisky D (2001) Putting tumours in context. Nat Rev Cancer 1:46–54

Blot E, Chen W, Vasse M, Paysant J, Denoyelle C, Pille JY, Vincent L, Vannier JP, Soria J, Soria C (2003) Cooperation between monocytes and breast cancer cells promotes factors involved in cancer aggressiveness. Br J Cancer 88:1207–1212

Bonecchi R, Sozzani S, Stine JT, Luini W, D'Amico G, Allavena P, Chantry D, Mantovani A (1998) Divergent effects of interleukin-4 and interferon-gamma on macrophage-derived chemokine production: an amplification circuit of polarized T helper 2 responses. Blood 92:2668–2671

Boring L, Gosling J, Chensue SW, Kunkel SL, Farese Jr., RV, Broxmeyer HE, Charo IF (1997) Impaired monocyte migration and reduced type 1 (Th1) cytokine responses in C-C chemokine receptor 2 knockout mice. J Clin Invest 100:2552–2561

Bosco MC, Reffo G, Puppo M, Varesio L (2004) Hypoxia inhibits the expression of the CCR5 chemokine receptor in macrophages. Cell Immunol 228:1–7

Burke B, Sumner S, Maitland N, Lewis CE (2002) Macrophages in gene therapy: cellular delivery vehicles and in vivo targets. J Leukoc Biol 72:417–428

Butterick CJ, Williams DA, Boxer LA, Jersild RA, Jr., Mantich N, Higgins C, Baehner RL (1981) Changes in energy metabolism, structure and function in alveolar macrophages under anaerobic conditions. Br J Haematol 48:523–532

Cardiff RD (2001) Validity of mouse mammary tumour models for human breast cancer: comparative pathology. Microsc Res Tech 52:224–320

Cardiff RD, Anver MR, Gusterson BA, Hennighausen L, Jensen RA, Merino MJ, Rehm S, Russo J, Tavassoli FA, Wakefield LM, Ward JM, Green JE (2000) The mammary pathology of genetically engineered mice: the consensus report and recommendations from the Annapolis meeting. Oncogene 19:968–988

Carta L, Pastorino S, Melillo G, Bosco MC, Massazza S, Varesio L (2001) Engineering of macrophages to produce IFN-gamma in response to hypoxia. J Immunol 166:5374–5380

Cazin M, Paluszezak D, Bianchi A, Cazin JC, Aerts C, Voisin C (1990) Effects of anaerobiosis upon morphology and energy metabolism of alveolar macrophages cultured in gas phase. Eur Respir J 3:1015–1022

Ceradini DJ, Kulkarni AR, Callaghan MJ, Tepper OM, Bastidas N, Kleinman ME, Capla JM, Galiano RD, Levine JP, Gurtner GC (2004) Progenitor cell trafficking is regulated by hypoxic gradients through HIF-1 induction of SDF-1. Nat Med 10:858–864

Chae P, Im M, Gibson F, Jiang Y, Graves DT (2002) Mice lacking monocyte chemoattractant protein 1 have enhanced susceptibility to an interstitial polymicrobial infection due to impaired monocyte recruitment. Infect Immun 70:3164–3169

Chang CI, Liao JC, Kuo L (2001) Macrophage arginase promotes tumor cell growth and suppresses nitric oxide-mediated tumor cytotoxicity. Cancer Res 61:1100–1106

Chang YC, Hsu TL, Lin HH, Chio CC, Chiu AW, Chen NJ, Lin CH, Hsieh SL (2003) Modulation of macrophage differentiation and activation by decoy receptor 3. J Leukoc Biol Dec 4 [Epub ahead of print]

Chen J, Bierhaus A, Schiekofer S, Andrassy M, Chen B, Stern DM, Nawroth PP (2001) Tissue factor–a receptor involved in the control of cellular properties, including angiogenesis. Thromb Haemost 86:334–345

Comoglio PM, Trusolino L (2002) Invasive growth: from development to metastasis. J Clin Invest 109:857–862

Cotterchio M, Kreiger N, Sloan M, Steingart A (2001) Nonsteroidal anti-inflammatory drug use and breast cancer risk. Cancer Epidemiol Biomarkers Prev 10:1213–1217

Coussens LM, Tinkle CL, Hanahan D, Werb Z (2000) MMP-9 supplied by bone marrow-derived cells contributes to skin carcinogenesis. Cell 103:481–490

Coussens LM, Werb Z (2001) Inflammatory cells and cancer: think different!. J Exp Med 193:F23–F26

Cramer T, Yamanishi Y, Clausen BE, Forster I, Pawlinski R, Mackman N, Haase VH, Jaenisch R, Corr M, Nizet V, Firestein GS, Gerber HP, Ferrara N, Johnson RS (2003) HIF-1alpha is essential for myeloid cell-mediated inflammation. Cell 112:645–657

Crowther M, Brown NJ, Bishop ET, Lewis CE (2001) Microenvironmental influence on macrophage regulation of angiogenesis in wounds and malignant tumors. J Leukoc Biol 70:478–490

Davies B, Miles DW, Happerfield LC, Naylor MS, Bobrow LG, Rubens RD, Balkwill FR (1993a) Activity of type IV collagenases in benign and malignant breast disease. Br J Cancer 67:1126–1131

Davies B, Waxman J, Wasan H, Abel P, Williams G, Krausz T, Neal D, Thomas D, Hanby A, Balkwill F (1993b) Levels of matrix metalloproteases in bladder cancer correlate with tumor grade and invasion. Cancer Res 53:5365–5369

De Palma M, Venneri MA, Galli R, Sergi Sergi L, Politi LS, Sampaolesi M, Naldini L (2005) Tie2 identifies a hematopoietic lineage of proangiogenic monocytes required for tumor vessel formation and a mesenchymal population of pericyte progenitors. Cancer Cell 8:211–226

Dinapoli MR, Calderon CL, Lopez DM (1996) The altered tumoricidal capacity of macrophages isolated from tumor-bearing mice is related to reduce expression of the inducible nitric oxide synthase gene. J Exp Med 183:1323–1329

Djemadji-Oudjiel N, Goerdt S, Kodelja V, Schmuth M, Orfanos CE (1996) Immunohistochemical identification of type II alternatively activated dendritic macrophages (RM 3/1 + 3, MS-1 + /-, 25F9-) in psoriatic dermis. Arch Dermatol Res 288:757–764

Dong Z, Kumar R, Yang X, Fidler IJ (1997) Macrophage-derived metalloelastase is responsible for the generation of angiostatin in Lewis lung carcinoma. Cell 88:801–810

Dupuy E, Hainaud P, Villemain A, Bodevin-Phedre E, Brouland JP, Briand P, Tobelem G (2003) Tumoral angiogenesis and tissue factor expression during hepatocellular carcinoma progression in a transgenic mouse model. J Hepatol 38:793–802

Duyndam MC, Hilhorst MC, Schluper HM, Verheul HM, van Diest PJ, Kraal G, Pinedo HM, Boven E (2002) Vascular endothelial growth factor-165 overexpression stimulates angiogenesis and induces cyst formation and macrophage infiltration in human ovarian cancer xenografts. Am J Pathol 160:537–548

Dvorak HF, Nagy JA, Berse B, Brown LF, Yeo KT, Yeo TK, Dvorak AM, van de Water L, Sioussat TM, Senger DR (1992) Vascular permeability factor, fibrin, and the pathogenesis of tumor stroma formation. Ann N Y Acad Sci 667:101–111

Edwards RL, Rickles FR (1992) The role of leukocytes in the activation of blood coagulation. Semin Hematol 29:202–212

Egeblad M, Werb Z (2002) New functions for the matrix metalloproteinases in cancer progression. Nat Rev Cancer 2:161–174

Engers R, Gabbert HE (2000) Mechanisms of tumor metastasis: cell biological aspects and clinical implications. J Cancer Res Clin Oncol 126:682–692

Fajardo LF, Kwan HH, Kowalski J, Prionas SD, Allison AC (1992) Dual role of tumor necrosis factor-alpha in angiogenesis. Am J Pathol 140:539–544

Fidler IJ (2001) Seed and soil revisited: contribution of the organ microenvironment to cancer metastasis. Surg Oncol Clin N Am 10:257–269

Fidler IJ (2002) Critical determinants of metastasis. Semin Cancer Biol 12:89–96

Fidler IJ (2003) Timeline: the pathogenesis of cancer metastasis: the 'seed and soil' hypothesis revisited. Nat Rev Cancer 3:453–458

Folkman J (1994) Tumor angiogenesis. Nat Med 1:206–232

Folkman J (1995) Tumor angiogenesis. In: Mendelsohn J, Howley PM, Israel MA, Liotta LA (eds) The molecular basis of cancer, Vol. 9. W. B. Saunders Company, Philadelphia, pp. 206–232

Folkman J, D'Amore PA (1996) Blood vessel formation: what is its molecular basis. Cell 87:1153–1155

Folkman J, Shing Y (1992) Angiogenesis. J Biol Chem 267:10931–10934

Funada Y, Noguchi T, Kikuchi R, Takeno S, Uchida Y, Gabbert HE (2003) Prognostic significance of CD8 + T cell and macrophage peritumoral infiltration in colorectal cancer. Oncol Rep 10:309–313

Gabrilovich DI, Chen HL, Girgis KR, Cunningham HT, Meny GM, Nadaf S, Kavanaugh D, Carbone DP (1996) Production of vascular endothelial growth factor by human tumors inhibits the functional maturation of dendritic cells. Nat Med 2:1096–1103

Gabrilovich DI, Velders MP, Sotomayor EM, Kast WM (2001) Mechanism of immune dysfunction in cancer mediated by immature Gr-1 + myeloid cells. J Immunol 166:5398–5406

Garcia-Rodriguez LA, Huerta-Alvarez C (2001) Reduced risk of colorectal cancer among long-term users of aspirin and nonaspirin nonsteroidal antiinflammatory drugs. Epidemiology 12:88–93

Ghezzi P, Dinarello CA, Bianchi M, Rosandich ME, Repine JE, White CW (1991) Hypoxia increases production of interleukin-1 and tumor necrosis factor by human mononuclear cells. Cytokine 3:189–194

Giraudo E, Inoue M, Hanahan D (2004) An amino-bisphosphonate targets MMP-9-expressing macrophages and angiogenesis to impair cervical carcinogenesis. J Clin Invest 114:623–633

Gonzalez-Perez A, Rodriguez L, Lopez-Ridaura R (2003) Effects of non-steroidal anti-inflammatory drugs on cancer sites other than the colon and rectum: a meta-analysis. BMC Cancer 3:1–12

Gordon S (2003) Alternative activation of macrophages. Nat Rev Immunol 3:23–35

Gorrin-Rivas MJ, Arii S, Mori A, Takeda Y, Mizumoto M, Furutani M, Imamura M (2000) Implications of human macrophage metalloelastase and vascular endothelial growth factor gene expression in angiogenesis of hepatocellular carcinoma. Ann Surg 231:67–73

Goswami S, Sahai E, Wyckoff JB, Cammer M, Cox D, Pixley FJ, Stanley ER, Segall JE, Condeelis JS (2005) Macrophages promote the invasion of breast carcinoma cells via a colony-stimulating factor-1/epidermal growth factor paracrine loop. Cancer Res 65:5278–5283

Griffiths L, Binley K, Iqball S, Kan O, Maxwell P, Ratcliffe P, Lewis C, Harris A, Kingsman S, Naylor S (2000) The macrophage – a novel system to deliver gene therapy to pathological hypoxia. Gene Ther 7:255–262

Grohmann U, Belladonna ML, Vacca C, Bianchi R, Fallarino F, Orabona C, Fioretti MC, Puccetti P (2001) Positive regulatory role of IL-12 in macrophages and modulation by IFN-gamma. J Immunol 167:221–227

Guida E, Stewart A (1998) Influence of hypoxia and glucose deprivation on tumour necrosis factor-alpha and granulocyte-macrophage colony-stimulating factor expression in human cultured monocytes. Cell Physiol Biochem 8:75–88

Gyetko MR, Todd 3rd, RF, Wilkinson CC, Sitrin RG (1994) The urokinase receptor is required for human monocyte chemotaxis in vitro. J Clin Invest 93:1380–1387

Hamada I, Kato M, Yamasaki T, Iwabuchi K, Watanabe T, Yamada T, Itoyama S, Ito H, Okada K (2002) Clinical effects of tumor-associated macrophages and dendritic cells on renal cell carcinoma. Anticancer Res 22:4281–4284

Hanahan D, Weinberg RA (2000) The hallmarks of cancer. Cell 100:57–70

Harmey JH, Dimitriadis E, Kay E, Redmond HP, Bouchier-Hayes D (1998) Regulation of macrophage production of vascular endothelial growth factor (VEGF) by hypoxia and transforming growth factor beta-1. Ann Surg Oncol 5:271–278

Harris AL (2002) Hypoxia–a key regulatory factor in tumour growth. Nat Rev Cancer 2:38–47

Hibbs Jr., JB, Taintor RR, Vavrin Z, Rachlin EM 1988 Nitric oxide: a cytotoxic activated macrophage effector molecule. Biochem Biophys Res Commun 157:87–94

Homey B, Muller A, Zlotnik A (2002) Chemokines: agents for the immunotherapy of cancer? Nat Rev Immunol 2:175–184

Horn D, van der Bosch J, Ruller S, Schlaak M (1991) Suppression of tumor cell susceptibility to monocyte-induced cell death by growth-inhibitory signals generated during monocyte/tumor cell interaction. J Cell Biochem 45:213–223

Ibe S, Qin Z, Schuler T, Preiss S, Blankenstein T (2001) Tumor rejection by disturbing tumor stroma cell interactions. J Exp Med 194:1549–1559

Ishigami S, Natsugoe S, Tokuda K, Nakajo A, Okumura H, Matsumoto M, Miyazono F, Hokita S, Aikou T (2003) Tumor-associated macrophage (TAM) infiltration in gastric cancer. Anticancer Res 23:4079–4083

Janeway CA, Travers P, Walport M, Shlomchik M (2001) Immunobiology, 5th edn. Garland Publishing, New York and London

Jenkins DC, Charles IG, Thomsen LL, Moss DW, Holmes LS, Baylis SA, Rhodes P, Westmore K, Emson PC, Moncada S (1995) Roles of nitric oxide in tumor growth. Proc Natl Acad Sci USA 92:4392–4396

Johnson WJ, Steplewski Z, Matthews TJ, Hamilton TA, Koprowski H, Adams DO (1986) Cytolytic interactions between murine macrophages, tumor cells, and monoclonal antibodies: characterization of lytic conditions and requirements for effector activation. J Immunol 136:4704–4713

Jussila L, Alitalo K (2002) Vascular growth factors and lymphangiogenesis. Physiol Rev 82:673–700

Kacinski BM (1995) CSF-1 and its receptor in ovarian, endometrial and breast cancer. Ann Med 27:79–85

Kacinski BM (1997) CSF-1 and its receptor in breast carcinomas and neoplasms of the female reproductive tract. Mol Reprod Dev 46:71–74

Kamate C, Baloul S, Grootenboer S, Pessis E, Chevrot A, Tulliez M, Marchiol C, Viguier M, Fradelizi D (2002) Inflammation and cancer, the mastocytoma P815 tumor model revisited: triggering of macrophage activation in vivo with pro-tumorigenic consequences. Int J Cancer 100:571–579

Kambayashi T, Alexander HR, Fong M, Strassmann G (1995) Potential involvement of IL-10 in suppressing tumor-associated macrophages. Colon-26-derived prostaglandin E2 inhibits TNF-alpha release via a mechanism involving IL-10. J Immunol 154:3383–3390

Kappel A, Ronicke V, Damert A, Flamme I, Risau W, Breier G (1999) Identification of vascular endothelial growth factor (VEGF) receptor-2 (Flk-1) promoter/enhancer sequences sufficient for angioblast and endothelial cell-specific transcription in transgenic mice. Blood 93:4284–4292

Kawase I, Komuta K, Ogura T, Fujiwara H, Hamaoka T, Kishimoto S (1985) Murine tumor cell lysis by antibody-dependent macrophage-mediated cytotoxicity using syngeneic monoclonal antibodies. Cancer Res 45:1663–1668

Keller R, Geiges M, Keist R (1990) L-arginine-dependent reactive nitrogen intermediates as mediators of tumor cell killing by activated macrophages. Cancer Res 50:1421–1425

Klassen DK, Sagone Jr., AL (1980) Evidence for both oxygen and non-oxygen dependent mechanisms of antibody sensitized target cell lysis by human monocytes. Blood 56:985–992

Ko SC, Chapple KS, Hawcroft G, Coletta PL, Markham AF, Hull MA (2002) Paracrine cyclooxygenase-2-mediated signalling by macrophages promotes tumorigenic progression of intestinal epithelial cells. Oncogene 21:7175–7186

Kusmartsev S, Gabrilovich DI (2003) Inhibition of myeloid cell differentiation in cancer: the role of reactive oxygen species. J Leukoc Biol 74:186–196

Lafleur MA, Handsley MM, Edwards DR (2003) Metalloproteinases and their inhibitors in angiogenesis. Expert Rev Mol Med 5:1–39

Lasek W, Basak G, Switaj T, Jakubowska AB, Wysocki PJ, Mackiewicz A, Drela N, Jalili A, Kaminski R, Kozar K, Jakobisiak M (2003) Complete tumour regressions induced by vaccination with IL-12 gene-transduced tumour cells in combination with IL-15 in a melanoma model in mice. Cancer Immunol Immunother 53(4):363–372, Epub 2003 Nov 7

Leek RD, Harris AL, Lewis CE (1994) Cytokine networks in solid human tumors: regulation of angiogenesis. J Leukoc Biol 56:423–435

Leek RD, Hunt NC, Landers RJ, Lewis CE, Royds JA, Harris AL (2000) Macrophage infiltration is associated with VEGF and EGFR expression in breast cancer. J Pathol 190:430–436

Leek RD, Landers RJ, Harris AL, Lewis CE (1999) Necrosis correlates with high vascular density and focal macrophage infiltration in invasive carcinoma of the breast. Br J Cancer 79:991–995

Leeper-Woodford SK, Mills JW (1992) Phagocytosis and ATP levels in alveolar macrophages during acute hypoxia. Am J Respir Cell Mol Biol 6:326–334

Leibovich SJ, Polverini PJ, Shepard HM, Wiseman DM, Shively V, Nuseir N (1987) Macrophage-induced angiogenesis is mediated by tumour necrosis factor-alpha. Nature 329:630–632

Leung SY, Wong MP, Chung LP, Chan AS, Yuen ST (1997) Monocyte chemoattractant protein-1 expression and macrophage infiltration in gliomas. Acta Neuropathol (Berl) 93:518–527

Lewis CE, Pollard JW (2006) Distinct role of macrophages in different tumor microenvironments. Cancer Res 66:605–612

Lewis JS, Landers RJ, Underwood JC, Harris AL, Lewis CE (2000) Expression of vascular endothelial growth factor by macrophages is up-regulated in poorly vascularized areas of breast carcinomas. J Pathol 192:150–158

Lewis JS, Lee JA, Underwood JC, Harris AL, Lewis CE (1999) Macrophage responses to hypoxia: relevance to disease mechanisms. J Leukoc Biol 66:889–900

Li C, Shintani S, Terakado N, Nakashiro K, Hamakawa H (2002) Infiltration of tumor-associated macrophages in human oral squamous cell carcinoma. Oncol Rep 9:1219–1223

Lin EY, Gouon-Evans V, Nguyen AV, Pollard JW (2002) The macrophage growth factor CSF-1 in mammary gland development and tumor progression. J Mammary Gland Biol Neoplasia 7:147–162

Lin EY, Li JF, Gnatovskiy L, Deng Y, Zhu L, Grzesik DA, Qian H, Xue XN, Pollard JW (2006) Macrophages regulate the angiogenic switch in a mouse model of breast cancer. Cancer Res 66:11238–11246

Lin EY, Nguyen AV, Russell RG, Pollard JW (2001) Colony-stimulating factor 1 promotes progression of mammary tumors to malignancy. J Exp Med 193:727–740

Lin EY, Pollard JW (2007) Tumor-associated macrophages press the angiogenic switch in breast cancer. Cancer Res 67:5064–5066

Liss C, Fekete MJ, Hasina R, Lingen MW (2002) Retinoic acid modulates the ability of macrophages to participate in the induction of the angiogenic phenotype in head and neck squamous cell carcinoma. Int J Cancer 100:283–289

Loercher AE, Nash MA, Kavanagh JJ, Platsoucas CD, Freedman RS (1999) Identification of an IL-10-producing HLA-DR-negative monocyte subset in the malignant ascites of patients with ovarian carcinoma that inhibits cytokine protein expression and proliferation of autologous T cells. J Immunol 163:6251–6260

Lorsbach RB, Murphy WJ, Lowenstein CJ, Snyder SH, Russell SW (1993) Expression of the nitric oxide synthase gene in mouse macrophages activated for tumor cell killing. Molecular basis for the synergy between interferon-gamma and lipopolysaccharide. J Biol Chem 268:1908–1913

Lu B, Rutledge BJ, Gu L, Fiorillo J, Lukacs NW, Kunkel SL, North R, Gerard C, Rollins BJ (1998) Abnormalities in monocyte recruitment and cytokine expression in monocyte chemoattractant protein 1-deficient mice. J Exp Med 187:601–608

Luboshits G, Shina S, Kaplan O, Engelberg S, Nass D, Lifshitz-Mercer B, Chaitchik S, Keydar I, Ben-Baruch A (1999) Elevated expression of the CC chemokine regulated on activation, normal T cell expressed and secreted (RANTES) in advanced breast carcinoma. Cancer Res 59:4681–4687

Luo Y, Zhou H, Krueger J, Kaplan C, Lee SH, Dolman C, Markowitz D, Wu W, Liu C, Reisfeld RA, Xiang R (2006) Targeting tumor-associated macrophages as a novel strategy against breast cancer. J Clin Invest 116:2132–2141

Lwaleed BA, Bass PS, Cooper AJ (2001) The biology and tumour-related properties of monocyte tissue factor. J Pathol 193:3–12

Maeda H, Kuwahara H, Ichimura Y, Ohtsuki M, Kurakata S, Shiraishi A (1995) TGF-beta enhances macrophage ability to produce IL-10 in normal and tumor-bearing mice. J Immunol 155:4926–4932

Maity A, Solomon D (2000) Both increased stability and transcription contribute to the induction of the urokinase plasminogen activator receptor (upar) message by hypoxia. Exp Cell Res 255:250–257

Mantovani A (1994) Tumor-associated macrophages in neoplastic progression: a paradigm for the in vivo function of chemokines. Lab Invest 71:5–16

Mantovani A, Bottazzi B, Colotta F, Sozzani S, Ruco L (1992) The origin and function of tumor-associated macrophages. Immunol Today 13:265–270

Mantovani A, Sica A, Locati M (2007) New vistas on macrophage differentiation and activation. Eur J Immunol 37:14–16

Mantovani A, Sozzani S, Locati M, Allavena P, Sica A (2002) Macrophage polarization: tumor-associated macrophages as a paradigm for polarized M2 mononuclear phagocytes. Trends Immunol 23:549–555

Martin JH, Edwards SW (1993) Changes in mechanisms of monocyte/macrophage-mediated cytotoxicity during culture. Reactive oxygen intermediates are involved in monocyte-mediated cytotoxicity, whereas reactive nitrogen intermediates are employed by macrophages in tumor cell killing. J Immunol 150:3478–3486

Meier CR, Schmitz S, Jick H (2002) Association between acetaminophen or nonsteroidal antiinflammatory drugs and risk of developing ovarian, breast, or colon cancer. Pharmacotherapy 22:303–309

Menetrier-Caux C, Montmain G, Dieu MC, Bain C, Favrot MC, Caux C, Blay JY (1998) Inhibition of the differentiation of dendritic cells from CD34(+) progenitors by tumor cells: role of interleukin-6 and macrophage colony-stimulating factor. Blood 92:4778–4791

Meyer-Siegler K, Hudson PB (1996) Enhanced expression of macrophage migration inhibitory factor in prostatic adenocarcinoma metastases. Urology 48:448–452

Mills CD, Shearer J, Evans R, Caldwell MD (1992) Macrophage arginine metabolism and the inhibition or stimulation of cancer. J Immunol 149:2709–2714

Monti P, Leone BE, Marchesi F, Balzano G, Zerbi A, Scaltrini F, Pasquali C, Calori G, Pessi F, Sperti C, Di Carlo V, Allavena P, Piemonti L (2003) The CC chemokine MCP-1/CCL2 in pancreatic cancer progression: regulation of

expression and potential mechanisms of antimalignant activity. Cancer Res 63:7451–7461

Mrowietz U, Schwenk U, Maune S, Bartels J, Kupper M, Fichtner I, Schroder JM, Schadendorf D (1999) The chemokine RANTES is secreted by human melanoma cells and is associated with enhanced tumour formation in nude mice. Br J Cancer 79:1025–1031

Munder M, Mallo M, Eichmann K, Modolell M (1998) Murine macrophages secrete interferon gamma upon combined stimulation with interleukin (IL)-12 and IL-18: a novel pathway of autocrine macrophage activation. J Exp Med 187:2103–2108

Murdoch C, Lewis CE (2005) Macrophage migration and gene expression in response to tumor hypoxia. Int J Cancer 117:701–708

Murphy PM (1996) Chemokine receptors: structure, function and role in microbial pathogenesis. Cytokine Growth Factor Rev 7:47–64

Mytar B, Woloszyn M, Szatanek R, Baj-Krzyworzeka M, Siedlar M, Ruggiero I, Wieckiewicz J, Zembala M 2003 Tumor cell-induced deactivation of human monocytes. J Leukoc Biol 74:1094–1101

Naylor MS, Stamp GW, Davies BD, Balkwill FR (1994) Expression and activity of MMPS and their regulators in ovarian cancer. Int J Cancer 58:50–56

Negus RP, Stamp GW, Hadley J, Balkwill FR (1997) Quantitative assessment of the leukocyte infiltrate in ovarian cancer and its relationship to the expression of C-C chemokines. Am J Pathol 150:1723–1734

Negus RP, Stamp GW, Relf MG, Burke F, Malik ST, Bernasconi S, Allavena P, Sozzani S, Mantovani A, Balkwill FR (1995) The detection and localization of monocyte chemoattractant protein-1 (MCP-1) in human ovarian cancer. J Clin Invest 95:2391–2396

Negus RP, Turner L, Burke F, Balkwill FR (1998) Hypoxia down-regulates MCP-1 expression: implications for macrophage distribution in tumors. J Leukoc Biol 63:758–765

Nesbit M, Schaider H, Miller TH, Herlyn M (2001) Low-level monocyte chemoattractant protein-1 stimulation of monocytes leads to tumor formation in nontumorigenic melanoma cells. J Immunol 166:6483–6490

Noguchi T, Wada S, Takeno S, Moriyama H, Kimura Y, Uchida Y (2003) Lymph node metastasis could be predicted by evaluation of macrophage infiltration and hsp70 expression in superficial carcinoma of the esophagus. Oncol Rep 10:1161–1164

O'Reilly MS, Holmgren L, Shing Y, Chen C, Rosenthal RA, Moses M, Lane WS, Cao Y, Sage EH, Folkman J (1994) Angiostatin: a novel angiogenesis inhibitor that mediates the suppression of metastases by a Lewis lung carcinoma. Cell 79:315–328

Ohno S, Inagawa H, Soma G, Nagasue N (2002) Role of tumor-associated macrophage in malignant tumors: should the location of the infiltrated macrophages be taken into account during evaluation? Anticancer Res 22:4269–4275

Ohno S, Ohno Y, Suzuki N, Kamei T, Koike K, Inagawa H, Kohchi C, Soma G, Inoue M (2004) Correlation of histological localization of tumor-associated macrophages with clinicopathological features in endometrial cancer. Anticancer Res 24:3335–3342

Ohta M, Kitadai Y, Tanaka S, Yoshihara M, Yasui W, Mukaida N, Haruma K, Chayama K (2003) Monocyte chemoattractant protein-1 expression correlates with macrophage infiltration and tumor vascularity in human gastric carcinomas. Int J Oncol 22:773–778

Onodera S, Suzuki K, Matsuno T, Kaneda K, Takagi M, Nishihira J (1997) Macrophage migration inhibitory factor induces phagocytosis of foreign particles by macrophages in autocrine and paracrine fashion. Immunology 92:131–137

Oosterling SJ, van der Bij GJ, Meijer GA, Tuk CW, van Garderen E, van Rooijen N, Meijer S, van der Sijp JR, Beelen RH, van Egmond M (2005) Macrophages direct tumour histology and clinical outcome in a colon cancer model. J Pathol 207:147–155

Patterson BC, Sang QXA (1997) Angiostatin-converting enzyme activities of human matrilysin (MMP-7) and gelatinase B/type IV collagenase (MMP-9). J Biol Chem 272:28823–28825

Payne AS, Cornelius LA (2002) The role of chemokines in melanoma tumor growth and metastasis. J Invest Dermatol 118:915–922

Poulter LW, Janossy G, Power C, Sreenan S, Burke C (1994) Immunological/physiological relationships in asthma: potential regulation by lung macrophages. Immunol Today 15:258–261

Pozzi LA, Weiser WY (1992) Human recombinant migration inhibitory factor activates human macrophages to kill tumor cells. Cell Immunol 145:372–379

Pullyblank AM, Guillou PJ, Monson JR (1995) Interleukin 1 and tumour necrosis factor alpha may be responsible for the lytic mechanism during anti-tumour antibody-dependent cell-mediated cytotoxicity. Br J Cancer 72:601–606

Ren Y, Tsui HT, Poon RT, Ng IO, Li Z, Chen Y, Jiang G, Lau C, Yu WC, Bacher M, Fan ST (2003) Macrophage migration inhibitory factor: roles in regulating tumor cell migration and expression of angiogenic factors in hepatocellular carcinoma. Int J Cancer 107:22–29

Riethdorf L, Riethdorf S, Gutzlaff K, Prall F, Loning T (1996) Differential expression of the monocyte chemoattractant protein-1 gene in human papillomavirus-16-infected squamous intraepithelial lesions and squamous cell carcinomas of the cervix uteri. Am J Pathol 149:1469–1476

Rofstad EK, Rasmussen H, Galappathi K, Mathiesen B, Nilsen K, Graff BA (2002) Hypoxia promotes lymph node metastasis in human melanoma xenografts by up-regulating the urokinase-type plasminogen activator receptor. Cancer Res 62:1847–1853

Roth SJ, Carr MW, Springer TA (1995) C-C chemokines, but not the C-X-C chemokines interleukin-8 and interferon-gamma inducible protein-10, stimulate transendothelial chemotaxis of T lymphocytes. Eur J Immunol 25:3482–3488

Ryffel B (1997) Interleukin-12: role of interferon-gamma in IL-12 adverse effects. Clin Immunol Immunopathol 83:18–20

Sato K, Kuratsu J, Takeshima H, Yoshimura T, Ushio Y (1995) Expression of monocyte chemoattractant protein-1 in meningioma. J Neurosurg 82:874–878

Sato N, Goto T, Haranaka K, Satomi N, Nariuchi H, Mano-Hirano Y, Sawasaki Y (1986) Actions of tumor necrosis factor on cultured vascular endothelial cells: morphologic modulation, growth inhibition, and cytotoxicity. J Natl Cancer Inst 76:1113–1121

Satoh T, Saika T, Ebara S, Kusaka N, Timme TL, Yang G, Wang J, Mouraviev V, Cao G, Fattah el MA, Thompson TC (2003) Macrophages transduced with an adenoviral vector expressing interleukin 12 suppress tumor growth and metastasis in a preclinical metastatic prostate cancer model. Cancer Res 63:7853–7860

Scannell G, Waxman K, Kaml GJ, Ioli G, Gatanaga T, Yamamoto R, Granger GA (1993) Hypoxia induces a human macrophage cell line to release tumor necrosis factor-alpha and its soluble receptors in vitro. J Surg Res 54:281–285

Schioppa T, Uranchimeg B, Saccani A, Biswas SK, Doni A, Rapisarda A, Bernasconi S, Saccani S, Nebuloni M, Vago L, Mantovani A, Melillo G, Sica A (2003) Regulation of the chemokine receptor CXCR4 by hypoxia. J Exp Med 198:1391–1402

Scholl SM, Pallud C, Beuvon F, Hacene K, Stanley ER, Rohrschneider L, Tang R, Pouillart P, Lidereau R (1994) Anti-colony-stimulating factor-1 antibody staining in primary breast adenocarcinomas correlates with marked inflammatory cell infiltrates and prognosis. J Natl Cancer Inst 86:120–126

Senger DR, Ledbetter SR, Claffey KP, Papadopoulos-Sergiou A, Peruzzi CA, Detmar M (1996) Stimulation of endothelial cell migration by vascular permeability factor/vascular endothelial growth factor through cooperative mechanisms involving the alphavbeta3 integrin, osteopontin, and thrombin. Am J Pathol 149:293–305

Seno H, Oshima M, Ishikawa TO, Oshima H, Taku K, Chiba T, Narumiya S, Taketo MM (2002) Cyclooxygenase 2- and prostaglandin E(2) receptor EP(2)-dependent angiogenesis in Apc(Delta716) mouse intestinal polyps. Cancer Res 62:506–511

Serafini P, De Santo C, Marigo I, Cingarlini S, Dolcetti L, Gallina G, Zanovello P, Bronte V (2004) Derangement of immune responses by myeloid suppressor cells. Cancer Immunol Immunother 53:64–72

Sharpe CR, Collet JP, McNutt M, Belzile E, Boivin JF, Hanley JA (2000) Nested case-control study of the effects of non-steroidal anti-inflammatory drugs on breast cancer risk and stage. Br J Cancer 83:112–120

Shaw GM, Levy PC, Lobuglio AF (1978) Human monocyte cytotoxicity to tumor cells. I. Antibody-dependent cytotoxicity. J Immunol 121:573–578

Shimizu T, Abe R, Nakamura H, Ohkawara A, Suzuki M, Nishihira J (1999) High expression of macrophage migration inhibitory factor in human melanoma cells and its role in tumor cell growth and angiogenesis. Biochem Biophys Res Commun 264:751–758

Shinohara H, Yano S, Bucana CD, Fidler IJ (2000) Induction of chemokine secretion and enhancement of contact-dependent macrophage cytotoxicity by engineered expression of granulocyte-macrophage colony-stimulating factor in human colon cancer cells. J Immunol 164:2728–2737

Sica A, Bronte V (2007) Altered macrophage differentiation and immune dysfunction in tumor development. J Clin Invest 117:1155–1166

Sica A, Saccani A, Bottazzi B, Bernasconi S, Allavena P, Gaetano B, Fei F, LaRosa G, Scotton C, Balkwill F, Mantovani A (2000) Defective expression of the monocyte chemotactic protein-1 receptor CCR2 in macrophages associated with human ovarian carcinoma. J Immunol 164:733–738

Sica A, Saccani A, Bottazzi B, Polentarutti N, Vecchi A, van Damme J, Mantovani A (2000) Autocrine production of IL-10 mediates defective IL-12 production and NF-kappa B activation in tumor-associated macrophages. J Immunol 164:762–767

Sica A, Schioppa T, Mantovani A, Allavena P (2006) Tumour-associated macrophages are a distinct M2 polarised population promoting tumour progression: potential targets of anti-cancer therapy. Eur J Cancer 42:717–727

Siegert A, Denkert C, Leclere A, Hauptmann S (1999) Suppression of the reactive oxygen intermediates production of human macrophages by colorectal adenocarcinoma cell lines. Immunology 98:551–556

Silzle T, Kreutz M, Dobler MA, Brockhoff G, Knuechel R, Kunz-Schughart LA (2003) Tumor-associated fibroblasts recruit blood monocytes into tumor tissue. Eur J Immunol 33:1311–1120

Skobe M, Hamberg LM, Hawighorst T, Schirner M, Wolf GL, Alitalo K, Detmar M (2001) Concurrent induction of lymphangiogenesis, angiogenesis, and macrophage recruitment by vascular endothelial growth factor-C in melanoma. Am J Pathol 159:893–903

Soiffer R, Hodi FS, Haluska F, Jung K, Gillessen S, Singer S, Tanabe K, Duda R, Mentzer S, Jaklitsch M, Bueno R, Clift S, Hardy S, Neuberg D, Mulligan R, Webb I, Mihm M, Dranoff G (2003) Vaccination with irradiated, autologous melanoma cells engineered to secrete granulocyte-macrophage colony-stimulating factor by adenoviral-mediated gene transfer augments antitumor immunity in patients with metastatic melanoma. J Clin Oncol 21:3343–3350

Sonoshita M, Takaku K, Sasaki N, Sugimoto Y, Ushikubi F, Narumiya S, Oshima M, Taketo MM (2001) Acceleration of intestinal polyposis through prostaglandin receptor EP2 in Apc(Delta 716) knockout mice. Nat Med 7:1048–1051

Sotomayor EM, DiNapoli MR, Calderon C, Colsky A, Fu YX, Lopez DM (1995) Decreased macrophage-mediated cytotoxicity in mammary-tumor-bearing mice is related to alteration of nitric-oxide production and/or release. Int J Cancer 60:660–667

Sozzani S, Sallusto F, Luini W, Zhou D, Piemonti L, Allavena P, Van Damme J, Valitutti S, Lanzavecchia A, Mantovani A (1995) Migration of dendritic cells in response to formyl peptides, c5a, and a distinct set of chemokines. J Immunol 155:3292–3295

Stamenkovic I (2000) Matrix metalloproteinases in tumor invasion and metastasis. Semin Cancer Biol 10:415–433

Sunderkotter C, Steinbrink K, Goebeler M, Bhardwaj R, Sorg C (1994) Macrophages and angiogenesis. J Leukoc Biol 55:410–422

Takahashi A, Kono K, Ichihara F, Sugai H, Fujii H, Matsumoto Y (2003) Vascular endothelial growth factor inhibits maturation of dendritic cells induced by lipopolysaccharide, but not by proinflammatory cytokines. Cancer Immunol Immunother 53(6):543–550, Epub 2003 Dec 10

Tang R, Beuvon F, Ojeda M, Mosseri V, Pouillart P, Scholl S (1992) M-CSF (monocyte colony stimulating factor) and M-CSF receptor expression by breast tumour cells: M-CSF mediated recruitment of tumour infiltrating monocytes? J Cell Biochem 50:350–356

Tang RP, Kacinski B, Validire P, Beuvon F, Sastre X, Benoit P, dela Rochefordiere A, Mosseri V, Pouillart P, Scholl S (1990) Oncogene amplification correlates with dense lymphocyte infiltration in human breast cancers: a role for hematopoietic growth factor release by tumor cells? J Cell Biochem 44:189–198

te Koppele JM, Keller BJ, Caldwell-Kenkel JC, Lemasters JJ, Thurman RG (1991) Effect of hepatotoxic chemicals and hypoxia on hepatic nonparenchymal cells: impairment of phagocytosis by Kupffer cells and disruption of the endothelium in rat livers perfused with colloidal carbon. Toxicol Appl Pharmacol 110:20–30

Thomsen LL, Miles DW, Happerfield L, Bobrow LG, Knowles RG, Moncada S (1995) Nitric oxide synthase activity in human breast cancer. Br J Cancer 72:41–44

Topoll HH, Zwadlo G, Lange DE, Sorg C (1989) Phenotypic dynamics of macrophage subpopulations during human experimental gingivitis. J Periodontal Res 24:106–112

Torisu H, Ono M, Kiryu H, Furue M, Ohmoto Y, Nakayama J, Nishioka Y, Sone S, Kuwano M (2000) Macrophage infiltration correlates with tumor stage and angiogenesis in human malignant melanoma: possible involvement of tnfalpha and IL- 1alpha. Int J Cancer 85:182–188

Trinchieri G (1995) Interleukin-12: a proinflammatory cytokine with immunoregulatory functions that bridge innate resistance and antigen-specific adaptive immunity. Annu Rev Immunol 13:251–276

Trinchieri G (1998) Interleukin-12: a cytokine at the interface of inflammation and immunity. Adv Immunol 70:83–243

Tsukamoto Y, Kuwabara K, Hirota S, Ikeda J, Stern D, Yanagi H, Matsumoto M, Ogawa S, Kitamura Y (1996) 150-kd oxygen-regulated protein is expressed in human atherosclerotic plaques and allows mononuclear phagocytes to withstand cellular stress on exposure to hypoxia and modified low density lipoprotein. J Clin Invest 98:1930–1941

Tsung K, Dolan JP, Tsung YL, Norton JA (2002) Macrophages as effector cells in interleukin 12-induced T cell-dependent tumor rejection. Cancer Res 62:5069–5075

Turner L, Scotton C, Negus R, Balkwill F (1999) Hypoxia inhibits macrophage migration. Eur J Immunol 29:2280–2287

Ueno T, Toi M, Saji H, Muta M, Bando H, Kuroi K, Koike M, Inadera H, Matsushima K (2000) Significance of macrophage chemoattractant protein-1 in macrophage recruitment, angiogenesis, and survival in human breast cancer. Clin Cancer Res 6:3282–3489

Urban JL, Shepard HM, Rothstein JL, Sugarman BJ, Schreiber H (1986) Tumor necrosis factor: a potent effector molecule for tumor cell killing by activated macrophages. Proc Natl Acad Sci USA 83:5233–5237

von Luettichau I, Nelson PJ, Pattison JM, van de Rijn M, Huie P, Warnke R, Wiedermann CJ, Stahl RA, Sibley RK, Krensky AM (1996) RANTES chemokine expression in diseased and normal human tissues. Cytokine 8:89–98

von Stebut E, Metz M, Milon G, Knop J, Maurer M (2003) Early macrophage influx to sites of cutaneous granuloma formation is dependent on MIP-1alpha /beta released from neutrophils recruited by mast cell-derived tnfalpha. Blood 101:210–215

Vrana JA, Stang MT, Grande JP, Getz MJ (1996) Expression of tissue factor in tumor stroma correlates with progression to invasive human breast cancer: paracrine regulation by carcinoma cell-derived members of the transforming growth factor beta family. Cancer Res 56:5063–5070

Watanabe K, Jose PJ, Rankin SM (2002) Eotaxin-2 generation is differentially regulated by lipopolysaccharide and IL-4 in monocytes and macrophages. J Immunol 168:1911–1918

Webb DS, Mostowski HS, Gerrard TL (1991) Cytokine-induced enhancement of ICAM-1 expression results in increased vulnerability of tumor cells to monocyte-mediated lysis. J Immunol 146:3682–3686

Weidner N, Folkman J, Pozza F, Bevilacqua P, Allred EN, Moore DH, Meli S, Gasparini G (1992) Tumor angiogenesis: a new significant and independent prognostic indicator in early-stage breast carcinoma. J Natl Cancer Inst 84:1875–1887

Williams CS, Mann M, DuBois RN (1999) The role of cyclooxygenases in inflammation, cancer, and development. Oncogene 18:7908–7916

Wong MP, Cheung KN, Yuen ST, Fu KH, Chan AS, Leung SY, Chung LP (1998) Monocyte chemoattractant protein-1 (MCP-1) expression in primary lymphoepithelioma-like carcinomas (lelcs) of the lung. J Pathol 186:372–377

Woodhouse EC, Chuaqui RF, Liotta LA (1997) General mechanisms of metastasis. Cancer 80:1529–1537

Wyckoff J, Wang W, Lin EY, Wang Y, Pixley F, Stanley ER, Graf T, Pollard JW, Segall J, Condeelis J (2004) A paracrine loop between tumor cells and macrophages is required for tumor cell migration in mammary tumors. Cancer Res 64:7022–7029

Xie QW, Cho HJ, Calaycay J, Mumford RA, Swiderek KM, Lee TD, Ding A, Troso T, Nathan C (1992) Cloning and characterization of inducible nitric oxide synthase from mouse macrophages. Science 256:225–228

Yokota J (2000) Tumor progression and metastasis. Carcinogenesis 21:497–503

Yoshimura T, Yuhki N, Moore SK, Appella E, Lerman MI, Leonard EJ (1989) Human monocyte chemoattractant protein-1 (MCP-1). Full-length cdna cloning, expression in mitogen-stimulated blood mononuclear leukocytes, and sequence similarity to mouse competence gene JE. FEBS Lett 244:487–493

Young HA, Hardy KJ (1995) Role of interferon-gamma in immune cell regulation. J Leukoc Biol 58:373–381

Yu AY, Frid MG, Shimoda LA, Wiener CM, Stenmark K, Semenza GL (1998) Temporal, spatial, and oxygen-regulated expression of hypoxia-inducible factor-1 in the lung. Am J Physiol 275:L818–L826

Yun JK, McCormick TS, Villabona C, Judware RR, Espinosa MB, Lapetina EG (1997) Inflammatory mediators are perpetuated in macrophages resistant to apoptosis induced by hypoxia. Proc Natl Acad Sci USA 94:13903–13908

Zhang L, Yoshimura T, Graves DT (1997) Antibody to Mac-1 or monocyte chemoattractant protein-1 inhibits monocyte recruitment and promotes tumor growth. J Immunol 158:4855–4861

Zhang Y, Deng Y, Luther T, Muller M, Ziegler R, Waldherr R, Stern DM, Nawroth PP (1994) Tissue factor controls the balance of angiogenic and antiangiogenic properties of tumor cells in mice. J Clin Invest 94:1320–1327

Zhou M, Zhang Y, Ardans JA, Wahl LM (2003) Interferon-gamma differentially regulates monocyte matrix metalloproteinase-1 and -9 through tumor necrosis factor-alpha and caspase 8. J Biol Chem 278:45406–45413

Zlotnik A, Yoshie O (2000) Chemokines: a new classification system and their role in immunity. Immunity 12:121–127

H.E. Kaiser and A. Nasir (eds.), Selected Aspects of Cancer Progression:
Metastasis, Apoptosis and Immune Response, 139–156.
© Springer Science + Business Media B.V. 2008

CHAPTER NINE

Classical and alternative activation of macrophages: different pathways of macrophage-mediated tumor promotion

Jo Van Ginderachter, Yuanqing Liu, Nick Devoogdt, Wim Noël, Lea Brys, Gholamreza Hassanzadeh Gh.,
Geert Raes, Anja Geldhof, Alain Beschin, Hilde Revets, and Patrick De Baetselier

Abstract: Macrophages often function as control switches of the immune system, securing the balance between pro- and anti-inflammatory reactions. For this purpose and depending on the activating stimuli, macrophages can develop into different subsets: classically (M1) or alternatively (M2) activated macrophages, the characterization of which is a current topic of investigation. Accumulating evidence suggests that both populations, using their own specific mechanisms, may influence the behaviour of cancer cells, shape the tumor microenvironment and subverte anti-tumor immunity, thereby contributing to tumor growth and progression.

Keywords: Classically activated macrophage (M1), Alternatively activated macrophage (M2), Tumor-associated macrophage, Myeloid-derived suppressor cells, Anti-tumor CTL response, Metastasis

Introduction

It is well established that cancer is a progressive disease, occurring in a series of well-defined steps, typically arising as a consequence of activating mutations (oncogenes) and/or deactivating mutations (tumor suppressor genes) in proliferating cells. In each step of the process – i.e. the primary tumor site, the lymph or blood circulation and in metastatic lesions – cancer cells are confronted with cells of the immune system. The extent to which the immune system is involved in controlling tumors, the "immune surveillance" theory (Burnet FM 1971), has long been a matter of debate. Only with the advent of targeted gene knock-out mice, evidence for the immune surveillance theory could be gathered (Smyth et al. 2001). Indeed, mice lacking functional IFN-γ, IFN-γR, IL-12, TRAIL, perforin, NKT cells, αβ T cells and γδ T cells all display a more rapid and frequent development of certain types of cancer (Kaplan et al. 1998; Girardi et al. 2001; Shankaran et al. 2001; Smyth et al. 2000; Van den Broek et al. 1996).

These findings boosted the belief in the potential power of tumor immunotherapy. Although innate immune effector cells, such as NK cells and macrophages, are endowed with the capacity to kill cancer cells and their metastases, most efforts were directed toward enforcing adaptive anti-tumor responses (Whiteside and Herberman 1995; Fidler 1985). According to current knowledge, adaptive immunity against tumors would require that signals from transformed cells cause activation of the antigen-presenting cells (APCs), in particular dendritic cells (DCs). Examples of such "danger" signals may

Cellular Immunology, Institute for Molecular Biology, Vrije Universiteit Brussel, Vlaams Interuniversitair Instituut voor Biotechnologie

include cytokines (e.g. IFN-α), heat shock proteins or NKG2D-ligands expressed by cells undergoing damage and necrotic death (Diefenbach et al. 2000; Gallucci and Matzinger 2001). Thus, DCs capture tumor-associated antigens that are secreted or shed by tumor cells or after cell lysis (Boon et al. 1997), become activated by concomitant "danger" signals and migrate to a secondary lymphoid organ to stimulate an appropriate T cell response.

Despite this knowledge, cancer immunotherapy trials faced limited success thus far. The main barriers to generating an effective anti-tumor response are dual. Tumors can either turn their genetic instability to advantage in order to evade innate and adaptive immune responses or can directly hamper an anti-tumor attack by establishing a tolerant environment or inducing immunosuppression (Khong and Restifo 2002; Chouaib et al. 1997; Sotomayor et al. 1996; Sakaguchi et al. 2001).

In this chapter, we will provide compelling data that one subset of immune cells, the macrophage, and the dynamic environment in which they live, facilitate malignant outgrowth and metastatic spread of neoplastic cells.

Function of classically (M1) versus alternatively (M2) activated macrophages

Macrophages are of crucial importance for host immune defenses. They are best well known for initiating an effective innate immune response against microbes by recognizing pathogen-associated molecular patterns (PAMPs) through pattern-recognition proteins (PRPs) (Medzhitov and Janeway 2000). Following phagocytosis, macrophages destroy most micro-organisms. In tumors, macrophages represent a major component of the leukocyte infiltrate, affecting diverse aspects of neoplastic tissues including vascularization, growth rate, stroma formation and dissolution (Mantovani et al. 1992). By producing diverse molecules and presenting antigens to T cells, these cells orient the adaptive immune response leading to the expansion and differentiation of lymphocytes specific for invaders or cancer cells (Akira et al. 2001; Belardelli and Ferrantini 2002).

Macrophage heterogeneity is well recognized. It arises as macrophages differentiate from myeloid progenitors, and is determined by the genetic background as well as by specific stimuli (Kuroda et al. 2002; Mills et al. 2000; Akagawa 2002). In this regard, microbial antigens, tumor products, as well as Th1 or Th2 effector T cells and their secretory products influence the heterogeneity and the state of activation of macrophage populations (Elgert et al. 1998; Munder et al. 1998). The better characterized response of macrophages to microbial molecules, cancer cells and host cytokines is the release of inflammatory/microbicidal/tumoricidal products. This "classical activation" profile occurs in a type I cytokine environment (IFN-γ, TNF) or upon recognition of PAMPs (LPS, lipoproteins, dsRNA, lipoteichoic acid, etc.) and endogenous "danger" signals (heat shock proteins, CD40L, etc.). As such, it plays an important role in protection against intracellular pathogens, and under certain conditions also cancer cells (Boehm et al. 1997; MacMicking et al. 1997). Classically activated macrophages (caMφ or M1) exert anti-proliferative and cytotoxic activities, resulting partly from their ability to secrete NO and pro-inflammatory cytokines (TNF, IL-1, IL-6) (Stuehr and Nathan 1989; Urban et al. 1986; Bonnotte et al. 2001). Although such inflammatory activity could be beneficial for the host in a tumor setting, the persistence of inflammatory processes often results in detrimental tissue damage during infections (Mueller 2002; Satoskar et al. 2000). Therefore, in the course of a response, inflammation is usually counteracted through the development of anti-inflammatory mechanisms. Ideally, this regulation must be spatially and temporally controlled.

Based on the observation that the development of M1 is inhibited by type II cytokines (IL-4, IL-13, IL-10) (Montaner et al. 1999; Suk et al. 1993), investigators have initiated the analysis of macrophages developing in a type II cytokine environment. Such macrophages perform a different activation program and were termed "alternatively activated" (aaMφ or M2) (Stein et al. 1992; Goerdt and Orfanos 1999). M2 express similar levels of CD11a, CD40, CD54, CD58, CD80 and CD86 co-stimulatory molecules as M1 (Schebesch et al. 1997). In addition, they exhibit enhanced endo- and phagocytic ability, increased expression of MHC class II molecules and can perform antigen presentation (Schebesch et al. 1997; Namangala et al. 2001). Therefore, M2, as M1, possess all the features required to drive an immune

response. Siamon Gordon (2003) proposed to restrict the definition of M2 to IL-4 and/or IL-13-elicited macrophages. It should be noted however that type II cytokines, PGE$_2$, glucocorticoids (Corraliza et al. 1995) and apoptotic cells (Mills 1991) induce a partly overlapping gene repertoire expression in macrophages, allowing these cells to fulfill specific functions.

Based on the observation that they are antagonistically regulated by type I cytokines, the present postulate is that M2, secreting anti-inflammatory molecules like IL-10 and TGF-β, down-regulate inflammatory processes initiated by M1. Accordingly, M2 exert selective immunosuppressive functions (Schebesch et al. 1997; Loke et al. 2000). Their presence in healthy individuals in the placenta, lungs and immune privileged sites (Mues et al. 1989), as well as in chronic inflammatory diseases like rheumatoid arthritis, atherosclerosis and psoriasis (Djemadji-Oudjiel et al. 1996; Goerdt et al. 1993; Szekanecz et al. 1994), further suggests that M2 protect organs and surrounding tissues against detrimental immune responses. They produce increased levels of factors involved in tissue remodeling (TGF-β1, PDGF-AA, PDGF-BB, and extracellular matrix proteins) and reduced levels of Matrix MetalloProtease-7 (MMP-7). Moreover, they increase the fibrinogenic activity of human fibroblasts (Song et al. 2000), promote angiogenesis (Kodelja et al. 1997) and wound repair during the healing phase of acute inflammatory reactions and in chronic inflammatory diseases (Djemadji-Oudjiel et al. 1996; Goerdt et al. 1993; Szekanecz et al. 1994). M2 also contribute to the induction and maintenance of peripheral tolerance (Flores Villanueva et al. 1994; Stevens et al. 1995).

Figure 1 summarizes the main properties of M1 and M2.

Molecular repertoire of M1 versus M2

The concept of M2 was first introduced in the human system by Goerdt and colleagues, by analyzing the phenotype of peripheral blood monocytes differentially activated *in vitro* in the presence of type I or type II cytokines (Goerdt and Orfanos 1999). M2 were characterized by the enhanced expression of receptors from the innate immune response displaying broad specificity for foreign antigens (i.e. PRPs), such as the macrophage mannose receptor (Stein et al. 1992), the β-glucan receptor (Mosser and Handman 1992) and the scavenger receptor type I (Geng and Hansson 1992). They express enhanced levels of MHC class II molecules like HLA-DR and HLA-DQ (Becker and Daniel 1990), the Fcε immunoglobulin receptor II (CD23) and aminopeptidase-N (CD13), a peptidase potentially involved in inactivating inflammatory mediators (Van Hal et al. 1992). Human M2 could be further characterized by the surface expression of the RM3/1 marker, a glucocorticoid-inducible splice variant of the scavenger receptor CD163 (Mues et al. 1989), and the high molecular weight IL-4/glucocorticoid-inducible protein MS-1 (Goerdt et al. 1991).

Interestingly, these macrophages specifically express the anti-inflammatory cytokine IL-1 receptor antagonist (IL-1Ra) and IL-10 (Fenton et al. 1992; Schebesch et al. 1997), but lack expression of M1-associated pro-inflammatory cytokines such as IL-1, TNF, IL-6 and IL-12 (Bonder et al. 1998; Cheung et al. 1990). In addition, M2 secrete the alternative macrophage activation associated CC-chemokine (AMAC-1) (Kodelja et al. 1998) that displays strong homology to MIP-1α, a chemokine possibly expressed by M1 (Standiford et al. 1993). It was postulated that AMAC-1 may bring together alternatively activated antigen presenting cells and naïve CD4+ T cells, leading to the induction of an anti-inflammatory (type II-based) immune response. It was also shown that human monocytes activated by IL-4 and IL-13 secrete the macrophage-derived chemokine (MDC), leading to preferential recruitment of type II cells through interaction with the CC chemokine receptor 4 (CCR4) (Bonecchi et al. 1998). Human M2 induced by IL-4, IL-10 or PGE$_2$ also express the prototype extracellular matrix proteins fibronectin and βIG-H3 (Gratchev et al. 2001) and secrete pro-fibrinogenic factors such as TGF-β, PDGF-AA and PDGF-BB (Song et al. 2000). 15-lipoxygenase activity, which is induced by IL-4 while inhibited by IFN-γ and hydrocortisone in cultured human monocytes, may represent another marker for M2 (Conrad and Lu 2000).

Studies reporting the enhanced expression of the mannose receptor on IL-4-treated murine macrophages have led to the concept of M2 in mice (Stein et al. 1992). Until recently, the discrimination between murine M1 and M2 was mainly documented at the biochemical

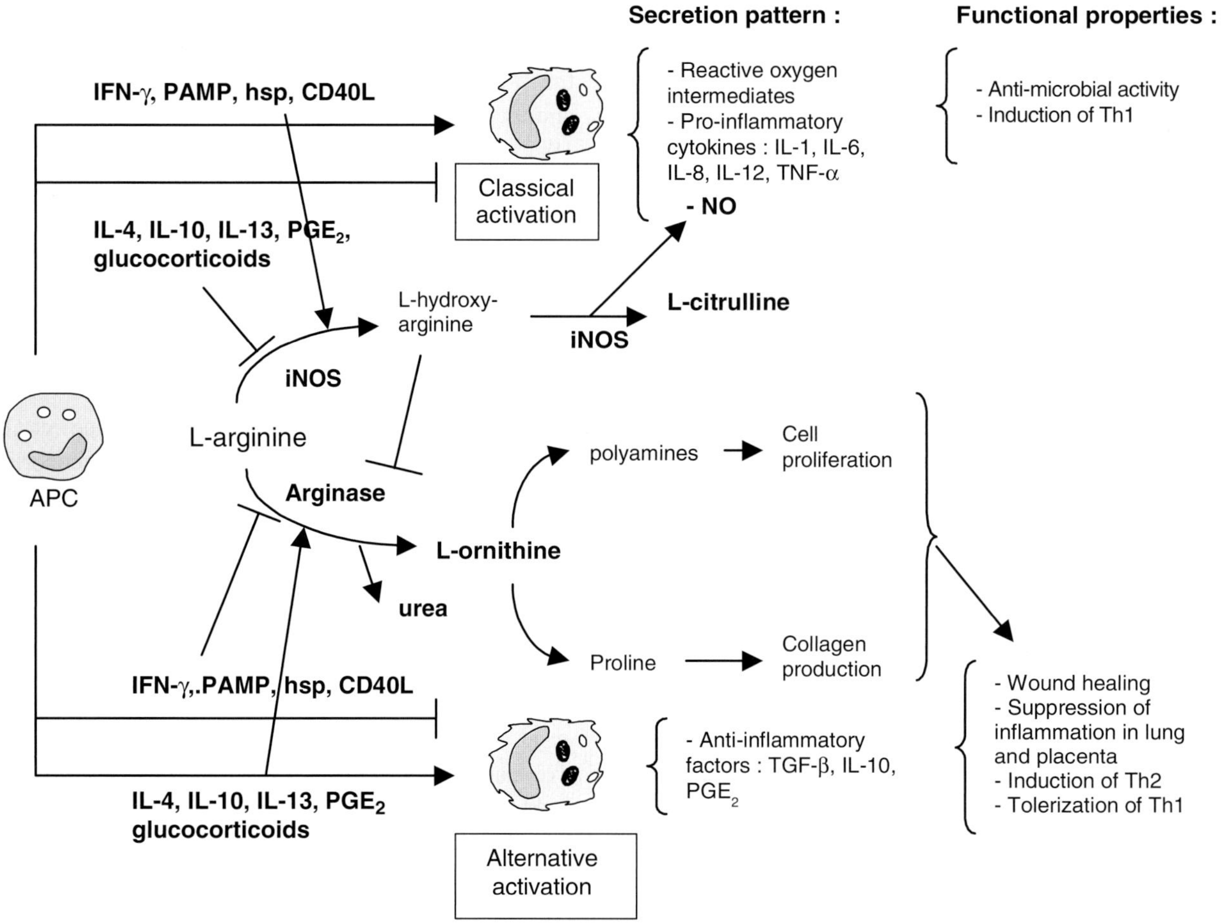

Secretion pattern :

Functional properties :

IFN-γ, PAMP, hsp, CD40L

- Reactive oxygen
  intermediates
- Pro-inflammatory
  cytokines : IL-1, IL-6,
  IL-8, IL-12, TNF-α
- **NO**

- Anti-microbial activity
- Induction of Th1

Classical
activation

IL-4, IL-10, IL-13, PGE₂,
glucocorticoids

L-hydroxy-
arginine

L-citrulline

iNOS

iNOS

L-arginine

Arginase

APC

polyamines → Cell
proliferation

L-ornithine

urea

IFN-γ,.PAMP, hsp, CD40L

Proline → Collagen
production

- Anti-inflammatory
  factors : TGF-β, IL-10,
  PGE₂

- Wound healing
- Suppression of
  inflammation in lung
  and placenta
- Induction of Th2
- Tolerization of Th1

IL-4, IL-10, IL-13, PGE₂
glucocorticoids

Alternative
activation

Fig. 1 Main properties of classically and alternatively activated macrophages. Classical activation of macrophages is stimulated by type I cytokines, PAMP, and endogenous danger signals such as heat-shock proteins and CD40L. In contrast, alternative activation is induced by type II cytokines, prostaglandinE₂, and glucocorticoids. Both activation states are characterized by different L-arginine metabolic pathways. Classically activating signals induce iNOS, converting L-arginine to NO and L-citrulline, while alternatively activating signals promote arginase, converting L-arginine to urea and L-ornithine

level, based on the L-arginine metabolism. Indeed, within the cytosol of macrophages, L-arginine can be metabolized by three different pathways (Wu and Morris 1998) resulting in the production of (i) L-citrulline and NO via iNOS, (ii) ureum and L-ornithine through the activity of arginase and/or (iii) agmatine via arginine decarboxylase (ADC). The iNOS pathway is induced by type I immune mediators, like IFN-γ or LPS, and suppressed by the type II cytokines IL-4 and IL-13 (Munder et al. 1999). In contrast, L-arginine metabolism through the arginase activity is induced by IL-4 and IL-13, but repressed by IFN-γ. The cross-regulation of the iNOS/arginase balance by type I/type II cytokines suggests that the measure of NO level and

of arginase activity in distinct macrophage populations reflects their activation state, classical or alternative (Munder et al. 1998) (Fig. 1).

Recent *in vitro* and *in vivo* studies have focussed on the differential gene expression in M2. Welch and colleagues have identified YM1, a chitinase-like secretory lectin, and arginase as the most highly upregulated genes in M2 by analyzing the transcriptional response in mouse peritoneal macrophages activated by IL-4 (Welch et al. 2002). The mannose receptor, the tissue inhibitor of metalloproteinase (TIMP) 1 and 2, the glucose-regulated heat shock protein 5, MIP-1γ and the IFN-γ-inducible lysosomal thioreductase Ifi 30 were also found to be induced by IL-4.

Loke et al. (2002) have established a gene expression profile of M2 elicited by IL-4 *in vivo* following the implantation of the nematode *Brugia malayi* in the peritoneal compartment of mice. YM1 and FIZZ1 (found in the inflammatory zone), a resistin-like secreted protein, were the most abundantly expressed genes in nematode-elicited M2 (10% and 2% of the transcripts, respectively). The expression of arginase was also induced, confirming the assignment of this enzyme with M2. Other genes that were abundantly expressed in *Brugia*-elicited M2 were serum amyloid A3, an acute phase protein; surfactant protein α, a secreted scavenger receptor cysteine-rich protein; fibronectin; C4 from complement and Fcγ receptor III. The CC chemokine C10 was also associated with filarial-activated M2. In another helminth infection model, *Taenia crassiceps*, alternatively activated macrophages were induced during the chronic stage of infection and characterized by high expression levels of CD23 and CCR5 (Rodriguez-Sosa et al. 2002).

Our group has focussed on the identification of genes that are differentially expressed in M2 versus M1 in experimental murine models of African trypanosomosis (Namangala et al. 2000; 2001). This study has confirmed the enhanced expression of FIZZ1 and YM genes in *in vivo*-elicited M2. The expression of the two genes was also up-regulated in M2 elicited *in vitro* following steering of macrophages with the type II cytokines IL-4 and IL-13. Moreover, IFN-γ counteracted the IL-4-mediated over-expression of FIZZ1 and YM genes *in vitro* (Raes et al. 2002). Hence, IFN-γ could possibly maintain low FIZZ1 and YM expression in M1. Although the differential expression of FIZZ1 and YM genes in M2 has been confirmed in several models, the role of the encoded proteins remains thus far elusive.

An overview of M1- and M2-associated genes, both in human and mouse is given in Table 1.

Tumor-associated macrophages

Tumor-associated macrophages were first investigated for their role in stimulating anti-tumor immune responses. However, more recent studies uncovered a tumor-promoting activity of macrophages (Elgert et al. 1998). This finding revealed the dual nature of

Table 1 List of type II cytokine-induced genes in macrophages

Human	Mouse
Membrane markers	
– Mannose receptor	– Mannose receptor
– β–glucan receptor	
– Scavenger receptor type I	
– MS-1	
– Fibronectin	– Fibronectin
– βIG-H3	
– CD23	– CD23
– CD13	
– RM3/1	
	– Surfactant protein-α
	– CCR5
Secreted products	
– IL-1Ra	
– IL-10	
– TGF-β	– TGF-β
– AMAC-1	
– MDC	
– PDGF-AA,BB	
– MMP-7	
	– C4 (complement)
	– C10 (chemokine)
	– TIMP1,2
	– MIP-1γ
Metabolic markers	
– Arginase	– Arginase
– 15-Lipoxygenase	– Serum amyloid A3
	– Glucose-regulated heat shock protein 5
	– Ym
	– FIZZ1
	– IFN-γ-inducible lysosomal thioreductase Ifi30

macrophages, which depending on their *in vivo* context can exert diametrically opposed activities.

Tumors can influence macrophage activity, both locally in the tumor microenvironment as in the periphery. Macrophages may contribute up to or more than half of a tumor's mass (Sunderkotter et al. 1994) and are actually required for the tumor to survive (Mantovani et al. 1992; Evans 1978; Fauve 1993). Therefore, tumors have developed strategies to attract monocytic precursors, the most important of which being the constitutive expression of chemokines (Mantovani 1999). For example, CCL2, also known as MCP-1, is probably the most frequently found CC chemokine in tumors (Mantovani et al. 2002) and has been shown to promote tumor formation by attracting monocytes to the tumor site (Nesbit et al. 2001). Similar functions are attributed to tumor-derived

CXCL1 and related molecules (CXCL2, CXCL3, CXCL8 or IL-8) (Haghnegahdar et al. 2000), VEGF (Duyndam et al. 2002), M-CSF (Lin et al. 2001; Nowicki et al. 1996) and TGF-β (Wahl et al. 1987).

In addition, the presence of a localized tumor can shape macrophage activation at tumor-distal sites, such as lymphoid organs. Such macrophages can mediate lymphocyte suppression and adversely affect host anti-tumor responses (Ting and Rodrigues 1982; Tsuchiya et al. 1988; Bluestone and Lopez 1983; Parhar and Lala 1988).

In this chapter, we will interpret literature data in terms of the M1/M2 dichotomy, and provide additional evidence from our lab for the tumor-promoting capacity of both types of macrophages.

Tumor-promoting effects of M1

The M1 phenotype of macrophages has been intensively studied over the past decades, including their role in cancer. Therefore, we will in first instance highlight the large body of evidence suggesting the tumor-promoting effects of M1.

Evidence for an M1 phenotype in tumor-associated macrophages

Matzinger's danger model argues that many cancers do not appear dangerous to the immune system because initially they grow as healthy cells and do not send out distress signals to activate dendritic cells and macrophages in a classical pro-inflammatory way (Fuchs and Matzinger 1996). This statement is indeed corroborated by the large body of evidence mentioned in the following chapter, illustrating that the tumor microenvironment is rather anti-inflammatory.

Therefore, many of the tumor-destructive actions of M1 are only seen upon experimental activation of macrophages *in vitro* or *in vivo*. Thus, tumoricidal activity can be detected in macrophages activated *in vitro* with LPS, polyI:C, lymphokines such as IFN-γ, and other pro-inflammatory agents (Pace et al. 1983; Meltzer 1981; Gifford and Lohmann-Matthes 1986; Remels et al. 1990). Likewise, *in vivo* eradication of tumors by macrophages requires prior injection of macrophage-activating products (Fidler et al. 1982; Akaza et al. 1995; Zhang et al. 2002; Auf et al.

2001; Tsung et al. 2002) or a massive activation of tumor-immune lymphocytes by immunisation, which then activate adjacent macrophages (Bonnotte et al. 2001). Almost invariably, macrophage cytotoxic mechanisms involve nitric oxide, hydrogen peroxide and/or TNF (Farias-Eisner et al. 1996; Klostergaard 1993). The cytotoxic activity of M1 can be targeted directly against the cancer cells or against the vascular endothelium, depriving the tumor of oxygen and nutrients.

Nevertheless, cancer cells can in theory produce endogenous danger signals (Gallucci and Matzinger 2001) including the expression of stress-induced ligands for the immunoactivating NKG2D receptor (Diefenbach et al. 2000; Bauer et al. 1999), and the release of intracellular nucleotides, intact double-stranded DNA or heat shock proteins from dying cancer cells (Tamura et al. 1997). Such signals might account for the occasional secretion of pro-inflammatory cytokines and chemokines in the tumor bed by infiltrating inflammatory cells, such as macrophages (i.e. M1), neutrophils and mast cells (Naylor et al. 1990, 1993; Vitolo et al. 1992; Burke et al. 1996; Di Carlo et al. 2001; Coussens et al. 1999). However, it should be noted that in some tumors, the cancer cells themselves are the source of pro-inflammatory mediators, the production of which is often regulated by inflammatory cells (Wigmore et al. 2002; Zhang et al. 1999; Hensley et al. 1998).

Conventional thinking suggests that inflammatory cells and cytokines mount an effective anti-tumor immune response. In recent years however, strong indications highlight a contribution of inflammation to tumor growth, progression and immunosuppression (Balkwill and Mantovani 2001; Coussens and Werb 2001; Cordon-Cardo and Prives 1999). Perhaps the strongest argument in favor of the notion that the inflammatory process is a cofactor in carcinogenesis, comes from the clinical observation that about 15% of the global cancer burden is attributable to chronic infections of which inflammation is a major determinant (Kuper et al. 2000). More specifically, epithelial cell turnover in the setting of chronic inflammation predisposes humans to carcinoma in the breast, liver, large bowel, urinary bladder, prostate, gastric mucosa, ovary and skin (Kuper et al. 2000; De Marzo et al. 1999; Ernst and Gold 2000; Ness and Cottreau 1999; Brocker et al. 1988). In addition, long-term use

of aspirin and nonsteroidal anti-inflammatory drugs reduces colon cancer risk by 40–50%, and may be preventive for lung, esophagus, and stomach cancer (Baron and Sandler 2000; Garcia Rodriguez and Huerta-Alvarez 2001).

Several tumor-promoting mechanisms have been suggested for inflammatory cytokines (Cordon-Cardo and Prives 1999). In first instance, they might provoke DNA damage, and consequently transformation, by inducing NO, H_2O_2, and/or O_2^- synthesis in macrophages or cancer cells. These reactive nitrogen or oxygen intermediates can directly oxidise DNA, resulting in mutagenic changes, and may damage some DNA repair proteins (Jaiswal et al. 2000). In the same vein, the inflammatory cytokine migration inhibitory factor (MIF) can inactivate the p53 tumor-suppressor protein, causing a deficient response to genetic damage (Hudson et al. 1999).

Secondly, the inflammatory cell infiltrate, particularly tumor-associated M1 macrophages, contribute to tumor angiogenesis (O'Byrne et al. 2000; Leek et al. 1999). The formation of new blood vessels is a crucial step in the growth of neoplastic tissues and provides the necessary conditions for dissemination at distant anatomical sites. M1 produce TNF, IL-1 and IL-6 which stimulate the production of angiogenic factors such as VEGF, IL-8 and the enzyme thymidine phosphorylase (Polverini and Leibovich 1984; Ono et al. 1999; Torisu et al. 2000; Leek et al. 1998; Schoppmann et al. 2002; Barbera-Guillem et al. 2002). However, it should be remarked that this angiogenic activity is not a prototypical feature of M1. Indeed, also M2 can promote angiogenesis, albeit via different mechanisms (Kodelja et al. 1997).

Thirdly, inflammation causes the induction of proteases involved in tissue remodeling. In order for invasion and metastasis to occur, the cancer cell must bypass the basement membrane. Therefore, excess matrix degradation is often a hallmark of cancer and could be an important component of the process of tumor progression (Fidler 1997). Matrix metalloproteinases or MMPs are a large family of over 20 proteins that together can degrade all the known components of the extracellular matrix. Certain individual MMP family members are only expressed in cancer cells, whereas some are only seen overexpressed in the stroma, typically by tumor-associated macrophages (McCawley and Matrisian 2000). For example, in one

skin tumor model, paracrine MMP-9 production by inflammatory cells was implicated in epithelial hyperproliferation, angiogenesis and increased malignant potential, and skin tumor development was reduced in MMP-9 "knock-out" mice (Coussens et al. 2000). However, the tumor-promoting effect of proteases is not generally applicable to all tumors. Our data illustrate that endogenous TNF, produced by tumor-infiltrating macrophages, has a direct pro-malignant activity on cancer cells. This activity was due to its stimulatory effect on the expression of *Secretory Leukocyte Protease Inhibitor* (*SLPI*) in these cells, a serine protease inhibitor with broad anti-inflammatory (Sallenave 2000) and as we have shown recently, malignancy-promoting properties (Devoogdt et al. 2003). The protease inhibiting capacity of SLPI was needed for its tumor-enhancing effect, since mutations in the protease-inhibiting domain abolished this effect (Devoogdt et al. 2003). Moreover, although some serine protease inhibitors confer protection to TNF-α-mediated cytotoxicity (Park et al. 2000; Ruggiero et al. 1987), SLPI expression in the cancer cells does not influence its lysis by TNF-α.

Fourthly, once the path is paved for cancer cells to spread, pro-inflammatory cytokines can promote seeding at distant anatomical sites by augmenting the expression of adhesion molecules on endothelial cells, recognizing their receptors on cancer cells (Garofalo et al. 1995; Okahara et al. 1994). In some instances, the cancer cells might even be embedded in emboli with migratory inflammatory cells, particularly macrophages, that alter the attachment of cancer cells to vascular endothelium (Starkey et al. 1984).

Suppression of anti-tumor CTL responses by M1

Finally, M1 can subvert an anti-tumor immune response by suppressing T cell activity. Indeed, Mills' group (1991) pinpointed nitric oxide, produced by the M1-associated inducible nitric oxide synthase, as an important effector molecule of suppressor macrophages. NO exerts heterogeneous and diverse phenotypic effects linked to its capacity to react with other inorganic molecules (such as oxygen, superoxide and transition metals), DNA (pyrimidine bases), prosthetic groups (such as heme) or proteins (leading to S-nitrosylation of thiol groups, nitration of tyrosine residues or disruption of metal-sulfide clusters) (Bogdan 2001). In the

context of a T lymphocyte-mediated response, NO has been shown to inhibit blastogenesis, by inhibiting IL-2 production, altering IL-2Rα expression and/or precluding IL-2R signalling (Blesson et al. 2002; Mazzoni et al. 2002). In tumor-bearers, splenic CD11b⁺ Gr-1⁺ myeloid-derived suppressor cells are often detected, which can suppress T cell responses by an NO-dependent mechanism (Mazzoni et al. 2002; Kusmartsev et al. 2000). Overall, NO-mediated suppression is regularly observed in lymphoid organs of tumor-bearing hosts, but only few reports demonstrate such activity in tumor-associated macrophages (Saio et al. 2001), most probably because the tumor microenvironment is most often anti-inflammatory (Elgert et al. 1998 and Chapter 4.2.1. of this manuscript). In any case, IFN-γ endogenously produced by activated T cells, is needed for the inhibitory activity of these myeloid cells. This leads to the conclusion that macrophage NO-mediated inhibition of the CTL response is a side effect of activating macrophages in a classical way rather than resulting from the action of a distinct subset of what have long been termed suppressor macrophages. These *in vitro* findings have recently been corroborated *in vivo*. Treatment of mice with the iNOS inhibiting agent N^G-monomethyl-L-arginine increases the subsequent anti-tumor response (Koblish et al. 1998; Medot-Pirenne et al. 1999). One important remark should be made in this context: NO is not an exclusivity of M1. Indeed, both Bronte's results (Bronte et al. 2003) and ours (Liu et al. 2003) demonstrate that alternatively activated myeloid suppressor cells (M2) in the spleen of tumor-bearers display at least two mechanisms of T-cell suppression: an NO-dependent suppression of mitogen-activated and memory T cells, and an NO-independent suppression of antigen-specific primary T cells. Finally, not all our attention should be focused on NO as the principal inhibitory molecule of M1. For instance, macrophages in tumor-draining lymph nodes can induce T-cell dysfunction by down-regulating CD3 ζ molecules and provoking T cell apoptosis via reactive oxygen intermediates (Otsuji et al. 1996; Takahashi et al. 2003).

Tumor-promoting effects of M2

M2 cells are still less well characterized than M1 cells. Therefore, investigations into the involvement of M2 cells in tumor progression have only been initiated

over the past few years. Nevertheless, this topic is gaining ever increasing interest from the scientific community, as illustrated by a recent review paper by Mantovani et al. (2002), identifying tumor-associated macrophages as a paradigm for polarized M2 mononuclear phagocytes.

Evidence for an M2 phenotype in tumor-associated macrophages

Based on what is known about tumor-associated macrophages (TAM), they often appear to have a phenotype and function similar to M2 macrophages (Mantovani et al. 2002).

Several indications for this statement came from scrutinizing the iNOS/arginase balance in TAM. It was proposed that macrophage arginine metabolism in the tumor bed via the NO synthase pathway (prototypically M1) would favor tumor inhibition because NO is tumoristatic/tumoricidal (Hibbs et al. 1987; Farias-Eisner et al. 1996). If instead, the predominant pathway of arginine metabolism is via arginase (prototypically M2), then tumor growth might be further promoted because ornithine is the precursor of polyamines that are required for cell replication (Chang et al. 2001; Pegg 1988). It turns out that macrophages isolated from tumor-bearing mice, especially the macrophages embedded within the tumor, often show a diminished tumoricidal capacity due to a reduced expression of the inducible nitric oxide synthase gene (DiNapoli et al. 1996; Klimp et al. 2001; Davel et al. 2002; Baskic et al. 2001; Mills et al. 1992; Alleva et al. 1994; Sotomayor et al. 1995). Rather, L-arginine metabolism via arginase is often increased in macrophages associated with progressing tumors (Davel et al. 2002; Baskic et al. 2001; Mills et al. 1992; Parajuli and Singh 1996). These *in vivo* characteristics of TAM mimic the *in vitro* effect of IL-4, which has been shown to preclude macrophage tumoricidal activation (Nishioka et al. 1990; Suk et al. 1993). In addition to a lowering in the production of tumoristatic and cytotoxic NO, TAM also produce fewer reactive oxygen intermediates (Tsunawaki and Nathan 1986), further allowing an undisturbed tumor growth. It is important to note that these findings are mainly applicable to macrophages derived from the tumor microenvironment, but not necessarily to peripheral macrophages, such as those in the spleen, which might still produce

significant levels of reactive nitrogen and oxygen intermediates.

Not only reactive nitrogen and oxygen intermediates can kill cancer cells, but also inflammatory cytokines such as TNF (Lejeune et al. 1998; Urban et al. 1986). Moreover, inflammatory cytokines and chemokines could guide the recruitment of both nonspecific and specific immune effector cells to the tumor bed. Evidence suggests however that TAM rather have an anti-inflammatory phenotype, secreting high amounts of IL-10 and TGF-β, but only limited amounts of TNF and IL-12 (Beissert et al. 1989; Sica et al. 2000; Kambayashi et al. 1995; Loercher et al. 1999; Maeda et al. 1995; Kim et al. 1995). IL-12 is a pro-inflammatory cytokine with immunoregulatory functions that bridge innate resistance and antigen-specific adaptive immunity (Trinchieri 1995) and lack of its production may be a serious impairment for effective anti-tumor responses.

Not many studies report on the expression of M1- or M2-related surface markers on TAM. Only Mantovani et al. (2002) mention high levels of macrophage mannose receptor, a marker which is rather associated with M2, on TAM. Clearly, this area of research needs to be further explored.

Tumor-associated M2 can be induced upon interactions with cancer cells, stromal fibroblasts and tumor-infiltrating T cells. Tumors may produce molecules that steer macrophages in a M2 direction, such as IL-4, IL-6, IL-10, TGF-β, prostaglandin E_2, M-CSF and p15E-related molecules, which have all been reported to block the cytotoxic functions of TAM (Elgert et al. 1998). Moreover, cancer cell apoptosis, a phenomenon which occurs quite frequently in tumors, results in the induction of M2 secreting IL-10 and TGF-β (Gough et al. 2001).

In tumors, the prevalence of Th2 cells is common, so that T-cell derived IL-4, IL-13 and IL-10 could reinforce the skewing of monocyte differentiation towards M2 (Sato et al. 1998; Maeurer et al. 1995; Sheu et al. 2001; Maeda and Shiraishi 1996; Pellegrini et al. 1996; Ibe et al. 2001). As a matter of fact, cancer cells have been shown to secrete chemokines that contribute to a type II tumor microenvironment. For instance, MCP-1, produced by tumors and TAM, can orient specific immunity in a Th2 direction (Gu et al. 2000). As a consequence, neutralization of MCP-1 boosted the generation of tumor-reactive T cells in a poorly immunogenic tumor model (Peng et al. 1997). Furthermore, Reed–Sternberg cells in Hodgkin's lymphoma can express CCL22 and CCL17 (van den Berg et al. 1999; Cossman et al. 1999). These chemokines recognize CCR4 that is preferentially expressed on polarized Th2 cells and on regulatory T cells, as well as on monocytes (Mantovani 1999; Mantovani et al. 2002; Iellem et al. 2001). In the same tumor, stromal cells produce CCL11, which attracts eosinophils and Th2 cells. Therefore, neoplastic elements and stroma use complementary tools to recruit cells associated with polarized type II responses (Mantovani et al. 2002). Also TAM can participate in the regulation of T cell responses by chemokines. CCL18, a chemokine constitutively produced by immature DC and macrophages exposed to IL-4, IL-13 or IL-10, is the most abundant chemokine in human ovarian ascites fluid (Schutyser et al. 2002). CCL18 attracts naïve T cells, which are then submerged in an environment dominated by M2 and immature DC.

Berrebi et al. (2003) have shed light on the molecular mechanism responsible for the anti-inflammatory effect of cytokines such as IL-10. IL-10, but also glucocorticoids, induce the synthesis of the leucin zipper GILZ in macrophages. GILZ inhibits the function of the key inflammatory transcription factor NF-κB. Most interestingly, while GILZ gene expression is down-regulated during delayed-type hypersensitivity reactions, its expression persists in macrophages infiltrating Burkitt lymphomas, adding additional evidence to the presence of M2-oriented TAM.

Suppression of anti-tumor CTL responses by M2

Immunosuppressive macrophages have been known for a long time (Oehler et al. 1977) and have most of the time been associated with a classically activated phenotype, mainly based on the secretion of nitric oxide. However, new evidence suggests that in some – or perhaps most – instances alternatively activated macrophage and suppressor macrophage populations may at least partially overlap (Goerdt and Orfanos 1999).

Indeed, lymphokines and LPS can stimulate cytotoxic activity in mouse peritoneal macrophages, but cause a reduction in the suppressor activity of these cells, linking suppression with M2 (Chang et al. 1988; Boraschi et al. 1983). In addition, IL-4 and

glucocorticoid-induced macrophages inhibit mitogen-mediated proliferation of CD4[+] T cells in a human co-culture system *in vitro* (Schebesch et al. 1997). Tzachanis et al. (2002) report that blockade of B7/CD28 facilitates the induction of T-cell unresponsiveness by generating suppressive M2. *In vivo*, both placental and alveolar macrophages have an alternatively activated phenotype (Kodelja et al. 1998; Mues et al. 1989) and are prototypes of naturally occurring suppressor macrophages (Chang et al. 1993; Holt et al. 1998). UVB-induced contact tolerance is another excellent model for the functional analysis of suppressive actions of alternatively activated macrophages (Stevens et al. 1995). Finally, M2-mediated immunosuppression has been reported in a number of parasitic infections, including schistosomiasis (Terrazas et al. 2001), tryanosomiasis (Namangala et al. 2001; De Baetselier et al. 2001) and *Brugia malayi* infections (Loke et al. 2000).

In cancer, the presence of immunosuppressive macrophages is amply documented, but fitting these cells into the M1/M2 model is not always straightforward. For example, prostaglandin E$_2$ is a well known macrophage-derived immunosuppressive agent, which is often upregulated in tumor-bearing hosts (Fujii et al. 1987; Pelus and Bockman 1979; Metzger et al. 1980). However, prostaglandin E$_2$ can be produced by either M1 macrophages receiving pro-inflammatory stimuli (Harris et al. 2002) or M2 macrophages receiving anti-inflammatory stimuli (Fadok et al. 1998). Likewise, ambiguous information comes from a population of immature Gr1[+]/CD11b[+] myeloid precursors found in the spleen of immunosuppressed tumor-bearing mice. While some reports identified NO as the immunosuppressive agent from Gr1[+]/CD11b[+] cells (Mazzoni et al. 2002; Kusmartsev et al. 2000), suggesting an M1 phenotype, other reports demonstrate that exposure of these Gr1[+]/CD11b[+] cells to IL-4 greatly increased the T-lymphocyte suppressive activity (Bronte et al. 2000; Apolloni et al. 2000; Gabrilovich et al. 2001). These myeloid-derived suppressor cells specifically target CD8[+] T cells, thereby precluding efficient anti-tumor CTL responses. A recent report by Bronte et al. (2003) might provide an explanation for these at first sight conflicting observations. These authors uncovered that IL-4 induced arginase is involved in the suppressive action of these cells, by increasing superoxide production through a pathway that likely

utilizes the reductase domain of inducible NO synthase. Hence, the simultaneous expression of arginase, inducible NO synthase and superoxide does not allow a straightforward classification of immunosuppressive Gr1[+]/CD11b[+] cells within the M1/M2 model.

Our work further elaborates on the role of splenic myeloid-derived suppressor cells in cancer. In one study we have analyzed the possible involvement of NK cells (via either cytokine production and/or cytolytic activity) in determining the cellular composition of myeloid cells in the spleen and evaluated the resulting effects on anti-tumor CTL responses (Geldhof et al. 2002). Hereby the BW-Sp3 lymphoma model was adopted since rejection of BW-Sp3 depends on eliciting strong CTL responses (VandenDriessche et al. 1994; Raes et al. 1998), and consequently this model allowed to examine the involvement of NK cells in early CD8[+]-dependent immune reactions. To address the importance of NK cells for the early anti-tumor defense, α-ASGM-1-treated mice were challenged with BW-Sp3 cells. Our results indicate that NK cells are a prerequisite for efficient CTL generation and that absence of NK cells favours the outgrowth of M2 cells. Subsequently these M2 cells, associated with NK depletion, suppress the restimulation of memory CTLs. Finally, preferential engagement of M2 by activated NK (LAK) cells reveal that cytolytic interactions might be involved in the regulatory function of NK cells to drive M1-dependent CTL responses.

In the same tumor model, a distinction can be made between hosts that spontaneously reject their tumor and hosts in which the tumor progresses (Raes et al. 1998; Van Ginderachter et al. 2000). We demonstrated that T cell activating antigen-presenting cells were induced in regressors, whereas T cell suppressive myeloid cells, characterized by a combined CD11b and Gr1 expression, predominated in the spleen of progressors (Liu et al. 2003). Indeed, *in vitro* depletion of CD11b[+] populations restored T cell cytotoxicity and proliferation in mice with progressing disease. This CTL inhibition was cell-to-cell contact-dependent but not mediated by nitric oxide. However, the same progressor suppressive cells prevented the activity of *in vitro* restimulated CTLs derived from regressors in a cell-to-cell contact and NO-dependent fashion. Thus, either NO-dependent or -independent suppressive pathways prevailed, depending on the

target CTL population. In addition, the suppressive population expressed a high arginase activity. However, in contrast to Bronte's results (2003), the high arginase activity is not directly involved in the suppressive process in this case. Nevertheless, these data provide yet another example of myeloid suppressor cells in the spleen of tumor-bearing mice, co-expressing arginase (M2 marker) and inducible NO synthase (M1 marker) (Fig.2).

Concluding remarks

Macrophages are in the first place versatile, regulatory cells able to respond to diverse environmental signals. It is remarkable that the same cell type, depending on the stimuli it gets, can have opposing pro-inflammatory (M1) or anti-inflammatory (M2) functions. While the molecular signature of M1 has been a subject of investigation for a long time, M2 cell gene expression

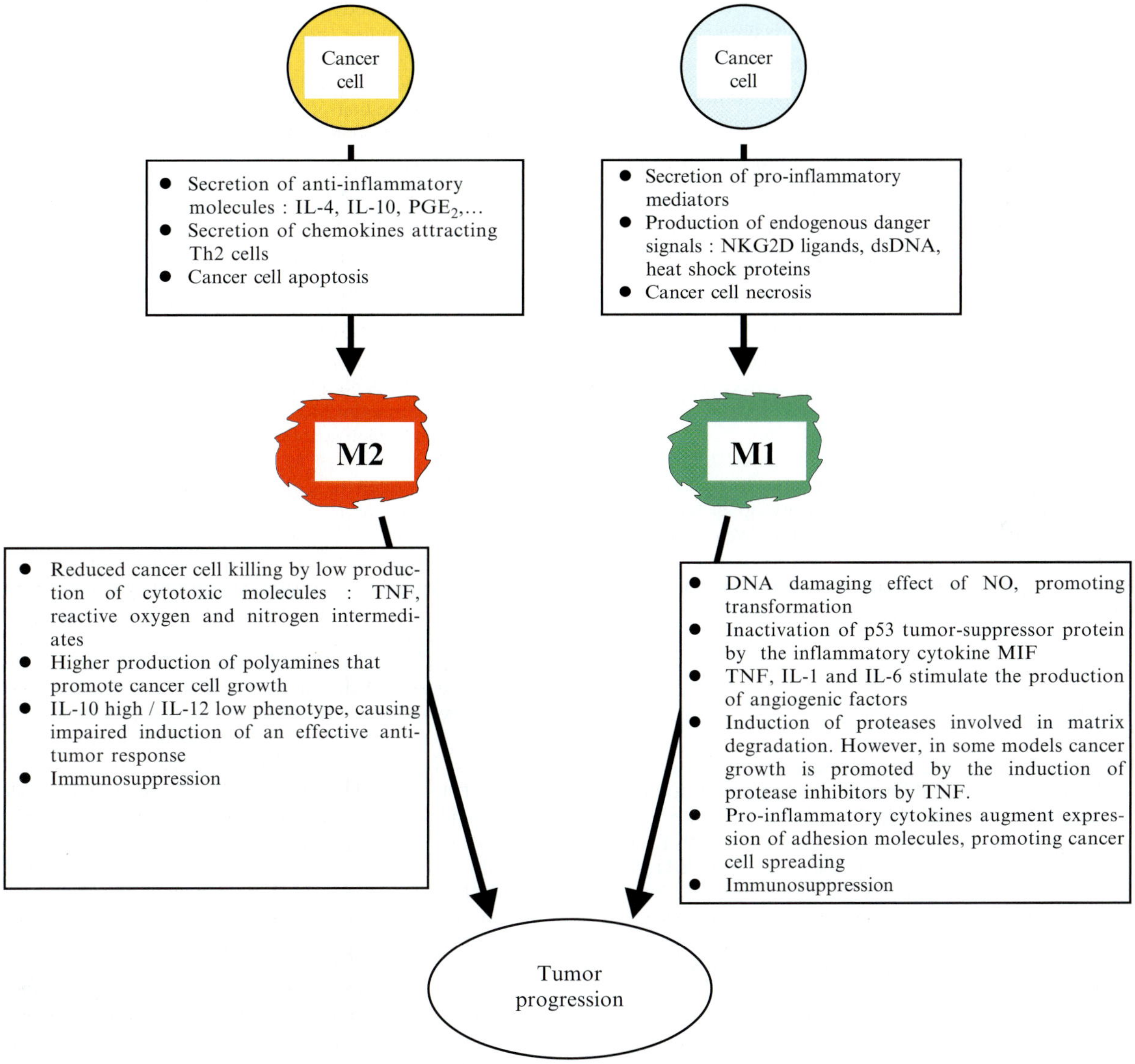

Cancer cell

- Secretion of anti-inflammatory molecules : IL-4, IL-10, PGE$_2$,...
- Secretion of chemokines attracting Th2 cells
- Cancer cell apoptosis

M2

- Reduced cancer cell killing by low production of cytotoxic molecules : TNF, reactive oxygen and nitrogen intermediates
- Higher production of polyamines that promote cancer cell growth
- IL-10 high / IL-12 low phenotype, causing impaired induction of an effective anti-tumor response
- Immunosuppression

Cancer cell

- Secretion of pro-inflammatory mediators
- Production of endogenous danger signals : NKG2D ligands, dsDNA, heat shock proteins
- Cancer cell necrosis

M1

- DNA damaging effect of NO, promoting transformation
- Inactivation of p53 tumor-suppressor protein by the inflammatory cytokine MIF
- TNF, IL-1 and IL-6 stimulate the production of angiogenic factors
- Induction of proteases involved in matrix degradation. However, in some models cancer growth is promoted by the induction of protease inhibitors by TNF.
- Pro-inflammatory cytokines augment expression of adhesion molecules, promoting cancer cell spreading
- Immunosuppression

Tumor progression

Fig. 2 Classical (M1) and alternative (M2) activation of macrophages: different pathways of macrophage-mediated tumor promotion. Cancer cells can steer the macrophage activation state into either M1 or M2. These M1 or M2 then employ their own mechanisms to promote tumor growth

is still largely unknown. It will be important for the future to fit the still scattered information on M2 into applicable models describing the modalities of these cells. Important aspects of M2 biology to consider are their anti-type I inflammatory activity, and wound healing and tissue repair promoting properties.

The involvement of macrophages in tumor growth is a long recognized fact. This review shows that tumor-mediated regulation of macrophage activities, either in the M1 or M2 direction, can favor tumor growth. However, considering the involvement of these cells in a multitude of functions – influencing immune competitiveness, cancer cell growth and invasive capacity, angiogenesis and tissue remodelling – it will not be straightforward to design therapies or drugs able to subvert macrophage tumor-promoting effects. Therefore, more data will be needed to describe the nature of macrophages in different locations – not only comparing lymphoid organs with the tumor site, but also different compartments within the tumor – and in cancers from different histological origin and at different stages, in order to obtain a clearer picture on the when and where of macrophage-mediated tumor progression. This knowledge might ultimately lead to a rationalisation of therapies aimed at tumor-associated macrophages.

Acknowledgements This work is supported by grants from the IWT (Instituut ter bevordering van Wetenschap en Technologie in de industrie) and FWO (Fonds voor Wetenschappelijk Onderzoek).

We thank Ella Omasta, Eddy Vercauteren, Martine Gobert and Roger Wijnants for excellent technical assistance.

References

Akagawa KS (2002) Functional heterogeneity of colony-stimulating factor-induced human monocyte-derived macrophages. Int J Hematol 76:27

Akaza H, Hinotsu S, Aso Y, Kakizoe T, Koiso K (1995) Bacillus Calmette-Guerin treatment of existing papillary bladder cancer and carcinoma in situ of the bladder. Four-year results. The Bladder Cancer BCG Study Group. Cancer 75:552

Akira S, Takeda K, Kaisho T (2001) Toll -like receptors: critical proteins linking innate and acquired immunity. Nat Immunol 2:675

Alleva DG, Burger CJ, Elgert KD (1994) Tumor-induced regulation of suppressor macrophage nitric oxide and TNF-α production: role of tumor-derived IL-10, TGF-β, and prostaglandin E$_2$. J Immunol 153:1674

Apolloni E, Bronte V, Mazzoni A, Serafini P, Cabrelle A, Segal DM, Young HA, Zanovello P (2000) Immortalized myeloid suppressor cells trigger apoptosis in antigen-activated T lymphocytes. J Immunol 165:6723

Auf G, Carpentier AF, Chen L, Le Clanche C, Delattre JY (2001) Implication of macrophages in tumor rejection induced by CpG-oligodeoxynucleotides without antigen. Clin Cancer Res 7:3540

Balkwill F, Mantovani A (2001) Inflammation and cancer: back to Virchow? Lancet 357:539

Barbera-Guillem E, Nyhus JK, Wolford CC, Friece CR, Sampsel JW (2002) Vascular endothelial growth factor secretion by tumor-infiltrating macrophages essentially supports tumor angiogenesis, and IgG immune complexes potentiate the process. Cancer Res 62:7042

Baron JA, Sandler RS (2000) Nonsteroidal anti-inflammatory drugs and cancer prevention. Annu Rev Med 51:511

Baskic D, Acimovic L, Samardzic G, Vujanovic NL, Arsenijevic NN (2001) Blood monocytes and tumor-associated macrophages in human cancer: differences in activation levels. Neoplasma 48:169

Bauer S, Groh V, Wu J, Steinle A, Phillips JH, Lanier LL, Spies T (1999) Activation of natural killer cells and T cells by NKG2D, a receptor for stress-inducible MICA. Science 285:727

Becker S, Daniel EG (1990) Antagonistic and additive effects of IL-4 and interferon-γ on human monocytes and macrophages: effects on Fc receptors, HLA-D antigens, and superoxide production. Cell Immunol 129:351

Beissert S, Bergholz M, Waase I, Lepsien G, Schauer A, Pfizenmaier K, Krönke M (1989) Regulation of tumor necrosis factor gene expression in colorectal adenocarcinoma: in vivo analysis by in situ hybridization. Proc Natl Acad Sci USA 86:5064

Belardelli F, Ferrantini M (2002) Cytokines as a link between innate and adaptive antitumor immunity. Trends Immunol 23:201

Berrebi D, Bruscoli S, Cohen N, Foussat A, Migliorati G, Bouchet-Delbos L, Maillot M-C, Portier A, Couderc J, Galanaud P, Peuchmaur M, Riccardi C, Emilie D (2003) Synthesis of glucocorticoid-induced leucine zipper (GILZ) by macrophages: an anti-inflammatory and immunosuppressive mechanism shared by glucocorticoids and IL-10. Blood 101:729

Blesson S, Thiery J, Gaudin C, Stancou R, Kolb J-P, Moreau JL, Theze J, Mami-Chouaib F, Chouaib S (2002) Analysis of the mechanisms of human cytotoxic T lymphocyte response inhibition by NO. Int Immunol 14:1169

Bluestone JA, Lopez C (1983) Suppression of the immune response in tumor-bearing mice. III. Induction of functionally suppressive antigen-driven macrophages. Cancer Invest 1:5

Boehm U, Klamp T, Groot M, Howard JC (1997) Cellular responses to IFN-γ. Annu Rev Immunol 15:749

Bogdan C (2001) Nitric oxide and the immune response. Nat Immunol 2:907

Bonder CS, Dickensheets HL, Finlay-Jones JJ, Donnelly RP, Hart PH (1998) Involvement of the IL-2 receptor gamma-chain in the control by IL-4 of human monocyte and macrophage proinflammatory mediator production. J Immunol 160:4048

Bonecchi R, Sozzani S, Stine JT, Luini W, D'Amico G, Allavena P, Chantry D, Mantovani A (1998) Divergent effects of interleukin-4 and interferon-γ on macrophage-derived chemokine production: an amplification circuit of polarized T helper 2 responses. Blood 92:2668

Bonnotte B, Larmonier N, Favre N, Fromentin A, Moutet M, Martin M, Gurbuxani S, Solary E, Chauffert B, Martin F (2001) Identification of tumor-infiltrating macrophages as the killers of tumor cells after immunization in a rat model system. J Immunol 167:5077

Boon T, Coulie PG, Van den Eynde B (1997) Tumour antigens recognized by T cells. Immunol Today 18:267

Boraschi D, Pasqualetto E, Ghezzi P, Salmona M, Bartalini M, Barbarulli G, Censini S, Soldateschi D, Tagliabue A (1983) Dissociation between macrophage tumoricidal capacity and suppressive activity: analysis with macrophage defective mouse strains. J Immunol 131:1707

Brocker EB, Zwaldo G, Holzmann B, Macher E, Sorg C (1988) Inflammatory cell infiltrates in human melanoma at different stages of tumor progression. Int J Cancer 41:562

Bronte V, Apolloni E, Cabrelle A, Ronca R, Serafini P, Zamboni P, Restifo NP, Zanovello P (2000) Identification of a CD11b+/ Gr-1+/CD31+ myeloid progenitor capable of activating or suppressing CD8+ T cells. Blood 96:3838

Bronte V, Serafini P, De Santo C, Marigo I, Tosello V, Mazzoni A, Segal DM, Staib C, Lowel M, Sutter G, Colombo MP, Zanovello P (2003) IL-4-induced arginase 1 suppresses alloreactive T cells in tumor-bearing mice. J Immunol 170:270

Burke F, Relf M, Negus R, Balkwill F (1996) A cytokine profile of normal and malignant ovary. Cytokine 8:578

Burnet FM (1971) Immunological surveillance in neoplasia. Transplant Rev 7:3

Chang C-I, Liao JC, Kuo L (2001) Macrophage arginase promotes tumor cell growth and suppresses nitric oxide-mediated tumor cytotoxicity. Cancer Res 61:1100

Chang M-DY, Pollard JW, Khalili H, Goyert SM, Diamond B (1993) Mouse placental macrophages have a decreased ability to present antigen. Proc Natl Acad Sci USA 90:462

Chang ZL, Bonvini E, Varesio L, Holden HT, Herberman RB (1988) Differential in vitro modulation of suppressor and antitumor functions of mouse macrophages by lymphokines and/or endotoxin. Cell Immunol 114:282

Cheung DL, Hart PH, Vitti GF, Whitty GA, Hamilton JA (1990) Contrasting effects of interferon-gamma and interleukin-4 on the interleukin-6 activity of stimulated human monocytes. Immunology 71:70

Chouaib S, Asselin-Paturel C, Mami-Chouaib F, Caignard A, Blay JY (1997) The host-tumor immune conflict: from immunosuppression to resistance and destruction. Immunol Today 18:493

Conrad DJ, Lu M (2000) Regulation of human 12/15-lipoxygenase by Stat6-dependent transcription. Am J Respir Cell Mol Biol 22:226

Cordon-Cardo C, Prives C (1999) At the crossroads of inflammation and tumorigenesis. J Exp Med 190:1367

Corraliza IM, Soler G, Eichmann K, Modolell M (1995) Arginase induction by suppressors of nitric oxide synthesis (IL-4, IL-10 and PGE2) in murine bone-marrow-derived macrophages. Biochem Biophys Res Commun 206:667

Cossman J, Annunziata CM, Barash S, Staudt L, Dillon P, He WW, Ricciardi_Castagnoli P, Rosen CA, Carter KC (1999) Reed-Sternberg cell genome expression supports a B-cell lineage. Blood 94:411

Coussens LM, Werb Z (2001) Inflammatory cells and cancer: think different! J Exp Med 193:F23

Coussens LM, Raymond WW, Bergers G, Laig-Webster M, Behrendtsen O, Werb Z, Caughey G, Hanahan D (1999) Inflammatory mast-cells up-regulate angiogenesis during squamous epithelial carcinogenesis. Genes Dev 13:1382

Coussens LM, Tinkle CL, Hanahan D, Werb Z (2000) MMP -9 supplied by bone marrow-derived cells contributes to skin carcinogenesis. Cell 103:481

Davel LE, Jasnis MA, de la Torre E, Gotoh T, Diament M, Magenta G, Sacerdote de Lustig E, Sales ME (2002) Arginine metabolic pathways involved in the modulation of tumor-induced angiogenesis by macrophages. FEBS Lett 532:216

De Baetselier P, Namangala B, Noël W, Brys L, Pays E, Beschin A (2001) Alternative versus classical macrophage activation during experimental African trypanosomosis. Int J Parasitol 31:575

De Marzo AM, Marchi VL, Epstein JI, Neson WG (1999) Proliferative inflammatory atrophy of the prostate: implications for prostatic carcinogenesis. Am J Pathol 155:1985

Devoogdt N, Hassanzadeh Ghassabeh G, Zhang J, Brys L, De Baetselier P, Revets H (2003) Secretory Leukocyte Protease Inhibitor promotes the tumorigenic and metastatic potential of cancer cells. Proc Natl Acad Sci USA 100:5778

Di Carlo E, Forni G, Lollini PL, Colombo MP, Modesti A, Musiani P (2001) The intriguing role of polymorphonuclear neutrophils in anticancer reactions. Blood 97:339

Diefenbach A, Jamieson AM, Liu SD, Shastri N, Raulet DH (2000) Ligands for the murine NKG2D receptor: expression by tumor cells and activation of NK cells and macrophages. Nat Immunol 1:119

DiNapoli MR, Calderon CL, Lopez DM (1996) The altered tumoricidal capacity of macrophages isolated from tumor-bearing mice is related to reduced expression of the inducible nitric oxide synthase gene. J Exp Med 183:1323

Djemadji-Oudjiel N, Goerdt S, Kodelja V, Schmuth M, Orfanos CE (1996) Immunohistochemical identification of type II alternatively activated dendritic macrophages (RM 3/1+, MS-1+/−, 25F9−) in psoriatic dermis. Arch Dermatol Res 288:757

Duyndam MC, Hilhorst MC, Schluper HM, Verheul HM, van Diest PJ, Kraal G, Pinedo HM, Boven E (2002) Vascular endothelial growth factor-165 overexpression stimulates angiogenesis and induces cyst formation and macrophage infiltration in human ovarian cancer xenografts. Am J Pathol 160:537

Elgert KD, Alleva DG, Mullins DW (1998) Tumor -induced immune dysfunction: the macrophage connection. J Leukoc Biol 64:275

Ernst P, Gold BD (2000) The disease spectrum of *Helicobacter pylori*: the immunopathogenesis of gastroduodenal ulcer and gastric cancer. Annu Rev Microbiol 54:615

Evans R (1978) Macrophage requirement for growth of a murine fibrosarcoma. Br J Cancer 37:1086

Fadok VA, Bratton DL, Konowal A, Freed PW, Westcott JY, Henson PM (1998) Macrophages that have ingested apoptotic cells in vitro inhibit proinflammatory cytokine production through autocrine/paracrine mechanisms involving TGF-β, PGE2, and PAF. J Clin Invest 101:890

Farias-Eisner R, Chaudhuri G, Aeberhard E, Fukuto JM (1996) The chemistry and tumoricidal activity of nitric oxide/

hydrogen peroxide and the implications to cell resistance/ susceptibility. J Biol Chem 271:6144

Fauve RM (1993) Macrophages and cancer: some aspects of the macrophage-cancer relationship. Res Immunol 144:265

Fenton MJ, Buras JA, Donnelly RP (1992) IL -4 reciprocally regulates IL-1 and IL-1 receptor antagonist expression in human monocytes. J Immunol 149:1283

Fidler IJ (1985) Macrophages and metastasis – a biological approach to cancer therapy. Cancer Res 45:4714

Fidler IJ (1997) Molecular biology of cancer: invasion and metastasis. In: Devita VT et al. (eds) Cancer: principles and practice of oncology, 5th edn. Lippincott-Raven, Philadelphia, pp 135–152

Fidler IJ, Barnes Z, Fogler WE, Kirsh R, Bugelski P, Poste G (1982) Involvement of macrophages in the eradication of established metastases following intravenous injection of liposomes containing macrophage activators. Cancer Res 42:496

Flores Villanueva PO, Harris TS, Ricklan DE, Stadecker MJ (1994) Macrophages from schistosomal egg granulomas induce unresponsiveness in specific cloned Th1 lymphocytes in vitro and down-regulate schistosomal granulomatous disease in vivo. J Immunol 152:1847

Fuchs EJ, Matzinger P (1996) Is cancer dangerous to the immune system? Semin Immunol 8:271

Fujii T, Igarashi T, Kishimoto S (1987) Significance of suppressor macrophages for immunosurveillance of tumor-bearing mice. J Natl Cancer Inst 78:509

Gabrilovich DI, Velders MP, Sotomayor EM, Kast WM (2001) Mechanism of immune dysfunction in cancer mediated by immature Gr-1$^+$ myeloid cells. J Immunol 166:5398

Gallucci S, Matzinger P (2001) Danger signals: SOS to the immune system. Curr Op Immunol 13:114

Garcia Rodriguez LA, Huerta-Alvarez C (2001) Reduced risk of colorectal cancer among long-term users of aspirin and nonaspirin nonsteroidal anti-inflammatory drugs. Epidemiology 12:88

Garofalo A, Chirivi R, Foglieni C, Pigott R, Mortarini R, Martin-Padura I, Anichini A, Gearing A, Sanchez-Madrid F, Dejana E (1995) Involvement of the very late antigen 4 integrin on melanoma in interleukin 1-augmented experimental metastasis. Cancer Res 55:414

Geldhof AB, Van Ginderachter JA, Liu YQ, Noël W, Raes G, De Baetselier P (2002) Antagonistic effect of NK cells on alternatively activated monocytes: a contribution of NK cells to CTL generation. Blood 100:4049

Geng YJ, Hansson GK (1992) Interferon -γ inhibits scavenger receptor expression and foam cell formation in human monocyte-derived macrophages. J Clin Invest 89:1322

Gifford GE, Lohmann-Matthes M-L (1986) The requirement for the continual presence of LPS for the production of TNF by thioglycolate induced peritoneal murine macrophages. Int J Cancer 38:135

Girardi M, Oppenheim DE, Steele CR, Lewis JM, Glusac E, Filler R, Hobby P, Sutton B, Tigelaar RE, Hayday AC (2001) Regulation of cutaneous malignancy by γδ T cells. Science 294:605

Goerdt S, Orfanos CE (1999) Other functions, other genes: alternative activation of antigen-presenting cells. Immunity 10:137

Goerdt S, Bhardwaj R, Sorg C (1993) Inducible expression of MS-1 high-molecular-weight protein by endothelial cells

of continuous origin and by dendritic cells/macrophages in vivo and in vitro. Am J Pathol 142:1409

Goerdt S, Walsh LJ, Murphy GF, Pober JS (1991) Identification of a novel high molecular weight protein preferentially expressed by sinusoidal endothelial cells in normal human tissues. J Cell Biol 113:1425

Gordon S (2003) Alternative activation of macrophages. Nat Rev Immunol 3:23

Gough MJ, Melcher AA, Ahmed A, Crittenden MR, Riddle DS, Linardakis E, Ruchatz AN, Emiliusen LM, Vile RG (2001) Macrophages orchestrate the immune response to tumor cell death. Cancer Res 61:7240

Gratchev A, Guillot P, Hakiy N, Politz O, Orfanos CE, Schledzewski K, Goerdt S (2001) Alternatively activated macrophages differentially express fibronectin and its splice variants and the extracellular matrix protein βIG-H3. Scand J Immunol 53:386

Gu L, Tseng S, Horner RM, Tam C, Loda M, Rollins BJ (2000) Control of TH2 polarization by the chemokine monocyte chemoattractant protein-1. Nature 404:407

Haghnegahdar H, Du J, Wang D, Strieter RM, Burdick MD, Nanney LB, Cardwell N, Luan J,Shattuck-Brandt R, Richmond A (2000) The tumorigenic and angiogenic effects of MGSA/GRO proteins in melanoma. J Leukoc Biol 67:53

Harris SG, Padilla J, Koumas L, Ray D, Phipps RP (2002) Prostaglandins as modulators of immunity. Trends Immunol 23:144

Hensley C, Spitzler S, McAlpine BE, Lynn M, Ansel JC, Solomon AR, Armstrong CA (1998) In vivo human melanoma cytokine production: inverse correlation of GM-CSF production with tumor depth. Exp Dermatol 7:335

Hibbs Jr., JB, Vavrin Z, Taintor RR (1987) L-arginine is required for expression of the activated macrophage effector mechanism causing selective metabolic inhibition in target cells. J Immunol 138:550

Holt PG, Schon-Hegard MA, Oliver J (1998) MHC class II antigen-bearing dendritic cells in pulmonary tissues of the rat. Regulation of antigen presentation activity by endogenous macrophage population. J Exp Med 167:262

Hudson JD, Shoaibi MA, Maestro R, Carnero A, Hannon GJ, Beach DH (1999) A proinflammatory cytokine inhibits p53 tumor suppressor activity. J Exp Med 190:1375

Ibe S, Qin Z, Schuler T, Preiss S, Blankenstein T (2001) Tumor rejection by disturbing tumor stroma-cell interactions. J Exp Med 194:1549

Iellem A, Mariani M, Lang R, Recalde H, Panina_Bordignon P, Sinigaglia F, D_Ambrosio D (2001) Unique chemotactic response profile and specific expression of chemokine receptors CCR4 and CCR8 by CD4(+)CD25(+) regulatory T cells. J Exp Med 194:847

Jaiswal M, LaRusso NF, Burgart LJ, Gores GJ (2000) Inflammatory cytokines induce DNA damage and inhibit DNA repair in cholangiocarcinoma cells by a nitric oxide-dependent mechanism. Cancer Res 60:184

Kambayashi T, Alexander HR, Fong M, Strassmann G (1995) Potential involvement of IL-10 in suppressing tumor-associated macrophages. Colon-26-derived prostaglandin E_2 inhibits TNF-α release via a mechanism involving IL-10. J Immunol 154:3383

Kaplan DH, Shankaran V, Dighe AS, Stockert E, Aguet M, Old LJ, Schreiber RD (1998) Demonstration of an

IFNγ-dependent tumor surveillance system in immuno-competent mice. Proc Natl Acad Sci USA 95:7556

Khong HT, Restifo NP (2002) Natural selection of tumor variants in the generation of "tumor escape" phenotypes. Nat Immunol 3:999

Kim J, Modlin RL, Moy RL, Dubinett SM, McHugh T, Nickoloff BJ, Uyemura K (1995) IL-10 production in cutaneous basal and squamous cell carcinomas. A mechanism for evading the local T-cell immune response. J Immunol 155:2240

Klimp AH, Hollema H, Kempinga C, van der Zee AG, de Vries EG, Daemen T (2001) Expression of cyclooxygenase-2 and inducible nitric oxide synthase in human ovarian tumors and tumor-associated macrophages. Cancer Res 61:7305

Klostergaard J (1993) Macrophage tumoricidal mechanisms. Res Immunol 144:274

Koblish HK, Hunter CA, Wysocka M, Trinchieri G, Lee WM (1998) Immune suppression by recombinant interleukin (rIL)-12 involves interferon γ induction of nitric oxide synthase 2 (iNOS) activity: inhibitors of NO generation reveal the extent of rIL-12 vaccine adjuvant effect. J Exp Med 188:1603

Kodelja V, Muller C, Tenorio S, Schebesch C, Orfanos CE, Goerdt S (1997) Differences in angiogenic potential of classically vs alternatively activated macrophages. Immunobiology 197:478

Kodelja V, Muller C, Politz O, Hakij N, Orfanos CE, Goerdt S (1998) Alternative macrophage activation-associated CC-chemokine-1, a novel structural homologue of macrophage inflammatory protein-1 α with a Th2-associated expression pattern. J Immunol 160:1411

Kuper H, Adami HO, Trichopoulos D (2000) Infections as a major preventable cause of human cancer. J Int Med 248:171

Kuroda E, Kito T, Yamashita U (2002) Reduced expression of STAT4 and IFN-γ in macrophages from BALB/c mice. J Immunol 168:5477

Kusmartsev SA, Li Y, Chan S-H (2000) Gr-1⁺ myeloid cells derived from tumor-bearing mice inhibit primary T cell activation induced through CD3/CD28 costimulation. J Immunol 165:779

Leek RD, Landers R, Fox SB, Ng F, Harris AL, Lewis CE (1998) Association of tumour necrosisfactor alpha and its receptors with thymidine phosphorylase expression in invasive breast carcinoma. Br J Cancer 77:2246

Leek RD, Landers RJ, Harris AL, Lewis CE (1999) Necrosis correlates with high vascular density and focal macrophage infiltration in invasive carcinoma of the breast. Br J Cancer 79:991

Lejeune FJ, Rüegg C, Liénard D (1998) Clinical applications of TNF-α in cancer. Curr Op Immunol 10:573

Lin EY, Nguyen AV, Russell RG, Pollard JW (2001) Colony-stimulating factor 1 promotes progression of mammary tumors to malignancy. J Exp Med 193:727

Liu YQ, Van Ginderachter JA, Brys L, De Baetselier P, Raes G, Geldhof AB (2003) NO-independent CTL suppression during tumor progression: association with arginase-producing (M2) myeloid cells. J Immunol 170:5064

Loercher AE, Nash MA, Kavanagh JJ, Platsoucas CD, Freedman RS (1999) Identification of an IL-10-producing HLA-DR-negative monocyte subset in the malignant ascites of patients with ovarian carcinoma that inhibits cytokine protein expression and proliferation of autologous T cells. J Immunol 163:6251

Loke P, MacDonald AS, Robb A, Maizels RM, Allen JE (2000) Alternatively activated macrophages induced by nematode infection inhibit proliferation via cell-to-cell contact. Eur J Immunol 30:2669

Loke P, Nair MG, Parkinson J, Guiliano D, Blaxter M, Allen JE (2002) IL -4 dependent alternatively-activated macrophages have a distinctive in vivo gene expression phenotype. BMC Immunol 3:7

MacMicking J, Xie QW, Nathan C (1997) Nitric oxide and macrophage function. Annu Rev Immunol 15:323

Maeda H, Shiraishi A (1996) TGF-beta contributes to the shift toward Th2-type responses through direct and IL-10-mediated pathways in tumor-bearing mice. J Immunol 156:73

Maeda H, Kuwahara H, Ichimura Y, Ohtsuki M, Kurakata S, Shiraishi A (1995) TGF-β enhances macrophage ability to produce IL-10 in normal and tumor-bearing mice. J Immunol 155:4926

Maeurer MJ, Martin DM, Castelli C, Elder E, Leder G, Storkus WJ, Lotze MT (1995) Host immune response in renal cell cancer: interleukin-4 (IL-4) and IL-10 mRNA are frequently detected in freshly collected tumor-infiltrating lymphocytes. Cancer Immunol Immunother 41:111

Mantovani A (1999) The chemokine system: redundancy for robust outputs. Immunol Today 20:254

Mantovani A, Bottazzi B, Colotta F, Sozzani S, Ruco L (1992) The origin and function of tumor-associated macrophages. Immunol Today 13:265

Mantovani A, Sozzani S, Locati M, Allavena P, Sica A (2002) Macrophage polarization: tumor-associated macrophages as a paradigm for polarized M2 mononuclear phagocytes. Trends Immunol 23:549

Mazzoni A, Bronte V, Visintin A, Spitzer JH, Apolloni E, Serafini P, Zanovello P, Segal DM (2002) Myeloid suppressor lines inhibit T cell responses by an NO-dependent mechanism. J Immunol 168:689

McCawley LJ, Matrisian LM (2000) Matrix metalloproteinases: multifunctional contributors to tumor progression. Mol Med Today 6:149

Medot-Pirenne M, Heilman MJ, Saxena M, McDermott PE, Mills CD (1999) Augmentation of an antitumor CTL response in vivo by inhibition of suppressor macrophage nitric oxide. J Immunol 163:5877

Medzhitov R, Janeway CA (2000) Innate immune recognition: mechanisms and pathways. Immunol Rev 173:89

Meltzer MS (1981) Macrophage activation for tumor cytotoxicity: characterization of priming and trigger signals during lymphokine activation. J Immunol 127:179

Metzger Z, Hoffeld JT, Oppenheim JJ (1980) Macrophage-mediated suppression. I. Evidence for participation of both hydrogen peroxide and prostaglandins in suppression of murine lymphocyte proliferation. J Immunol 124:983

Mills CD (1991) Molecular basis of "suppressor" macrophages. Arginine metabolism via the nitric oxide synthetase pathway. J Immunol 146:2719

Mills CD, Shearer J, Evans R, Caldwell MD (1992) Macrophage arginine metabolism and the inhibition or stimulation of cancer. J Immunol 149:2709

Mills CD, Kincaid K, Alt JM, Heilman MJ, Hill AM (2000) M -1/M-2 macrophages and the Th1/Th2 paradigm. J Immunol 164:6166

Montaner LJ, da Silva RP, Sun J, Sutterwala S, Hollinshead M, Vaux D, Gordon S (1999) Type 1 and type 2 cytokine regulation of macrophage endocytosis: differential activation by IL-4/IL-13 as opposed to IFN-γ or IL-10. J Immunol 162:4606

Mosser DM, Handman E (1992) Treatment of murine macrophages with interferon-γ inhibits their ability to bind leishmania promastigotes. J Leukoc Biol 52:369

Mueller C (2002) Tumour necrosis factor in mouse models of chronic intestinal inflammation. Immunology 105:1–8

Mues B, Langer D, Zwadlo G, Sorg C (1989) Phenotypic characterization of macrophages in human term placenta. Immunology 67:303

Munder M, Eichmann K, Modolell M (1998) Alternative metabolic states in murine macrophages reflected by the nitric oxide synthase/arginase balance: competitive regulation by CD4+ T cells correlates with Th1/Th2 phenotype. J Immunol 160:5347

Munder M, Eichmann K, Moran JM, Centeno F, Soler G, Modolell M (1999) Th1/Th2-regulated expression of arginase isoforms in murine macrophages and dendritic cells. J Immunol 163:3771

Namangala B, De Baetselier P, Brijs L, Stijlemans B, Noël W, Pays E, Carrington M, Beschin A (2000) Attenuation of *Trypanosoma brucei* is associated with reduced immunosuppression and concomitant production of Th2 lymphokines. J Infect Dis 181:1110

Namangala B, De Baetselier P, Noël W, Brys L, Beschin A (2001) Alternative versus classical macrophage activation during experimental African trypanosomosis. J Leukoc Biol 69:387

Naylor MS, Stamp GW, Balkwill FR (1990) Investigation of cytokine gene expression in human colorectal cancer. Cancer Res 50:4436

Naylor MS, Stamp GW, Foulkes WD, Eccles D, Balkwill FR (1993) Tumor necrosis factor and its receptors in human ovarian cancer. Potential role in disease progression. J Clin Invest 91:2194

Nesbit M, Schaider H, Miller TH, Herlyn M (2001) Low - level monocyte chemoattractant protein-1 stimulation of monocytes leads to tumor formation in nontumorigenic melanoma cells. J Immunol 166:6483

Ness RB, Cottreau C (1999) Possible role of ovarian epithelial inflammation in ovarian cancer. J Natl Cancer Inst 91:1459

Nishioka Y, Sone S, Okubo A, Ogura T (1990) Suppression by interleukin-4 of activation of human blood monocytes to the tumoricidal state. Jpn J Cancer Res 81:936

Nowicki A, Szenajch J, Ostrowska G, Wojtowicz A, Wojtowicz K, Kruszewski AA, Maruszynski M, Aukerman SL, Wiktor-Jedrzejczak W (1996) Impaired tumor growth in colony-stimulating factor 1 (CSF-1)-deficient, macrophage-deficient *op/op* mouse: evidence for a role of CSF-1-dependent macrophages in formation of tumor stroma. Int J Cancer 65:112

O'Byrne KJ, Dalgleish AG, Browning MJ, Steward WP, Harris AL (2000) The relationship between angiogenesis and the immune response in carcinogenesis and the progression of malignant disease. Eur J Cancer 36:151

Oehler JR, Hernerman RB, Campbell Jr., DA, Djeu JY (1977) Inhibition of rat mixed lymphocyte cultures by suppressor macrophages. Cell Immunol 29:238

Okahara H, Yagita H, Miyake K, Okumura K (1994) Involvement of very late antigen 4 (VLA-4) and vascular cell adhesion molecule 1 (VCAM-1) in tumor necrosis factor alpha enhancement of experimental metastasis. Cancer Res 54:3233

Ono M, Torisu H, Fukushi J, Nisjie A, Kuwano M (1999) Biological implication of macrophage infiltration in human tumor angiogenesis. Cancer Chemother Pharmacol 43:69

Otsuji M, Kimura Y, Aoe T, Okamoto Y, Saito T (1996) Oxidative stress by tumor-derived macrophages suppresses the expression of CD3 ζ chain of T-cell receptor complex and antigen-specific T-cell responses. Proc Natl Acad Sci USA 93:13119

Pace JL, Russell SW, Torres BA, Johnson HM, Gray PW (1983) Macrophage activation to cytotoxicity by recombinant IFN-γ. J Immunol 130:2011

Parajuli P, Singh SM (1996) Alteration in IL-1 and arginase activity of tumor-associated macrophages: a role in the promotion of tumor growth. Cancer Lett 107:249

Parhar RS, Lala PK (1988) Prostaglandin E$_2$-mediated inactivation of various killer lineage cells by tumor-bearing host macrophages. J Leukoc Biol 44:474

Park IC, Park MJ, Choe TB, Jang JJ, Hong SI, Lee SH (2000) TNF-alpha induces apoptosis mediated by AEBSF-sensitive serine protease(s) that may involve upstream caspase-3/CPP32 protease activation in a human gastric cancer cell line. Int J Oncol 16:1243

Pegg AE (1988) Polyamine metabolism and its importance in neoplastic growth and as a target for chemotherapy. Cancer Res 48:759

Pellegrini P, Berghella AM, Del Beato T, Cicia S, Adorno D, Casciani CU (1996) Disregulation in TH1 and TH2 subsets of CD4+ T cells in peripheral blood of colorectal cancer patients and involvement in cancer establishment and progression. Cancer Immunol Immunother 42:1

Pelus LM, Bockman RS (1979) Increased prostaglandin synthesis by macrophages from tumor-bearing mice. J Immunol 123:2118

Peng L, Shu S, Krauss JC (1997) Monocyte chemoattractant protein inhibits the generation of tumor-reactive T cells. Cancer Res 57:4849

Polverini PJ, Leibovich J (1984) Induction of neovascularization in vivo and endothelial proliferation in vitro by tumor-associated macrophages. Lab Invest 51:635

Raes G, Van Ginderachter J, Liu YQ, Brys L, Thielemans K, De Baetselier P, Geldhof A (1998) Active antitumor immunotherapy, with or without B7-mediated costimulation, increases tumor progression in an immunogenic murine T cell lymphoma model. Cancer Immunol Immunother 45:257

Raes G, De Baetselier P, Noël W, Beschin A, Brombacher F, Hassanzadeh Gh G (2002) Differential expression of FIZZ1 and Ym1 in alternatively versus classically activated macrophages. J Leukoc Biol 71:597

Remels L, Fransen L, Huygen K, De Baetselier P (1990) Poly I: C activated macrophages are tumoricidal for TNF-α-resistant 3LL tumor cells. J Immunol 144:4477

Rodriguez-Sosa M, Satoskar AR, Caldcron R, Gomcz-Garcia L, Saavedra R, Bojalil R, Terrazas LI (2002) Chronic helminth infection induces alternatively activated macrophages expressing high levels of CCR5 with low interleukin-12 production and Th2-biasing ability. Infect Immun 70:3656

Ruggiero V, Johnson SE, Baglioni C (1987) Protection from tumor necrosis factor cytotoxicity by protease inhibitors. Cell Immunol 107:317

Saio M, Radoja S, Marino M, Frey AB (2001) Tumor-infiltrating macrophages induce apoptosis in activated CD8+ T cells by a mechanism requiring cell contact and mediated by both the cell-associated form of TNF and nitric oxide. J Immunol 167:5583

Sakaguchi S, Sakaguchi N, Shimizu J, Yamazaki S, Sakihama T, Itoh M, Kuniyasu Y, Nomura T, Toda M, Takahashi T (2001) Immunologic tolerance maintained by CD25+ CD+ regulatory T cells: their common role in controlling autoimmunity, tumor immunity, and transplantation tolerance. Immunol Rev 182:18

Sallenave JM (2000) The role of secretory leukocyte proteinase inhibitor and elafin (elastase-specific inhibitor/skin-derived antileukoprotease) as alarm antiproteinases in inflammatory lung disease. Respir Res 1:87

Sato M, Goto S, Kaneko R, Ito M, Sato S, Takeuchi S (1998) Impaired production of Th1 cytokines and increased frequency of Th2 subsets in PBMC from advanced cancer patients. Anticancer Res 18:3951

Satoskar AR, Rodig S, Telford 3rd, SR, Satoskar AA, Ghosh SK, von Lichtenberg F, David JR (2000) IL -12 gene-deficient C57BL/6 mice are susceptible to *Leishmania donovani* but have diminished hepatic immunopathology. Eur J Immunol 30:834

Schebesch C, Kodelja V, Müller C, Hakij N, Bisson S, Orfanos CE, Goerdt S (1997) Alternatively activated macrophages actively inhibit proliferation of peripheral blood lymphocytes and CD4+ T cells in vitro. Immunology 92:478

Schoppmann SF, Birner P, Stöckl J, Kalt R, Ullrich R, Caucig C, Kriehuber E, Nagy K, Alitalo K, Kerjaschki D (2002) Tumor -associated macrophages express lymphatic endothelial growth factors and are related to peritumoral lymphangiogenesis. Am J Pathol 161:947

Schutyser E, Struyf S, Proost P, Opdenakker G, Laureys G, Verhasselt B, Peperstraete L, Van de Putte I, Saccani A, Allavena P, Mantovani A, Van Damme J (2002) Identification of biologically active chemokine isoforms from ascitic fluid and elevated levels of CCL18/pulmonary and activation-regulated chemokine in ovarian carcinoma. J Biol Chem 277:24584

Shankaran V, Ikeda H, Bruce AT, White JM, Swanson PE, Old LJ, Schreiber RD (2001) IFN γ and lymphocytes prevent primary tumour development and shape tumour immunogenicity. Nature 410:1107

Sheu BC, Lin RH, Lien HC, Ho HN, Hsu SM, Huang SC (2001) Predominant Th2/Tc2 polarity of tumor-infiltrating lymphocytes in human cervical cancer. J Immunol 167:2972

Sica A, Saccani A, Bottazzi B, Polentarutti N, Vecchi A, van Damme J, Mantovani A (2000) Autocrine production of IL-10 mediates defective IL-12 production and NF-κB activation in tumor-associated macrophages. J Immunol 164:762

Smyth MJ, Thia KY, Street SE, Cretney E, Trapani JA, Taniguchi M, Kawano T, Pelikan SB, Crowe NY, Godfrey DI (2000) Differential tumour surveillance by natural killer (NK) and NKT cells. J Exp Med 191:661

Smyth MJ, Godfrey DI, Trapani JA (2001) A fresh look at tumor immunosurveillance and immunotherapy. Nat Immunol 2:293

Song E, Ouyang N, Horbelt M, Antus B, Wang M, Exton MS (2000) Influence of alternatively and classically activated macrophages on fibrogenic activities of human fibroblasts. Cell Immunol 204:19

Sotomayor EM, DiNapoli MR, Calderon C, Colsky A, Fu YX, Lopez DM (1995) Decreased macrophage-mediated cytotoxicity in mammary-tumor-bearing mice is related to alteration of nitric-oxide production and/or release. Int J Cancer 60:660

Sotomayor EM, Borrello I, Levitsky HI (1996) Tolerance and cancer: a critical issue in tumor immunology. Crit Rev Oncog 7:433

Standiford TJ, Kunkel SL, Liebler JM, Burdick MD, Gilbert AR, Strieter RM (1993) Gene expression of macrophage inflammatory protein-1 α from human blood monocytes and alveolar macrophages is inhibited by interleukin-4. Am J Respir Cell Mol Biol 9:192

Starkey JR, Liggitt HD, Jones W, Hosick HL (1984) Influence of migratory blood cells on the attachment of tumor cells to vascular endothelium. Int J Cancer 34:535

Stein M, Keshav S, Harris N, Gordon S (1992) Interleukin 4 potently enhances murine macrophage mannose receptor activity: a marker of alternative immunologic macrophage activation. J Exp Med 176:287

Stevens SR, Shibaki A, Meunier L, Cooper KD (1995) Suppressor T cell-activating macrophages in ultraviolet-irradiated human skin induce a novel, TGF-β-dependent form of T cell activation characterized by deficient IL-2r α expression. J Immunol 155:5601

Stuehr DJ, Nathan CF (1989) Nitric oxide, a macrophage product responsible for cytostasis and respiratory inhibition in tumor target cells. J Exp Med 169:1543

Suk K, Somers SD, Erickson KL (1993) Regulation of murine macrophage function by IL-4: IL-4 and IFN-γ differentially regulate macrophage tumoricidal activation. Immunology 80:617

Sunderkotter C, Steinbrink K, Goebeler M, Bhardwaj R, Sorg C (1994) Macrophages and angiogenesis. J Leukoc Biol 55:410

Szekanecz Z, Haines GK, Lin TR, Harlow LA, Goerdt S, Rayan G, Koch AE (1994) Differential distribution of intercellular adhesion molecules (ICAM-1, ICAM-2, and ICAM-3) and the MS-1 antigen in normal and diseased human synovia. Their possible pathogenetic and clinical significance in rheumatoid arthritis. Arthritis Rheum 37:221

Takahashi A, Kono K, Ichihara F, Sugai H, Amemiya H, Iizuka H, Fujii H, Matsumoto Y (2003) Macrophages in tumor-draining lymph node with different characteristics induce T-cell apoptosis in patients with advanced stage-gastric cancer. Int J Cancer 104:393

Tamura Y, Peng P, Liu K, Daou M, Srivastava PK (1997) Immunotherapy of tumors with autologous tumor-derived heat shock protein preparations. Science 278:117

Terrazas LI, Walsh KL, Piskorska D, McGuire E, Harn Jr., DA (2001) The schistosome oligosaccharide lacto-*N*-neotetraose expands Gr-1+ cells that secrete anti-inflammatory cytokines and inhibit proliferation of naïve CD4+ cells: a potential mechanism for immune polarization in helminth infections. J Immunol 167:5294

Ting CC, Rodrigues D (1982) Tumor cell-triggered macrophage-mediated suppression of the T-cell cytotoxic response to tumor-associated antigens. I. Characterization of the cell components for induction of suppression. J Natl Cancer Inst 69–867

Torisu H, Ono M, Kiryu H, Furue M, Ohmoto Y, Nakayama J, Nishioka Y, Sone S, Kuwano M (2000) Macrophage infiltration correlates with tumor stage and angiogenesis in human malignant melanoma: possible involvement of TNF-a and IL-1a. Int J Cancer 85:182

Trinchieri G (1995) Interleukin-12: a proinflammatory cytokine with immunoregulatory functions that bridge innate resistance and antigen-specific adaptive immunity. Annu Rev Immunol 13:251

Tsuchiya Y, Igarashi M, Suzuki R, Kumagai K (1988) Production of colony-stimulating factor by tumor cells and the factor-mediated induction of suppressor cells. J Immunol 141:699

Tsunawaki S, Nathan CF (1986) Macrophage deactivation. Altered kinetic properties of superoxide-producing enzyme after exposure to tumor cell-conditioned medium. J Exp Med 164:1319

Tsung K, Dolan JP, Tsung YL, Norton JA (2002) Macrophages as effector cells in interleukin 12-induced T cell-dependent tumor rejection. Cancer Res 62:5069

Tzachanis D, Berezovskaya A, Nadler LM, Boussiotis VA (2002) Blockade of B7/CD28 in mixed lymphocyte reaction cultures results in the generation of alternatively activated macrophages, which suppress T-cell responses. Blood 99:1465

Urban JL, Shepard HM, Rothstein JL, Sugarman BJ, Schreiber H (1986) Tumor necrosis factor: a potent effector molecule for tumor cell killing by activated macrophages. Proc Natl Acad Sci USA 83:5233

van den Berg A, Visser L, Poppema S (1999) High expression of the CC chemokine TARC in Reed-Sternberg cells. A possible explanation for the characteristic T-cell infiltrate in Hodgkin's lymphoma. Am J Pathol 154:1685

Van den Broek ME, Kagi D, Ossendorp F, Toes R, Vamvakas S, Lutz WK, Melief CJ, Zinkernagel RM, Hengartner H (1996) Decreased tumor surveillance in perforin-deficient mice. J Exp Med 184:1781

Van Ginderachter JA, Liu YQ, Geldhof AB, Brys L, Thielemans K, De Baetselier P, Raes G (2000) B7-1, IFN-gamma and anti-CTLA-4 co-operate to prevent T-cell tolerization during immunotherapy against a murine T-lymphoma. Int J Cancer 87:539

Van Hal PT, Hopstaken-Broos JP, Wijkhuijs JM, Te Velde AA, Figdor CG, Hoogsteden HC (1992) Regulation of aminopeptidase-N (CD13) and Fcε RIIb (CD23) expression by IL-4 depends on the stage of maturation of monocytes/macrophages. J Immunol 149:1395

VandenDriessche T, Bakkus M, Toussaint-Demylle D, Thielemans K, Verschueren H, De Baetselier P (1994) Tumorigenicity of mouse T lymphoma cells is controlled by the level of major histocompatibility complex class I H-2K^k antigens. Clin Exp Metastasis 12:73

Vitolo D, Zerbe T, Kanbour A, Dahl C, Herberman RB, Whiteside TL (1992) Expression of mRNA for cytokines in tumor-infiltrating mononuclear cells in ovarian adenocarcinoma and invasive breast cancer. Int J Cancer 51:573

Wahl SM, Hunt DA, Wakefield LM, McCartney-Francis N, Wahl LM, Roberts AB, Sporn MB (1987) Transforming growth factor type β induces monocyte chemotaxis and growth factor production. Proc Natl Acad Sci USA 84:5788

Welch JS, Escoubet-Lozach L, Sykes DB, Liddiard K, Greaves DR, Glass CK (2002) Th2 cytokines and allergic challenge induce Ym1 expression in macrophages by a STAT6-dependent mechanism. J Biol Chem 277:42821

Whiteside TL, Herberman RB (1995) The role of natural killer cells in immune surveillance of cancer. Curr Op Immunol 7:704

Wigmore SJ, Fearon KC, Sangster K, Maingay JP, Garden OJ, Ross JA (2002) Cytokine regulation of constitutive production of interleukin-8 and -6 by human pancreatic cancer cell lines and serum cytokine concentrations in patients with pancreatic cancer. Int J Oncol 21:881

Wu G, Morris Jr., SM, (1998) Arginine metabolism: nitric oxide and beyond. Biochem J 336(Pt 1):1

Zhang F, Lu W, Dong Z (2002) Tumor -infiltrating macrophages are involved in suppressing growth and metastasis of human prostate cancer cells by IFN-beta gene therapy in nude mice. Clin Cancer Res 8:2942

Zhang Y, Khoo HE, Esuvaranathan K (1999) Effects of bacillus Calmette-Guèrin and interferon alpha-2B on cytokine production in human bladder cancer cell lines. J Urol 161:977

H.E. Kaiser and A. Nasir (eds.), Selected Aspects of Cancer Progression:
Metastasis, Apoptosis and Immune Response, 157–167.
© Springer Science + Business Media B.V. 2008

CHAPTER TEN

Characterization of tumor-directed cellular immune responses in humans

Dirk Nagorsen, MD, Vladia Monsurro, PhD, and Francesco M. Marincola, MD

Abstract: Understanding tumor/host immune interactions may help to fight cancer. Growing knowledge about T cell responses and increasing success of immunotherapeutic approaches have created the need for methods to characterize tumor-directed cellular immune responses. The spectrum of methods reaches from protein-based methods, including tetramers or intracellular flow cytometry, to genetic assays, such as TCR analysis or microarray techniques, further on to functional assays analysing proliferation and microtoxicity. Here, we describe these and further methods and explain their respective application in human tumor immunology.

Introduction

The host's cellular immune system can fight autologous tumor (Dunn et al. 2002) through a tumor-directed T cell response. Tumor-specific T cell responses can develop either spontaneously (Nagorsen et al. 2003) or after specific immunotherapy (Rosenberg 2001). The growing knowledge about natural T cell responses and increasing success of cellular immunotherapy against cancer have created the need for efficient methods to identify and characterize tumor-directed cellular immune responses. It is well established that malignancies express tumor-associated antigens (TAA), such as differentiation antigens, shared antigens overexpressed on cancers, cancer germline antigens, or mutated antigens, which can be recognized by TAA-specific CD8$^+$ T cells in a major histocompatibility complex (MHC) class I restricted fashion (Boon et al. 1997; Renkvist et al. 2001). Therefore, the main focus of technologic development since the early 1990s has been on monitoring these TAA-specific CD8$^+$ T cell responses in peripheral blood (Keilholz et al. 2002). New technologies allow *ex vivo* characterization of immune system/tumor interactions at the actual tumor site, and the monitoring could be extended to other relevant cell populations (e.g. CD4$^+$ T cells, monocytes, dendritic cells [DC], or natural killer [NK] cells), that may play an additional role in defense against tumors.

Based on their underlying principles, methods for the characterization of tumor-directed cellular immune responses can be divided into three major groups: (1) protein-based assays, (2) functional assays, and (3) molecular genetics-based methods (see Fig. 1). Depending on the specific question asked, each

Immunogenetics Section, Department of Transfusion Medicine, Clinical Center, National Institutes of Health, Bethesda, MD.
Address correspondence to Dr. F.M. Marincola at the Department of Transfusion Medicine, Clinical Center, NIH, Building 10, Room 1C711, 10 Center Drive, Bethesda, MD, 20892-1502, U.S.A., Tel: (301) 496-3098, FAX (301) 4021360, e-mail: fmarincola@mail.cc.nih.gov

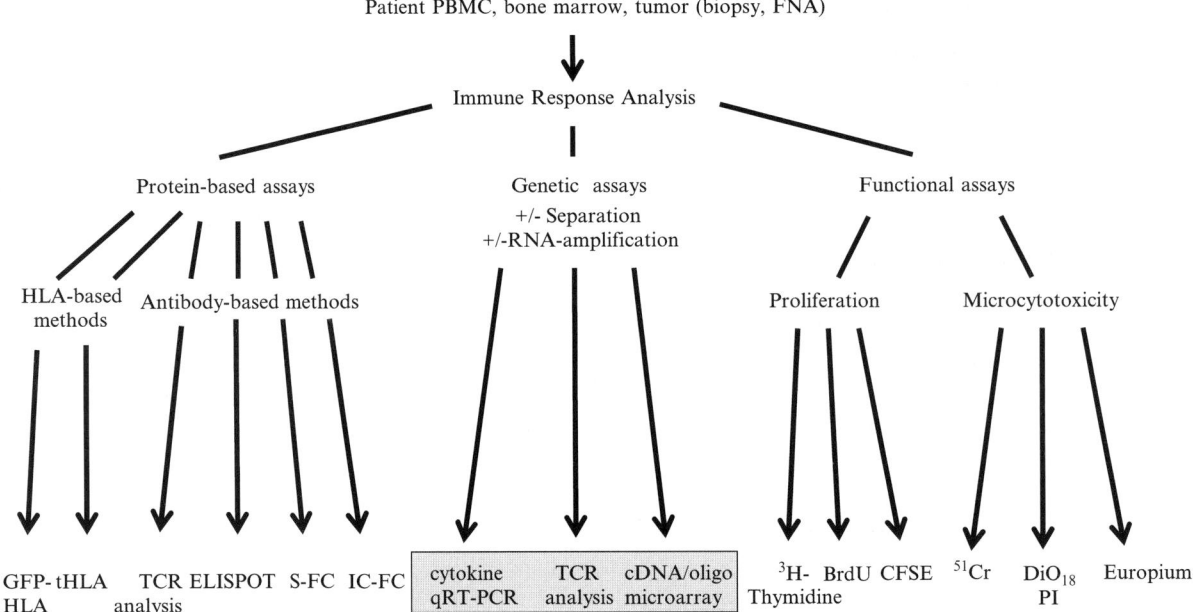

Fig. 1 The expanding range of methods for the characterization of tumor-directed cellular immune responses. In recent years, genetic-based methods (gray background) have started to complement protein-based and functional methods. For several genetic approaches a meaningful cell subset has to be separated. This can be achieved by flow cytometric separation, microbeads, microdissection, or under specific culture conditions (Modified from Nagorsen et al. 2002)

category has its respective advantages and disadvantages, and a combined approach will be most useful.

Protein-based methods

Protein-based methods can be divided into antibody-based and human leukocyte antigen (HLA)-based methods. Antibody-based T cell assays are classified as ELISPOT and flow cytometry methods. The ELISPOT assay is a relatively easy-to-perform and sensitive enzyme-linked immunosorbent spot assay detecting specific T cells by their antigen (Ag)-induced secretion of cytokines (mainly IFNγ) (Schmittel et al. 2000). Peripheral mononuclear cells (PBMC) – or other cell populations of interest – are put onto 96-well nitrocellulose plates coated with specific first anti-cytokine (IFNγ) monoclonal antibodies (mAb). During 24 h of coincubation with Ag (e.g. by peptide or tumor cells), specific T cells secrete cytokines, which are bound by the surrounding mAb. Then a second, biotinylated anti-cytokine-mAb is added to the plate, and later streptavidin-alkaline phosphatase is added. Finally, the plates can be analyzed through

a microscope, or better, by using an automated image analysis system for spots representing cytokine-releasing T cells. Several groups have shown a good correlation between the number of IFNγ secreting T cells in the ELISPOT assay and cytotoxicity determined by the traditional chromium release assay (Di Fabio et al. 1994; Miyahira et al. 1995; Scheibenbogen et al. 2000; see also *Functional assays*). This ELISPOT assay can be used to detect T cell responses not only against peptides but also against tumor cells or soluble proteins (CD4+ T cell responses) (Schmittel et al. 2001). Automated image analysis systems have overcome the subjectivity of spot counting. The major disadvantage of the ELISPOT is that further phenotyping is laborious and possible only by indirect methods, such as using blocking antibodies or depleting cell subsets.

Intracellular IFNγ flow cytometric staining (IC-FC) is almost as sensitive as the ELISPOT assay and has the major advantage of allowing further phenotyping of Ag-specific T cells. A good correlation was shown between IFNγ ELISPOT assay and IC-FC (Asemissen et al. 2001). IC-FC is based on flow cytometric measurement of intracellular cytokine levels (also most

common: IFNγ) after Ag-specific stimulation of T cells, blocking of cytokine secretion to allow intracellular accumulation of cytokines, and permeabilization of cells (Suni et al. 1998). A further flow cytometric approach for detecting IFNγ-secreting T cells is the use of a bispecific antibody trapping secreted IFNγ at the cell surface (Brosterhus et al. 1999). This so-called IFNγ-catch-reagent is a bispecific anti-CD45 and anti-IFNγ antibody binding with its anti-CD45 portion to the leukocyte surface and "catching" secreted IFNγ with the other side. Then a labeled second anti-IFNγ antibody (the so-called detection-antibody) binds to a different site of the IFNγ molecule and allows flow cytometric analysis. This IFNγ-secretion assay is particularly useful for separation of live, specific T cells because it does not require permeabilization and fixation. However, the specificity of this assay for monitoring T cell responses remains to be elucidated (Asemissen et al. 2001). Like the ELISPOT assay, both flow cytometric methods can detect reactivity against unknown antigens (for example, presented by autologous tumor cells) and provide basic information about the functionality of the detected cells.

The above methods are used for monitoring T cell responses *ex vivo* from peripheral blood or bone marrow. At the tumor site, so far, only less specific antibody-based methods can be used to survey immune responses, such as staining for infiltrating T cells (CD3, CD4, CD8), NK cells (CD56) or phagocytes/antigen-presenting cells (CD14, CD80, CD83, CD86). Antibodies have been used to detect T cell receptor (TCR) distributions in peripheral blood as well as at the tumor site, where a significant increase of a given TCR subgroup is likely to correspond to the clonal expansion of a TAA-specific clone. Antibodies that specifically recognize a V gene subgroup allow a quantitation of the V region expression of αβ T cells and provide an opportunity to further study the phenotype of relevant T cells. However, sensitivity and accuracy of these data must be interpreted cautiously, since antibodies cover only about 70% of T cells and the identification of a V gene subgroup is no direct proof for clonotypic identity of a given TCR (please see also *Genetic methods*).

Two HLA-based methods are in use for determining Ag-specific T cells *ex vivo*: multimerized HLA class I molecules carrying a specific peptide and labeled with a fluorescent marker (tHLA or tetramers) (Altman et al. 1996) and peptide-HLA-green-fluorescence-protein

(GFP) complexes (Tomaru et al. 2003). To produce tHLA, recombinant HLA heavy chains containing a biotinylation site and recombinant β2-microglobulin are synthesized in *E. coli*. The respective epitope peptide is added to the HLA heavy chain and β2-microglobulin. The refolded complex is then isolated and biotinylated. Fluorescent streptavidin is added to induce a tetramer formation. Once established, tHLA are time saving and relatively easy-to-use. The tetramers are added to the cells and coincubated for 15 min to bind to complementary T cell receptors; then antibodies (e.g. anti-CD3 or anti-CD8) can be added. tHLA is the only method that allows an estimation of the avidity between TCR and peptide-loaded HLA class I molecules (Dutoit et al. 2002). This is especially important because the avidity of Ag recognition by tumor-specific CD8+ T cells seems to correlate with the efficiency of tumor recognition. This parameter may determine the potency of a tumor rejection Ag and correlate in adoptive transfer models with protection against tumor and viral infection (Alexander-Miller et al. 1996; Yee et al. 1999; Zeh et al. 1999). Furthermore, tHLA (and other flow cytometric approaches as well) allow a detailed phenotyping of specific T cells. However, no direct conclusions on the functional activity of T cells are possible because tetramer and intracellular IFNγ double-positive T cells are difficult to detect due to down-regulation of TCR tetramer complexes during the staining procedure (Lee et al. 1999; Nielsen et al. 2000). Two main staining methods have been developed for *in situ* tetramer staining (Haanen et al. 2000; Skinner et al. 2000). Skinner and Haase (2002) give a comprehensive overview. There is, however, not much experience with these techniques in humans, and it has been difficult to establish tetramer staining in human tissues (personal experience). Whether these difficulties are due to TCR engagement in the tumor environment is unclear. It remains to be seen what role these important and promising techniques will play. A second HLA-based technique is the detection of Ag-specific T cells by peptide-HLA-GFP complexes (Tomaru et al. 2003). Antigen presenting cells (APC) expressing HLA-A*0201 coupled to green fluorescent protein (GFP) are pulsed with an Ag-derived peptide. During stimulation of Ag-specific T cells, these T cells incorporate the peptide-HLA-GFP complexes. Thus, Ag-specific T cells can be identified and analyzed by fluorescence-based detection methods.

The incremental understanding of the function of several Ag-specific T cell subtypes led to a greater appreciation of the importance of characterizing Ag-specific T cells directly *ex vivo*. Several subpopulations of TAA-specific T cells have been defined by correlating specific surface markers with T cell functions, such as cytotoxicity or homing. Hamann et al. (1999) used CD45RA and CD27 expression to describe three subsets of CD8$^+$ T cells: CD27$^+$CD45RA$^+$ T cells represent naive, CD27$^+$CD45RA$^-$ memory, and CD27$^-$CD45RA$^+$ effector subpopulations. Sallusto et al. (1999) defined T cells based on the expression of the lymph node-homing chemokine receptor CCR7 as naive T cells (CD45RA$^+$CCR7$^+$), central memory T cells (CD45RA$^-$CCR7$^+$), effector memory T cells (CD45RA$^-$CCR7$^-$) and terminally differentiated effector T cells (CD45RA$^+$CCR7$^-$). A similar distinction of T cell subsets was made using CD27/CD28 (Appay et al. 2002). We know now that these classifications represent different cells with different characteristics that may be important for cancer defense, such as apoptosis-inducing ligands, cytotoxic effector functions, cytokines, and homing behavior. Thus, a determination of T cell characteristics is a very helpful tool for a realistic reflection of their capabilities *in vivo*. In many studies, however, functional aspects of Ag-specific T cells have been described after *in vitro* culture, with repeated Ag restimulation and/or exogenous cytokines to increase the sensitivity. Although T cell stimulation *in vitro* over a few days is probably too short to generate specific T cells from their precursors and, thus, comparative quantitative analyses are reliable, even short-term culture leads to dramatic changes in T cells. Besides a huge increase of the frequency of Ag-specific T cells in PBMC from 3% to more than 60% after *in vitro* sensitization and addition of IL-2 (Monsurro et al. 2001), culture causes alterations of phenotype and cytotoxic functions. A strong increase of cytolytic activity was shown after *in vitro* restimulation of Ag-specific T cells (Pittet et al. 2001). After *in vitro*-sensitization and *in vitro*-culture for 10 days, Ag-specific CD8$^+$ T cells expressed more perforin (more than 80% of CD8$^+$ tHLA$^+$ cells compared to 17% before culture) and up-regulated CD27 (from less than one-third to more than two-thirds of CD8$^+$tHLA$^+$ cells) (Monsurro et al. 2002). This strongly suggests that a characterization representing the *in vivo* state of specific T cells is inaccurate after *in*

vitro treatment of Ag-specific T cells. Thus, methods treating cells with cytokines and/or peptides (*in vitro* sensitization or short time culture) are not useful for *ex vivo* monitoring and characterization of specific T cell responses. A clear distinction between directly *ex vivo*, short-term *in vitro*, and long-term *in vitro* T cell responses should be used in publications.

In summary, protein-based methods represent the current first line tools for the analysis of Ag-specific T cell responses (Keilholz et al. 2002). ELISPOT, tHLA, and IC-FC are particularly suitable for monitoring immunotherapy trials. Flow cytometric assays can be used to further characterize Ag-specific T cells. However, there are limitations when monitoring systemic immune responses for the purpose of predicting clinical outcome. So far, no clear correlation has been found between immune response and clinical effect. Obviously, circulating T cells represent only one of various factors interacting during an immune response. Other components determining homing of T cells to the tumor site, their survival and function at the tumor site, and adjustment of cancer cells trying to evade T cell recognition may make it difficult to find such a correlation.

Functional assays

Although cytokine production after Ag stimulation also reflects T cell function, a traditional understanding is that the microtoxicity and T cell proliferation assays are tests of T cell functionality in a narrower sense. Assays to analyze the ability of T cells to destroy cells have been known for many years. The classical assay for detection of target cell destruction uses lysis of radiolabeled target cells by lymphocyte populations usually after limiting dilution (Brunner et al. 1968). Target cells (e.g. tumor cells or Ag-presenting cells) are labeled with ^{51}chromium (^{51}Cr) and then incubated with effector cells at several effector/target ratios. Radioactivity caused by ^{51}Cr release is measured in the supernatant of these cultures using a gamma counter. Valmori et al. (2002b) demonstrated direct *ex vivo* tumor cell lysis by peripheral TAA-specific T cells using this assay. Several other methods have been developed to substitute for the ^{51}Cr release assay. Comparable results to the ^{51}Cr assay can be obtained by europium release assay (Zons et al. 1997). This method is based on the

measurement of the nonradioactive, fluorescent marker europium by labeled target cells after coincubation with effector cells (e.g. T cells or NK cells) (Blomberg et al. 1986). In another method, target cells are stained with a green fluorescent membrane dye (DiO_{18}) and the nuclear dye propidium iodide (PI) (Mattis et al. 1997). While all target cells are stained green, only lysed cells are stained with PI. However, these alternative methods have not been used widely so far.

Although these cytotoxicity assays are rather time-consuming and allow only semiquantitative estimation of Ag-specific T cells, they are important because they allow actual testing of whether effector cells are able to lyse cells. One disadvantage is that these methods often require culturing and addition of cytokines, such as interleukin (IL)-2, which may induce major alterations in the functional status of T cells and, thus, may not reflect the actual T cell activity *in vivo* (Monsurro et al. 2002). Which test of this variety will develop for immune monitoring purposes remains to be determined.

Proliferation assays measure the ability of T cells to proliferate in response to a given antigen. Thymidine incorporation is the most common assay. For this method, T cells are coincubated with an antigen (and cytokines) for several days. Then, cells are pulsed with titrated thymidine. Finally, the incorporation of radioactive thymidine into the DNA of proliferating cells is measured. As a nonradioactive alternative, a good correlation was shown between the traditional thymidine assay and bromo-2′-deoxyuridine (BrdU)-incorporation assay (Messele et al. 2000). Although these methods are relatively sensitive, they do not give data on cell viability and do not indicate which populations of cells are proliferating after Ag exposure. A more modern approach uses carboxyfluorescein diacetate, succinimidyl ester (CFSE), a molecule that diffuses into cells and is intracellularly converted to a membrane-binding fluorescent (Lyons and Parish 1994). With each cell division, the fluorescence intensity is halved. Thus, after an Ag exposure, cycles of division can be analyzed by flow cytometry. This method allows a further phenotyping of the dividing/proliferating cells by simultaneous staining with differentially fluorescent antibodies (e.g. CD4 or CD8).

Proliferation assays are rated as comparatively imprecise and are not recommended for monitoring vaccination studies (Keilholz et al. 2002). Nevertheless

they are pivotal for certain tasks. Obviously, these culture assays cannot provide direct data about *ex vivo* characteristics, but their read-out, proliferation in response to an antigenic signal, seems to be crucial for an efficient immune defense.

Taken together, functional assays are not useful for first-line monitoring of TAA-specific T cells. However, they can provide important information about the proliferation and killing abilities of cells.

Genetic methods

Genetic methods for the study of specific immune cell function are based mainly on the measurement of transcript levels. Quantitative (real-time) reverse transcriptase polymerase chain reaction (qRT-PCR), TCR clonality analyses, and complementary DNA (cDNA) or oligonucleotide microarrays are molecular genetics-based methods that may extend the analysis of Ag-specific T cells. We isolated three major contributions of genetic-based methods for the analysis of cellular immune responses (Nagorsen et al. 2002): (1) sensitivity, (2) globality, and (3) plasticity, respectively referring to (1) necessary improvement of sensitivity (qRT-PCR) to analyze particular T cell subsets, (2) an expanded analysis yielding simultaneous information about thousands of factors (microarrays) rather than only a few dozen (protein-based methods), and (3) the changes of T cell frequency, clonality, and localization (TCR analysis) in response to environmental stimulation in natural or therapeutic conditions. Genetic methods allow high-throughput measurements with high sensitivity and a detailed look into phenotype, function, and genotypic variation of Ag-specific T cells. However, these methods have their limitations, including lack of discrimination between individual cells and a possible dissociation of messenger RNA (mRNA) expression and its corresponding protein level, which might be especially important in cancer patients with alterations of gene translation. Genetic methods complement rather than replace protein-based methods.

Kammula et al. (1999) were the first to apply qRT-PCR to the analysis of TAA-specific T cells. They observed immunization-induced T cell reactivity by measuring IFNγ transcript levels in PBMC of peptide-vaccinated melanoma patients exposed *ex vivo*

to relevant epitope. qRT-PCR is based on a simple concept (Heid et al. 1996; Kruse et al. 1997). Since the amount of cDNA after amplification will be directly proportional to the log2 of the original amount, RNA copies can be calculated using a standard curve from samples with a known amount of RNA. This logarithmic amplification determines one of the major advantages of qRT-PCR: its high sensitivity. In comparative studies, detection of vaccine-induced responses in circulating T cells by qRT-PCR correlated with results obtained by *in vitro* sensitization, T cell phenotyping with tHLA, and IC-FC (Monsurro et al. 2001; Kammula et al. 2000; Nielsen and Marincola 2000). The use of qRT-PCR for analysis of specific T cells has some specific problems. Standard "housekeeping" genes, such as β-actin mRNA, cannot be used because they do not consider variations in target cell frequency. Since CD8[+] T cells are the responding cells to HLA class I restricted TAA-derived epitope, CD8 mRNA has been proposed as a reference to normalize variations in the ratio of various T cell subsets in a given population (i.e. PBMC) (Kammula et al. 1999). Also, the determination of a threshold for "positive results" is under discussion. Kammula et al. (1999) proposed a two- to three-fold increase in transcript levels above baseline conditions as a positive result based on extensive analyses of variance in samples stimulated with relevant and irrelevant epitope. In addition to the study of peripheral blood, analysis of tumor specimens obtained from tumor fine needle aspirates (FNA) is an important application of qRT-PCR. Combining qRT-PCR with FNA, dynamic changes in expression of antigens, cytokines, and other markers were documented during immunization (Ohnmacht et al. 2001; Mocellin et al. 2001). The limited amount of material obtainable with FNA would have allowed the study of only a few markers, while the RNA extracted from individual FNA and amplified according to Wang's method allowed the study of a virtually unlimited number of genes (Wang et al. 2000; Wang and Marincola 2002). In conclusion, although qRT-PCR requires more validation, it is a sensitive tool for monitoring immune activities in tumor patients and offers singular advantages.

Analyses of TCR clonality make it possible to follow the adjustment of T cell frequency, clonality, and homing behavior to various natural or therapeutic conditions. T cells recognize Ag through engagement of their TCR made of αβ or γδ chain heterodimers. Most T cells express TCR consisting of α and β chains, which mediate binding to epitope-loaded HLA molecules. Genes encoding α and β TCR chains rearrange in the thymus during maturation using germline variable (V), diversity (D), and junction (J) segments (Davis and Bjorkman 1988). The TCR loci contain 44–46 (α) and 40–42 (β) functional V regions belonging to 32–34 (α) and 21–23 (β) subgroups, 2 D (only β) genes, and 50 (α) and 13 (β) J regions (Rowen et al. 1996; Lefranc and Lefranc 2001). During rearrangement in the thymus, α and β germline genes combine (combinatorial diversity), and a flexible process involving endonucleases, exonucleases, and terminal transferase creates V(D)J junctions (junctional diversity) of different lengths coding for different sequences. Compared to antibody-based analysis, molecular detection of specific V genes/subgroups containing several genes has the advantage of covering 100% of the αβ TCR repertoire. However, it can be performed only on extracted RNA or by in situ hybridization on fixed cells and tissues. TCR analysis is mainly based on RT-PCR using V region-specific primers paired to a constant region primer. Low-resolution methods using agarose gels and high-resolution methods using polyacrylamide gels are deployed. For low-resolution methods, after agarose gel separation, PCR products are either detected directly in the gel or on Southern blots. Such methods are semiquantitative at best. However, the large variations in expression levels in tissue infiltrates allow the identification of subgroups likely to contain expanded TAA-specific T cell clones. High-resolution methods are based on clone-specific recognition of functional α and β V(D)J junctions that vary in size and sequence. V(D)J regions correspond to the complementarity-determining region 3 (CDR3) that contains the only non-germline encoded diversity of the TCR and is responsible for the specificity of the HLA-peptide complex (i.e. Ag recognition). CDR3 can be used to identify specific clones and to detect them in different locations at different times in a clonotypic fashion (Puisieux et al. 1994). The most commonly used method analyzes CDR3 size patterns on denaturing sequencing gels using either radioactively or fluorescence-labeled DNA analyzed on automated sequencers (Cochet et al. 1992; Hingorani ct al. 1993; Kissela et al. 1994). Well-established variations of this approach

are spectratyping (Kissela et al. 1994) and immunoscope (Pannetier et al. 1995). The CDR3 lengths vary over 8–10 codons, resulting in Gaussian-like curve of peaks spaced by three nucleotides, with the most intense band representing the largest number of transcripts. The sensitivity of the detection of an expanded clone depends on the primers used to obtain the amplified DNA. At best, a given clone can be detected at a resolution of one cell in 10^5. In the further developed TC landscape technique (Guillet et al. 2001), CDR3 size patterns can be represented graphically based on expression levels assessed by real-time qPCR and the degree of perturbation of patterns. In addition to CDR3 size analysis, single stranded conformational polymorphism (SSCP) (Andrews et al. 1997) and heteroduplexing (Wack et al. 1996) can be used to analyze TCR repertoires. In both techniques, PCR products are migrated on non-denaturing polyacrylamide gels as single-stranded DNAs or heteroduplexed to a clonal reference DNA belonging to the same subfamily. Both methods allow the detection of PCR products obtained from expanded clones on the basis of their sequence, that is, a band detected above the background corresponds to a given CDR3 sequence that determines the migration pattern of the amplified clonal DNA. The advantage of both methods over CDR3 size patterns is that a band identifies a clone, whereas a size peak on a denaturing sequencing gel may correspond to one or several clones. Heteroduplexing and CDR3 size pattern analysis have been compared and heteroduplexing has been found to be slightly more sensitive (Maini et al. 1998). Applying the above TCR-based techniques to subsets of Ag-specific T cells together with molecular cloning and sequencing throughout a disease process represents a powerful complement to protein-based methods by analyzing T cell clonality and tracking the kinetics of T cell responses.

Functional genomic methods document the expression of thousands of genes in a given cell population in one microarray simultaneously ("molecular fingerprint"). This technique has literally revolutionized molecular biology by adding multiplication, automation, and high throughput to conventional gene expression analysis (Schena et al. 1995). Two different methods have been developed: cDNA- and oligonucleotide-based arrays. For cDNA-based arrays, thousands of genes (cDNA, 600–1,000 base pairs long) are spotted in high density but separately on a solid surface, such as a glass slide. Test RNA (such as immune cells) and reference RNA (e.g. pooled healthy donor PMBC) are separately converted into cDNA with simultaneous incorporation of different reporter fluorochrome molecules (e.g. test Cy5 red and reference Cy3 green). Labeled test and reference cDNA are co-hybridized to the immobilized arrayed genes. Genes equally expressed by both test and reference samples will fluoresce with both colors digitally portrayed as yellow. Genes present only in the test material fluoresce red, and those present in the reference material fluoresce green. By measuring both fluorescence intensities, the relevant individual gene expression level can be calculated as a ratio of test fluorescence intensity over reference fluorescence intensity. Oligo-microarrays represent an alternative technology for high-throughput gene expression analysis. This technology is available in two variations, which are based either on short (approximately 25 bases) or long (50–80 bases) oligonucleotides synthesized *in situ*. Both, cDNA- and oligoarray, have specific advantages and disadvantages. cDNA microarrays have a high sensitivity and a slightly reduced specificity due to the use of comparatively long pieces of cDNA. They permit the analysis of either sense or antisense RNA because of printed double-stranded cDNA. Oligo-microarrays are often more specific, allowing even, if appropriately designed, the detection of single nucleotides polymorphisms. Production-related requirements raise oligoarray manufacturing costs and make this technology more expensive than cDNA microarrays. Both techniques can cause technical difficulties related to array quality (spot printing), labeling efficiency, and statistical error caused by the large number of parameters analyzed (Wang and Marincola 2002). Normalization and quality control of microarrays are the basis for generating meaningful data. Repetition and reciprocal labeling especially distinguish actual gene expression data from experimental variation and labeling bias (Wang et al. 2000). Powerful statistical analysis software, such as hierarchical cluster analysis (Eisen et al. 1998), helps to avoid errors and identify true patterns of gene expression. "Fingerprints" of gene expression have been linked to various functions of immune cells, including dendritic cells and T cells (Huang et al. 2001; Granucci et al. 2001; Chtanova et al. 2001; Hamalainen et al. 2001). After an initial effort to

extensively investigate immune cells by microarrays *in vitro* to identify differentiation pathways or tolerance/immunity decision processes (Glynne and Watson 2001), the effect of immunotherapeutic treatments was analyzed directly *ex vivo*. Panelli et al. (2002) analyzed RNA from PBMC and FNA of melanoma metastases of patients undergoing IL-2 therapy. This study suggests that the immediate effect of IL-2 on the tumor microenvironment is transcriptional activation of genes associated with monocyte function. Wang et al. (2002) used cDNA microarrays to analyze RNA from FNA samples from melanoma metastases prior to immunotherapy. A gene pattern was identified that predicted a clinical response. About half of these genes were related to T cell regulation, suggesting that immune responsiveness might be predetermined by the tumor microenvironment.

Since genetic-based methods cannot distinguish between RNA from different cell subsets, cells of interest must be separated prior to analysis. This can be done by several methods including laser microdissection, FNA over a time period, as the above examples (Panelli et al. 2002; Wang et al. 2002) show, or protein-based methods. Live Ag-specific T cells can be labeled by tHLA or using the cytokine-secretion assay. In a second step, cells thus labeled can be separated using magnetic beads, laser microdissection, or a flow cytometric cell sorter. Several studies describe promising examples of combining protein-based separation and subsequent molecular analysis of Ag-specific T cells. Douek et al. (2002) used qRT-PCR and TCR clonotyping to characterize HIV-specific T cells separated by cytokine-secretion assay. This study has demonstrated a high variability in the number of clonotypes specific for HIV epitopes. Valmori et al. (2002a) and Lim et al. (2002) have enriched melanoma-Ag-specific T cells with tHLA and subsequently analyzed CDR3 lengths/sequences and clonotypes by qRT-PCR. Jäger et al. (2002) separated a subset of tHLA-positive T cells in melanoma patients and further characterized these cells by TC landscape and sequencing.

Several approaches allow the separation of cell subsets, and tumor samples can be obtained by FNA during a time course. Often, however, only small numbers of cells can be obtained using these techniques. The amount of RNA extracted might not be sufficient for extensive analyses, such as microarrays or TCR analy-

ses. In these circumstances, RNA amplification can be a very helpful tool. Amplification can increase RNA 10^5-fold without changing the original proportion of transcript levels and allows complex gene expression analyses when only a limited number of cells is available (Wang et al. 2000). RNA amplification was used to monitor tumor–host interaction in melanoma patients undergoing immunotherapy (Panelli et al. 2002; Wang et al. 2002). The studies used an optimized procedure on low-abundance RNA samples by combining antisense RNA amplification (Phillips and Eberwine 1996) with template switch effect (for more technical details about the amplification methods, please see Wang and Marincola 2002). The addition of high-fidelity RNA amplification allows analyses of huge numbers of genes present in any tissue sample without causing significant alteration of their relative expression. Routine mRNA amplification was recommended even for all cDNA microarray-based analysis of gene expression (Feldman et al. 2002).

Conclusions

Appropriate and effective immune monitoring must rely on a variety of methods to embrace the complex of tumor-directed cellular immune responses. The choice of a particular method depends on the specific question and the features of the technique. ELISPOT, IC-FC, and tHLA represent the current first-line monitoring of TAA-specific T cell responses (Keilholz et al. 2002) since they are relatively easy to perform, reliable, have been used often in clinical trials, and allow in part (IC-FC and tHLA) a detailed characterization of T cell subsets based on markers such as CD45RA, CD27, and CCR7. Especially for such subset analysis, performing direct *ex vivo* analysis is important to reflect the actual *in vivo* state of immune cells. The additional application of molecular methods can augment and deepen the immune response analysis. qRT-PCR can increase the sensitivity of detection of specific T cell responses. cDNA microarrays allow high-throughput analysis of cellular immune responses by portraying the expression of thousands of genes simultaneously. Alterations of T cell responses can be followed *ex vivo* at the clonal level using different TCR-based approaches. Furthermore, genetic-based methods allow analyses

of the expression of all genes for which the sequence is known, thus eliminating the need to develop antibodies or other biological markers, as is required for protein-based methods. However, molecular methods do not allow phenotypic analyses of individual cells. This limitation can be circumvented by prior cell separation followed by RNA amplification if necessary. Furthermore, although protein and gene expression levels correlate for some cytokines, for example, IFNγ (Favre et al. 1997), this association is not known for most genes associated with immune responses. Genetic-based assays do not provide information about amount and biochemical characteristics of a gene product. Nevertheless, high-throughput molecular genetic methods establish a basis for subsequent confirmatory studies, including further gene-specific methods and proteomic approaches. Proteomic methods hold promise for gaining additional large-scale information about isoforms or post-translational alterations of immune response-related proteins (Naour 2001; Mosca et al. 2003).

References

Alexander-Miller MA, Leggatt GR, Sarin A, Berzofsky JA (1996) Role of antigen, CD8, and cytotoxic T lymphocyte (CTL) avidity in high dose antigen induction of apoptosis of effector CTL. J Exp Med 184:485–492

Altman JD, Moss PA, Goulder PJ, Barouch DH, McHeyzer-Williams MG, Bell JI, McMichael AJ, Davis MM (1996) Phenotypic analysis of antigen-specific T lymphocytes. Science 274:94–96

Andrews DM, Leary CP, Hishii M, Shen J, Kurnick JT (1997) Use of single stranded conformational polymorphism (SSCP) for analysis of the T cell receptor. In: Oksenberg JR (ed) The antigen T cell receptor: selected protocols and applications. Chapman & Hall, New York, pp 373–406

Appay V, Dunbar PR, Callan M, Klenerman P, Gillespie GM, Papagno L, Ogg GS, King A, Lechner F, Spina CA, Little S, Havlir DV, Richman DD, Gruener N, Pape G, Waters A, Easterbrook P, Salio M, Cerundolo V, McMichael AJ, Rowland-Jones SL (2002) Memory CD8+ T cells vary in differentiation phenotype in different persistent virus infections. Nat Med 8:379–385

Asemissen AM, Nagorsen D, Keilholz U, Letsch A, Schmittel A, Thiel E, Scheibenbogen C (2001) Flow cytometric determination of intracellular or secreted ifngamma for the quantification of antigen reactive T cells. J Immunol Methods 251:101–108

Blomberg K, Granberg C, Hemmila I, Lovgren T (1986) Europium-labelled target cells in an assay of natural killer cell activity. I. A novel non-radioactive method based on time-resolved fluorescence. J Immunol Methods 86:225–229

Boon T, Coulie PG, Van den Eynde B (1997) Tumor antigens recognized by T cells. Immunol Today 18:267–268

Brosterhus H, Brings S, Leyendeckers H, Manz RA, Miltenyi S, Radbruch A, Assenmacher M, Schmitz J (1999) Enrichment and detection of live antigen-specific CD4 and CD8 T cells based on cytokine secretion. Eur J Immunol 29:4053

Brunner KT, Mauel J, Cerottini JC, Chapuis B (1968) Quantitative assay of the lytic action of immune lymphoid cells on 51-Cr-labelled allogeneic target cells in vitro; inhibition by isoantibody and by drugs. Immunology 14:181–196

Chtanova T, Kemp RA, Sutherland AP, Ronchese F, Mackay CR (2001) Gene microarrays reveal extensive differential gene expression in both CD4(+) and CD8(+) type 1 and type 2 T cells. J Immunol 167:3057–3063

Cochet M, Pannetier C, Regnault A, Darche S, Leclerc C, Kourilsky P (1992) Molecular detection and in vivo analysis of the specific T cell response to a protein antigen. Eur J Immunol 22:2639–2647

Davis MM, Bjorkman PJ (1988) T-cell antigen receptor genes and T-cell recognition. Nature 334:395–402

Di Fabio S, Mbawuike IN, Kiyono H, Fujihashi K, Couch RB, McGhee JR (1994) Quantitation of human influenza virus-specific cytotoxic T lymphocytes: correlation of cytotoxicity and increased numbers of IFN-gamma producing CD8+ T cells. Int Immunol 6:11–19

Douek DC, Betts MR, Brenchley JM, Hill BJ, Ambrozak DR, Ngai KL, Karandikar NJ, Casazza JP, Koup RA (2002) A novel approach to the analysis of specificity, clonality, and frequency of HIV-specific T cell responses reveals a potential mechanism for control of viral escape. J Immunol 168:3099–3104

Dunn GP, Bruce AT, Ikeda H, Old LJ, Schreiber RD (2002) Cancer immunoediting: from immunosurveillance to tumor escape. Nat Immunol 3:991–998

Dutoit V, Rubio-Godoy V, Doucey MA, Batard P, Lienard D, Rimoldi D, Speiser D, Guillaume P, Cerottini JC, Romero P, Valmori D (2002) Functional avidity of tumor antigen-specific CTL recognition directly correlates with the stability of MHC/peptide multimer binding to TCR. J Immunol 168:1167–1171

Eisen MB, Spellman PT, Brown PO, Botstein D (1998) Cluster analysis and display of genome-wide expression patterns. Proc Natl Acad Sci USA 95:14863–14868

Favre N, Bordmann G, Rudin W (1997) Comparison of cytokine measurements using ELISA, ELISPOT and semi-quantitative RT-PCR. J Immunol Methods 204:57–66

Feldman AL, Costouros NG, Wang E, Qian M, Marincola FM, Alexander HR, Libutti SK (2002) Advantages of mRNA amplification for microarray analysis. Biotechniques 33:906–912, 914

Glynne RJ, Watson SR (2001) The immune system and gene expression microarrays – new answers to old questions. J Pathol 195:20–30

Gorski J, Yassai M, Zhu X, Kissela B, Kissella B [corrected to Kissela B, Keever C, Flomenberg N]: (1994) Circulating T cell repertoire complexity in normal individuals and bone marrow recipients analyzed by CDR3 size spectratyping. Correlation with immune status. J Immunol 152:5109–5119

Granucci F, Vizzardelli C, Pavelka N, Feau S, Persico M, Virzi E, Rescigno M, Moro G, Ricciardi-Castagnoli P (2001) Inducible IL-2 production by dendritic cells

revealed by global gene expression analysis. Nat Immunol 2:882–888

Guillet M, Sebille F, Soulillou J (2001) TCR usage in naive and committed alloreactive cells: implications for the understanding of TCR biases in transplantation. Curr Opin Immunol 13:566–571

Haanen JB, van Oijen MG, Tirion F, Oomen LC, Kruisbeek AM, Vyth-Dreese FA Schumacher TN (2000) In situ detection of virus- and tumor-specific T-cell immunity. Nat Med 6:1056–1060

Hamalainen H, Zhou H, Chou W, Hashizume H, Heller R, Lahesmaa R (2001) Distinct gene expression profiles of human type 1 and type 2 T helper cells. Genome Biol 2(7): RESEARCH0022

Hamann D, Roos MT, van Lier RA (1999) Faces and phases of human CD8 T-cell development. Immunol Today 20:177–180

Heid CA, Stevens J, Livak KJ, Williams PM (1996) Real time quantitative PCR. Genome Res 6:986

Hingorani R, Choi IH, Akolkar P, Gulwani-Akolkar B, Pergolizzi R, Silver J, Gregersen PK (1993) Clonal predominance of T cell receptors within the CD8+ CD45RO+ subset in normal human subjects. J Immunol 151:5762–5769

Huang Q, Liu D, Majewski P, Schulte LC, Korn JM, Young RA, Lander ES, Hacohen N (2001) The plasticity of dendritic cell responses to pathogens and their components. Science 294:870–875

Jäger E, Hohn H, Necker A, Förster R, Karbach J, Freitag K, Neukirch C, Castelli C, Salter RD, Knuth A, Maeurer MJ (2002) Peptide-specific CD8+ T-cell evolution in vivo: response to peptide vaccination with Melan-A/MART-1. Int J Cancer 98:376–388

Kammula US, Lee KH, Riker AI, Wang E, Ohnmacht GA, Rosenberg SA, Marincola FM (1999) Functional analysis of antigen-specific T lymphocytes by serial measurement of gene expression in peripheral blood mononuclear cells and tumor specimens. J Immunol 163:6867–6875

Kammula US, Marincola FM, Rosenberg SA (2000) Real-time quantitative polymerase chain reaction assessment of immune reactivity in melanoma patients after tumor peptide vaccination. J Natl Cancer Inst 92:1336–1344

Keilholz U, Weber J, Finke JH, Gabrilovich DI, Kast WM, Disis ML, Kirkwood JM, Scheibenbogen C, Schlom J, Maino VC, Lyerly HK, Lee PP, Storkus W, Marincola F, Worobec A, Atkins MB (2002) Immunologic monitoring of cancer vaccine therapy: results of a workshop sponsored by the Society for Biological Therapy. J Immunother 25:97–138

Kruse N, Pette M, Toyka K, Rieckmann P (1997) Quantification of cytokine mrna expression by RT PCR in samples of previously frozen blood. J Immunol Methods 210:195

Lee KH, Wang E, Nielsen MB, Wunderlich J, Migueles S, Connors M, Steinberg SM, Rosenberg SA, Marincola FM (1999) Increased vaccine-specific T cell frequency after peptide-based vaccination correlates with increased susceptibility to in vitro stimulation but does not lead to tumor regression. J Immunol 163:6292–6300

Lefranc MP, Lefranc G (2001) The T cell receptor facts book. Academic, London

Lim A, Baron V, Ferradini L, Bonneville M, Kourilsky P, Pannetier C (2002) Combination of MHC-peptide multimer-based T cell sorting with the Immunoscope permits sensitive ex vivo quantitation and follow-up of human CD8+ T cell immune responses. J Immunol Methods 261:177–194

Lyons AB, Parish CR (1994) Determination of lymphocyte division by flow cytometry. J Immunol Methods 171:131–137

Maini MK, Wedderburn LR, Hall FC, Wack A, Casorati G, Beverley PC (1998) A comparison of two techniques for the molecular tracking of specific T-cell responses; CD4+ human T-cell clones persist in a stable hierarchy but at a lower frequency than clones in the CD8+ population. Immunology 94:529–535

Mattis AE, Bernhardt G, Lipp M, Forster R (1997) Analyzing cytotoxic T lymphocyte activity: a simple and reliable flow cytometry-based assay. J Immunol Methods 204: 135–142

Messele T, Roos MT, Hamann D, Koot M, Fontanet AL, Miedema F, Schellekens PT, Rinke de Wit TF (2000) Nonradioactive techniques for measurement of in vitro T-cell proliferation: alternatives to the [(3)H]thymidine incorporation assay. Clin Diagn Lab Immunol 7:687–692

Miyahira Y, Murata K, Rodriguez D, Rodriguez JR, Esteban M, Rodrigues MM, Zavala F (1995) Quantification of antigen specific CD8+ T cells using an ELISPOT assay. J Immunol Methods 181:45–54

Mocellin S, Ohnmacht GA, Wang E, Marincola FM (2001) Kinetics of cytokine expression in melanoma metastases classifies immune responsiveness. Int J Cancer 93:236–242

Monsurro V, Nielsen MB, Perez-Diez A, Dudley ME, Wang E, Rosenberg SA, Marincola FM (2001) Kinetics of TCR use in response to repeated epitope-specific immunization. J Immunol 166:5817–5825

Monsurro V, Nagorsen D, Wang E, Provenzano M, Dudley ME, Rosenberg SA, Marincola FM (2002) Functional heterogeneity of vaccine-induced CD8(+) T cells. J Immunol 168:5933–5942

Mosca PJ, Lyerly HK, Ching CD, Hobeika AC, Clay TM, Morse MA (2003) Proteomics for monitoring immune responses to cancer vaccines. Curr Opin Mol Ther 5:39–43

Nagorsen D, Wang E, Marincola FM, Even J (2002) Transcriptional analysis of tumor-specific T-cell responses in cancer patients. Crit Rev Immunol 22:449–462

Nagorsen D, Scheibenbogen C, Marincola FM, Letsch A, Keilholz U (2003) Natural T cell immunity against cancer. Clin Cancer Res 9:4296–303

Naour F (2001) Contribution of proteomics to tumor immunology. Proteomics 1:1295–1302

Nielsen MB, Marincola FM (2000) Melanoma vaccines: the paradox of T cell activation without clinical response. Cancer Chemother Pharmacol 46:s62–s66

Nielsen MB, Monsurro V, Migueles SA, Wang E, Perez-Diez A, Lee KH, Kammula U, Rosenberg SA, Marincola FM (2000) Status of activation of circulating vaccine-elicited CD8+ T cells. J Immunol 165:2287–2296

Ohnmacht GA, Wang E, Mocellin S, Abati A, Filie A, Fetsch P, Riker AI, Kammula US, Rosenberg SA, Marincola FM (2001) Short-term kinetics of tumor antigen expression in response to vaccination. J Immunol 167:1809–1820

Panelli MC, Wang E, Phan G, Puhlmann M, Miller L, Ohnmacht GA, Klein HG, Marincola FM (2002) Gene-expression profiling of the response of peripheral blood mononuclear cells and melanoma metastases to systemic IL-2 administration. Genome Biol 3(7):RESEARCH0035

Pannetier C, Even J, Kourilsky P (1995) T-cell repertoire diversity and clonal expansions in normal and clinical samples. Immunol Today 16:176–181

Phillips J, Eberwine JH (1996) Antisense RNA amplification: a linear amplification method for analyzing the mrna population from single living cells. Methods 10:283–288

Pittet MJ, Speiser DE, Lienard D, Valmori D, Guillaume P, Dutoit V, Rimoldi D, Lejeune F, Cerottini JC, Romero P (2001) Expansion and functional maturation of human tumor antigen-specific CD8+ T cells after vaccination with antigenic peptide. Clin Cancer Res 7:796s–803s

Puisieux I, Even J, Pannetier C, Jotereau F, Favrot M, Kourilsky P (1994) Oligoclonality of tumor-infiltrating lymphocytes from human melanomas. J Immunol 153:2807–2818

Renkvist N, Castelli C, Robbins PF, Parmiani G (2001) A listing of human tumor antigens recognized by T cells. Cancer Immunol Immunother 50(1):3–15

Rosenberg SA (2001) Progress in human tumour immunology and immunotherapy. Nature 411:380–384

Rowen L, Koop BF, Hood L (1996) The complete 685-kilobase DNA sequence of the human beta T cell receptor locus. Science 272:1755–1762

Sallusto F, Lenig D, Forster R, Lipp M, Lanzavecchia A (1999) Two subsets of memory T lymphocytes with distinct homing potentials and effector functions. Nature 401:708–712

Scheibenbogen C, Romero P, Rivoltini L, Herr W, Schmittel A, Cerottini JC, Wolfel T, Eggermont AM, Keilholz U (2000) Quantitation of antigen-reactive T cells in peripheral blood by ifngamma-ELISPOT assay and chromium-release assay: a four-centre comparative trial. J Immunol Methods 244:81–89

Schena M, Shalon D, Davis RW, Brown PO (1995) Quantitative monitoring of gene expression patterns with a complementary DNA microarray. Science 270:467–470

Schmittel A, Keilholz U, Thiel E, Scheibenbogen C (2000) Quantification of tumor-specific T lymphocytes with the ELISPOT assay. J Immunother 23:289–295

Schmittel A, Keilholz U, Bauer S, Kuhne U, Stevanovic S, Thiel E, Scheibenbogen C (2001) Application of the IFN-gamma ELISPOT assay to quantify T cell responses against proteins. J Immunol Methods 247:17–24

Skinner PJ, Haase AT (2002) In situ tetramer staining. J Immunol Methods 268:29–34

Skinner PJ, Daniels MA, Schmidt CS, Jameson SC, Haase AT (2000) Cutting edge: In situ tetramer staining of antigen-specific T cells in tissues. J Immunol 165:613–617

Suni MA, Picker LJ, Maino VC (1998) Detection of antigen-specific T cell cytokine expression in whole blood by flow cytometry. J Immunol Methods 212:89–98

Tomaru U, Yamano Y, Nagai M, Maric D, Kaumaya PT, Biddison W, Jacobson S (2003) Detection of virus-specific T cells and CD8(+) T-cell epitopes by acquisition of peptide-HLA-GFP complexes: analysis of T-cell phenotype and function in chronic viral infections. Nat Med 9:469–476

Valmori D, Dutoit V, Schnuriger V, Quiquerez AL, Pittet MJ, Guillaume P, Rubio-Godoy V, Walker PR, Rimoldi D, Lienard D, Cerottini JC, Romero P, Dietrich PY (2002a) Vaccination with a Melan-A peptide selects an oligoclonal T cell population with increased functional avidity and tumor reactivity. J Immunol 168:4231–4240

Valmori D, Scheibenbogen C, Dutoit V, Nagorsen D, Asemissen AM, Rubio-Godoy V, Rimoldi D, Guillaume P, Romero P, Schadendorf D, Lipp M, Dietrich PY, Thiel E, Cerottini JC, Lienard D, Keilholz U (2002b) Circulating Tumor-reactive CD8(+) T cells in melanoma patients contain a CD45RA(+)CCR7(−) effector subset exerting ex vivo tumor-specific cytolytic activity. Cancer Res 62:1743–1750

Wack A, Montagna D, Dellabona P, Casorati G (1996) An improved PCR-heteroduplex method permits high-sensitivity detection of clonal expansions in complex T cell populations. J Immunol Methods 196:181–192

Wang E, Marincola FM (2002) Amplification of small quantities of mrna for expression analysis. In: David Bowtell, Joseph Sambrook (eds) DNA microarrays. A molecular cloning manual. Cold Spring Harbor Laboratory Press, New York, pp 204–213

Wang E, Miller LD, Ohnmacht GA, Liu ET, Marincola FM (2000) High-fidelity mrna amplification for gene profiling. Nat Biotechnol 18(4):457–459

Wang E, Miller LD, Ohnmacht GA, Mocellin S, Perez-Diez A, Petersen D, Zhao Y, Simon R, Powell JI, Asaki E, Alexander HR, Duray PH, Herlyn M, Restifo NP, Liu ET, Rosenberg SA, Marincola FM (2002) Prospective molecular profiling of melanoma metastases suggests classifiers of immune responsiveness. Cancer Res 62:3581–3586

Yee C, Savage PA, Lee PP, Davis MM, Greenberg PD (1999) Isolation of high avidity melanoma-reactive CTL from heterogeneous populations using peptide-MHC tetramers. J Immunol 162:2227–2234

Zeh 3rd., HJ Perry-Lalley D, Dudley ME, Rosenberg SA, Yang JC (1999) High avidity ctls for two self-antigens demonstrate superior in vitro and in vivo antitumor efficacy. J Immunol 162:989–994

Zons P von, Crowley-Nowick P, Friberg D, Bell M, Koldovsky U, Whiteside TL (1997) Comparison of europium and chromium release assays: cytotoxicity in healthy individuals and patients with cervical carcinoma. Clin Diagn Lab Immunol 4:202–207

H.E. Kaiser and A. Nasir (eds.), Selected Aspects of Cancer Progression:
Metastasis, Apoptosis and Immune Response, 169–191.
© *Springer Science + Business Media B.V.* 2008

CHAPTER ELEVEN

Immunological aspects of Marek's disease virus (MDV)-induced lymphoma progression

Immune Suppression and Modulation

Mark S. Parcells[1]* and Shane C. Burgess[2]

Abstract: Marek's disease is a highly transmissible T-cell lymphoma of chickens caused by the only known acutely transforming alphaherpesvirus, Marek's disease virus (MDV). Losses due to MDV-induced tumors (lymphomas, skin leukoses, etc.) have been minimized in poultry production since the early 1970s through the use of non-sterilizing vaccines. Initial lymphoma development in MDV-infected chickens is dependent on the challenge strain of MDV, the genetic susceptibility of the exposed chickens, the level of challenge and relative exposure to other adventitious agents. In this chapter, we examine the factors affecting the progression of MDV-induced, CD4+ T-lymphomas as a consequence of the immune insult incurred by the host during early lytic and latent phases of MDV infection, and as a consequence of the factors expressed by the transformed T-lymphoblasts. A special emphasis has been placed on the role of de-regulated host surface antigen expression on lymphoma progression. Several of these antigens (CD29/CD49e, CD30, CD44) are common to invasive and metastatic human lymphomas, suggesting common mechanisms of immune modulation in lymphoma progression.

Keywords: Marek's disease, lymphoma progression, herpesvirus oncology, immunophenotype, lymphomagenesis

Introduction

Marek's disease (MD) is a highly transmissible herpesvirus infection associated with paralysis, immunosuppression and T-cell lymphomas of chickens. MD has served as a model for herpesvirus oncology for over three decades. Despite recent insight into the molecular mechanism of T-cell transformation in MD, the mechanisms of lymphoma development and progression in susceptible lines of chickens remain enigmatic. As the causative agent, Marek's disease virus (MDV) is the only known acute-transforming alphaherpesvirus. Moreover, lymphomas induced by MDV have been largely controlled in commercial poultry production through the use of non-sterilizing, cell-associated vaccines (i.e. vaccines do not prevent persistent superinfection of the host). MDV can not only cause visceral T-cell lymphomas, paralysis, blindness, and neurological dysfunction, but can also induce a less clinically obvious, yet profound immunosuppression. In this chapter, we explore MDV-induced T-lymphoma formation as a result of the combination of the immunosuppression incurred by the host during MDV infection and the distinctive expression changes in

[1]O-404, Center of Excellence for Poultry Science, University of Arkansas, Fayetteville, AR USA.
[2]Department of Basic Sciences, College of Veterinary Medicine, Mississippi State University, Starkville, MS USA.
*Current address: 052 Townsend Hall, University of Delaware Newark. DE, USA

MDV-transformed T-lymphoblasts as they contribute to further lymphoma progression and metastasis.

Marek's disease

Pathogenesis

The pathogenesis of MD is dependent on virus-, host- and environmental-specific conditions (Calnek and Witter 1997). Virus-specific conditions include the virulence level of the MDV (pathotype) to which the chicken is exposed and challenge dose. Host-specific conditions include the inherent genetic susceptibility of the line challenged and the level of protection provided by vaccination. Environmental factors include the physical conditions within the facility (temperature, ammonia levels, relative humidity, etc.) as well as the level of challenge from other adventitious agents (bacteria, other viruses, fungi, parasites, etc.).

The current paradigm is that MDV infection occurs via the inhalation of infectious dander (Calnek and Witter 1997). The dander-associated MDV particles are then and phagocytosed, presumably by macrophages of the trachea or lung epithelium which then transmits MDV infection to macrophages, though this remains to be proven. Most of our understanding regarding MDV pathogenesis comes from genetically defined specified-pathogen-free chickens, raised in isolation and challenged with defined MDV strains, at standard doses and by standard routes of exposure. This brief introduction will focus on such defined models.

Upon entry, four stages of MDV infection are recognized; namely, (1) early cytolytic infection, (2) latent infection, (3) secondary lytic infection and (4) transformation (tumor development and progression) (Calnek and Witter 1997). With strains of high virulence, these stages become somewhat blurred with the tumor development seen in stages 2 through 4. Shedding of MDV is from the feather follicle epithelium from 10 days-post-infection (dpi) (Beasley et al. 1970; Calnek et al. 1970).

During the early cytolytic infection, MDV replicates in B-cells of the bursa of Fabricius. MDV replication in B-cells is lytic and productive/restrictive, in that infectious virus is strictly cell-associated (Calnek et al. 1984; Payne et al. 1976). Nearly concomitant with B-cell infection, T-lymphocytes become infected,

presumably through interactions with B-cells and/or infected antigen presenting cells (APCs). Replication is then detected in the thymus, spleen and finally, as a cell-associated viremia in peripheral blood leukocytes (PBLs). Recently, macrophages have been demonstrated to be infected by MDV during this initial phase and the level of macrophage infection is greater in MDV strains of high virulence (Barrow et al. 2003).

Replication in the bursa initiates 1–3 days post-inoculation in 3 week old or older chickens, infected via intra-abdominal injection of lytically infected chicken kidney cells (CKC) or embryo fibroblasts (CEF) (Calnek and Witter 1997). Replication in the thymus and spleen follows shortly thereafter. Peak replication in the spleen, at levels much higher than in the bursa and thymus, occurs approximately 7 days post infection (dpi). In vaccinated, resistant chickens, the cell-associated viremia in PBL peaks at about 14 dpi and falls thereafter to a low but persistent level of latent infection. In susceptible chickens, the pattern of viremia is similar, but the level of latent infection in PBLs remains high and persists into the period of lymphoma development (28–42 dpi).

In chickens infected at 1 day-of-age, the progression of MDV viremia is delayed. Peaks of bursal, thymic and splenic infections occur approximately 7 days later than the infections in older chickens (M.S. Parcells, 1996, unpublished data). This delay is most likely due to the low numbers of target cells in the primary and secondary immune organs of the chicken at this time. Despite the delay in viremia onset, chickens exposed at day of age are more susceptible to MDV infection and generally exhibit more severe expression of MD (Heier and Jarp 2000). This phenomenon of viremia in older chickens and subsequent lesser MD clinical expression has been called age-associated resistance.

Clinical signs, lesions and mortality

The clinical signs specific to "classical MD" are paralysis (due to peripheral neural lymphomas) (Marek 1907). At post-mortem examination, thickening of the peripheral nerves and enlargement of gonads are seen. As MDV has evolved greater virulence, clinical signs have increased to include blindness, cachexia, diarrhoea, skin lesions, and torticollis (Witter et al. 1979). This "acute MD" is due to increased distribution of lymphomas in all tissue of the body. Highly virulent MDVs (vv⁺ MDVs) also cause death during the cytolytic phase of disease in young chickens (Witter 1997; Witter et al. 1999).

MD lesions are edema and cytolysis during the cytolytic phase and both inflammatory and proliferative in later stages (Calnek and Witter 1997). These later lesions, which are lymphomas involving infiltration and proliferation of lymphoblasts, have been graded A-C based on classical histopathological examination (Payne 1985). A-type lesions are primarily inflammatory, being composed of macrophages, heterophils and CD8⁺ T-lymphocytes. C-type lesions are primarily proliferating, transformed CD4⁺ T-lymphocytes.

The transformed cells within MD lymphomas, like many human lymphomas, express high levels of CD30, a member of the tumor necrosis receptor family (TNFRII) (Burgess et al. 2004). A more detailed description of these lymphoma cells is given below. MD lymphomas range in size from microscopic to frank and can be found in all visceral organs, muscle, nerves eyes and skin. Recent MDV isolates cause early proliferative lesions in the spleen by 5–7 days post-infection (Rosenberger et al. 1997; Witter 1997). These become encysted and necrotic within 2–3 weeks, as they appear to outgrow their blood supply.

Mortality caused by MDV can occur at various times post-infection depending on the virulence and dosage of the challenge virus. A period of early death occurs with highly virulent strains of MDV within 10–14 dpi (Witter et al. 1999). This early mortality has been associated with monocytosis, bursal and thymic atrophy, proliferative neurological lesions and paralysis (Barrow et al. 2003). A second peak of mortality is seen in the proliferative phase (4–6 weeks post-infection) and is primarily lymphoma-associated. Finally, in chickens infected later in life, there is a late phase of sporadic mortality which is also associated with tumor formation. One common clinical presentation is for hens to develop lymphomas as they go into lay (19–21 weeks). Recent isolates of highly virulent MDVs cause distinct neurological signs in chickens of various ages, as well as a high incidence of spleen tumors (Gimeno et al. 2002).

Mortality attributable to MD can also be caused by secondary viruses and bacteria due to the profound immune suppression caused by MDV (see below). This additional MD-associated loss is often unappreciated and may go undetected. Thus the cost of MD to the poultry industry is almost certainly much higher than the losses recognized due to lymphomas and other direct mortality alone.

Moreover, MD clinical expression can be exacerbated by coinfection with avian leukosis/sarcoma viruses (ALV/ASV), chicken infectious anemia virus (CIAV), infectious bursal disease virus (IBDV) and reticuloendotheliosis virus (REV) (Jakovleva and Mazurenko 1979; Miles et al. 2001; Takagi et al. 1996; von Bulow et al. 1986; Witter et al. 1979).

Target cells

MDV lytic (albeit productive/restrictive) replication *in vivo* is primarily in B-cells, some T-cells, epitheliod cells (feather follicle epithelium, Schwann cells, etc.), and macrophages (Barrow et al. 2003; Calnek et al. 1984; Shek et al. 1983). In the feather follicle epithelium (FFE), MDV undergoes fully productive infection and this is the only recognized site of glycoprotein D (gD) expression (Calnek et al. 1970; Niikura et al. 1999). Although nonessential for pathogenesis and horizontal transmission (Anderson et al. 1998), the maintenance of gD in the MDV genome suggests a functional role for this expression. Exogenous expression of gD in CEF during MDV lytic infection does not stimulate the expression of cell-free virus, and there are likely to be other cellular factors required for the production of cell-free MDV (Zelnik et al. 1999).

The primary site of MDV latent infection is the activated CD4⁺ T-lymphocyte (Shek et al. 1983). The identification of the CD4⁺ T-lymphocyte as the major latency target is based on: (1) the proliferative components of an MD lymphomas are CD4⁺ T-lymphocytes (Nazerian and Sharma 1975), (2) cell lines established from MDV-induced lymphomas are almost exclusively CD4⁺ T-lymphocytes (Parcells et al. 1999; Schat et al. 1991), (3) MDV-transformed cell lines contain the MDV genome in a latent state that can be reactivated upon co-cultivation with susceptible cell types (CEF, CKC), and (4) CD4⁺ T-lymphocytes are the major site of infectivity in the peripheral blood from 3 weeks post-infection when lytically infected lymphocytes cannot be detected in the circulation (Baigent et al. 1998; Calnek et al. 1984).

The status of MDV genome during latency has been debated. Early biochemical studies suggested an episomal form (Hirai et al. 1981; Rziha and Bauer 1982), but subsequent studies of lymphomas and derived cell lines suggest integration of MDV at the ends of chromosomes (telomeres) is the predominant form of the latent MDV genome (Delecluse and Hammerschmidt 1993; Delecluse et al. 1993). The recent finding of an MDV-encoded telomerase templating RNA, vTR, as well as telomeric repeats at the junctions between genome segments (Fragnet et al. 2003; Kishi et al. 1988, 1991), suggest that these may be involved in the targeting of the MDV genome to telomeric sites. Reactivation of the MDV genome from these sites, therefore, would require some excision or targeted replication mechanism yielding monomeric genomes. These mechanisms are theoretical at this point. Conversely, minute quantities of episomal genomes may be present that serve as replication templates during reactivation.

Oncogenesis

Meq

The mechanism of MDV-mediated transformation of chicken CD4+ T-lymphocytes is currently unknown, but several virus genes have been implicated. The most likely and best characterized candidate oncogene is *meq*, the gene encoding the **M**arek's **E**coRI-**Q**-encoded protein, Meq (Jones et al. 1992). Meq is a nuclear b-ZIP protein of 339 amino acids that is expressed in MDV-induced lymphomas and established cell lines. Meq is localized to the nucleolus and nucleoplasmic coiled bodies in chicken T-lymphocytes and in cells transfected with Meq expression vectors (Liu et al. 1997). Splice variants of Meq, termed Meq-sp (Meq/vIL8, 202 a.a., and Meq/vIL8Δexon 3, 154 a.a.), are expressed primarily during lytic infection, whereas full length Meq (339 a.a.) is expressed primarily in the latent/transforming infection (Peng and Shirazi 1996a; Peng et al. 1995; M.S. Parcells, 2003, unpublished data). Meq expression is also regulated by antisense expression, but the regulation and kinetics of these Meq products have not been fully characterized (Peng and Shirazi 1996b; Arumugaswami et al. in preparation). Interestingly, Meq/vIL8 localizes to PML-like bodies as well as the nucleoplasm, but does not interact with the Meq full length protein (Schmidt, personal communication).

Mcq expression confers anti-apoptotic, proliferative and morphological transformation to rat and chicken fibroblast cell lines (Liu et al. 1998; Kung, personal communication). Meq blocks apoptosis induced by TNF-α, UV-irradiation, and ceramide treatment, blocking death receptor signalling, p53-mediated and lysosphingolipid-mediated apoptotic pathways (Liu et al. 1998). Thus, pleiotropic pro-survival signals mediated by Meq are likely to be downstream and common to each of these pathways. Transfection of *meq* alone into activated T-lymphocytes, however, is not sufficient for neoplastic transformation (Kung et al. 2001).

Insight into the proliferative functions of Meq has been restricted to the non-lymphoid cells described above. A number of proteins are currently known to interact with Meq (Kung et al. 2001). Cyclin-dependent kinase 2, the activity of which is required for the G1- > S transition, associates with Meq and phosphorylates it at serine 42 (Liu et al. 1999). The function of this association is presently unknown, but since Meq also contains a retinoblastoma (Rb) family protein-binding motif (LxCxE), the binding of Meq to both of these proteins, may serve to aid in the phosphorylation of Rb. Moreover, Meq binds p53 and contains a C-terminal binding protein motif (ctBP), both factors associated with cell cycle progression (Liu et al. 1999; Kung et al. 2001, Brown et al. 2006).

The DNA-binding sequences of Meq are termed MEREs (Meq response elements) (Qian et al. 1996). Binding to MEREs is dependent on Meqs dimerization partner. Meq/c-Jun heterodimers, however, bind to MERE-I sequences (AP-1-like motif, GTGA<u>TGACTCA</u>, where AP-1 is underlined), and are potent transactivators in this context *in vitro* (Qian et al. 1995; Kung, personal communication). Thus, Meq has the ability to bind different DNA sequences and either activate or repress gene expression based on its dimerization partners. As full length Meq is expressed primarily during latent phases of infection in T-lymphocytes, roles in anti-apoptosis, proliferation and latency maintenance can be envisioned. In contrast, Meq homdimerizes and binds to MERE-II sequences (CACA-motif), and repress transcription in this context (Qian et al. 1996; Levy et al. 2003). As MERE-II sequences are present at the MDV origin of lytic replication, ICP4 promoter and early antigen pp38 promoters, it is likely that Meq functions at one level as a repressor of MDV lytic genes (Levy et al. 2003).

The identification of Meq target genes is yielding very important insight into the Meq's proliferative

effects. Meq has recently been found to upregulate genes that are also upregulated by v-Jun. These include v-Jun target proteins JIP, JTAP, and heparin-binding epidermal growth factor (Hb-EGF) (H.-J. Kung, personal communication).

Interestingly, the transactivator protein, Tax, of human T-lymphotrophic virus I (HTLVI), also degregulates Jun signaling (Mori et al. 2000). In addition to the upregulation of Jun-responsive genes, Meq also upregulates anti-apoptotic genes bcl-2 (in CEF line, DF-1) and bcl-x$_L$ in MDV-transformed T-cells (Ohashi et al. 1999; H.-J. Kung, personal communication). Also upregulated by Meq is Ski, a regulator of TGF-ß receptor signaling, and inhibitory sMADs, proteins involved in blocking the gene repression and apoptosis induced by TGF-ß (Pessah et al. 2002). The blocking of TGFß-mediated apoptosis is seen in EBV-transformed B-cell lines and HTLV-I-transformed T-cell lines (Arnulf et al. 2002; Inman and Allday 2000). As one function of TGF-ß is the downregulation of the host acute inflammatory response and the induction of apoptosis in activated lymphocytes after inflammation, the protection of latently infected, activated T-cells from this apoptosis would be of prime importance to the maintenance of MDV in its host.

Other potential oncogenes

Despite the fact that Meq has all the hallmarks of a transformation antigen, Meq in and of itself does not appear capable of full transformation of chicken T-cells. Several other MDV genes have been implicated in contributing to oncogenesis by their expression in cell lines and tumors and their loss of expression during attenuation. These genes map predominantly to a block of approximately 11 kbp within the repeats flanking the unique-long region of the genome. This block of genes is largely unique to MDV-1 (oncogenic) strains and are not found, or are represented in drastically divergent forms in MDV-2 and HVT (Afonso et al. 2001; Izumiya et al. 2001; Kingham et al. 2001; Lee et al. 2000; Tulman et al. 2000).

These genes include phosphoproteins pp38/pp24, pp14, a 1.8 kb family of RNAs and most recently, a telomerase RNA (vTR) (Bradley et al. 1989a, b; Chen et al. 1992; Cui et al. 1990; Fragnet et al. 2003). Initially thought to be tumor antigens, pp38/pp24 are now known to be associated with the MDV lytic infection, in fact, pp38 expression within lymphocytes is a currently used as a measure for the early cytolytic infection *in vivo* (Baigent et al. 1998; Burgess and Davison 2002). The apparent requirement for pp38 expression for the maintenance of latency in an antisense cell-line based model has never been mechanistically defined, so its role as a factor affecting lymphoma progression is questionable. Recent analysis of a pp38 deletion mutant suggests that pp38 may be a factor affecting the permissivity of lymphocytes to MDV lytic infection (Reddy et al. 2002).

Likewise putative oncoproteins pp14 and p7, also mapping to this region, are expressed in both lytic and latent infection, and the lack of any additional functional analysis of these gene products have hindered our understanding of its possible contributory role to MDV oncogenesis (Hong and Coussens 1994; Hong et al. 1995; Peng et al. 1992). The 1.8 kb family of RNAs map adjacent to the lytic origin of replication and are disrupted during attenuation of the virus by serial cell culture passage (Bradley et al. 1989b). Since no protein products have been associated with this region, it is difficult to likewise envision this region as being important to MDV oncogenicity. As BAC-based and cosmid-based infectious clones of the MDV genome have recently been constructed, this region can be dissected and functionally analyzed (Schumacher et al. 2000).

A more promising candidate oncogene is the recently discovered telomerase RNA, vTR (Fragnet et al. 2003). This gene is expressed in MDV cell lines but not in lytic infection (Parcells, unpublished data). Moreover, vTR has been shown to functionally substitute for the cellular TR in a mouse cell line (Fragnet et al. 2003). As telomerase activity is greatly upregulated in activated T-cells (Weng et al. 1997), the maintenance of telomere length in the sustained proliferation associated with MDV-mediated transformation would be critical to the viruses long-term survival. An interesting observation regarding vTR is that it apparently is capable of out-competing the cellular TR for the reverse transcriptase portion of the protein (telomerase reverse transcriptase, TERT) (Fragnet et al. 2003). The vTR is also partially antisense to the ICP0 homolog of MDV, suggesting that this gene not only contributes to sustained telomerase activity but that it may be directly associated with downregulating the MDV lytic infection (Parcells, unpublished data).

The effect of vTR on cellular proliferation as well as on the regulation of MDV lytic infection await further characterization, but mechanistically, this gene is as exciting as Meq in terms of its potential contribution to transformation.

Given the efficiency of MDV-mediated transformation, it is presumed that lymphomas are polyclonal in origin, but fluorescent *in situ* hybridization (FISH) analysis of MDV lymphomas and derived cell lines suggests that lymphomas are essentially monoclonal in origin (Delecluse et al. 1993). However, that once a frank lymphoma develops from a lymphoproliferative lesion, it is highly likely to metastasize and become the dominant event present. The rapidity of MD lymphoma progression at this stage makes the predominant monoclonal effect most likely.

Pathotypes of MDV

Since the identification of MDV as the etiologic agent of MD and the advent of large scale commercial vaccination for MD, MDV field strains have increased in virulence (Witter 1983, 1997). In the US, many strains of MDV have been ranked according to their virulence by lesion scores in unvaccinated, HVT-vaccinated and bivalently vaccinated chickens of a specific genetic background (line $7_2 \times 15I_5$) (Witter 1997). Although a continuum, the strains of MDV have been grouped as mild (mMDV), virulent (vMDV), very virulent (vvMDV) and very virulent plus (vv+ MDV). In addition, according to specific neurological disease expression, strains have been further characterized according to a letter scheme (A–C) that roughly corresponds to the v, vv and vv+ (Gimeno et al. 2002). The expression of these pathotypes *in vivo* can be described as: (1) vMDV strains cause some paralysis, nerve lesions and relatively few lymphomas, (2) vvMDV strains cause more visceral lymphomas, greater immune suppression, and more neurologic lesions, (3) vv+ MDVs cause profound stunting in young chickens, usually spleen lymphomas in older chickens and early proliferative lesions that outgrow their blood supply (Rosenberger et al. 1997), becoming necrotic. vv+ MDVs also cause the most profound neurologic signs and lesions (blindness, paralysis, torticollis, brain infiltrations, etc.) (Gimeno et al. 1999).

In addition to these classifications, MDV strains also can be categorized according to their immuno-

suppressive potential. Using these criteria (lymphoid organ atrophy, duration of lytic infection, lymphoid organ lesion scores), strains of v, vv, and vv+ segregate according to pathotype (Calnek et al. 1998). To date, however, the molecular basis for MDV pathotype is unknown.

Immunity to MDV

MD has presented a very provocative paradigm: the ability to vaccinate for the prevention of a very rapid and aggressive cancer. The widespread vaccination for MD, in fact, has been one of the greatest success stories for vaccine-based prophylaxis. To date, most "breaks" of MD in vaccinated flocks are due more to management issues (over-dilution of vaccine, exposure of vaccine to heat, contamination of vaccination apparatus) than to the evolution of MDVs of greater virulence. In summarizing the current understanding of immunity to MDV, the literature falls primarily into the study of two major aspects of immunity: (1) genetics of MDV resistance and (2) vaccine responses. This assertion is somewhat different than the innate vs acquired immune responses, as genetic differences may include inherent aspects of both innate susceptibility/resistance of particular target cells, as well as inherent differences in surveillance (i.e., CTL responses).

Genetic resistance to MDV

Even prior to the identification of MDV as the etiologic agent of MD, the resistance of specific lines of chickens to lymphoma formation was described in some detail, reviewed in (Calnek 1985). The most well described loci affecting MDV resistance are those encoding the major histocompatability genes (B-F/B-L, B-G) encoding the classical major histocompatability genes, MHC class I and class II (Bacon et al. 1981, 1983; Bacon and Witter 1993, 1994; Briles et al. 1983).

In addition to classical MHC genes, which are located in the B gene cluster, another set of MHC genes are present in the separate gene cluster known as "R*fp*Y" (Briles et al. 1993; Miller et al. 1994). The R*fp*Y genes are separated by the nucleolar organizing region of chicken chromosome 16 from the MHC B locus (Miller et al. 1996). MD lymphoma incidence

is significantly influenced by gene(s) in the R*fp*-Y region in the stock in which the R*fp*Y was originally identified (Wakenell et al. 1996). However, a common role for R*fp*Y in MD resistance remains controversial. Three independent publications found that R*fp*Y is not genetically linked to MD lymphoma resistance. The first group investigated the issue in two MHC B^2 congenic lines (one susceptible, one resistant) and in lines N (B^{21}; resistant) and P (B^{19}; susceptible) (Bacon et al. 1996; Pharr et al. 1997; Vallejo et al. 1997). The second group used four lines of chickens selected for multi-trait immuno-competence (Lakshmanan and Lamont 1998). The third group, used another MHC B^2 congenic and differentially susceptible system (Bumstead 1998).

MD resistant genotypes of chickens have decreased MDV loads (Baigent et al. 1998; Bumstead et al. 1997). Certainly CTL specific for MDV latent and lytic genes have been demonstrated (Markowski-Grimsrud and Schat 2002; Omar and Schat 1996, 1997; Omar et al. 1998; Schat and Xing 2000; Uni et al. 1994). Current data also supports a role for NK cells in maintaining low MDV loads. The chicken homologue of the mouse cytomegalovirus resistance gene (Cmv-1) (an NK receptor) was mapped using MDV-load differences between inbred B2 congenic chicken lines (Bumstead 1998). However, down-regulation of MHC class I has been reported, using cell-line cells in which MDV was reactivated out of latency (Hunt et al. 2001), but not lymphoma cells. It is likely that the effect of this form of MDV resistance functions during the cytolytic phase of MD. Regardless, differences in MDV-loads after infection between MD-resistant and MD-susceptible chickens are postulated to play a major role in resistance to MD lymphomas. The proposed mechanism is by decreasing the numbers of T lymphocytes that become latently infected and thus eligible for neoplastic transformation.

Beyond the B-locus, several quantitative trait loci (QTLs) have been identified that affect specific phenotypic traits of MD expression (Vallejo et al. 1998; Yonash et al. 1999). The phenotypes associated with these QTLs were similar in aspects of disease presentation (nerve inflammation, tumor development, viremia level, etc.) resulting in the conclusion that some QTLs showed epistatic relationships. Although specific genes associated with these QTLs have yet to be identified, a recent mapping of a susceptibil-

ity QTL to the growth axis genes (GH1) and the interaction of an MDV gene product (SORF-2) with chicken growth hormone suggest that complementary approaches may augment mapping of these traits (Liu et al. 2001). These results suggest that the variability of genetic resistance and relative disease expression seen with MDV infection may not only be affected by variations in host target cell susceptibility or immune surveillance, but by polymorphisms in MDV gene products and the cellular targets with which they interact.

Observations regarding differences in the immune responses between MDV-resistant and susceptible lines of chickens have been somewhat counter-intuitive: Susceptible lines of chickens tend to have more robust and prolonged immune responses (Calnek et al. 1989a), as measured by lectin assays, MLR and antibody levels (Pevzner et al. 1981a, b; Yamamoto et al. 1991). Immune parameters that do appear to correlate with genetic resistance to MDV are associated with innate immune responses (nitric oxide production, NK cell activity, IFN responses) (Djeraba et al. 2002b, c; Xing and Schat 2000b).

MD vaccines

Despite the great success of MD vaccines, the means by which they elicit protective immunity is only partially understood, at best. Presumably, the representation of common virus lytic antigens by vaccine strains such as CVI-988 (attenuated non-oncogenic MDV serotype 1) (Rispens et al. 1972a, b), SB-1 (a naturally occurring non-oncogenic serotype -2 MDV) (Schat and Calnek 1978) and FC-126 (HVT) (Witter et al. 1970, 1976) to the immune system elicits potent CTL and antibody responses that limit MDV-1 lytic infection upon challenge. This limitation of the MDV-1 lytic infection is believed to limit the pool of latently infected CD4+ T-cells, thereby limiting the number of cells capable of undergoing transformation (Calnek 1982). This understanding is therefore based on the similarity of MDV-1 and vaccine strain antigens in terms of epitopes presented, and to some extent, on the similarities of interactions between vaccine strains and the target tissues in which MDV-1 replicates (Schat and Xing 2000).

Anti-MD-lymphoma immunity in MD is more controversial than the immunity to lytic antigens

introduced above. In 1976, Payne et al. (1976) introduced a "two-step model" for MD immunity. In this model, immunity is directed towards both MDV antigens and antigens expressed by neoplastically transformed MD lymphoma cells. Although a comprehensive review of tumour immunity in MD is beyond the scope of this chapter and has recently be reviewed elsewhere (Burgess and Venugopal 2002).

One point is worth emphasizing in the context of MD vaccines, however. The current bivalent vaccine used to prevent MD in the USA (SB-1 and FC-126) is comprised of viruses that do not encode the *meq* oncogene (see above). Consequently, no CTL could be primed to this prevalent latency and transformation-associated antigen through MDV-2/HVT vaccination, yet this vaccine prevents MDV-1-induced lymphoma formation. Only attenuated non-oncogenic MDV-1 vaccines such as CVI-988, used more commonly in Europe, contain the *meq* gene, albeit a mixture of heterogeneous *meq* isoforms (Chang et al. 2002). CTL specific to Meq peptides have been demonstrated in SB-1-vaccinated chickens (Omar and Schat 1996), and vaccines such as CVI-988 may potentially promote anti-lymphoma immunity via this mechanism. In the case of the most commonly used MD vaccines in the USA (SB-1/HVT), such a mechanism cannot be the key to any anti-tumor response, as many latently expressed gene products of MDV-1 strains are not encoded by either MDV-2 or HVT (Afonso et al. 2001; Izumiya et al. 2001; Kingham et al. 2001).

As our assumption is that vaccines must present antigens common with a pathogen, at least at the level of T-cell epitopes, to elicit a primary immune response. The ability of MDV-2 and HVT to elicit protection to MDV-1-induced lymphomas is very interesting. As the transformation phase arises presumably, from latently infected T-cells, and the genes expressed during MDV latency are largely unique to MDV-1-strains, then how is an anti-tumor response elicited by MDV-2/HVT vaccination?

Several different lines of evidence suggest that components important to vaccinal protection are directed to MDV-transformed cells, e.g. are elicited to MDV-transformed cells. Firstly, proteins encoded by MDV and expressed in transformed cells are apparently recognized by vaccine induced CTL, despite the fact that the most highly expressed antigens in MDV-1 transformed lymphocytes (Meq, etc.) (Ross 1997

#115) are not encoded by the vaccine viruses SB-1 (MDV-2) and HVT. Even in the absence of encoding this protein, SB-1-vaccination is apparently capable of inducing CTL to this antigen, albeit at a low level of specific lysis (Omar and Schat 1996). The most conserved virus structural antigens (gB, gC, gE) among the three serotypes are not expressed in the context of the transformed cell (Ui et al. 1998), and hence are not likely to be involved in an anti-tumor response. Secondly, some level of vaccinal protection can be elicited through the inoculation of glutaraldehye-fixed MDV-1-transformed cell lines (Nazerian and Witter 1984; Powell 1975). Although protection is elicited by these cell line vaccines, the level of immunity is insufficient to employ this method commercially (Nazerian and Witter 1984). Thirdly, similar lymphoproliferative responses, with accompanying upregulation of a Marek's-associated tumor surface antigen (MATSA), are elicited by SB-1 and HVT vaccination (Calnek et al. 1979; Witter et al. 1976). MATSA, or at least one MATSA (Lee et al. 1983), has been determined to be a cellular antigen upregulated on activated chicken T-lymphocytes and thrombocytes (McColl et al. 1987). Antibodies and CTL to MATSA, as well as to MDV lymphoblastoid cells can be detected post vaccination (Quere and Dambrine 1988; Sharma 1977; Sharma et al. 1978), and these responses, and or responses to other virus-encoded or deregulated host-encoded antigens may be important to the vaccine-induced anti-tumor response.

As vaccination largely controls MD clinical expression through the development of protective immune responses, the development of MDV-induced lymphomas in susceptible chickens must contend with the development of this protective response. The role of the immune system in keeping tumor progression in check has been well established. Recent experiments involving immunosuppressive drugs in humans, suggest that the inability of the host to mount an effective immune response against abnormal cells (i.e. those expressing tumor antigens, altered cellular antigens or improperly processed antigens, etc.) contributes to lymphoma progression (Kirby et al. 2002; Lim and Bertouch 2002; Nathanson et al. 2002; Tremblay et al. 2002). Similarly, the immunosuppressive nature of MDV, likely contributes to lymphoma progression.

MDV and immunosuppression

MDV is not only neoplastically transforming, but also immunosuppressive. The immune parameters affected by MDV are described here according to the phase of infection with which they are associated; namely, cytolytic, latent and transformation phases.

Early cytolytic infection

During the early cytolytic infection, a key mechanism of immune suppression is the loss of lymphocytes from the primary and secondary lymphoid organs (Calnek and Witter 1997). The edema, hypertrophy and following atrophy of these organs accompanies measurable losses of both humoral and cell mediated responses as measured by decreased antibody responses to sheep red blood cells, and decreased lectin (Con-A, PHA), and mixed lymphocyte-induced proliferation (Confer et al. 1980; Theis et al. 1975; Yamamoto et al. 1995). Early responses to MDV infection suggested the elicitation of "suppressor macrophage" populations that mediated immunosuppressive effects (Lee et al. 1978b; Sharma 1988; Theis 1981). The factors induced during this stage of infection have not been fully defined, but include IFN-I (α/β), IFN-II (γ) and reactive oxygen and nitrogen species (Buscaglia and Calnek 1988; Levy et al. 1999; Volpini et al. 1995; Xing and Schat 2000b). Notably however, it is this MDV-induced host response that is also associated with the onset of MDV latency (see below).

In addition to the targeted loss of lymphocytes in the primary and secondary immune organs, there is an apparent loss of both CD4[+] and CD8[+] T-cells in the circulation and a decrease in the level of CD8 expression in CD8[+] T-cells (Morimura et al. 1996, 1997, 1998). Loss of these cells is through apoptosis and may be due to the direct cytolytic infection of these cells, or through indirect effects (i.e. induction of TGF-ß, etc.).

The relative contribution of this early immunosuppression on lymphoma development is controversial or, at the very least, unclear. Pathotypic analyses of MDV strains suggests that higher virulence MDVs induce more profound immunosuppression (Calnek et al. 1998). A characteristic of these strains is a marked increase viremia and duration of the cytolytic phase of infection (Barrow et al. 2003). Interestingly, the highest virulence MDV isolates tend not to induce the greatest incidence of lymphomas, but tend to induce greater incidence of neurological lesions, stunting, inflammatory lesions and mortality (Gimeno et al. 1999; Rosenberger et al. 1997; Witter 1997; Witter et al. 1999). Moreover, some recombinant MDV strains that established relatively low levels of viremia and were mildly immunosuppressive, were still lymphomagenic (Parcells et al. 1995, 2001). Conversely a recombinant MDV (called RM-1) containing an LTR insertion upstream of the SORF 2 gene showed high early viremia with profound thymic atrophy, but did not induce tumors – even when injected into young chickens (Jones et al. 1996; Witter et al. 1997).

Mechanistically, RM-1 may block lymphoma formation through the ablation of the transformation target cells (i.e. CD4[+] T-lymphocytes). The efficacy of RM-1 as a vaccine against vv[+] MDV challenge certainly suggests such a mechanism. However, other viruses of equal or greater replication levels (RB1B, MD-5), also induce profound thymic atrophy yet induce high incidence of lymphomas (Schat et al. 1982). Alternatively, RM-1 may have deficiencies that impair its entry into latency. RM-1 could also have alterations in *meq* or other transformation-associated genes that affect its ability to transform T-cells.

From the standpoint of the current paradigm; namely, that the greater the level of MDV-1 lytic infection, the greater the level of latent infection and hence, the greater the tumor incidence, the RM-1 example demonstrates that cytolytic infection and neoplastic-transformation are not absolutely linked. Likewise, the low level viremias established by RB1B-based deletion viruses (Parcells et al. 1995), which retain oncogenicity, also question the relative role of early immune suppression on lymphoma development.

MDV latency

Herpesviruses have evolved the capacity to undergo two distinct types of replication, lytic (replicative) and latent (quiescent). Latency in MD was initially a clinical description of chickens that had recovered from the clinical signs associated with the cytolytic phase of MDV infection, yet had not reached the tumor phase. Once more of MDV virology was understood, the description of this latency was defined as

the presence of the MDV genome in the absence of expression of virus internal antigen (VIA) and virus membrane antigen (VMA) (Calnek et al. 1984; Shek et al. 1983). Although defined by polyclonal antibodies of unknown specificity, these antigens are likely to be MDV proteins of early and late kinetic classes, respectively; due to the greater abundance of VIA-positive latently infected lymphocytes.

The definition of latency differs depending on whether the disease, or the state of the virus, is being discussed. Clinical latency is that period of the infection in which there is no clinical manifestation of disease. MDV latency is now defined, using molecular biological techniques, according to the MDV genome state in terms of its structure and pattern of gene expression. Such methods for defining herpesvirus latency are used for all heperviruses. For MDV, virus latency has most recently been described as the presence of the DNA genome and a limited set of gene transcripts (*meq*, latency-associated RNAs [LATs or MSRs]) antisense to the immediate early regulatory protein, ICP4 (Cantello et al. 1994, 1997; Li et al. 1994, 1998; McKie et al. 1995). MDV latency is defined, moreover, by an even more limited set of translated proteins (Meq, pp14) (Hong et al. 1995; Jones et al. 1992), and the absence of early proteins (e.g. pp38) (Baigent et al. 1998).

In chickens supporting MDV latent infection, there is likely a constant low percentage of cells undergoing reactivation and lytic infection at peripheral sites. Even in lymphomas and MDCC, in which by definition most cells must maintain MDV in a latent state, a small percentage (typically 5%) will be undergoing spontaneous reactivation (Burgess and Davison 2002; Calnek et al. 1981; Parcells et al. 1999).

In the study of latency and the response of latent MDV to IFN-γ treatment, there appears to be at least two levels of quiescence (Volpini et al. 1996). One level of quiescence, in which VIA can be induced within 24h of withdrawal of conditioned medium (CM) or purified IFN-γ, occurs within 1–2 weeks of *in vivo* infection (Volpini et al. 1995). A more quiescent phase of latency, seen in circulating T-cells after 2 weeks post infection requires sustained, longer-term culture for reactivation of virus and these cells are less responsive to withdrawal of CM or IFN-γ. The acquisition of this highly restricted latency, may provide advantage to MDV as a long-term maintenance strategy

within a particular host or within a flock. From our understanding of human herpesvirus latency, this long-term maintenance strategy has been highly successful.

Immunosuppressive effects of MDV latency have not been defined during this phase of infection, as any immunosuppression observed at this time is likely to be the consequence of damage sustained during the cytolytic phase of infection. Moreover, effects of serum from birds at this stage on *in vitro* cell proliferation assays are largely thought to be mediated by the macrophage/reticuloendothelial response to cytolytic infection (Carpenter and Sevoian 1983; Lee et al. 1978b; Sharma 1988). The effects of this serum can be inhibitory or excitatory to lymphocyte assays, depending on the time post cytolytic infection and the level of cytolytic infection supported by the host (Lee et al. 1978a). These effects can be explained by the cytokine response to cytolytic infection and the regeneration of the primary and secondary immune organs, respectively.

Modulation of MDV latency by host cytokines has been studied prior to the identification of exactly what factors were present (Buscaglia and Calnek 1988; Buscaglia et al. 1988). These factors, particularly those limiting MDV lytic infection are now more defined. The effects of IFN-γ, nitric oxide, and myelomonocytic growth factor (MCG) on MDV lytic infection *in vitro* and *in vivo* have been well characterized (Djeraba et al. 2002a, b, c; Levy et al. 1999; Xing and Schat 2000b), suggesting that any modulatory response by MDV latent infection is likely to be mediated, to a large extent, by the host response to the early cytolytic phase of infection. Immunosuppressive effects of latently infected T-cells have been therefore limited to the transformation phase of infection and are discussed below.

MDV-induced transformation

Depending on the susceptibility of the host chicken and the strain and challenge level of MDV, the transformation phase of MDV infection can commence as early as 2 weeks post infection, but typically is seen in chickens infected at 1-day-of-age by 4–6 weeks dpi (Calnek 1986; Calnek and Witter 1997). In breeding and laying chickens, the tumor phase can begin as chickens go into lay, during peak lay, or upon challenge with another immunosuppressive agent.

Commencing with the onset of lymphoma formation is profound immunosuppression caused by factors secreted by transformed lymphocytes, as well as by antigens expressed on their surface. Since the target for transformation is the CD4$^+$ T-cell, the key regulator of acquired immunity, dysfunction of these cells is particularly devastating to the host.

Little is known of the soluble factors expressed by MDV lymphomas. Most of what is known regarding the expression of MDV-transformed cells has come from the study of MDCCs and is discussed below (see "Factors expressed by MDV-transformed cell lines"). As the transformed component of MDV-induced lymphomas has recently been identified (Burgess and Davison 2002), and a chicken expressed sequence tag (EST)-based microarray has recently been developed (Morgan et al. 2001), the characterization of MDV lymphoma expression is likely imminent. Some insight into factors expressed during the tumor phase of MDV infection can be gleaned from recent cytokine expression analysis of MDV-susceptible and resistant lines of chickens. During the course of MDV early cytolytic infection, there are noted increases in inflammatory cytokines/chemokines (IL-2, IL-6 and IL-8) followed by sustained IFN-γ and inducible nitric oxide synthase (iNOS) expression (Xing and Schat 2000a). This study focused on the initial phases of infection and terminated at day 15, so any shifts in expression into the latent and transforming phases were not reported.

In genetically susceptible lines of chickens (lines 6 and N), there was noted a consistent increase in IL-6 and IL-18 in infected splenocytes as opposed to resistant chickens (lines 7 and P) (Kaiser et al. 2003). Another group noted increases in arginase activity in macrophage/monocytes from susceptible chickens (B$^{13/13}$) with concomitant low levels of nitric oxide production, whereas a resistant line (B$^{19/19}$) showed only a transient increase in arginase activity followed by a larger and more sustained level of nitric oxide production (Djeraba et al. 2002c). Arginase and inducible nitric oxide synthase (iNOS) expression in macrophages are regulated differentially by T$_{h2}$ and T$_{h1}$ cytokines, with TGF-β and prostaglandin E$_2$ (PGE$_2$) inducing primarily arginase (Djeraba et al. 2002c). These studies, however, examined essentially the early phase of MDV infection, but the similarities in findings; namely, that the immune response in susceptible chickens is more inflammatory and skewed towards a T$_{h2}$ response is provocative. A modulatory effect of PGE$_2$ on macrophage expression is of particular interest given the finding that MDCCs express this immune modulator. In light of the surface expression of MDCC and lymphomas described below, the findings of differences in cytokine expression profiles from resistant and susceptible chickens are particularly interesting, suggesting that the skewing of the immune response during MDV latent/transforming phases of infection may be fundamentally different in resistant vs susceptible chickens. Such a shift in the T$_{h1}$ to T$_{h2}$ response to transformed cells has also recently been identified in other malignancies including renal carcinoma, with the induction of PGE$_2$ found to be largely responsible for this shift in response (Smyth et al. 2003).

Early work on MDV lymphomas was focussed on surface antigens expressed on lymphoma cells. MATSA, an activated leukocyte antigen was described by several groups using polyclonal and monoclonal antibodies (Lee et al. 1983; McColl et al. 1987; Witter et al. 1975). This highly glycosylated antigen, was found to be induced during the early cytolytic phase of infection but remained upregulated on MDV-transformed lymphoblasts (Murthy and Calnek 1979). In addition to MATSA, glycolipid antigens were described on MDV lymphomas (Forssman antigen and Hanganutziu and Deicher antigen) (Ikuta et al. 1981a, b; Kawai et al. 1991; Naiki et al. 1982). The significance of changes in glycolipid surface expression is unknown, but has been noted in other neoplastic diseases, suggesting a functional difference in these changes.

MDV lymphomas and cell lines

Antigens expressed on MDV-transformed cell lines

In order to determine the functional lineage and potential signalling pathways de-regulated during MDV transformation, we and others have studied the surface antigens present on MDV-transformed cell lines (Burgess and Davison 2002; Dienglewicz and Parcells 1999; Nazerian and Sharma 1975; Parcells et al. 1999; Schat et al. 1991). This endeavor has been limited by a paucity of chicken lymphocyte antigen-specific monoclonal antibodies. A list of these antibodies, their specificities and their expression on two MDV- and one REV-transformed cell lines is given in Table 1. Of the cell lines examined, most MDV-transformed lines show a T-helper immunophenotype and express antigens involved in extravasation (CD29, CD44, CD49e). The REV-transformed cell line RECC-CU91 expresses CD166 (ligand of CD6, ALCAM, activated leukocyte cellular adhesion molecule), another antigen associated with lymphocyte activation, extravasation and tumor metastasis (Corbel et al. 1996; Skonier et al. 1997).

Given these immunophenotypes, two key aspects of this surface antigen characterization are discernible: (1) MDV-transformed cell lines express antigens indicating that they are at least partially activated (CD28, CD30, CD44, MATSA, GRL-2, MHC-II), and (2) MDV-transformed cell lines express antigens suggesting that they are highly invasive (CD29, CD44, CD49e, and CD164, a.k.a., sialomucin). The activation status of MDV-transformed cell lines is inferred by the MHC-II expression and the level of auxilliary activation antigen expression, but a key determinant of T-cell activation, CD25 (IL-2Rα) expression is not definitively known. The monoclonal antibody developed for chicken CD25 (INNCH.16) (Schauenstein et al. 1988), shows marked variation in the ability to stain MDV-transformed cell lines (Hrdlickova et al. 1994; Kaplan et al. 1992; Quere 1992). Another monoclonal reported to detect chicken CD25, B337, appears to recognize the non-induced low affinity subunit, CD132 (IL-2γ) (Lee and Tempelis 1992) and (Parcells, unpublished data). As IL-2Rα expression is the hallmark of T-cell activation, the equivocal nature of these results has suggested that MDV-transformed cells represent a unique, quasi-activated cell population. Expression of CD166 (ALCAM/BEN/GRASP), was found in avian T-cells to be induced simultaneously with CD25 expression (Corbel et al. 1992). The

Table 1 Surface antigens identified on MDV-transformed cell lines

Antigen	UA01[a]	UD14[b]	CU91[c]	Function	Ab source
CD3	++	+++	+	T-cell activation	SBT[d]
CD4	+++	++++	+	MHC-II binding	SBT
CD25 (IL2R-α)	−	−	−	T-cell activation	K. Hala[e]
CD28	+++	+++	−	B-7-binding (activation)	SBT
CD29 (§1 integrin)	++++	++++	++++	BCM-binding	(CSAT) DSHB[f]
CD30 (RS antigen)	++++	++++	++++	Activation/immune evasion	F. Davison[g]
CD44 (HA-binding)	++++	++++	+++	Extravasation/activation	SBT
CD45	+	+	++	PTPase/activation	SBT
CD49e (α-5 integrin)	++	++	++	ECM-binding	(A21F7) DSHB
CD132 (IL2R-γ)	+++	+++	+++	IL-2R subunit, constitutive	C. Tempelis[h]
CD164 (sialomucin)	+++	+++	+	Migration (?)	F. Davison[g]
CD 166 (ALCAM)	−	−	+++	Binds CD6/NgCAM	(BEN) DSHB
MATSA[i]	++++	++++	++	Activation Antigen (CD69?)	L.F.Lee
GRL-1	−	−	−	Activation antigen	DSHB
GRL-2	+	+	++	Activation antigen	(GRL-2) DSHB
TCR-2	++	+++	−	T-cell receptor (α/β)	SBT
TCR-3	−	+	+	T-cell receptor (α/β2)	SBT
MHC-I	+++	+++	++++	Self/non-self determinant	SBT
MHC-II	+++	+++	++++	Self/non-self determinant	SBT

[a]*RB 1B-transformed T-cell line (MDCC-UA01) (Dienglewicz & Parcells 1999).*
[b]*RB 1B-transformed T-cell line (MDCC-UD14) (Parcells Anderson, & Morgan, 1995; Parcells, Dienglew icz, Anderson, & Morgan, 1999).*
[c]*REV -transformed T-cellline (RECC -CU91) (Pratt, Morgan, & Schat, 1992).*
[d]*Southern Biotechnology Associates, Inc., Birmingham, AL.*
[e]*Provided by Dr. Kani Hala, Innsbruck, Austria (mAb INNC H.16) (Schauenstein et al. 1988).*
[f]*Developmental Studied Hybridoma Bank, University of Iowa, Iowa City, IA*
[g]*Provided by Dr. Fred Davison, Institute of Animal Health, Compton, Berkshire, UK (mAb AV37) (Burgess and Davison 1999).*
[h]*Provided by Dr. Constantine Tempelis, UC-Berkeley, CA (mAb, B 337) (Lee and Tempelis 1992).*
[i]*Marek's Disease-Associated Tumor Surface Antigen, (mAb 14B 367), provided by Dr. Lucy F. Lee, USDA-ADOL, EastLansing, MI (Lee et al. 1983).*

apparent absence of this from MDV-transformed cell lines, therefore, is also suggestive of a quasi-activation state and apparently differs from the MDV-transformed component of lymphomas examined *ex vivo* (see below).

In addition to the above antigens, MDV-transformed cell lines were found to express altered glycolipids on their cell surface (Forssman antigen, Hanganutziu and Deicher heterophil antigen), and chicken fetal antigen (CFA) (Ikuta et al. 1981b; Kawai et al. 1991; Naiki et al. 1982; Ohashi et al. 1986). CFA is expressed on erythrocytes (Qureshi et al. 1986), and in the context of MDCC-MSB-1 cells, CFA has been found to be immunosuppressive to NK killing (Ohashi et al. 1987). Given the importance of NK responses to anti-tumor immunity (Soloski 2001), the expression of fetal and developmentally regulated antigens by MDV-transformed cells may be directed to circumvent these responses.

As MDV has not been found to transform T-cells *in vitro*, the unique or aberrant immunophenotypes of many MDV-transformed cell lines suggests that some *in vivo* lineages, normally ablated during T-cell ontogeny, may be the actual targets for MDV transformation. This notion is supported by the generation of T-cell lines using a wing-web model of MDV-transformation (Calnek et al. 1989b). In this model, allogeneic chicken kidney cells were infected with MDV and injected into the wing web of chickens. At various times post infection, these sections of wing web were excised and the infiltrating lymphocytes cultured. By this method, cell lines could be established and the immunophenotypes of these lines varied from double negative (CD4-/CD8-), double positive (CD4+/CD8+) to single CD4+ and CD8+ T-cells (Schat et al. 1991). In general, most established MDV-transformed T-cell lines are single positive CD4+ T-helper cells. There are very few CD8+ MDV-transformed cell lines, as would be expected given the results or Morimura et al. showing that CD8+ T-cells undergo apoptosis and have CD8+ expression downregulated during the lytic phase of infection (Morimura et al. 1996, 1997). Interestingly, we have found that cell lines established from vv+ MDV-induced lymphomas tend to have a higher incidence of aberrant immunophenotype (Shamblin et al. 2004). As these viruses are more immunosuppressive (Calnek et al. 1998), an aberrant T-cell expansion mechanism is suggested. Conversely,

alterations of T-cell expression may be mediated by MDV gene expression.

Cell lines established from MDV-lymphomas have been found to be tumorigenic when re-injected *in vivo* (Coleman and Schierman 1982; DiFronzo and Schierman 1990), suggesting that they represent a transformed element, but how closely they reflect the initial lymphoma or even the transformed component of the initial lymphoma is discussed below.

Antigens expressed on lymphoma cells ex vivo

An alternative to studying MDCCs for understanding MDV-induced lymphomagenesis has been through analysis of lymphomas directly *ex vivo*. Such analyses provide insight into the development, progression and architecture of MDV lymphomas. As the number of mAbs specific for avian lymphoid and myeloid populations increases, so will the insight into the actual cell populations contributing to MD lymphomas. Lymphomas, like other tissues, presumably follow the structure-function paradigm.

Compared with MDCC, very little work has been done to characterize surface antigen expression by cells within MD lymphomas. On one hand this is surprising because the technologies used to immunophenotype MDCC can equally well be applied to MD lymphomas. On the other hand MDCC have been assumed to be, and were used as, models for the actual neoplastically transformed cells in MD lymphomas. The approach was reasonable because until recently the neoplastically transformed cells in MD lymphomas were unidentified. MD lymphomas have been known to be heterogeneous prior to mAb technology and MDCC were a practical way to try to separate the neoplastically transformed cells from the rest of the lymphoma. Now that the neoplastically transformed cells have been identified directly *ex vivo* (Burgess and Davison 2002), the assumption that MDCC would be reasonable models of the neoplastically transformed phenotype has been proven sound.

The neoplastically transformed MD lymphoma cells, which occur in clusters scattered throughout the lymphoma mass, are CD4+, MHC class I[hi], MHC class II[hi], CD25[hi] and CD28[lo/-] (Burgess and Davison 2002). Distinctive to these cells is the gross over expression of CD30 (recognized by the mAb AV37;

Burgess et al. 2004), which is two orders of magnitude greater than CD30 expression on physiologically activated T lymphocytes, the next highest expressors of CD30 (Burgess and Davison 2002). The implications of this phenotype are that the neoplastically transformed cells in MD lymphomas are activated CD4[+] T helper cells. However, these neoplastically transformed cells have down-regulated CD28 possibly reflecting an immune escape mechanism mediated by MDV or a host mechanism to attempt to promote cell death. As CD28 expression can be quite high on MDCCs (Table 1), one inference is that in the lack of an immune response in culture, the expression of this co-activator is unchecked.

In "normal physiology" CD30 is expressed by sub-populations of memory T helper lymphocytes that secrete T-helper 2 (T_{H2}) cytokines (Bengtsson et al. 2000; Harlin et al. 2002; Lucey et al. 1996). T_{H2} cytokines antagonize CMI, suggesting another potential and likely important mechanism of immune evasion by the neoplastically transformed cells in MD.

The implication of the neoplastically transformed cells in MD over expressing CD30 goes beyond MD. The neoplastically transformed cells in very many human lymphoma types, of both herpesviral and non-herpesviral etiology, grossly over-express CD30 (Lucey et al. 1996). Because it was first recognized as a marker of neoplastic-transformation of human B lymphocytes after Epstein–Barr virus infection, the common name for CD30 in human medicine is the "Hodgkin's disease" or "Reed–Sternberg" antigen (Falini et al. 1987). CD30 is a lymphocyte lineage specific antigen and is over-expressed by both B and T lymphocyte derived human lymphomas (Lucey et al. 1996). MD is the only known natural herpesviral infection causing CD30 over-expressing lymphomas in its natural host outside of humans (Burgess et al. 2004).

In one area phenotyping of MDCC could never be useful. Direct *ex vivo* analysis is critical to define the structure–function relationships of both different cells within heterogeneous MD lymphomas and also for defining those cells involved in MD lymphoma regression. This regression is at least part of the reason that gross lymphomas do not occur in resistant chicken genotypes (Biggs et al. 1968; Burgess and Venugopal 2002; Morimura et al. 1999; Quere and Dambrine 1988). Notably, in both MD susceptible and MD resistant genotypes, CD30[hi] cells are present in the blood of genetically susceptible and resistant chickens from 3 dpi and infiltrate tissues from this time (Burgess and Davison 1999; Burgess et al. 2001).

Using a 14-day-old inbred chick model, lines N & 61 (resistant) and 15I & 7$_2$ (susceptible) chickens were infected with the HPRS-16 (virulent) MDV. Clinically, all line 72 chickens developed progressive MD; line 15I chickens generally developed fluctuating MD clinical signs and some individuals even recovered (Burgess et al. 2001). Lines 61 and 72 had a similar patterns of lymphoid infiltration into non-lymphoid tissues during cytolytic infection, although the numbers of infiltrating lymphocytes were greater in line 72. By the time lymphomas were visible in line 72 (the tumor phase), histological lesions in line 61 were regressing. In the cytolytic and latent phases, in all genotypes, most infiltrating cells were CD4[+]. After this time, line 72 and 15I lesions increased in size and CD4[+] lymphocytes were predominant whereas line 61 and N lesions decreased in size and CD8[+] cells predominated (Burgess et al. 2001).

Notably CD30[hi] cells were present in similar numbers in all genotypes in the cytolytic phase (Burgess and Davison 1999). This suggests neoplastically transformed cells were present in all genotypes regardless of susceptibility to "full-blown" MD. However, after the cytolytic phase, CD30[hi] cell numbers increased in lines 7^2 and [15]I but decreased in lines 61 and N. Regardless of the chicken genotype, AV37 immunostaining for CD30 was weak in lesions with many CD8[+] cells suggesting CD30 expression, or CD30[hi] cells themselves, were controlled by CD8[+] cells (Burgess and Davison 1999).

To this point we have discussed only host cell antigens in MD lymphomas. However MDV is present in the lymphoma cells. Firstly, CD30[hi] lymphoma cells carry more latent MDV genomes than any other cell type in the lymphomas.(Burgess and Davison 1999). In "mature lymphomas", the few CD8[+] and TCRγδ[+] killer T lymphocytes are cytolytically infected, expressing the early pp38 and the late, gB antigens (Burgess and Davison 2002). Most importantly, from the point of view of neoplastic transformation, the CD30[hi] cells express most Meq protein. In fact Meq expression is directly proportional to CD30 expression (Burgess et al. 2004). As this is unlikely to be coincidental, the transactivation of the CD30 promoter by

Meq/Jun heterodimers is a clear possibility and currently under study (see below).

Factors expressed by MDV-transformed cell lines

MDV-transformed cell lines express immunosuppressive factors, as measured by decreased ConA blast formation and mixed lymphocyte reaction (MLR) (Quere 1992; Theis 1981). Moreover, the work of Quere (1992) has shown that most of this immunosuppressive activity is through secretion of soluble factors. As mentioned above, the expression of CFA by MSB-1 cells is apparently immunosuppressive to NK cells, even when given in soluble form *in vivo* (Ohashi et al. 1987). At least one soluble factor expressed by an MDV-transformed cell line has been identified as prostaglandin (PGE_2) (Bumstead and Payne 1987). The significance of PGE_2, as discussed above, is that as a product of cycloxygenase-2 (COX-2) activity (Chilton et al. 1997), it regulates the expression of (iNOS) by macrophage (Djeraba et al. 2002c). In addition, we have found that supernatant medium from MDV-transformed cell lines induces surface MHC-II expression on chicken primary macrophages whereas IFN-γ induces both MHC-I and MHC-II (M.S. Parcells, 2003, unpublished data). The significance of these data are not clear, but suggest that MDV-transformed cell lines express factors that are distinct from IFN-γ in their activity. Thus, like several other factors seen in MDV lymphoma formation and ostensibly their progression, secreted factors of MDCC appear to modulate the immune response towards a T_{h2}-like response.

CD30, TNFRs and their role in lymphoma progression

Like the transformed component of MDV lymphomas and the derived cell lines, chicken lymphocytes transformed by other viruses also over-express CD30. CD30 is also over-expressed by chicken cell lines neoplastically transformed by Rous-associated virus-2 (RAV-2) and by the acute-transforming reticuloendotheliosis virus type T (REV-T) (Burgess et al. 2004). CD30 is also over-expressed by MD lymphomas from turkeys. MDV, ALV, RAV-2 and REV-T all neoplastically transform lymphocytes, yet their target specificities and mechanisms differ (Burgess et al. submitted. ALVs preferentially transform B- and pre-B-cells, whereas MDV transforms only T-cells. REV-T transforms both B- and and T-cells. Yet lymphomas caused by each these viruses over-express CD30. This suggests that either each virus is subverting the CD30 signalling pathway and/or CD30 over-expression is a conserved response in neoplastically transformed chicken lymphocytes (and probably neoplastically transformed human lymphocytes) (Burgess et al. 2004).

$CD30^{hi}$ cells from MDV-induced lymphomas also express high levels of the Meq oncoprotein (Burgess et al. submitted). As Meq expression is associated with MDV lymphomas and cell lines, and $CD30^{hi}$ cells are the transformed component of MDV-induced lymphomas, this association may be coincidental. However, the promoter region of the chicken CD30 gene contains several MERE-I sites (AP-1-like sites that bind Meq/Jun heterodimers), as well as a MERE-II site (CACA-motif that binds Meq/Meq homodimers) (Burgess et al. 2004). As c-Jun signaling (either through transcriptional activation or protein stablization) is induced in v-*myc* and c-*myc*-disrupted transformed B-cells (Gavine et al. 1999), in v-*rel* transformed T-cells and MDV-transformed T-cells (Kralova et al. 1998), the induction of this pathway may be directly or indirectly involved in CD30 upregulation. Although the pathways involved in CD30 upregulation have not been elucidated, CD30 expression peaks 4–5 days after activation in normal T-cells (Chiarle et al. 1999). The ligand for membrane-bound CD30 is CD153 (a.k.a. CD30 ligand) and the stimulation of CD30 on mammalian T-cells has been linked with the release of IL-13 and T_{H2} cytokines IL-4 and IL-10 (Bengtsson et al. 2000; Harlin et al. 2002). As these cytokines have yet to be cloned from the chicken, and no antibody reagents are currently available for these proteins, the expression profiles of MDV-transformed cells remains unknown.

Other surface antigens

The expression of homophilic and extracellular matrix (ECM)-binding antigens on MDV-transformed cell lines is indicative of their metastatic and lymphoma forming potential (Berg et al. 2002). $CD49e^+$ /$CD29^+$ (α-5/β-1 integrin) expression is directly associated with dissemination and metastasis of lymphomas in SCID mice (Blase et al. 1995). In addition to this metastatic potential, integrin signaling through AKT (PKB) induces cell survival, thus providing not only

adhesion, but additional anti-apoptotic signals to the proliferating cell (Pankov et al. 2003). Expression of CD49e/CD29 by HTLV-I-infected T-lymphomcytes and is likely to contribute to HTLV-I lymphomagenesis (Hasegawa et al. 1998).

Expression of CD44 by lymphomas is indicative of metastasis as well (Walter et al. 1995). CD44 binds hyaluronate and is a key determinant of T-cell extravasation (Haynes et al. 1991). Signaling via the CD44 receptor depends on the splice variant of CD44 that is expressed (Akisik et al. 2002). Lymphoma progression has been associated with specific splice variants of CD44 but the chicken form recognized by the anti-CD44 mAb has not been thoroughly characterized.

Expression of CD164 by MDV-transformed cells is intriguing, as this antigen is associated with $CD34^+$ stem cells (Watt et al. 1998). Its role in development of hematopoeitic progenitor cells suggests an involvement in cell migration (Zannettino et al. 1998). Expression of GRL-2 is found on granule containing leukocytes, including some activated T-cells. The role of this molecule in terms of lymphoma development, however, is uncharacterized, but suggests that MDCC may be enhanced in their secretion of soluble factors (Thomas et al. 1993, 1994).

Overall effects on lymphoma development

Given the level of immune suppression and modulation that occurs during and as a consequence of MDV-lymphomagenesis, it is difficult to separate those factors rendering the host incapable of mounting an effective anti-viral response from those that affect lymphoma development and progression. Conversely, the immunity elicited by MDV vaccination is still relatively undefined in terms of delineation between anti-virus and anti-tumor responses. What does seem clear, however, is that upon transformation by MDV, a common immunoevasive and immunomodulatory set of pathways is activated that appears common to many human lymphomas. At least one of these immunoevasive pathways may involve CD30 surface expression and presumably the modulation of the host immune response to more of a T_{H2} effector response. Interestingly, the level of genetic susceptibility to MDV lymphoma formation and progression appears to be determined by the extent to which MDV is successful in shifting the immune response to a T_{h2}

expression profile. The upregulation of specific antigens by MDV in transformed T-cells, and the secretion of PGE_2 and perhaps other factors, may have a net effect on the immune response within the lymphoma. The result would be escape of the transformed cells from CTL-, NK- and macrophage-mediated responses. Interestingly, on of the cellular genes upregulated during MDV infection of chicken embryo fibroblasts is the IL-13 receptor (IL-13R) (Morgan et al. 2001), a cytokine activated by CD30 signaling (Harlin et al. 2002).

Intracellular adhesion molecules expressed by MDV lymphomas may explain their tissue dissemination and metastatic potential. The expression of CD49e/CD29 in particular is highly indicative of such a lymphoma as upregulation of this integrin is associated with late stage metastasis. In addition to mediating ECM-binding, the proliferative/pro-survival signaling of integrins is essential for tumor progression as well as the blocking of anoikis during anchorage-independent growth (Pankov et al. 2003). Expression of other developmentally regulated (CD164) and activation-associated (MATSA, CD44, GRL-2) may also contribute to migration or proliferation of these cells.

A key element in MD lymphoma development and progression is the expression of Meq. This b-ZIP protein appears to de-regulate c-Jun signaling, a common pathway altered in other human T-lymphomas. The contribution of this oncoprotein to the alteration of surface antigen expression, although inferred by the coincidental expression of CD30, has yet to be shown.

Summary

MDV-induced T-lymphomagenesis occurs rapidly (within 4–6 weeks) and like other herpesviral lymphomas, involves the expression of latent gene products. The most consistently detected of the MDV latent gene products is the Meq oncoprotein, a b-ZIP protein capable of altering c-Jun signaling and affecting cell survival and proliferation. Mechanisms affecting lymphoma progression are currently unknown but are characterized by the upregulation of surface antigens on the transformed cells within lymphomas. These antigens include fetal and developmentally regulated antigens, and CD30, a TNFRII molecule expressed on activated

lymphocytes and associated with a T_{H2} expression profile. As anti-tumor responses are largely cell mediated, the expression of CD30 by the transformed cells, and potential subsequent expression of T_{H2} cytokines could disrupt the CMI to the transformed cell. In addition, several other antigens (CD44, CD49e/CD29, CD164, MATSA and GRL-2) are likely to promote the angiogenesis, extravasation, proliferation and pro-survival of transformed T-cells. The aggressive nature of MDV-induced lymphomas, coupled with their rapid and reproducible induction in genetically defined populations continue to make MD lymphomagenesis a compelling model. MD is the only known model of naturally occurring CD30 over-expressing lymphoma. With the completion and annotation of the chicken genome sequence, the transformed component of MD-lymphomas will be able to be completely described. The MD model promises to yield important insight into basic and broadly applicable mechanisms of lymphomagenesis in a general biomedical context. Moreover, a clearer understanding of the immunity elicited by MDV vaccination to lymphomagenesis promises to provide hints for novel immunotherapeutic targets for the treatment of human lymphomas and leukemias.

Acknowledgements The authors wish to thank Hsing-Jien Kung, Karel A. Schat, Jagdev M. Sharma and Richard L. Witter for helpful discussions and the sharing of unpublished data in the preparation of this work.

References

Afonso CL, Tulman ER, Lu Z, Zsak L, Rock DL, Kutish GF (2001) The genome of turkey herpesvirus. J Virol 75(2):971–978

Akisik E, Bavbek S, Dalay N (2002) CD44 variant exons in leukemia and lymphoma. Pathol Oncol Res 8(1):36–40

Anderson AS, Parcells MS, Morgan RW (1998) The glycoprotein D (US6) homolog is not essential for oncogenicity or horizontal transmission of Marek's disease virus. J Virol 72(3):2548–2553

Anobile JM, Arumugaswami V, Downs D, Czymmek K, Parcells M, and Schmidt CJ (2006) Nuclear Localization and Dynamic Properties of the Marek's Disease Virus Oncogene Products Meq and Meq/vIL8. J Virol 80:1160–1166

Arnulf B, Villemain A, Nicot C, Mordelet E, Charneau P, Kersual J, Zermati Y, Mauviel A, Bazarbachi A, Hermine O (2002) Human T-cell lymphotropic virus oncoprotein Tax represses TGF-beta 1 signaling in human T cells via c-Jun activation: a potential mechanism of HTLV-I leukemogenesis. Blood 100(12):4129–4138

Bacon LD, Witter RL (1993) Influence of B-haplotype on the relative efficacy of Marek's disease vaccines of different serotypes. Avian Dis 37(1):53–59

Bacon LD, Witter RL (1994) Serotype specificity of B-haplotype influence on the relative efficacy of Marek's disease vaccines. Avian Dis 38(1):65–71

Bacon LD, Witter RL, Crittenden LB, Fadly A, Motta J (1981) B-haplotype influence on Marek's disease, Rous sarcoma, and lymphoid leukosis virus-induced tumors in chickens. Poult Sci 60(6):1132–1139

Bacon LD, Crittenden LB, Witter RL, Fadly A, Motta J (1983) B5 and B15 associated with progressive Marek's disease, Rous sarcoma, and avian leukosis virus-induced tumors in inbred 15I4 chickens. Poult Sci 62(4):573–578

Bacon LD, Vallejo RL, Cheng HH, Witter RL (1996) Failure of RFP-Y genes to influence resistance to Marek's disease. Paper presented at the 5th International Symposium on Marek's Disease. East Lansing, MI

Baigent SJ, Ross LJ, Davison TF (1998) Differential susceptibility to Marek's disease is associated with differences in number, but not phenotype or location, of pp38[+] lymphocytes. J Gen Virol 79(Pt 11):2795–2802

Barrow AD, Burgess SC, Howes K, Nair VK (2003) Monocytosis is associated with the onset of leukocyte and viral infiltration of the brain in chickens infected with the very virulent Marek's disease virus strain C12/130. Avian Pathol 32(2):183–191

Beasley JN, Patterson LT, McWade DH (1970) Transmission of Marek's disease by poultry house dust and chicken dander. Avian Dis 14(1):45–53

Bengtsson A, Scheynius A, Avila-Carino J (2000) Crosslinking of CD30 on activated human Th clones enhances their cytokine production and downregulates the CD30 expression. Scand J Immunol 52(6):595–601

Berg LP, James MJ, Alvarez-Iglesias M, Glennie S, Lechler RI, Marelli-Berg FM (2002) Functional consequences of non-cognate interactions between CD4[+] memory T lymphocytes and the endothelium. J Immunol 168(7):3227–3234

Biggs PM, Thorpe RJ, Payne LN (1968) Studies on genetic resistance to Marek's disease in the domestic chicken. Br Poult Sci 9(1):37–52

Blase L, Daniel PT, Koretz K, Schwartz-Albiez R, Moller P (1995) The capacity of human malignant B-lymphocytes to disseminate in SCID mice is correlated with functional expression of the fibronectin receptor alpha 5 beta 1 (CD49e/CD29). Int J Cancer 60(6):860–866

Bradley G, Hayashi M, Lancz G, Tanaka A, Nonoyama M (1989a) Structure of the Marek's disease virus bamhi-H gene family: genes of putative importance for tumor induction. J Virol 63(6):2534–2542

Bradley G, Lancz G, Tanaka A, Nonoyama M (1989b) Loss of Marek's disease virus tumorigenicity is associated with truncation of rnas transcribed within bamhi-H. J Virol 63(10):4129–4135

Briles WE, Briles RW, Taffs RE, Stone HA (1983) Resistance to a malignant lymphoma in chickens is mapped to sub-region of major histocompatibility (B) complex. Science 219(4587):977–979

Briles WE, Goto RM, Auffray C, Miller MM (1993) A polymorphic system related to but genetically independent of the chicken major histocompatibility complex. Immunogenetics 37(6):408–414

Brown AC, Baigent SJ, Smith LP, Chattoo JP, Petheridge LJ, Hawes P, Allday MJ, Nair V (2006) Interaction of MEQ protein and C-terminal-binding protein is critical for induction of lymphomas by Marek's disease virus. Proc Natl Acad Sci 103(6):1687–1692

Bumstead JM, Payne LN (1987) Production of an immune suppressor factor by Marek's disease lymphoblastoid cell lines. Vet Immunol Immunopathol 16(1–2):47–66

Bumstead N (1998) Genetic resistance to avian viruses. Rev Sci Tech 17(1):249–255

Bumstead N, Sillibourne J, Rennie M, Ross N, Davison F (1997) Quantification of Marek's disease virus in chicken lymphocytes using the polymerase chain reaction with fluorescence detection. J Virol Methods 65(1):75–81

Burgess SC, Davison TF (1999) A quantitative duplex PCR technique for measuring amounts of cell-associated Marek's disease virus: differences in two populations of lymphoma cells. J Virol Methods 82(1):27–37

Burgess SC, Davison TF (2002) Identification of the neoplastically transformed cells in Marek's disease herpesvirus-induced lymphomas: recognition by the monoclonal antibody AV37. J Virol 76(14):7276–7292

Burgess SC, Venugopal KN (2002) Anti-tumor immune responses after infection with the Marek's disease and avian leukosis oncogenic viruses of poultry. In: Mathew T (ed) Modern concept of immunology in veterinary medicine – poultry immunology. Thajema Publishing, pp 236–291

Burgess SC, Basaran BH, Davison TF (2001) Resistance to Marek's disease herpesvirus- induced lymphoma is multiphasic and dependent on host genotype. Vet Pathol 38(2):129–142

Burgess SC, Young JR, Baaten BJ, Hunt L, Ross LN, Parcells MS, Kumar PM, Tregaskes CA, Lee LF, and Davison TF (2004) Marek's disease is a natural model for lymphomas overexpressing Hodgkin's disease antigen (CD30). Proc Natl Acad Sci USA 101:13879–13884

Buscaglia C, Calnek BW (1988) Maintenance of Marek's disease herpesvirus latency in vitro by a factor found in conditioned medium. J Gen Virol 69(Pt 11):2809–2818

Buscaglia C, Calnek BW, Schat KA (1988) Effect of immunocompetence on the establishment and maintenance of latency with Marek's disease herpesvirus. J Gen Virol 69(Pt 5):1067–1077

Calnek BW (1982) Marek's disease vaccines. Dev Biol Stand 52:401–405

Calnek BW (1985) Genetic resistance. In Payne LN (ed) Marek's disease. Scientific basis and methods of control. Martinus Nijhoff Publishing, Boston, pp 293–328

Calnek BW (1986) Marek's disease – a model for herpesvirus oncology. Crit Rev Microbiol 12(4):293–320

Calnek BW, Witter RL (1997) Marek's disease. In Calnek BW (ed) Diseases of poultry. 10th edn. Iowa State University Press, Ames, IA, pp 367–413

Calnek BW, Adldinger HK, Kahn DE (1970) Feather follicle epithelium: a source of enveloped and infectious cell- free herpesvirus from Marek's disease. Avian Dis 14(2):255–267

Calnek BW, Carlisle JC, Fabricant J, Murthy KK, Schat KA (1979) Comparative pathogenesis studies with oncogenic and nononcogenic Marek's disease viruses and turkey herpesvirus. Am J Vet Res 40(4):541–548

Calnek BW, Shek WR, Schat KA (1981) Spontaneous and induced herpesvirus genome expression in Marek's disease tumor cell lines. Infect Immun 34(2):483–491

Calnek BW, Schat KA, Ross LJ, Shek WR, Chen CL (1984) Further characterization of Marek's disease virus-infected lymphocytes. I. In vivo infection. Int J Cancer 33(3):389–398

Calnek BW, Adene DF, Schat KA, Abplanalp H (1989a) Immune response versus susceptibility to Marek's disease. Poult Sci 68(1):17–26

Calnek BW, Lucio B, Schat KA, Lillehoj HS (1989b) Pathogenesis of Marek's disease virus- induced local lesions. 1. Lesion characterization and cell line establishment. Avian Dis 33(2):291–302

Calnek BW, Harris RW, Buscaglia C, Schat KA, Lucio B (1998) Relationship between the immunosuppressive potential and the pathotype of Marek's disease virus isolates. Avian Dis 42(1):124–132

Cantello JL, Anderson AS, Morgan RW (1994) Identification of latency-associated transcripts that map antisense to the ICP4 homolog gene of Marek's disease virus. Virology 204(1):242–250

Cantello JL, Parcells MS, Anderson AS, Morgan RW (1997) Marek's disease virus latency- associated transcripts belong to a family of spliced RNAs that are antisense to the ICP4 homolog gene. J Virol 71(2):1353–1361

Carpenter SL, Sevoian M (1983) Cellular immune response to Marek's disease: listeriosis as a model of study. Avian Dis 27(2):344–356

Chang KS, Lee SI, Ohashi K, Ibrahim A, Onuma M (2002) The detection of the meq gene in chicken infected with Marek's disease virus serotype 1. J Vet Med Sci 64(5):413–417

Chen XB, Sondermeijer PJ, Velicer LF (1992) Identification of a unique Marek's disease virus gene which encodes a 38-kilodalton phosphoprotein and is expressed in both lytically infected cells and latently infected lymphoblastoid tumor cells. J Virol 66(1):85–94

Chiarle R, Podda A, Prolla G, Gong J, Thorbecke GJ, Inghirami G (1999) CD30 in normal and neoplastic cells. Clin Immunol 90(2):157–164

Chilton FH, Fonteh AN, Johnson MM, Surette ME (1997) Metabolism of Arachidonic Acid. In: Crystal RG, West JB, Weibel ER, Barnes PJ (eds) The lung – scientific foundations, Vol. 1. Lippincott-Raven, Philadelphia, pp 77–88

Coleman RM, Schierman LW (1982) Transplantable Marek's disease lymphomas. I. Growth characteristics during development in two inbred lines of chickens. Avian Dis 26(2):245–256

Confer AW, Adldinger HK, Buening GM (1980) Cell-mediated immunity in Marek's disease: correlation of disease-related variables with immune responses in age-resistant chickens. Am J Vet Res 41(3):313–318

Corbel C, Bluestein HG, Pourquie O, Vaigot P, Le Douarin NM (1992) An antigen expressed by avian neuronal cells is also expressed by activated T lymphocytes. Cell Immunol 141(1):99–110

Corbel C, Pourquie O, Cormier F, Vaigot P, Le Douarin NM (1996) BEN/SC1/DM-GRASP, a homophilic adhesion molecule, is required for in vitro myeloid colony formation by avian hemopoietic progenitors. Proc Natl Acad Sci USA 93(7):2844–2849

Cui ZZ, Yan D, Lee LF (1990) Marek's disease virus gene clones encoding virus-specific phosphorylated polypeptides and serological characterization of fusion proteins. Virus Genes 3(4):309–322

Delecluse HJ, Hammerschmidt W (1993) Status of Marek's disease virus in established lymphoma cell lines: herpesvirus integration is common. J Virol 67(1):82–92

Delecluse HJ, Schuller S, Hammerschmidt W (1993) Latent Marek's disease virus can be activated from its chromosomally integrated state in herpesvirus-transformed lymphoma cells. EMBO J 12(8):3277–3286

Dienglewicz RL, Parcells MS (1999) Establishment of a lymphoblastoid cell line using a mutant MDV containing a green fluorescent protein expression cassette. Acta Virol 43:106–112

DiFronzo NL, Schierman LW (1990) Transplantable Marek's disease lymphomas. IV. Differences in lethality of lymphoma cell lines determined by route of inoculation. J Immunol 144(12):4883–4887

Djeraba A, Kut E, Rasschaert D, Quere P (2002a) Antiviral and antitumoral effects of recombinant chicken myelomonocytic growth factor in virally induced lymphoma. Int Immunopharmacol 2(11):1557–1566

Djeraba A, Musset E, Lowenthal JW, Boyle DB, Chausse AM, Peloille M, Quere P (2002b) Protective effect of avian myelomonocytic growth factor in infection with Marek's disease virus. J Virol 76(3):1062–1070

Djeraba A, Musset E, van Rooijen N, Quere P (2002c) Resistance and susceptibility to Marek's disease: nitric oxide synthase/arginase activity balance. Vet Microbiol 86(3):229–244

Falini B, Stein H, Pileri S, Canino S, Farabbi R, Martelli MF, Grignani F, Fagioli M, Minelli O, Ciani C et al. (1987) Expression of lymphoid-associated antigens on Hodgkin's and Reed- Sternberg cells of Hodgkin's disease. An immunocytochemical study on lymph node cytospins using monoclonal antibodies. Histopathology 11(12):1229–1242

Fragnet L, Blasco MA, Klapper W, Rasschaert D (2003) The RNA subunit of telomerase is encoded by Marek's disease virus. J Virol 77(10):5985–5996

Gavine PR, Neil JC, Crouch DH (1999) Protein stabilization: a common consequence of mutations in independently derived v-Myc alleles. Oncogene 18(52):7552–7558

Gimeno IM, Witter RL, Reed WM (1999) Four distinct neurologic syndromes in Marek's disease: effect of viral strain and pathotype. Avian Dis 43(4):721–737

Gimeno IM, Witter RL, Neumann U (2002) Neuropathotyping: a new system to classify Marek's disease virus. Avian Dis 46:909–918

Harlin H, Podack E, Boothby M, Alegre ML (2002) TCR-independent CD30 signaling selectively induces IL-13 production via a TNF receptor-associated factor/p38 mitogen-activated protein kinase-dependent mechanism. J Immunol 169(5):2451–2459

Hasegawa H, Nomura T, Kishimoto K, Yanagisawa K, Fujita S (1998) SFA-1/PETA-3 (CD151), a member of the transmembrane 4 superfamily, associates preferentially with alpha 5 beta 1 integrin and regulates adhesion of human T cell leukemia virus type 1-infected T cells to fibronectin. J Immunol 161(6):3087–3095

Haynes BF, Liao HX, Patton KL (1991) The transmembrane hyaluronate receptor (CD44): multiple functions, multiple forms. Cancer Cells 3(9):347–350

Heier BT, Jarp J (2000) Risk factors for Marek's disease and mortality in white Leghorns in Norway. Prev Vet Med 44(3–4):153–165

Hirai K, Ikuta K, Kitamoto N, Kato S (1981) Latency of herpesvirus of turkey and Marek's disease virus genomes in a chicken T-lymphoblastoid cell line. J Gen Virol 53(Pt 1):133–143

Hong Y, Coussens PM (1994) Identification of an immediate-early gene in the Marek's disease virus long internal repeat region which encodes a unique 14-kilodalton polypeptide. J Virol 68(6):3593–3603

Hong Y, Frame M, Coussens PM (1995) A 14-kDa immediate-early phosphoprotein is specifically expressed in cells infected with oncogenic Marek's disease virus strains and their attenuated derivatives. J Vet Med Sci 57(1):157–160

Hrdlickova R, Nehyba J, Humphries EH (1994) V-rel induces expression of three avian immunoregulatory surface receptors more efficiently than c-rel. J Virol 68(1):308–319

Hunt HD, Lupiani B, Miller MM, Gimeno I, Lee LF, Parcells MS (2001) Marek's disease virus down-regulates surface expression of MHC (B Complex) Class I (BF) glycoproteins during active but not latent infection of chicken cells. Virology 282(1):198–205

Ikuta K, Kitamoto N, Shoji H, Kato S, Naiki M (1981a) Expression of Forssman antigen of avian lymphoblastoid cell lines transformed by Marek's disease virus or avian leukosis virus. J Gen Virol 52(Pt 1):145–151

Ikuta K, Kitamoto N, Shoji H, Kato S, Naiki M (1981b) Hanganutziu and Deicher type heterophile antigen expressed on the cell surface of Marek's disease lymphoma-derived cell lines. Biken J 24(1–2):23–37

Inman GJ, Allday MJ (2000) Resistance to TGF-beta1 correlates with a reduction of TGF-beta type II receptor expression in Burkitt's lymphoma and Epstein-Barr virus-transformed B lymphoblastoid cell lines. J Gen Virol 81 (Pt 6):1567–1578

Izumiya Y, Jang HK, Ono M, Mikami T (2001) A complete genomic DNA sequence of Marek's disease virus type 2, strain HPRS24. Curr Top Microbiol Immunol 255:191–221

Jakovleva LS, Mazurenko NP (1979) Increased susceptibility of leukemia-infected chickens to Marek's disease. Neoplasma 26(4):393–396

Jones D, Lee L, Liu JL, Kung HJ, Tillotson JK (1992) Marek disease virus encodes a basic-leucine zipper gene resembling the fos/jun oncogenes that is highly expressed in lymphoblastoid tumors [published erratum appears in Proc Natl Acad Sci USA 1993 Mar 15;90 (6):2556]. Proc Natl Acad Sci USA 89(9):4042–4046

Jones D, Brunovskis P, Witter R, Kung HJ (1996) Retroviral insertional activation in a herpesvirus: transcriptional activation of US genes by an integrated long terminal repeat in a Marek's disease virus clone. J Virol 70(4):2460–2467

Kaiser P, Underwood G, Davison F (2003) Differential cytokine responses following Marek's disease virus infection of chickens differing in resistance to Marek's disease. J Virol 77(1):762–768

Kaplan MH, Dhar A, Brown TR, Sundick RS (1992) Marek's disease virus-transformed chicken T-cell lines respond to lymphokines. Vet Immunol Immunopathol 34(1–2):63–79

Kawai T, Kato A, Higashi H, Kato S, Naiki M (1991) Quantitative determination of N-glycolylneuraminic acid expression in human cancerous tissues and avian lymphoma cell lines as a tumor- associated sialic acid by gas chromatography-mass spectrometry. Cancer Res 51(4):1242–1246

Kingham BF, Zelnik V, Kopacek J, Majerciak V, Ney E, Schmidt CJ (2001) The genome of herpesvirus of turkeys: comparative analysis with Marek's disease viruses. J Gen Virol 82(Pt 5):1123–1135

Kirby B, Owen CM, Blewitt RW, Yates VM (2002) Cutaneous T-cell lymphoma developing in a patient on cyclosporin therapy. J Am Acad Dermatol 47(2 Suppl):S165–167

Kishi M, Harada H, Takahashi M, Tanaka A, Hayashi M, Nonoyama M, Josephs SF, Buchbinder A, Schachter F, Ablashi DV et al. (1988) A repeat sequence, GGGTTA, is shared by DNA of human herpesvirus 6 and Marek's disease virus. J Virol 62(12):4824–4827

Kishi M, Bradley G, Jessip J, Tanaka A, Nonoyama M (1991) Inverted repeat regions of Marek's disease virus DNA possess a structure similar to that of the a sequence of herpes simplex virus DNA and contain host cell telomere sequences. J Virol 65(6):2791–2797

Kralova J, Liss AS, Bargmann W, Bose Jr., HR, (1998) AP-1 factors play an important role in transformation induced by the v-rel oncogene. Mol Cell Biol 18(5):2997–3009

Kung HJ, Xia L, Brunovskis P, Li D, Liu JL, Lee LF (2001) Meq: an MDV-specific bzip transactivator with transforming properties. Curr Top Microbiol Immunol 255:245–260

Lakshmanan N, Lamont SJ (1998) Rfp-Y region polymorphism and Marek's disease resistance in multitrait immunocompetence-selected chicken lines. Poult Sci 77(4):538–541

Lee LF, Sharma JM, Nazerian K, Witter RL (1978a) Suppression and enhancement of mitogen response in chickens infected with Marek's disease virus and the herpesvirus of turkeys. DTW Dtsch Tierarztl Wochenschr 85(8):325–328

Lee LF, Sharma JM, Nazerian K, Witter RL (1978b) Suppression of mitogen-induced proliferation of normal spleen cells by macrophages from chickens inoculated with Marek's disease virus. Veterinarii a(5):99–101

Lee LF, Liu X, Sharma JM, Nazerian K, Bacon LD (1983) A monoclonal antibody reactive with Marek's disease tumor-associated surface antigen. J Immunol 130(2):1007–1011

Lee LF, Wu P, Sui D, Ren D, Kamil J, Kung HJ, Witter RL (2000) The complete unique long sequence and the overall genomic organization of the GA strain of Marek's disease virus. Proc Natl Acad Sci USA 97(11):6091–6096

Lee TH, Tempelis CH (1992) Possible 110kDa receptor for interleukin 2 in the chicken. Dev Comp Immunol 16(6):463–472Levy AM, Heller ED, Leitner G, Davidson I (1999) Effect of native chicken interferon on MDV replication. Acta Virol 43(2–3):121–127

Levy AM, Lzumiya Y, Brunovskis P, Xia L, Parcells MS, Reddy SM, Lee L, Chen HW, Kung HJ (2003) Characterization of the chromosomal binding sites and dimerization partners of the viral oncoprotein Meq in Marek's disease virus-transformed T cells. J Virol 77:12841–12851

Levy AM, Gilad O, Xia L, Lzumiya Y, Choi J, Tsalenko A, Yakhini Z, Witter R, Lee L, Cardona CJ, Kung HJ (2005) Marek's disease virus Meq transforms chicken cells via the v-jun transcriptional cascade: a converging transforming pathway for avian oncoviruses. Proc Natl Acad Sci USA 102:14831–14836

Li D, O'Sullivan G, Greenall L, Smith G, Jiang C, Ross N (1998) Further characterization of the latency-associated transcription unit of Marek's disease virus. Arch Virol 143(2):295–311

Li DS, Pastorek J, Zelnik V, Smith GD, Ross LJ (1994) Identification of novel transcripts complementary to the Marek's disease virus homologue of the ICP4 gene of herpes simplex virus. J Gen Virol 75(Pt 7):1713–1722Lim IG, Bertouch JV (2002) Remission of lymphoma after drug withdrawal in rheumatoid arthritis. Med J Australia 177(9):500–501

Liu HC, Kung HJ, Fulton JE, Morgan RW, Cheng HH (2001) Growth hormone interacts with the Marek's disease virus SORF2 protein and is associated with disease resistance in chicken. Proc Natl Acad Sci USA 98(16):9203–9208

Liu JL, Lee LF, Ye Y, Qian Z, Kung HJ (1997) Nucleolar and nuclear localization properties of a herpesvirus bzip onco-protein, MEQ. J Virol 71(4):3188–3196

Liu JL, Ye Y, Lee LF, Kung HJ (1998) Transforming potential of the herpesvirus oncoprotein MEQ: morphological transformation, serum-independent growth, and inhibition of apoptosis. J Virol 72(1):388–395

Liu JL, Ye Y, Qian Z, Qian Y, Templeton DJ, Lee LF, Kung HJ (1999) Functional interactions between herpesvirus oncoprotein MEQ and cell cycle regulator CDK2. J Virol 73(5):4208–4219

Lucey DR, Clerici M, Shearer GM (1996) Type 1 and type 2 cytokine dysregulation in human infectious, neoplastic, and inflammatory diseases. Clin Microbiol Rev 9(4):532–562

Marek J (1907) Multiplenervenetzundung bei Huehnern. Dtsh Tieraerztl Wochenschr 15:417–421

Markowski-Grimsrud CJ, Schat KA (2002) Cytotoxic T lymphocyte responses to Marek's disease herpesvirus-encoded glycoproteins. Vet Immunol Immunopathol 90(3–4):133–144

McColl KA, Calnek BW, Harris WV, Schat KA, Lee LF (1987) Expression of a putative tumor-associated surface antigen on normal versus Marek's disease virus-transformed lymphocytes. J Natl Cancer Inst 79(5):991–1000

McKie EA, Ubukata E, Hasegawa S, Zhang S, Nonoyama M, Tanaka A (1995) The transcripts from the sequences flanking the short component of Marek's disease virus during latent infection form a unique family of 3′-coterminal rnas. Virology 207(1):205–216

Miles AM, Reddy SM, Morgan RW (2001) Coinfection of specific-pathogen-free chickens with Marek's disease virus (MDV) and chicken infectious anemia virus: effect of MDV pathotype. Avian Dis 45(1):9–18

Miller MM, Goto R, Zoorob R, Auffray C, Briles WE (1994) Regions of homology shared by Rfp-Y and major histocompatibility B complex genes. Immunogenetics 39(1):71–73

Miller MM, Goto RM, Taylor Jr., RL, Zoorob R, Auffray C, Briles RW, Briles WE, Bloom SE (1996) Assignment of Rfp-Y to the chicken major histocompatibility complex/NOR microchromosome and evidence for high-frequency recombination associated with the nucleolar organizer region. Proc Natl Acad Sci USA 93(9):3958–3962

Morgan RW, Sofer L, Anderson AS, Bernberg EL, Cui J, Burnside J (2001) Induction of host gene expression following infection of chicken embryo fibroblasts with oncogenic Marek's disease virus. J Virol 75(1):533–539

Mori N, Fujii M, Iwai K, Ikeda S, Yamasaki Y, Hata T, Yamada Y, Tanaka Y, Tomonaga M, Yamamoto N (2000) Constitutive activation of transcription factor AP-1 in primary adult T-cell leukemia cells. Blood 95(12):3915–3921

Morimura T, Ohashi K, Kon Y, Hattori M, Sugimoto C, Onuma M (1996) Apoptosis and CD8-down-regulation in the thymus of chickens infected with Marek's disease virus. Arch Virol 141(11):2243–2249

Morimura T, Ohashi K, Kon Y, Hattori M, Sugimoto C, Onuma M (1997) Apoptosis in peripheral CD4+ T cells and thymocytes by Marek's disease virus-infection. Leukemia 11(Suppl 3):206–208

Morimura T, Ohashi K, Sugimoto C, Onuma M (1998) Pathogenesis of Marek's disease (MD) and possible mechanisms of immunity induced by MD vaccine. J Vet Med Sci 60(1):1–8

Morimura T, Cho KO, Kudo Y, Hiramoto Y, Ohashi K, Hattori M, Sugimoto C, Onuma M (1999) Anti-viral and anti-tumor effects induced by an attenuated Marek's disease virus in CD4- or CD8-deficient chickens [In Process Citation]. Arch Virol 144(9):1809–1818

Murthy KK, Calnek BW (1979) Marek's disease tumor-associated surface antigen (MATSA) in resistant versus susceptible chickens. Avian Dis 23(4):831–837

Naiki M, Fujii Y, Ikuta K, Higashi H, Kato S (1982) Expression of Hanganutziu and Deicher type heterophile antigen on the cell surface of Marek's disease lymphoma. Adv Exp Med Biol 152:445–456

Nathanson S, Lucidarme N, Landman-Parker J, Deschenes G (2002) Long-term survival of renal graft complicated with Burkitt lymphoma. Pediatr Nephrol 17(12):1066–1068

Nazerian K, Sharma JM (1975) Detection of T-cell surface antigens in a Marek's disease lymphoblastoid cell line. J Natl Cancer Inst 54(1):277–279

Nazerian K, Witter RL (1984) Immunization against Marek's disease transplantable cell lines. Avian Dis 28(1):160–167

Niikura M, Witter RL, Jang HK, Ono M, Mikami T, Silva RF (1999) MDV glycoprotein D is expressed in the feather follicle epithelium of infected chickens. Acta Virol 43 (2–3):159–163

Ohashi K, Mikami T, Higashihara T, Kodama H, Izawa H (1986) Monoclonal antibody to chicken fetal antigen on Marek's disease lymphoblastoid cell line MDCC-MSB1. Cancer Res 46(11):5858–5863

Ohashi K, Mikami T, Kodama H, Izawa H (1987) Suppression of NK activity of spleen cells by chicken fetal antigen present on Marek's disease lymphoblastoid cell line cells. Int J Cancer 40(3):378–382

Ohashi K, Morimura T, Takagi M, Lee SI, Cho KO, Takahashi H, Maeda Y, Sugimoto C, Onuma M (1999) Expression of bcl-2 and bcl-x genes in lymphocytes and tumor cell lines derived from MDV-infected chickens. Acta Virol 43(2–3):128–132

Omar AR, Schat KA (1996) Syngeneic Marek's disease virus (MDV)-specific cell-mediated immune responses against immediate early, late, and unique MDV proteins. Virology 222(1):87–99

Omar AR, Schat KA (1997) Characterization of Marek's disease herpesvirus-specific cytotoxic T lymphocytes in chickens inoculated with a non-oncogenic vaccine strain of MDV. Immunology 90(4):579–585

Omar AR, Schat KA, Lee LF, Hunt HD (1998) Cytotoxic T lymphocyte response in chickens immunized with a recombinant fowlpox virus expressing Marek's disease herpesvirus glycoprotein B. Vet Immunol Immunopathol 62(1):73–82

Pankov R, Cukierman E, Clark K, Matsumoto K, Hahn C, Poulin B, Yamada KM (2003) Specific beta1 integrin site selectively regulates Akt/protein kinase B signaling via local activation of protein phosphatase 2A. J Biol Chem 278(20):18671–18681

Parcells MS, Anderson AS, Morgan TW (1995) Retention of oncogenicity by a Marek's disease virus mutant lacking six unique short region genes. J Virol 69(12):7888–7898

Parcells MS, Dienglewicz RL, Anderson AS, Morgan RW (1999) Recombinant Marek's disease virus (MDV)-derived lymphoblastoid cell lines: regulation of a marker gene within the context of the MDV genome. J Virol 73(2):1362–1373

Parcells MS, Lin SF, Dienglewicz RL, Majerciak V, Robinson DR, Chen HC, Wu Z, Dubyak GR, Brunovskis P, Hunt HD, Lee LF, Kung HJ (2001) Marek's disease virus (MDV) encodes an interleukin-8 homolog (vil-8): characterization of the vil-8 protein and a vil-8 deletion mutant MDV. J Virol 75(11):5159–5173

Payne LN (1985) Pathology. In Payne LN (ed) Marek's disease: scientific basis and methods of control Vol. xiii. Martinus Nijhoff Publishing, Boston, pp 43–67

Payne LN, Frazier JA, Powell PC (1976) Pathogenesis of Marek's disease. Int Rev Exp Pathol 16:59–154

Peng F, Bradley G, Tanaka A, Lancz G, Nonoyama M (1992) Isolation and characterization of cDNAs from bamhi-H gene family rnas associated with the tumorigenicity of Marek's disease virus. J Virol 66(12):7389–7396

Peng Q, Shirazi Y (1996a) Characterization of the protein product encoded by a splicing variant of the Marek's disease virus Eco-Q gene (Meq). Virology 226(1):77–82

Peng Q, Shirazi Y (1996b) Isolation and characterization of Marek's disease virus (MDV) cDNAs from a MDV-transformed lymphoblastoid cell line: identification of an open reading frame antisense to the MDV Eco-Q protein (Meq). Virology 221(2):368–374

Peng Q, Zeng M, Bhuiyan ZA, Ubukata E, Tanaka A, Nonoyama M, Shirazi Y (1995) Isolation and characterization of Marek's disease virus (MDV) cDNAs mapping to the bamhi-I2, bamhi-Q2, and bamhi-L fragments of the MDV genome from lymphoblastoid cells transformed and persistently infected with MDV. Virology 213(2):590–599

Pessah M, Marais J, Prunier C, Ferrand N, Lallemand F, Mauviel A, Atfi A (2002) C-Jun associates with the oncoprotein Ski and suppresses Smad2 transcriptional activity. J Biol Chem 277(32):29094–29100

Pevzner IY, Kujdych I, Nordskog AW (1981a) Immune response and disease resistance in chickens. II. Marek's disease and immune response to GAT. Poult Sci 60(5):927–932

Pevzner IY, Stone HA, Nordskog AW (1981b) Immune response and disease resistance in chickens. I. Selection for high and low titer to Salmonella pullorum antigen. Poult Sci 60(5):920–926

Pharr GT, Vallejo RL, Bacon LD (1997) Identification of Rfp-Y (Mhc-like) haplotypes in chickens of Cornell lines N and P. J Hered 88(6):504–512

Powell PC (1975) Immunity to Marek's disease induced by glutaraldehyde-treated cells of Marek's disease lymphoblastoid cell lines. Nature 257(5528):684–685

Qian Z, Brunovskis P, Rauscher 3rd., F, Lee L, Kung HJ (1995) Transactivation activity of Meq, a Marek's disease herpesvirus bzip protein persistently expressed in latently infected transformed T cells. J Virol 69(7):4037–4044

Qian Z, Brunovskis P, Lee L, Vogt PK, Kung HJ (1996) Novel DNA binding specificities of a putative herpesvirus bzip oncoprotein. J Virol 70(10):7161–7170

Quere P (1992) Suppression mediated in vitro by Marek's disease virus-transformed T-lymphoblastoid cell lines: effect on lymphoproliferation. Vet Immunol Immunopathol 32(1–2):149–164

Quere P, Dambrine G (1988) Development of anti-tumoral cell-mediated cytotoxicity during the course of Marek's disease in chickens. Ann Rech Vet 19(3):193–201

Qureshi MA, Trembicki KA, Dietert RR, Bacon LD (1986) Chicken developmental antigens in 15I5-B-congenic lines. J Hered 77(6):435–440

Reddy SM, Lupiani B, Gimeno IM, Silva RF, Lee LF, Witter RL (2002) Rescue of a pathogenic Marek's disease virus with overlapping cosmid DNAs: use of a pp38 mutant to validate the technology for the study of gene function. Proc Natl Acad Sci USA 99(10):7054–7059

Rispens BH, Vloten HV, Mastenbroek N, Maas HJ, Schat KA (1972a) Control of Marek's disease in the Netherlands. I. Isolation of an avirulent Marek's disease virus (strain CVI 988) and its use in laboratory vaccination trials. Avian Dis 16(1):11–19

Rispens BH, Vloten HV, Mastenbroek N, Maas JL, Schat KA (1972b) Control of Marek's disease in the Netherlands. II. Field trials on vaccination with an avirulent strain (CVI 988) of Marek's disease virus. Avian Dis 16(1):139–152

Ross N, O'Sullivan G, Rothwell C, Smith G, Burgess SC, Rennie M, Lee LF, Davison TF (1997) Marek's disease virus EcoRI-Q gene (meq) and a small RNA antisense to ICP4 are abundantly expressed in CD4+ cells and cells carrying a noval lymphoid marker, AV37, in Marek's disease lymphomas. J Gen Virol. 78:2191–2198

Rosenberger JK, Cloud SS, Olmeda-Miro N (1997, July 21) Epizootiology and adult transmission of Marek's disease. Paper presented at the Avian Tumor Virus Symposium. NV, Reno

Rziha HJ, Bauer B (1982) Circular forms of viral DNA in Marek's disease virus-transformed lymphoblastoid cells. Arch Virol 72(3):211–216

Schat KA, Calnek BW (1978) Characterization of an apparently nononcogenic Marek's disease virus. J Natl Cancer Inst 60(5):1141–1146

Schat KA, Xing Z (2000) Specific and nonspecific immune responses to Marek's disease virus. Dev Comp Immunol 24(2–3):201–221

Schat KA, Calnek BW, Fabricant J (1982) Characterization of two highly oncogenic strains of Marek's disease virus. Avian Pathol 11:593–605

Schat KA, Chen CL, Calnek BW, Char D (1991) Transformation of T-lymphocyte subsets by Marek's disease herpesvirus. J Virol 65(3):1408–1413

Schauenstein K, Kromer G, Hala K, Bock G, Wick G (1988) Chicken-activated-T- lymphocyte-antigen (CATLA) recognized by monoclonal antibody INN-CH 16 represents the IL- 2 receptor. Dev Comp Immunol 12(4):823–831

Schumacher D, Tischer BK, Fuchs W, Osterrieder N (2000) Reconstitution of Marek's disease virus serotype 1 (MDV-1) from DNA cloned as a bacterial artificial chromosome and characterization of a glycoprotein B-negative MDV-1 mutant. J Virol 74(23):11088–11099

Shamblin, CE, Greene N, Arumugaswami V, Dienglewicz RL, Parcells MS (2004) Comparative analysis of Marek's disease virus (MDV) glycoprotein-, lytic antigen pp38- and transformation antigen Meq-encoding genes: association of meq mutations with MDVs of high virulence. Vet Microbiol 102:147–167.

Sharma JM (1977) Role of tumor antigen in vaccine protection in Marek's disease. J Biol Stand 5(4):333–339

Sharma JM (1988) Presence of natural suppressor cells in the chicken embryo spleen and the effect of virus infection of the embryo on suppressor cell activity. Vet Immunol Immunopathol 19(1):51–66

Sharma JM, Witter RL, Coulson BD (1978) Development of cell-mediated immunity to Marek's disease tumor cells in chickens inoculated with Marek's disease vaccines. J Natl Cancer Inst 61(5):1273–1280

Shek WR, Calnek BW, Schat KA, Chen CH (1983) Characterization of Marek's disease virus-infected lymphocytes: discrimination between cytolytically and latently infected cells. J Natl Cancer Inst 70(3):485–491

Skonier JE, Bowen MA, Aruffo A, Bajorath J (1997) CD6 recognizes the neural adhesion molecule BEN. Protein Sci 6(8):1768–1770

Smyth GP, Stapleton PP, Barden CB, Mestre JR, Freeman TA, Duff MD, Maddali S, Yan Z, Daly JM (2003) Renal cell carcinoma induces prostaglandin E2 and T-helper type 2 cytokine production in peripheral blood mononuclear cells. Ann Surg Oncol 10(4):455–462

Soloski MJ (2001) Recognition of tumor cells by the innate immune system. Curr Opin Immunol 13(2):154–162

Takagi M, Ishikawa K, Nagai H, Sasaki T, Gotoh K, Koyama H (1996) Detection of contamination of vaccines with the reticuloendotheliosis virus by reverse transcriptase polymerase chain reaction (RT-PCR). Virus Res 40(2):113–121

Theis GA (1981) Subpopulations of suppressor cells in chickens infected with cells of a transplantable lymphoblastic leukemia. Infect Immun 34(2):526–534

Theis GA, McBride RA, Schierman LW (1975) Depression of in vitro responsiveness to phytohemagglutinin in spleen cells cultured from chickens with Marek's disease. J Immunol 115(3):848–853

Thomas JL, Pourquie O, Coltey M, Vaigot P, Le Douarin NM (1993) Identification in the chicken of GRL1 and GRL2: two granule proteins expressed on the surface of activated leukocytes. Exp Cell Res 204(1):156–166

Thomas JL, Stieber A, Gonatas N (1994) Two proteins associated with secretory granule membranes identified in chicken regulated secretory cells. J Cell Sci 107(Pt 5):1297–1308

Trapp S, Parcells MS, Kamil JP, Schumacher D, Tischer BK, Kumar PM, Nair VK, Osterrieder N (2006) A virus-encoded telomerase RNA promotes malignant T cell lymphomagenesis. J Exp Med 203:1307–1317

Tremblay F, Fernandes M, Habbab F, de BEMD, Loertscher R, Meterissian S (2002) Malignancy after renal transplantation: incidence and role of type of immunosuppression. Ann Surg Oncol 9(8):785–788

Tulman ER, Afonso CL, Lu Z, Zsak L, Rock DL, Kutish GF (2000) The genome of a very virulent Marek's disease virus. J Virol 74(17):7980–7988

Ui M, Endoh D, Cho KO, Kon Y, Iwata A, Maki Y, Sato F, Kuwabara M (1998) Transcriptional analysis of Marek's disease virus (MDV) genes in MDV- transformed lymphoblastoid cell lines without MDV-activated cells. J Vet Med Sci 60(7):823–829

Uni Z, Pratt WD, Miller MM, O'Connell PH, Schat KA (1994) Syngeneic lysis of reticuloendotheliosis virus-transformed cell lines transfected with Marek's disease virus genes by virus-specific cytotoxic T cells. J Virol 68(12):8239–8253

Vallejo RL, Pharr GT, Liu HC, Cheng HH, Witter RL, Bacon LD (1997) Non- association between Rfp-Y major histocompatibility complex-like genes and susceptibility to Marek's disease virus-induced tumours in 6(3) × 7(2) F2 intercross chickens. Anim Genet 28(5):331–337

Vallejo RL, Bacon LD, Liu HC, Witter RL, Groenen MA, Hillel J, Cheng HH (1998) Genetic mapping of quantitative trait loci affecting susceptibility to Marek's disease virus induced tumors in F2 intercross chickens. Genetics 148(1):349–360

Volpini LM, Calnek BW, Sekellick MJ, Marcus PI (1995) Stages of Marek's disease virus latency defined by variable sensitivity to interferon modulation of viral antigen expression. Vet Microbiol 47(1–2):99–109

Volpini LM, Calnek BW, Sneath B, Sekellick MJ, Marcus PI (1996) Interferon modulation of Marek's disease virus genome expression in chicken cell lines. Avian Dis 40(1):78–87

von Bulow V, Rudolph R, Fuchs B (1986) [Enhanced pathenogicity of chicken anemia agent (CAA) in dual infections with Marek's disease virus (MDV), infectious bursal disease virus (IBDV) or reticuloendotheliosis virus (REV)]. Zentralbl Veterinarmed [B] 33(2):93–116

Wakenell PS, Miller MM, Goto RM, Gauderman WJ, Briles WE (1996) Association between the Rfp-Y haplotype and the incidence of Marek's disease in chickens. Immunogenetics 44(4):242–245

Walter J, Schirrmacher V, Mosier D (1995) Induction of CD44 expression by the Epstein-Barr virus latent membrane protein LMP1 is associated with lymphoma dissemination. Int J Cancer 61(3):363–369

Watt SM, Buhring HJ, Rappold I, Chan JY, Lee-Prudhoe J, Jones T, Zannettino AC, Simmons PJ, Doyonnas R, Sheer D, Butler LH (1998) CD164, a novel sialomucin on CD34(+) and erythroid subsets, is located on human chromosome 6q21. Blood 92(3):849–866

Weng N, Levine BL, June CH, Hodes RJ (1997) Regulation of telomerase RNA template expression in human T lymphocyte development and activation. J Immunol 158(7): 3215–3220

Witter RL (1983) Characteristics of Marek's disease viruses isolated from vaccinated commercial chicken flocks: association of viral pathotype with lymphoma frequency. Avian Dis 27(1):113–132

Witter RL (1997) Increased virulence of Marek's disease virus field isolates. Avian Dis 41(1):149–163

Witter RL, Nazerian K, Purchase HG, Burgoyne GH (1970) Isolation from turkeys of a cell-associated herpesvirus antigenically related to Marek's disease virus. Biken J 13(1):53–57

Witter RL, Stephens EA, Sharma JM, Nazerian K (1975) Demonstration of a tumor- associated surface antigen in Marek's disease. J Immunol 115(1):177–183

Witter RL, Sharma JM, Offenbecker L (1976) Turkey herpesvirus infection in chickens: induction of lymphoproliferative lesions and characterization of vaccinal immunity against Marek's disease. Avian Dis 20(4):676–692

Witter RL, Lee LF, Bacon LD, Smith EJ (1979) Depression of vaccinal immunity to Marek's disease by infection with reticuloendotheliosis virus. Infect Immun 26(1):90–98

Witter RL, Li D, Jones D, Lee LF, Kung HJ (1997) Retroviral insertional mutagenesis of a herpesvirus: a Marek's disease virus mutant attenuated for oncogenicity but not for immunosuppression or in vivo replication. Avian Dis 41(2):407–421

Witter RL, Gimeno IM, Reed WM, Bacon LD (1999) An acute form of transient paralysis induced by highly virulent strains of Marek's disease virus. Avian Dis 43(4):704–720

Xing Z, Schat KA (2000a) Expression of cytokine genes in Marek's disease virus-infected chickens and chicken embryo fibroblast cultures. Immunology 100(1):70–76

Xing Z, Schat KA (2000b) Inhibitory effects of nitric oxide and gamma interferon on in vitro and in vivo replication of Marek's disease virus. J Virol 74(8):3605–3612

Yamamoto Y, Okada I, Matsuda H, Okabayashi H, Mizutani M (1991) Genetic resistance to a Marek's disease transplantable tumor cell line in chicken lines selected for different immunological characters. Poult Sci 70(7):1455–1461

Yamamoto H, Hattori M, Ohashi K, Sugimoto C, Onuma M (1995) Kinetic analysis of T cells and antibody production in chickens infected with Marek's disease virus. J Vet Med Sci 57(5):945–946

Yonash N, Bacon LD, Witter RL, Cheng HH (1999) High resolution mapping and identification of new quantitative trait loci (QTL) affecting susceptibility to Marek's disease. Anim Genet 30(2):126–135

Zannettino AC, Buhring HJ, Niutta S, Watt SM, Benton MA, Simmons PJ (1998) The sialomucin CD164 (MGC-24v) is an adhesive glycoprotein expressed by human hematopoietic progenitors and bone marrow stromal cells that serves as a potent negative regulator of hematopoiesis. Blood 92(8):2613–2628

Zelnik V, Majerciak V, Szabova D, Geerligs H, Kopacek J, Ross LJ, Pastorek J (1999) Glycoprotein gd of MDV lacks functions typical for alpha-herpesvirus gd homologues. Acta Virol 43(2–3):164–168

H.E. Kaiser and A. Nasir (eds.), Selected Aspects of Cancer Progression:
Metastasis, Apoptosis and Immune Response, 193–222.
© *Springer Science + Business Media B.V.* 2008

CHAPTER TWELVE

Abnormal variation of the immune response as related to cancer

Gerhard R.F. Krueger[1] and L. Maximilian Buja[2]

Abstract: This overview chapter reviews the complex interrelationship between the immune system, cell proliferation and cancer. There exist, in essence, two opposite actions of the immune system with regard to tumor development: immune suppression of neoplasia (commonly: immune surveillance) and immune stimulation of tumor growth. In addition, there exists experimental evidence that certain tumors of the immune system itself (e.g. malignant lymphomas) may be caused by inadequate function of the immune system itself, i.e. by dysregulation. On the basis of examples for the latter, and outlook is finally provided of how techniques of computational simulation of such regulatory pathways can contribute to the study of immunology and tumor induction and growth.

Keywords: Cell proliferation, cell differentiation, cell inhibition, cancer, immune system, computational modeling

[1]Department of Pathology & Laboratory Medicine and of Internal Medicine, The University of Texas – Houston Medical School, 6431 Fannin St., Houston, Texas 77030, U.S.A. Dedicated to Robert Fischer. University of Cologne, Germany, in appriciation of his continued support
[2]Department of Pathology and Laboratory Medicine, University of Texas – Houston Medical School, 6431 Fannin St., Houston, Texas 77030, U.S.A.

Introduction

As studies of phylogeny and ontogeny of multicellular organisms and man have shown, any control mechanism of cell replication and cell differentiation develop in close relationship to parts of the immune system (Dawe 1969; Good and Finstad 1969). The latter apparently serves to guarantee cell and tissue homeostasis clearly beyond it's well defined defense reactions against foreign intruders including cancer cells. Many of the diverse functions of the immune system are covered in other parts of the book, and we will outline here only the interplay of immune system and neoplasia. Such *immune regulation of tumor growth* as summarized in Fig. 1 must consider both, stimulatory mechanisms (growth facilitation, tumor induction) and inhibitory actions (cell rejection). The first part of this chapter is devoted to an overview of the interplay between neoplasia and immune system, while the second part will present some exemplary computational modeling of such an interplay.

Growth facilitation of preexisting tumor cells

The concept is based on the reasonable assumption that neoplastic cells arise continuously in the living organism secondary to a variety of carcinogenic and co-carcinogenic influences, and that such atypical cells are

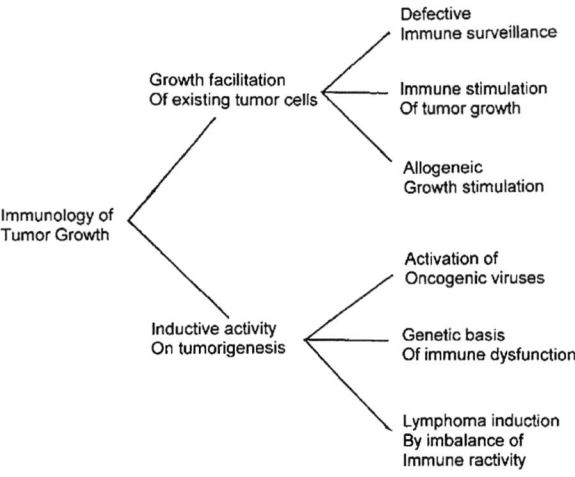

Fig. 1 Immune regulation of tumor growth

Table 1 Immunogenic changes in cancer cells (Adapted from D Pardoll 2003)

Cancer cells	Immunogenic changes
Genetic alterations	*Multiple neoantigens*
Epigenetic lability	*Altered antigen density*
Expression of growth factors and cytokines	*Growth stimulation and local immune inhibition*
Invasion and metastasis	*Induction of inflammation and of adaptive immunity*

eliminated by non-immune and immune mechanisms. A most powerful non-immune mechanism is connected to the cell cycle, recognizing chromosomal mutations and inducing repair or cell elimination by apoptosis (Buja et al. 1993; Holland et al. 1978; Kaufman and Kaufman 1993). Consequently, individuals with defective repair mechanisms such as in *Xeroderma pigmentosum* or with defective apoptosis as in *Canale–Smith syndrome* experience a higher incidence of neoplasia than their healthy siblings (Gelehrter and Collins 1997; Holland et al. 1978; Peters et al. 1999). Surviving atypical cells are thought to be immunogenic and prone to be recognized by cells of the immune system. Many atypical cells are probably destroyed by an effective host defense, i.e. *immune surveillance* in the broadest sense, and only those escaping destruction will grow to form a clinically manifest tumor. Paul Ehrlich was the first to formulate the immune surveillance theory at the beginning of the last century (Himmelweit 1957), and ever since it has repeatedly been confirmed and rejected. Currently, immune surveillance of tumor growth constitutes an accepted mechanism again (Kleist 1980; Melief and Schwartz 1975; Pardoll 2003). Table 1 summarizes some of the major changes which distinguish a tumor cell from its normal counterpart and thus render it immunogenic. It should be noted, however, that uncontrolled cell proliferation ultimately causing a clinically relevant tumor may arise from differentiation blockage of normal, untransformed cells (Krueger 1972a, 1973). Such cells may not possess classical tumor associated antigens, yet only retain the characteristic antigen composition of their respective developmental stage.

A classical example for such uncontrolled leukemia-like cell proliferation was the erythroblastosis in what later became known as *pernicious anemia*.

Another more recent example is the above mentioned Canale–Smith syndrome with leukemia-like lymphoblastosis (Krueger et al. 2002a).

Defective immune surveillance

Thomas (1959) and McFarlane Burnet (1970) revived Ehrlich's idea of an immunologic host defense against malignant tumors and formulated the immune surveillance theory (Burnet 1970; Pardoll 2003). It was supported by observations of increased tumor incidences in immunodeficient animals and man (Allison 1970; Fahey 1971; Gatti and Good 1970; Penn and Starzl 1972) as well as by the fact that oncogenic viruses and chemical carcinogens are commonly immunosuppressive besides their potentials to transform cells (Gabrielsen and Good 1967; Allison 1970; Chieco-Bianchi et al. 1974; Cremer 1967; Baldwin 1973; Friedman and Kateley 1975; Gilette and Fox 1977; Prehn 1963; Rubin 1971; Stjernswaerd 1965; Stutman 1969; Wedderburn and Salaman 1968). *Vice versa*, non-specific stimulation of the immune system by BCG (*Bilie Calmette Guerin*), Echinacea and other substances decreased the incidences of primary tumors in experiments or were tentatively applied for the adjuvant treatment of malignant tumors (Bast et al. 1974). There were a number of unexplained findings, however, that questioned the general validity of the immune surveillance theory (Gleichmann and Gleichmann 1973; Kripke and Borsos 1974). For instance, immune-deficient nude mice did not appear to exhibit an increased tumor incidence (Rygaard and Poulsen 1976) and human patients with inherited or acquired immune deficiencies did not reveal an increased incidence of the tumors in an age-adjusted normal population (Krueger et al. 2004; Penn 1978), but rather a more complex pattern of tumor types

with preferences of malignant lymphomas and tumors of the skin (Table 2).

However, more detailed investigations of nude mice did show increased tumor development at older ages (Rygaard and Poulsen 1976), and the significantly increased malignant lymphomas and certain skin tumors in immunodeficient human patients were explained by oncogenic activities of "opportunistic" infectious agents such as Epstein-Barr virus and papillomavirus or by human herpesvirus-8 (Krueger 1993; List et al. 1987). The immune surveillance theory thus is still valid.

Table 2 Malignant tumors in renal allograft recipients

Incidence 5–7%	
Types of tumors	
Malignant lymphomas	20.2%
Skin tumors (carcinoma, basalioma)	39.8%
Female genital tract carcinomas	8.2%
Pulmonary carcinomas	4.3%
Gastrointestinal carcinomas	2.5%
Mammary carcinoma	2.5%
Sarcomas	1.3%
Non-lymphatic leukemias	1.3%
Others	19.8%

Total of 766 cases

Immune surveillance is to a large part, yet not exclusively lymphocyte-controlled (Chism et al. 1976; Geldhof et al. 2002; Felzmann et al. 2002). Antibody-dependent killer cells (K cells), natural killer cells (NK cells) and macrophages participate in surveillance mechanisms (Alexander et al. 1976; Domzig and Lohmann-Matthes 1979; Kaplan and Morahan 1976; Actor 2007; Kiessling et al. 1976; Klein 1989; West et al. 1977; Russell et al. 1980). Non-immune cells as carriers of *innate immunity* cooperate with immunocompetent lymphocytes, the carriers of *adaptive immunity*, to produce an efficient and specific host response (Fig. 2; for review see Paul 1998). Such a response involves several types of lymphocytes and it's outcome may differ according to the participating cell populations as well as to it's locality. The CD8+ T cell population, for instance, includes cytotoxic T cells (CTL) which specifically may destroy foreign cells, as well as CD25+ regulatory T cells (Treg) which may be counteractive (Treg cells are not separately identified in Fig. 2). Similarly, the CD4+ T cell population comprises CD25+ suppressor cells (also Treg) supporting tolerance induction, Th1 cells producing interferon gamma (IFNγ) and tumor necrosis

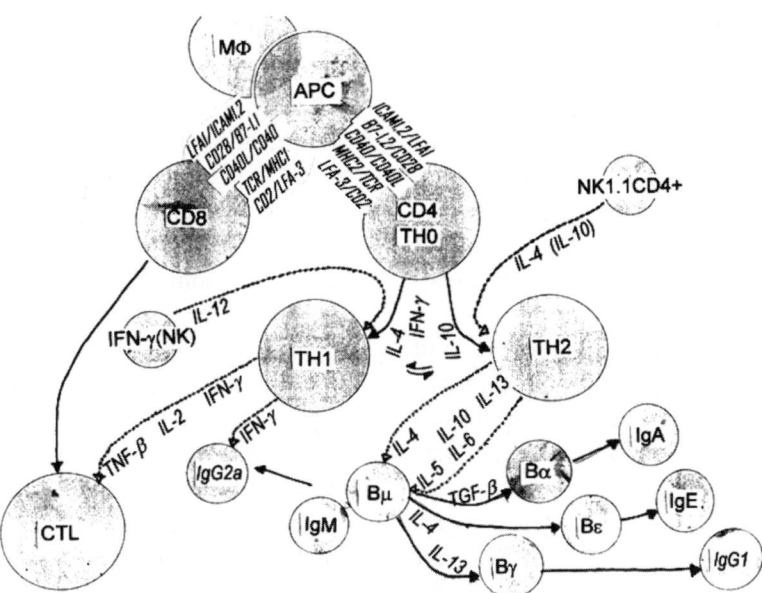

Fig. 2 Cell cooperation in immune response: _____ transformation into; .... action upon.
Abbreviations: MΦ macrophage; APC antigen presenting cell; CD8 and CD4 suppressor, helper or regulator cells; TH0, TH1 and TH2 forms of regulator cells; CTL cytotoxic T lymphocyte; IgM, IgA, IgE and IgG cells immunoglulin-producing cells; Bμ,α,γ,ε cells B cell intermediates; NK1.1CD4+ NK-type regulator cell; IFNγ(NK) NK cell and regulator cell; IL and number are different types of interleukins (cytokies); TNF tumor necrosis factor; IFN interferon. Links between APC/CD8/CD4 receptors and their ligands.

factor (TNF) adding to the toxic antitumor effect, as well as Th2 cells producing inflammatory cytokines (Gallin and Snyderman 1999). The latter can inhibit cytotoxicity by Th1 cells and support local invasion by atypical cells, yet also activate NK cell function and thus innate immunity. Enhanced NK cell activity then appears a necessary precursor for efficient adaptive immunity (Paul 1998). The local and systemic antitumor response thus appears as a complex network regulation in which the quantitative expression of tumor antigens, T cell receptors (TCR), recognition and activation factors play a decisive role such as MHC (major histocompatibility complex) receptors MHC I or MHC II, cytokines and chemokines (Krueger et al. 2002b, 2003a). There appear to exist thus four major ways permitting tumor cells to grow despite of immune surveillance:

(a) Defectiveness of effector cells for immune surveillance
(b) Imbalance of the immune response to the tumor
(c) Inadequate immunogenicity of tumor cells
(d) Production of blocking factors for the immune defense

Defectiveness of effector cells for immune surveillance

Defectiveness of one or several cell populations engaged in destruction of foreign cells can lead to escape of tumor cells from host defense and thus facilitate neoplastic growth. Nature has presented us with an adequate experiment for this phenomenon in the pathology of Epstein-Barr virus (EBV) infection of human patients: healthy individuals will suffer a limited proliferative disease of EBV-transformed B-lymphocytes upon infection known as infectious mononucleosis.

Immunocompromised patients, however, will either succumb from lethal infectious mononucleosis or will develop a malignant lymphoma of transformed B-lymphocytes, i.e. Burkitt's or Burkitt's-type lymphoma (Bar et al. 1974; Krueger 1985; Look et al. 1981; Kleindienst and Brocker 2003). Defectiveness of immune effector cells can result from a variety of exogenous influences or endogenous conditions such as immunosuppressive chemicals (carcinogens, chemotherapeutics, and immune suppressive agents), ionizing radiation, infections with certain viruses or other organisms, as well as genetically determined

immune deficiency syndromes (IDS) (Krueger 1985). Depending upon the kind of immunosuppressant, its dose, route of administration, and the condition of the patient, the sequelae can be unspecific affecting all kinds of proliferating cells such as in high dose radiation, carcinogen administration and cancer chemotherapy, or be rather selective interfering with the function of only certain cell populations of the immunocompetent tissues. In order to more clearly identify the possible targets for such immunosuppressive effects, the immunocompetent tissues are dissected into their individual cooperating cell populations as shown in Fig. 2.

Various cell members of the immune system exhibit a selective sensitivity towards exogenous immunosuppressive effects. Most *carcinogenic chemicals* exert a measurable depression of both antibody production (B-cell response) and cell mediated immunity (T-cell response) during the latent period of tumor induction while non-carcinogenic analogues are not immunosuppressive (Ball 1970; Krueger 1985; Malmgren et al. 1952; Prehn 1963; Rowland and Hurd 1970; Stjernswaerd 1965, 1969).

There are well over 3,000 pure chemical carcinogens known of widely diverse structures (Coombs 1980). Their means of interference with immunocompetent cells appears similar to their interference with other cells. These substances cause genetic mutations, and their uptake in cells is apparently related to the cellular metabolic activity. Cells of the immunocompetent tissues possess a quite active nucleic acid metabolism especially during antigenic stimulation and cell proliferation; they should thus be especially susceptible to the action of carcinogenic compounds.

There is some indication, however, that immunosuppression by carcinogens might also be caused by nonspecific toxic influences, since immune reactivity will return to normal after removal of the carcinogen.

For instance, treatment with 7,12-dimethyl benzanthracene (DMBA) and with N-nitroso butylurea (NBU) caused only a transient depression of antibody formation during the latent period of tumor development while cellular immunity remained suppressed persistently (Kraus and Krueger 1981; Szakal and Hanna 1972). Thus carcinogen induced immunosuppression can be selective to some extent; selectivity in the most simple way being determined by the different degree of metabolic activity of various cell populations

(e.g. rapidly dividing short-lived lymphocytes versus slowly dividing long-lived lymphocytes).

Increased incidences of second primary malignancies are reported after extended anti-tumor chemotherapy. Incidence rates are given, for instance, for primary treated Hodgkin's disease of 2–7.7%. The types of secondary neoplasms observed are shown in Table 3. The chemotherapeutic agents used in these patients are summarized in Tables 4 and 5.

As in the allograft recipients (see Table 2), the incidence in the various types of secondary neoplasms varies considerably as compared to the age-adjusted population (being about 20–26% (Coltman and Dixon 1982; Glicksman et al. 1982), for leukemia within 7 years after initiation of primary chemotherapy). Thus a complex mechanism of tumor promotion is expected. By far the highest risk of developing a second primary exists in patients on combination radio- and chemotherapy or on radiotherapy alone (Table 4).

Arseneau et al. (1972) state for Hodgkin's patients that the risk of developing second primaries is increased each three-fold in radiation or chemotherapy alone, and 23-fold in combined radio- and chemotherapy.

The respective data observed by others vary, however (Coltman and Dixon 1982; Glicksman et al. 1982).

The tumor promoting mechanism again probably results from a combination of direct carcinogenic and supportive immunosuppressive effects. Many cancer chemotherapeutic agents are overtly carcinogenic. Their application form may also include carcinogenic impurities such as heavy metals (chromium, nickel) and drug additives.

The possible mode of interference with the host's immune responses has been summarized elsewhere (Krueger 1975): functional depression of immunocompetent tissues, decrease of immunogenicity of tumor cells, increased immunogenicity of tumor cells (such as by pyrazole-4-carboxamide and derivatives), immunostimulation by such drugs as hapten or immunogen, and induction of autoantibody formation and immune complexes. Immunosuppression by individual agents was discussed in several reviews (Camiener and Wechter 1972; Henderson and Krueger 1977; Krueger 1972b, 1985; Spreafico and Anaclerio 1977). Their action on cells is polyfunctional and includes inhibition of protein and nucleic acid synthesis, deletion of genetic information (false templates by incorporation into DNA of wrong nucleic acid bases), blocking enzymes, and interference with cell replication.

Like the carcinogens, the action is quite unselective affecting various kinds of proliferating cells including lymphoid cells; hormones also show certain selectivities such as the cortisone sensitivity of thymic

Table 3 Second primary tumors in Hodgkin's lymphoma (From Krueger GRF: Cologne University Tumor Conference on Hodgkin's Disease, October 1982)

Total number of cases evaluated	8,592
Total number of second primaries	160
Incidence of second primaries	1.88%
Types of tumors	
Non-lymphatic leukemias	40 (25.0%)
Skin carcinomas	17 (10.62%)
Sarcomas	17 (10.62%)
Gastrointestinal carcinomas	15 (9.37%)
Pulmonary carcinomas	9 (5.62%)
Non-Hodgkin's lymphomas and lymphocytic leukemias	6 (3.75%)
Female genital tract carcinomas	8 (5.0%)
Mammary carcinomas	8 (5.0%)
Urinary bladder carcinomas	8 (5.0%)
Head and neck carcinomas	5 (3.12%)
Prostatic carcinomas	4 (2.5%)
Thyroid carcinoma	3 (1.87%)
Malignant melanoma	3 (1.87%)
Brain tumor	1 (0.62%)
Renal carcinoma	1 (0.62%)
Malignant tumors NOS	15 (9.37%)

Table 4 Therapeutic regimens and second primaries in Hodgkin's disease

Radiotherapy only	34 (21.0%)
Chemotherapy only	11 (7.0%)
Combined radio- and chemotherapy	115 (71.9%)

Table 5 Chemotherapeutic agents used in Hodgkin's disease with second primary tumors

Total case number 160

Chemotherapeutic agents	
Vincristine	89 (55.5%)
Procarbazine	79 (49.2%)
Corticosteroids	53 (33.3%)
N-lost	51 (31.7%)
Chlorambucil	40 (25.39%)
Bleomycin	23 (14.28%)
Methotrexate	20 (12.69%)
Cyclophosphamide	20 (12.69%)
Adriamycin	15 (9.52%)
BCNU	8 (5.0%)
L-asparaginase	5 (3.2%)
Other	3 (1.6%)

lymphocytes and peripheral T-cells for cytotoxic tumor cell lysis (Ahmed et al. 1979).

A more systematic investigation is still necessary, however, of the effect of chemotherapy drugs on immunocompetent cell populations before a definite evaluation of therapeutic immunosuppression and cancer can be attempted. Orsini et al. (1980) tested a number of anti-cancer agents for their suppressive activity an anti-tumor cell-mediated immunity in vitro introducing a test system which may be useful as first step for screening procedures.

Immunosuppressive agents used for the treatment of allograft recipients, graft-versus-host disease (GVHD) and autoimmune diseases consist of a few substances: corticosteroids, azathioprine, methotrexate, cyclophosphamide, antilymphocyte sera and cyclosporin. Cyclophosphamide and methotrexate are also cancer chemotherapeutics (see above).

Intravenous prednisone injection (1 gr) causes a transient fall in T-cell levels accompanied by a depression of T-cell reactivity without apparent selectivity an the T-cell subpopulations in the peripheral blood. Low dose maintenance therapy with prednisone or with azathioprine had but little influence an these cell populations. Similarly, both substances depressed killer cell (K-cell) functions only at high doses.

Antithymocyte globulin (ATG), however, caused a profound fall in peripheral T-cells associated by a rise of Ig membrane receptor carrying B-lymphocytes. Also K-cell cytotoxicity was significantly inhibited (Thomas et al. 1982).

In accordance with these findings, Yu et al. (1975) describe no significant lymphocyte changes after azathioprine administration in terms of cell numbers, density gradient distribution of peripheral lymphocytes, T- and B-cell ratios, and response to phytohemagglutinin.

Corticosteroids caused only transient lymphocytopenia (primarily T-cell depletion) and ATG induced a more persistent lymphocytopenia.

Cyclosporin A has been shown to ameliorate GVHD, an acute cytotoxic T-cell response in bone marrow allograft recipients, and also to cause leukocytopenia and thrombocytopenia (Atkinson et al. 1982; Miller et al. 1982).

In essence, there appears to be a certain selectivity of immunosuppressive drugs for T-cells and K-cells that interferes with above described immune surveillance.

Ionizing radiation induces marked diffuse atrophic and necrotic changes in lymphoid tissues (Zollinger 1960). Most sensitive appear small lymphocytes which undergo apoptosis and rhexis at doses as low as 20–50 rads (DeBruyns 1948; Pizon 1955). Apoptosis in 50% of the cells was noted at doses of 150 rads (Zollinger 1960).

Significant lymphocytopenia occurs after total body irradiation (TBI) at doses as low as 25 rads. Recovery is quite delayed and normal peripheral lymphocyte counts may not be reached until 5–10 years after radiation treatment.

B-cell depression, although initially more profound, recovers more rapidly than T-cell reduction, so that in later phases of radiation recovery a selective T-cell defect may result.

According to Dixon et al. (1952) and Parker and Vavra (1969) the effects on the humoral immune response are more prominent than on the cellular response provided radiation occurs shortly before antigenic stimulation.

Engers and Louis (1979) demonstrated an enhanced cytolytic T-cell activity after 500 rads TBI in mice while the humoral antibody response in terms of complement-mediated cell lysis was completely abrogated. Depending upon the radiation dose administered, certain T-cell populations are affected selectively (T-suppressor cells being more sensitive). This pertains similarly to radiomimetic chemotherapeutic substances (Krueger 1985, 1972b).

These observations are restricted, however, to the earlier postradiation period and more detailed investigations of the function of lymphocyte subpopulations are necessary during later periods (i.e. even after several years post irradiation when second primary tumors tend to occur.).

Infections with certain viruses and other organisms can induce various degrees of immune deficiencies and thus interfere with immune surveillance. The immunosuppressive effect of leukemogenic RNA viruses is well-known (Bendinelli and Nardini 1973; Cremer 1973; Cremer et al. 1971; Dent et al. 1965; Klein and Klein 1965, 1966) in terms of T-lymphocyte depletion, delayed graft rejection, disturbance of lymphocyte homing, and decreased antibody formation against third antigens.

Not all oncogenic viruses are overtly immunosuppressive *per se* (e.g. many oncogenic DNA viruses). When individual cell populations of the immune

system are investigated including non-specific natural killer cells (NK cells), selective defects, however, are demonstrated during the course of many viral infections. Persistent Epstein-Barr virus, cytomegaly virus, HTLV (strain I, II) and HIV infections are currently the best known examples (Purtilo 1980; Purtilo et al. 1981, 1982). There exists broad evidence that a defective immune reactivity is frequently a necessary precondition for oncogenesis (e.g. infections of newborn animals with polyoma virus or development of Burkitt's lymphoma in immunocompromised individuals).

Besides, viral infections (especially when persistent) may induce substances that can interfere with the immune response such as blocking factors and immune complexes (see later).

Lymphocytes from patients with long-standing EBV infection, for instance, have lost their specific cytotoxic activity for transformed cells (Moss et al. 1977).

There are a number of non-oncogenic viruses which may become immunosuppressive under certain conditions (Notkins et al. 1970): cytomegalovirus (CMV), Newcastle disease virus (NDV), lymphocytic choriomeningitis virus (LCM) Aleutian mink disease virus (AMDV) and to some extent lactic dehydrogenase elevating virus (LDV).

Tuberculin skin reactivity was reported to be depressed by measles virus, Influenza virus, chicken pox and polio virus.

Blastic transformation of lymphocytes in culture is abolished by the following viruses, many of which are common pathogens and also found in tumor patients: measles, rubella, NDV, polio, ECHO, rheo, vesicular stomatitis virus, mumps, Influenza A, Sendai, adeno, herpes simplex, vaccinia and human wart virus.

Similarly, other chronic infections such as tuberculosis, schistosomiasis and malaria are often accompanied by some immunologic defects (see e.g. Burkitt's lymphoma in African children with persistent malaria).

One can well imagine that all those viral infections especially when persistent – may serve as "cocarcinogens" by interfering with immune surveillance. We have demonstrated repeatedly, for instance, the frequent persistence of rubella virus in patients with angioimmunoblastic lymphadenopathy, a prelymphomatous condition that may progress into overt malignant lymphoma (Krueger and Konorza 1977; Krueger et al. 1979).

The increased risk of developing a malignant lymphoma is well known in patients with inherited and acquired immune deficiency syndromes (Good and Finstad 1968; Louie and Schwartz 1978; Penn 1974). The types of tumors observed in this population consist nearly exclusively of different kinds of malignant lymphomas.

In essence, the activity of effector cells for immune surveillance, can be depressed by a variety of influences, and in many instances this depression can be followed by increased susceptibility to malignant lymphomas.

Any discussion of effector cell defectiveness must also include quantitative and functional deficiencies of antigen-presenting cells (i.e. for instance macrophages and dendritic reticulum cells; DRC). Tumor draining lymph nodes and tumor tissue itself are apparently depleted of such cells (Blohm et al. 2002; Laguens et al. 2002), which may severely affect the amplification of an eventual immune response against the tumor. Antitumor immune therapy should thus consider the addition of functioning DRCs (Kleindienst and Brocker 2003; Nelson et al. 1975; Waller and Ernstoff 2003).

Imbalance of the immune response to the tumor

In our early experiments to induce malignant lymphomas by coincident persistent immunosuppression and immunostimulation (Krueger 1971; Krueger and Heine 1972; Krueger et al. 1971a) it was noted that severe combined T- and B-cell depression caused the death of the animal (by infection) rather than an increased tumor development.

Suppression of the adequate T-cell response, however, combined with persistent immunostimulation gave rise to a high incidence of malignant lymphomas. From this it appears that a dissociated immunologic responsiveness rather than a complete defectiveness may favor the growth of malignant neoplasias. Similarly, in viral and chemically induced malignant lymphomas, a significant decrease in T-lymphocytes was observed during the latent period of tumor development (Kraus and Krueger 1981; Mertens and Krueger 1976), while B-lymphocytes were not so obviously changed initially.

Depressed T-cell numbers have been observed in many tumor patients but this could be secondary to the tumor itself and not necessarily related to its development.

Besides, quantitative changes in T- and B-cell populations preceding malignant tumors, quite selective decreases and increases were described in certain T-cell subpopulations. Seeley et al. (1981), for instance, observed that the development of malignant lymphomas in children with X-linked lymphoproliferative disease (XLP) was preceded or accompanied by a dramatic inversion of the natural CD4/CD8 cell ratio with significant increase in suppressor cells. Acquired immune deficiency syndrome (AIDS) observed predominantly in homosexual patients is characterized by a CD4+ cell deficiency with subsequent opportunistic infections, Kaposi's sarcoma, lymphoproliferative disorders and squamous cell carcinoma. Lymphoproliferation includes all kinds from polyclonal reactive lymphoproliferation to malignant lymphoma (Krueger 1993). Similarly, in common variable immune deficiency syndrome and other immune deficiency diseases which carry an increased risk for lymphoma, regulatory T-cells (Treg) exhibit increased activities (Reinherz et al. 1981; Siegal et al. 1976; Waldmann et al. 1974, 1975).

There is now solid evidence that T regulator cells play a major role in the potentiation of tumor growth (Broder et al. 1979). Such cells with suppressor activities include not only Treg cells, but also macrophages and possibly B-cells depending upon the respective phase of the immune response.

Treg cell activities in the widest sense are antigen-specific (immune regulator cells) or non-specific (suppressor macrophages). Specific Treg cells regulate the production of immunoglobulins (isotypes and idiotypes) and of cytotoxic T lymphocytes (CTL). They may also interfere with the function of killer- and natural killer cells; thus all immune and non immune functions essential for surveillance of tumor growth can be counteracted by suppressor cells (Broder et al. 1979).

Naor (1980) suggested in his "sneaking through concept" that low doses or a weakly antigenic tumor may stimulate suppressor cells rather than an effective host defense, allowing the tumor to escape immune destruction.

In fact, experimental transplantation of suppressor cells together with tumor cells is followed by an enhanced tumor growth as compared to tumor cell grafting alone (Greenberg and Greene 1976).

In contrast, destruction of suppressor cells by specific antisera retards growth. In the mouse, suppressor cells carry antigens encoded by the IJ-sublocus of the *H2* complex, and anti IJ antisera are able to specifically react against suppressor cells (Broder et al. 1979).

Although the negative effect on the immune responsiveness of various chemotherapeutic and radiotherapeutic measures has been stressed in the last paragraph, it must be mentioned in this context that this general effect is obviously dose dependent. In a number of instances an augmentation of host immunity was demonstrated after conventional chemotherapy thought to be the result of the destruction of selectively vulnerable suppressor cells. Cyclophosphamide and ionizing radiation have proven especially useful in this regard in experimental systems (Askenase et al. 1975; Turk et al. 1972).

Inadequate immunogenicity of malignant cells

Malignant tumors apparently do not consist of an entirely homogeneous cell population and these cells are not completely stable in their antigenic composition. This information comes in large part from studies of the mechanism of tumor metastasis (Grundmann 1980).

Recognition and destruction of malignant cell populations apparently can be avoided by several vital reactions such as deletion, modulation, or differential expression of surface antigens. For instance, antigens on the cytoplasmic membrane of tumor cells can be shed (Alexander 1975; Bystryn and Smalley 1977; Ting and Rogers 1977) thus rendering the tumor non-antigenic.

Shedding can be induced by reaction of the surface antigen with specific antibody. Deletion of surface antigens furthermore can result from proteolysis associated with the tumor (Lakshmi et al. 1982; Latner and Sherbet 1979). Tumor cells can exhibit a differential immunogenicity because of various amounts of the glycocalyx coating the cell membrane. Kim et al. (1975) were able to correlate the thickness of the glycocalyx of rat mammary carcinoma cells with their immunogenicity and the potential to metastasize. Immunogenicity and thickness of glycocalyx were directly related while metastasizing capacity showed an inverse relationship.

As shown in Fig. 2, MHC antigens are necessary for effective antigen recognition and induction of the immune response. These antigens support the recognition of tumor antigens by macrophages and by

T lymphocytes, and it is especially the class I MHC who are necessary for cytotoxic cell mediated immune responses (CTL cell response). There are a considerable number of tumors with major alterations in MHC class I phenotypes (Cabrera et al. 2003) and profound influence on immune recognition. MHC class I losses, for instance, were identified in 90% of carcinomas of the cervix uteri, in 88% of carcinoma of the breast, in 66% of laryngeal cancer and in 51% of malignant melanoma (Cabrera et al. 2003).

Also in other systems, non-metastatic tumors were generally more immunogenic than metastatic ones (Currie and Alexander 1974; Davey et al. 1976).

Furthermore, the cellular composition of a tumor appears to be inhomogeneous including cells with different degrees of antigen expression. Specific host defense mechanisms thus can select in favor of less immunogenic cell populations which are able to escape immune destruction. Evidence for this is provided by antibody binding studies to cells from metastatic and non-metastatic (highly immunogenic) tumors; the binding rates varied percentagewise but not in +/− pattern indicating a mixed cell composition in both metastatic and non-metastatic neoplasms (Sherbet 1982). Tumor cells must be seen as highly responsive cells able to adjust to environmental differences by antigenic modulation. This can result in variation in antigen densities in the surface membrane (Fenyoe et al. 1968; Ioachim et al. 1972, 1974), loss of the tumor antigens or transplantation antigens (the latter ones being necessary for immune recognition) and also acquisition of new antigenic structures related to the environment (Fogel et al. 1979; Schirrmacher et al. 1975).

The lipid bilayer of the cytoplasmic membrane of a cell is in no way a stable structure. Instead, various environmental and internal influences are able to change the lipid fluidity of the cell membrane and thus control also the availability of substances (including antigens) embedded in the membrane (Muller 1979; Nicolson 1976; Shinitzky 1976).

Drugs or metabolic disturbances which raise or lower the cholesterol/phospholipid ratio of the interstitial fluid or interfere with the organization of the microfilament/microtubule system of the cell will also affect the expression and mobility of cell membrane receptors.

From this it appears reasonable also to assume that the antigenicity of a given cell population will not be stable over a period of time.

Production of blocking factor for the immune response

As has been known for a long time, the immunocompetent host may react in a dual response against its immunogenic tumor: tumor rejection by cytotoxic T-cells, K-cells, NK-cells, macrophages and complement-activating antibodies, as well as by specific enhancement of tumor growth (Sundar et al. 1982; Gershon et al. 1974).

The Hellströms were the first to direct our attention to 'blocking factors' which enable an immunogenic tumor to evade immune destruction by the host's defense system (Hellstroem and Hellstroem 1969, 1974, 1976; Hellstroem et al. 1968; Sjoegren 1973).

Such blocking factors for T-cell or K-cell induced tumor cell killing were identified as free tumor antigens or immune complexes, made up of these tumor antigens with their specific antibodies (Hellstroem and Hellstroem 1974). They can be removed from the Serum by passages through the appropriate immunoadsorbants (Tamerius et al. 1975, 1976), and eluted from their binding to neoplastic cells or to lymphatic cells (Hellstroem and Hellstroem 1976, 1969; Sjoegren et al. 1971) with glycin buffer (pH 2.8) or 0.2M urea (Tamerius et al. 1975, 1976).

Specific blocking antibodies are synthesized by T-cell dependent B-lymphocytes (e.g. splenic cells in mice) after only brief stimulation by antigen (Kall and Hellstroem 1975; Nelson et al. 1975).

Serum blocking activity is observed relatively early during tumor development and is usually readily detectable at the time of the initial diagnosis of the malignancy. It is maintained during progressive tumor growth and rapidly disappears after surgical removal of the neoplasm (Sjoegren and Bansal 1971).

Evidence for the existence of serum blocking factors has been provided in various tumors such as bronchogenic carcinomas, Hodgkin's disease, colorectal cancer, nasopharyngeal carcinoma, and breast cancer (Cohen 1976; Currie 1973; Jose and Seshadri 1974; Sundar et al. 1982).

Sundar et al. (1982; Karamaju et al. 1983) were able to identify this blocking factor in patients with nasopharyngeal carcinoma (NPC) as specific IgA anti VCA of EBV-infected cells. It interfered with the antibody-dependent cellular cytotoxicity (K-cell

response) against EBV transformed cells and with the lymphocyte stimulation by EBV antigens. Others were able to show that IgG antibodies in infectious mononucleosis blocked cell mediated immunity (Lai et al. 1974; Wainrwight et al. 1979).

The presence of immune complexes and of carcinoembryonic antigens (CEA) was found to correlate with tumor extent and progression in patients with bronchogenic carcinoma thus also probably interfering with the host's anti tumor response (Dent et al. 1980). Similarly, immune complexes were detected in a number of experimental and human cancers (Heimer and Klein 1976; Jerry et al. 1976; Robins et al. 1979). In addition, non-immunoglobulin components were identified with immune suppressive activities. These include α-globulin components (Cooperband et al. 1972, 1976), inhibiting lymphocyte stimulation, α-fetoprotein (AFP) (Bacacs et al. 1977; Cerottini and Brunner 1974), carcinoembryonal antigen (CEA) (Warnatz 1979) and αl-antitrypsin (Arora et al. 1978). Also non-proteinaceous substances such as prostaglandins (produced by macrophages) were shown to be immunosuppressive (Plescia et al. 1976).

As shown in Fig. 2 and discussed in detail in two review papers (Krueger et al. 2002b, 2003a), there are a rather large number immune-modulating cytokines, chemokines and growth factors which serve to improve the response against tumor-associated antigens (Waller and Ernstoff 2003; Miller et al. 2003; Redlinger et al. 2003). At least indirect blocking mechanisms of effective anti-tumor immunity can arise from abnormal patterns of immune cytokines. Elevated levels of IL-6 and IL-10, for instance, may cause suppression of cellular immunity and consequently deteriorate the prognosis of such tumors as carcinoma of colon, liver, breast and leukemia (Belluco et al. 2000; Chau et al. 2000; Fayad et al. 2001; Yokoe et al. 2000).. This reaction suggests an increased activity of Th2 T-helper cells over the Th1 cells as actually demonstrated in some cutaneous T cell lymphomas (Yoo et al. 2001). Other factors such Fas (CD95) and Fas ligands assist in the control of metastatic spread (Owen-Schaub et al. 2000) along with a number of local non-immune mechanisms (Mareel and Leroy 2003). Various clinical trials consequently included attempts to use cytokines and primed dendritic cells for improving anti-tumor immunity (for review see Waller and Ernstoff 2003).

Immune stimulation of tumor growth

Prehn and Lappe (1971) first proposed in their immunostimulation hypothesis that an immunogenic tumor may not only activate immune surveillance but also promote the growth of a nascent tumor. Very low titers of "lymphotoxin", for instance, stimulate protein synthesis and DNA replication of the target cell and thus enhance tumor cell proliferation. This proposition was experimentally supported by Shearer and Parker (1975) who selected a fast growing permanent L-cell line variant by treatment with cytostimulatory doses (1:200) of specific anti-L-cell antiserum. This tumor cell line was almost completely resistant to usually cytotoxic doses of this antiserum in the presence of complement. The immunostimulatory effect in this system was a consequence of immunoselection in favor of cells resistant to the destructive effect of specific antibodies. Selected L-cell variants were less susceptible to both immunostimulatory and immunoinhibitory effects and obviously carried less surface antigens than sensitive clones.

The basic mechanism leading to antibody resistant cell lines is presently not well understood; several hypotheses have been ventured:

(a) Selective immunostimulation of certain cell populations in the tumor, which then will overgrow other populations.
(b) Destruction of highly antigenic tumor cells by antibodies allowing unaffected cell clones to proliferate more rapidly.
(c) Modulation of cell metabolism by antibody eventually causing permanent alterations in cell behavior.

Loss of cell surface antigen and dedifferentiation rather than selection could be one of the possible sequelae of antibody action. The process thus may be similar to antigenic modulation of tumor cells as mentioned before.

Selective effects in antigenic tumors of a similar kind have been demonstrated before (Bartlett 1972; Bubenik et al. 1967). Under special circumstances such as the development of malignant lymphomas, proliferation and growth factors of immunocompetent cells may enhance the proliferation of tumor cells themselves (see later).

Allogeneic growth stimulation

In their hypothesis of immunostimulation of malignant tumors, Prehn and Lappe (1971) and others (Treves et al. 1974; Umiel and Trainin 1974) point to several analogies between an antigenic tumor growing in a responsive host and allografts or fetuses growing in equally immunocompetent individuals. There is some indication, in fact, that in embryogenesis mild antigenic differences between ovum and mother enhance implantation, placentation, and embryonal growth.

Similarly, certain antigenic differences between allograft and host may contribute to enhanced growth of the graft. In tissue culture studies, immunologically sensitized or con-A stimulated lymphocytes enhanced tumor cell colony grow up to certain ratios (Fidler 1973); at 1:10,000 or above (tumor cell to lymphocyte) growth was inhibited. Similar observations were made with lymphocytes from lymph nodes draining a B16 melanoma while splenic lymphocytes inhibited *in vitro* tumor growth. This inhibition was reversible, however, by serum from the mice collected during tumor growth (Bartholomaeus et al. 1974).

Lymphocytes collected from lymph nodes and spleen during the first 2–3 weeks after tumor transplantation stimulated target cell proliferation in culture, despite the tumor being highly immunogenic (Prehn 1976).

Selective immune stimulation of suppressor T cells by tumor antigens, such as CD4+CD25+ regulatory cells (Treg as described before), support tumor growth by downregulating an efficient cellular immune response (Golger et al. 2002; Jones et al. 2003; Peng et al. 2002).

Thus, again, there exists some indication for a dual immunologic response to a growing malignant tumor: immunologic growth inhibition and tumor cell destruction as well as growth stimulation. Which of the two reactions may predominate in the individual case appears dependent upon the "antigenic strength" of the tumor, the tumor cell (i.e. antigen) dose, and upon the number of immune effector cells available at the site of the growing tumor.

Prehn's hypothesis of immunostimulation of a malignancy, without doubt deserves serious consideration, especially since immunotherapy programs have already been established in many places.

Inductive activity of the immune system in tumorigenesis

As mentioned before and shown in Table 2, there is an unexplained high incidence of certain tumor types, especially malignant lymphomas, in human allograft recipients.

Similarly, a high percentage of patients with genetically determined or acquired immune deficiency syndromes suffer from malignant tumors of the immune system (Gatti and Good 1970; Page et al. 1963; Peterson et al. 1964; Ten Bensel et al. 1966).

The lymphoma incidence in the latter patients is roughly *5–15%* as compared to about *12/100,000* (non-Hodgkin's and Hodgkin's lymphomas) in the age-matched average population. Thus there apparently exists a condition in these individuals which favors the preferential development of malignant lymphomas (besides skin cancer which so far widely escapes interpretations) (Penn 1978; Walder et al. 1971).

In very general terms, the basic condition common to all such syndromes is a combined immunostimulation and immune deficiency, immunostimulation by allografts by environmental pathogens or even nonpathogenic inhabitants of the body (Krueger 1970a,b; Krueger 1972c, 1979; O'Conor 1970).

A number of experimental models making use of a similar combined immunosuppression and immunostimulation resulted in significantly elevated incidences in malignant lymphomas (Table 6).

A number of different hypotheses have been offered to explain the increased incidences of cancer and malignant lymphomas in transplant recipients and in immune deficiency states:

(a) Activation of latent oncogenic viruses (Hanto et al. 1981; Schwartz 1972)
(b) Genetic susceptibility and disturbance of immune surveillance
(c) Carcinogenic action in certain cases (and disturbed immune surveillance)
(d) Lymphoma induction by chronic graft-versus-host disease (GVHD; Schwarz and Andre-Schwarz 1968; Schwartz and Beldotti 1965)
(e) Lymphoma induction by an unbalanced immune response ("dysregulative theory") (Krueger 1972a, c, 1974, 1979).

Table 6 Experimental models of lymphoma induction secondary to combined immunosuppression and immunostimulation (Krueger 1989)

Species	Immunosuppression	Immunostimulation	Lymphoma
Mouse	Azathioprine, methotrexate, cyclophosphamide	LDH-,virus, vaccinia virus, BSA Freund's complete adjuvant	Lymphoblastic T cell lymphoma
	Murine leukemia virus (MoLV, GLV)	Murine leukemia virus	Lymphoblastic T cell lymphoma
	Chronic graft-versus-host disease (GVHD)	Allograft	"Reticulum cell lymphoma", immunoblastic lymphoma
	Endogenous GVHD-like syndrome	Immunogenic autologous cells	Immunoblastic B cell lymphoma
	Endogenous "active" RNA virus	Endogenous "active" RNA virus	SJL-disease, "type B reticulum cell sarcoma" immunoblastic lymphoma and plasmacytosis
Rat	Tolerance induction	Tumor allograft	Immunoblastic lymphoma
Hamster	Tolerance induction	Xenograft	"Reticulum cell sarcoma"

According to our own investigations, there appear to be three basic disturbances which can lead to uncontrolled cell proliferation in the immune system and thus to malignant lymphoma: an imbalance of proliferation and differentiation factors, an unresponsiveness of lymphoid cells to regulatory factors (see later in part II of this chapter) and a deficient apoptosis.

Activation of latent oncogenic viruses

Activation of persistent (latent) oncogenic viruses was demonstrated after irradiation, chemical immunosuppression, administration of anti-lymphocyte sera and induction of a graft-versus-host disease (GVHD). The two latter manipulations include immunostimulation as well as immunosuppression (some chemotherapeutics can probably act as haptens and thus become immunostimulatory).

Certain immune reactions as such (e.g. allograft rejection) can be accompanied by an activation of viral infection (Hirsch et al. 1970, 1973), and these viruses can become oncogenic (Armstrong et al. 1970).

Virus infections are known to occur quite frequently in immunocompromised individuals including man such as with CMV, EBV, herpes simplex I and II, HHV-6, HHV-8 and human papilloma virus; some of which possess oncogenic potentials.

The most common types of malignancies in human allograft recipients, lymphoma and squamous cell carcinoma, are at least in part suspected to have a viral etiology (Melnick et al. 1974; Rapp and Reed 1977; Zur Hausen 1975).

Hanto et al. (1981) were actually able to show anti EBV-antibodies, virus replication and EBV DNA in tumor biopsies from 4 out of 6 renal allograft recipients with lymphoproliferative disorders. Similarly, EBV was successfully linked to fatal lymphoproliferation in a genetically determined immune defect, the x-linked lymphoproliferative disease, in other immune deficiency syndromes including AIDS as well as in transplant recipients (Krueger 1993; Purtilo 1980; Purtilo et al. 1982; Sakamoto et al. 1982; Fauci and Lane 2000).

There is thus good evidence for a viral etiology of a number of tumors in immunocompromised individuals. Susceptible cells are transformed to tumor cells by oncogenic viruses and a deficient immune surveillance allows them unrestricted proliferation. Immune stimulation at the same time (e.g. by the allograft or by infectious organisms including the virus itself) will provide increased numbers of blasts in the immune system allowing enhanced replication of lymphotropic oncogenic viruses (such as EBV).

This can explain the predominance of lymphoid malignancies in the described conditions (Krueger 1985) and may result from combined and coincident immunostimulatory and immunosuppressive activities of such persistent virus infections (Krueger 1972a, 1975, 1989). This mechanism probably operates independently from the various etiology of viral oncogenesis, i.e. of insertional mutagenesis, acute transformation, transactivation of protooncogens or inactivation of tumor suppressor genes (Nathanson 2002). Immune dysregulation as pathogenic principle for lymphomagenesis is not contradicting todays genetic theories of cancer, but rather supplementing and supporting it, as we will see later (Krueger et al. 2001).

Genetic susceptibility and disturbance of immune surveillance

There are more than 240 inherited neoplastic syndromes; among these are breast cancer, gastric carcinoma, retinoblastoma, neurofibromatosis with malignant complications, familial polyposis and colon cancer, xeroderma pigmentosum and malignancies of the skin, multiple endocrine adenomatosis, uterine cancer and certain lymphoproliferative diseases. Some of these are accompanied by chromosomal aberrations.

In many of such malignant diseases, however, the exact nature of genetic influences on tumor development is only partly explained and needs further elucidation. It is frequently not a matter of a simple Mendelian relationship. Only exceptional familial malignancies exhibit a simple one-locus trait, in most of the other tumors it appears rather a matter of genetic control of "disposition" with a large number of gene pairs acting additively.

The notion of a genetic susceptibility for neoplastic growth is supported by the increased occurrence of some genes controlled by the major histocompatibility complex in immunologic and in neoplastic diseases (Abramova et al. 1981; Chan et al. 1981).

Genetic determinants such as the major histocompatibility complex (MHC) can influence tumor development in a variety of ways: genetic control of oncogenesis itself (i.e. "malignant transformation"), tumor antigenicity and host response against the tumor (Williams 1977).

As far as the patients discussed in this chapter are concerned at least the ones which inherited immune deficiency syndromes suffer from a hereditary disorder of their immune system which may also influence later lymphoma development. It may thus be difficult to distinguish between some endogenous tendency for malignant cell proliferation (such as probably exists in patients with Down's syndrome and subsequent leukemia) or disturbed repair (such as in xeroderma pigmentosum) or exogenous influences (e.g. by viruses, chemicals).

No obvious correlation exists in the inhomogeneous group of allograft recipients and a possible genetic susceptibility for tumor development. However, of interest in this regard was the observation of Hoover and Fraumeni (1973) that in kidney allograft recipients there was an excessively high incidence of cancers other than lymphomas when the donor was a sibling. (This phenomenon could probably also be interpreted immunologically, but further evaluation must await the accumulation of more data. Law et al. (1977) published the interesting observation that some healthy family members of patients with familial cancer of the colon and uterus as well as with lymphoproliferative malignancies may show mild chromosomal abnormalities and a defective lymphocyte response to mitogens and foreign cells (MLC test).

Carcinogenic action and disturbed immune surveillance

Some of the chemotherapeutic substances and radiation used for immunosuppression also possess apparent carcinogenic potentials as mentioned before. Thus considering the possible pathogenesis of tumors in human transplant patients, chemical carcinogenesis should not be excluded.

There exists, however, an apparently different pattern of neoplasms induced by carcinogens as compared to those in transplant recipients (see Tables 2 and 3). Reflecting chemical and radiological carcinogenesis are patients with second malignancies occurring after cancer chemotherapy or after immunosuppressive chemotherapy of autoimmune diseases (Gruenwald and Rosner 1970; Kahn et al. 1971). In both situations an exceptionally large number of myeloproliferative malignancies is observed. This pertains especially to neoplasms following the use of alkylating agents.

In allograft recipients, non-lymphocytic leukemias occur only in about 1.3% (Krueger 1979) while lymphoproliferative neoplasms make up some 20% of the tumors.

Immune deficiency diseases are usually not treated by additional immunosuppression or cancer chemotherapy. In these patients the overwhelming part of the tumors are malignant lymphomas.

Thus, if carcinogenesis plays a part in some of the allograft recipients, then probably only in a certain low percentage, yet a defective immune response and/or immunostimulation may rather contribute to the selectively high incidence of malignant lymphomas in this population.

Lymphoma induction by chronic graft-versus-host disease

In a series of elegant experiments, several authors (Armstrong et al. 1970; Gleichmann and Gleichmann

1976; Schwartz and Beldotti 1965; Schwarz and Andre-Schwarz 1968) have obtained malignant lymphomas by inducing chronic graft-versus-host disease (GVHD) in mice. It was theorized that chronic GVHD by interfering with the host's defense will activate oncogenic viruses and so initiate malignant lymphoproliferation (Armstrong et al. 1973; Hirsch et al. 1970). Gleichmann and Gleichmann (1976, 1980) have further developed this system to include GVHD-like reactions of host lymphocytes against antigenically modified syngeneic cells.

Antigenic stimulation was achieved by the anti-epileptic drug diphenyl hydantoin (DPH). This substance was chosen because of previous reports of an increased incidence in lymphomas in epileptic patients treated with DPH and of lymphoproliferative disorders in the mouse after DPH administration (Chiari 1951; Krueger 1970c; Krueger and Bedoya 1978; Krueger and Heine 1972; Saltzstein and Ackerman 1959). The mechanisms of an autologous GVHD-like syndrome could probably also be induced by a number of exogenous influences able to change cell surface antigenicities such as viral infection (EBV, vaccinia virus), radiation and certain chemicals (haptens for instance). The spectrum of ensuing clinical syndromes is summarized in Fig. 3.

Lymphoproliferation which occurs in the B-cell region of the lymphatic system is apparently T-cell dependent since it does not occur in T-cell deficient nude mice (Gleichmann and Gleichmann 1976).

In essence, chronic GVHD or GVHD-like syndromes cause an apparent disturbance of immunologic homeostatic mechanisms which *per se* favor virologic or dysregulative lymphoma development. This theory sheds some light on the pathogenesis for lymphoma development without explaining the tumor etiology itself.

Lymphoma induction following an unbalanced immune response

The concept of a dysregulative theory of lymphoma development is based on the previously mentioned observations of increased lymphoma incidences in human allograft recipients and in patients with immune deficiency diseases. The basic mechanism was seen in the coincident persistent immunostimulation and immunosuppression (Krueger 1970a, b).

A number of experiments to this effect apparently supported this assumption (Table 6). The working hypothesis was that antigens physiologically stimulate proliferation of responsive cells and that this cell proliferation will also occur in deficient immune reactivity and in tolerant individuals provided cell division itself is not blocked (Krueger 1972a, 1973). Antigen-induced physiological cell proliferation is controlled by the antigen itself and will come to a standstill when the antigen is neutralized and eliminated. It is also controlled by cells and cell factors in the lymphoid tissues (e.g. suppressor cells and factors).

In the case of a persistent immune deficiency causing the antigen to persistently affect the proliferating control system, cell proliferation will continue with production of the clinically manifest malignant lymphoma.

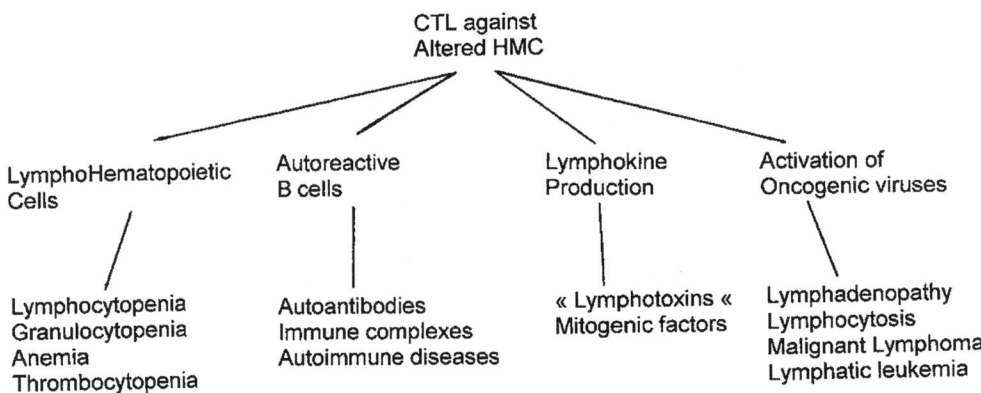

Fig. 3 Endogenous GVHD-like reactions causing autoimmune disorders and lymphomas (From Gleichmann and Gleichmann 1976)

There are three basic disturbances which actually can embrace all the theories of lymphoma development discussed so far:

(a) An imbalance of proliferation and differentiation factors controlling the immune system
(b) Unresponsiveness of lymphoid cells to such regulatory factors
(c) Inherited or acquired deficiencies of apoptosis

Before discussing these three basic defects we shall briefly attempt to summarize the currently known immunoregulatory circuits. There are, so far, *3–4* cell populations identified as responsible for regulating cell proliferation and maturation in the immune system: thymic epithelial cells (and theoretically similar cells in the, bursa equivalent'), T-inducer and suppressor cells, and macrophages. Thymic epithelial cells regulate the differentiation of preT-cells to T-cells (thymic and peripheral) through the production of thymic hormones (thymosin, thymopoietin, serum thymus factor; (Bach and Carnaud 1976; Goldstein 1975; Hadden and Stewart 1981; Marshall et al. 1978; Low et al. 1979; Krueger et al. 2003b). Similarly, a 'bursa epithelial factor' (e.g. ubiquitin, bursapoietin; Goldstein 1977) may control the transition of pre-B-cells to B-cells. T-inducer and T-suppressor cells regulate T-cell proliferation and differentiation, T-T-cell cooperation, T-Bcell cooperation, T-cell-macrophage cooperation and also hematopoietic stem cell proliferation, fibroblast and osteoclast proliferation, probably through similar helper or suppressor factors (lymphocyte mitogenic factor, proliferation inhibition factor, immunoglobulin-like molecule IgT, immunoregulatory α-globulins IRA, macrophage migration inhibition factor MIF, macrophage activation factor MAF and newer factors) (Cantor and Boyse 1977; Greene et al. 1977; Tada et al. 1976a, b; Krueger et al. 2002b, 2003a).

Macrophages themselves interact with T-lymphocytes to enhance their immune reactivity through the production of interleukin I (Oppenheim et al. 1980; Rosenstreich and Mitzel 1978). A tentative synopsis of these immunoregulatory circuits is shown in Fig. 4 (Cantor and Boyse 1975; Reinherz and Schlossman 1980; Reinherz et al. 1979a; see also Krueger et al. 2002b).

Based on clinical observations of selective regulator cell defects in immune deficiency diseases and in malignant lymphomas (Moretta et al. 1977; Reinherz and Rosen 1981; Siegal et al. 1978; Waldmann et al. 1978), as well as in a few experimental models (Heine et al. 1983a, b; Krueger et al. 1983), my collaborator A. Karpinski has mimicked in an early computer-assisted model various cellular defects in the T-cell immune system and obtained a large number of different T-lymphocyte neoplasms and immune defects (Krueger 1989). This model could also be expanded to include similar lesions of the B-cell system. All were a consequence of either regulatory imbalances or unresponsiveness of target cells.

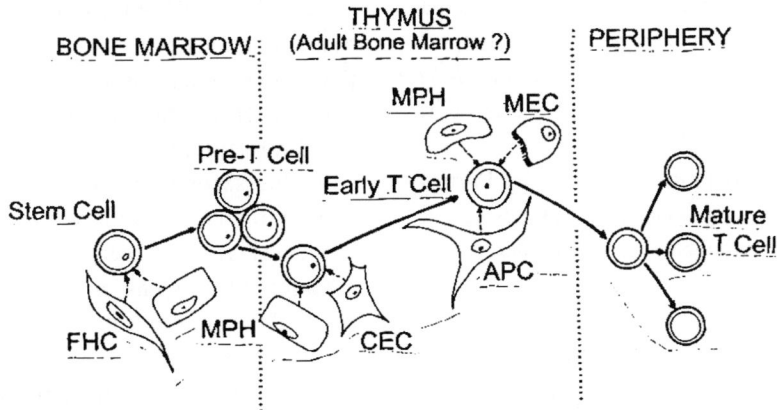

Fig. 4 Immunoregulatory events of T cell maturation and factorial influences from FHC (fibrohistiocytes), MPH (macrophages), CEC (cortical epithelial cells), MEC (medullary epithelial cells) and APC (antigen-presenting cells)

Imbalance of proliferation and differentiation factors resp. the representative cell populations

In the year 1971 Fudenberg (1971) concluded from his clinical experience that 'patients with generally determined T-cell defects, if they survive long enough, also have a high incidence of malignant diseases, at least ten times that of the general population. Besides malignant lymphomas, autoimmune diseases of various kinds are more frequent in T-cell deficient individuals similar to the findings in NZB mice (whose T-cell function also is defective; Leventhal and Talal 1970). These observations suggest an apparent deficient regulatory function of certain T-lymphocyte populations.

T-cell defects were also demonstrated in many malignant tumors including lymphomas (Costanzi et al. 1977; Kenady et al. 1977; Zembala et al. 1977), although it was not always clear whether this was secondary to the tumor itself or preceded it.

In angioimmunoblastic lymphadenopathy, a human prelymphomatous condition, and in some experimental models, T-cell decrease was also found preceding malignant lymphoma while B-lymphocytes may exhibit a polyclonal hyperplasia (Kosmidis et al. 1978; Kraus 1981; Krueger et al. 1978, 1981). The latter effect was tentatively interpreted as a consequence of hyperactive helper cell populations (Kosmidis et al. 1978).

Graft-versus-host disease (GVHD), another disturbance that conditions the diseased for lymphoma development, is accompanied by aberrations of suppressor T-cells (Reinherz et al. 1979b). We investigated similar prelymphomatous conditions, lymphoepithelioid cellular lymphoma (Lennert's lymphoma) which later progressed to B-immunoblastic lymphoma or to pleomorphic T-cell lymphoma. These cases revealed a prelymphomatous increase in suppressor cells (helper/suppressor cells = *0.3–0.8* instead of normally *1.3–2.4;* total percent T-cell in the blood *59%)* (Krueger, unpublished data from our laboratory).

This concurs with observations in XLP by Purtilo et al. (1982), that an inversion of the helper/suppressor cell ratio usually preceded lymphoma development, as well as of Kaposi sarcoma in homosexuals (Editorial, Lancet (Editorial 1981; Krueger 1985)) and of increased suppressor cell number in patients with chronic lymphocytic leukemia (Mills and Cawley 1982) and myeloma (Mills and Cawley 1983). Finally, there exist a number of reports documenting similar changes of prelympho-

matous cell proliferation with CD4/CD8 cell inversion during persistent active infection with lymphotropic herpesviruses (Krueger et al. 1988, 1989, 1992). Such prelymphomatous cell proliferations that can progress to overt malignant lymphoma, were termed "atypical polyclonal lymphoproliferation" or "APL". Jaffe and coworkers have called such changes – without referring to the original publications – "autoimmune lymphoproliferative disorder" (Lim et al. 1998).

Thus, there are apparently a certain number of B-cell lymphomas and prelymphomatous conditions associated with elevated T-suppressor cells.

In addition, T-memory cells against EBV antigens were deficient in XLP (Harada et al. 1984).

A similar T-suppressor cell defect is known from murine SJL disease, a B-cell lymphoproliferative disorder progressing to malignant lymphoma (Bergholz et al. 1983; Nakano and Cinader 1980).

Besides the B-cell prelymphomatous proliferations and the B-cell lymphomas with increased suppressor T-cell activity, there are a number of T-cell prelymphomas (Nichols et al. 1982; Rosenthal et al. 1982) described and T-cell lymphomas and leukemias with various functional activities (helper- and suppressor cell function, interleukin II production; (Friedman et al. 1982; Kaur et al. 1982; Moriya et al. 1981; Saxon et al. 1979).

There are two possible ways to interpret described changes in the B-cell proliferative diseases:
(a) depression of immune surveillance, or (b) dysregulation of B-cell response.

In the first case (a), hyperactive T-suppressor cells should interfere with the T-cell and B-cell immune surveillance to destroy proliferating foreign cells.

Atypical cells which arise (by mutation or by transformation) will show continued growth under these conditions. These cells should exhibit signs of atypia such as chromosomal changes, increased cell membrane fluidity, etc.

In the second case (b), hyperactive T-suppressor cells will interact with helper T-cells and antibody producing B-cells keeping the immune response to all kinds of antigens or to specific antigens insufficient. This represents an over-activity of Treg cells as described before.

Persistent antigen stimulation will be followed by progressive immunocyte proliferation in an attempt to make up for the deficiency.

Under such conditions, proliferating lymphocytes should not be atypical *per se*. The proliferation should be polyclonal in many cases responding to a variety of antigens, or monoclonal responding to only one specific antigen.

This second explanation (b) would follow the dysregulative theory of lymphomagenesis Figs. 5–7).

T-cell lymphomas consist of a variety of functioning and nonfunctioning neoplasms of different degrees of differentiation. There are helper cell lymphomas (like Sezary's Syndrome and T-cell immunoblastoma with hypergamma globulinemia (Broder et al. 1976; Friedman et al. 1982), suppressor cell leukemias (subacute T-cell leukemia; (Uchiyama et al.

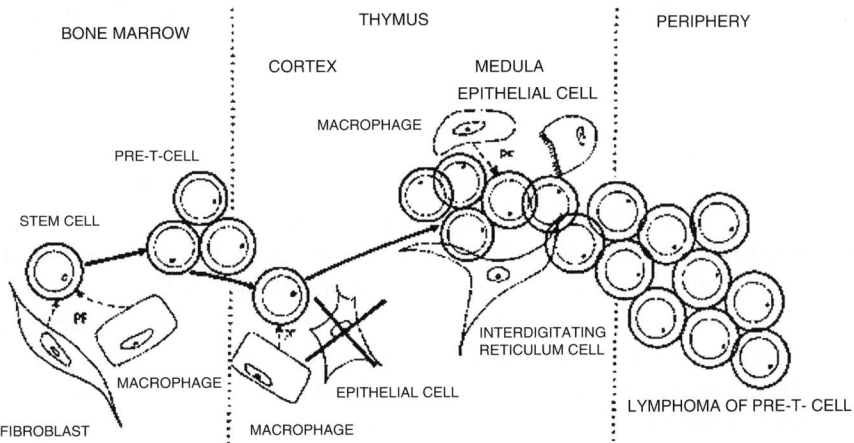

Fig. 5 Pre-Te cell lymphoma secondary to thymic epithelial defect and thymopoietin loss, i.e. proliferation of non-transformed undifferentiated cells responding to proliferation factors

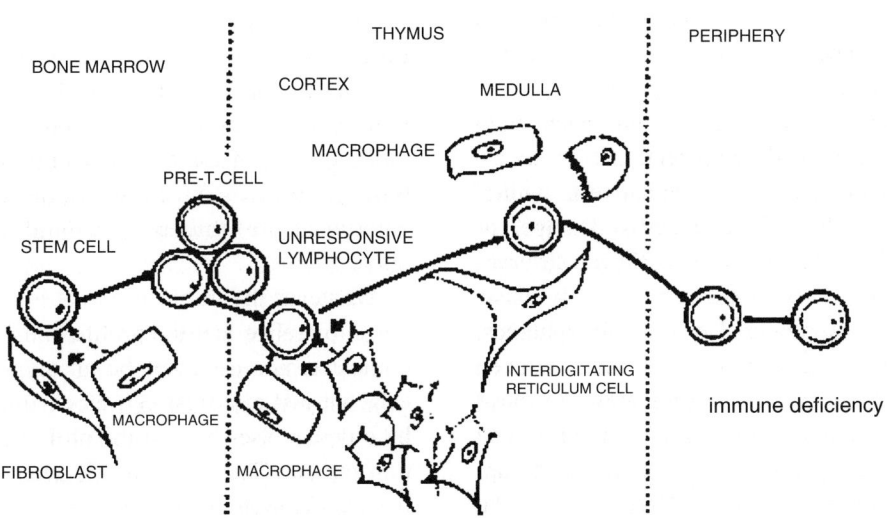

Fig. 6 Thymoma and immune deficiency secondary to unresponsive thymic lymphocytes. Pseudo-compentatory reactive cortical epithelial cells may eventually cause an epithelial thymoma

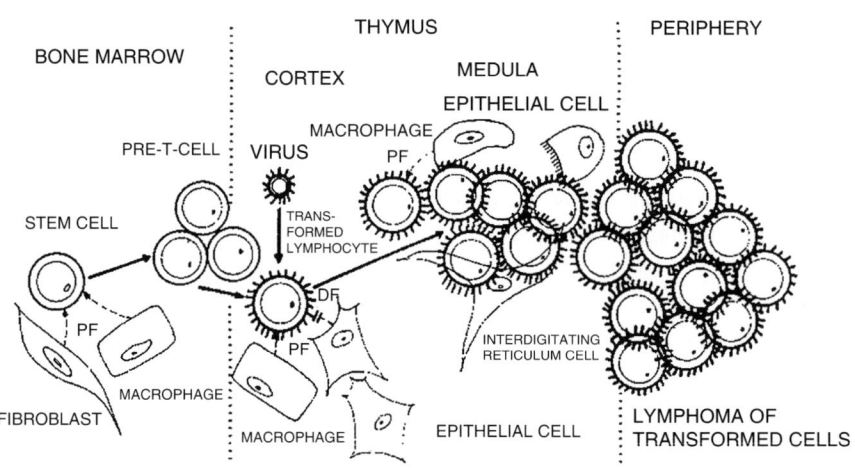

Fig. 7 Cortical T cell lymphoma secondary to malignant cell transformation

1977), and non-helper/non-suppressor T-cell or 0-cell lymphomas.

The theoretical basic defect in the pathogenesis of these tumors could be either an intrinsic disturbance of the proliferating cells themselves (i.e. transformed cells producing their own growth factor in analogy to the sarcoma growth factor of virus-induced murine sarcomas; (DeLarco et al. 1980); in this case, lymphoma cells again should exhibit signs of atypia. The defect could be in the target cell being unresponsive for produced helper or suppressor factors and thus causing continuous proliferation of T-cells. Here, proliferating cells should not be atypical but the non-proliferating target cell should be either 'transformed' or exhausted.

This mechanism again would be consistent with dysregulative lymphoma development.

The third possibility, also dysregulative in nature, consists in the production of defective T-helper or T-suppressor cells secondary to insufficient differentiation factors. The latter possibility probably exists in non-helper/non-suppressor T-or 0-cell lymphomas, as we were able to show in Moloney virus induced mouse lymphomas. In these experiments we have demonstrated the development of malignant pre-T-cell lymphomas associated with a defect of thymic epithelial cells to produce thymic hormones (Heine et al. 1983a, b; Krueger et al. 1983).

Thymic epithelial cells actively replicated the virus and degenerated, while non-transformed prethymic

T-lymphocytes continued to proliferate without further maturation.

Interestingly enough, deficiencies of thymic hormones have been demonstrated in a variety of immune deficiency diseases, i.e. diseases prone to later lymphoma development (Astaldi et al. 1978; Atkinson et al. 1982; Iwata et al. 1981; Pardoll 2003; Turk et al. 1972; Uchiyama et al. 1977); systematic studies to this effect in prelymphomatous conditions and in malignant lymphomas still need to be done (especially with regard to the exact lymphoma cell identification related to the type of immune defect).

The etiology of all the different lymphoproliferative syndromes can be genetically inherited (chromosomal abnormalities, structural or functional), malignant transformation (virus, carcinogen, radiation), as well as eradication of certain cell types (lytic virus infection, antigen exposure during embryonal life) thus imbalancing the immunoregulative circuit.

Whatever the initial insult, it is of utmost importance to realize that the proliferating cell causing the clinical tumor need not be an atypical transformed cell, but just a normal cell blocked in differentiation by a defect elsewhere in the proliferation/differentiation circuit. Chemotherapeutic intervention to block the proliferation of cells can not have ultimate effects unless the proliferating cell itself is atypical.

In this context it should be recalled that the erythropoietic cell proliferation in pernicious anemia was

once thought to be neoplastic, before the basic metabolic defect was identified.

Unresponsiveness of the target cell for immunoregulative influences

This concept has its parallel in the development of endocrine tumors secondary to a target cell unresponsiveness, for instance, the appearance of parathyroid adenomas in chronic renal disease. It is conceivable that similar helper- or suppressor cell lymphomas occur if the target cell is unresponsive for their factors or missing (see before).

Target cell transformation with loss of surface receptors, autoantibody reaction, and target cell elimination by lytic virus infection, other damage or by premature antigen exposure could be the causative effect.

Besides, metabolic disturbances may make this cell transiently unresponsive. Once reactive proliferation of regulatory cells has produced more than a critical number of cells, the imbalance in various cell types may support further proliferation of the regular cell independently of continuing stimulation (Metcalf 1971).

Transient metabolic disturbances can arise, for instance, by hormonal effects, zinc deficiency (Cunningham-Rundles et al. 1981; Fernandez et al. 1979), toxic influences, etc. Zinc deficiency is followed by defective thymic hormone production with subsequent T-cell impairment (Gryboski 1975; Lombeck et al. 1975) and by lymphoma development in rats.

Correcting zinc deficiency during the state of immune deficiency will normalize immune reactivity, but supplementing zinc in an already existing lymphoma will not reverse the malignant disease.

There is a number of immune deficiency syndromes in which a stem cell defect is believed to be the basic disturbance (Griscelli et al. 1978; Waldmann et al. 1980); these cells apparently do not respond to environmental stimuli.

It will be quite interesting to examine the developing lymphomas in such cases with respect to their cellular origin: stem cell versus regulator cell?

Transplantation of cultured thymus (i.e. regulator cells) in 30 cases with combined immune deficiency disease resulted in a lethal polyclonal B-cell proliferation in 3 cases (Borzy et al. 1979).

Apparently B-cells had a selective defect that influenced thymic control mechanisms responding to proliferative influences (inducer cells) but not to repressive influences (repressor cells); the possibility can not be excluded, however, that suppressor cell differentiation in the transplanted thymus was defective.

A similar dysregulative mechanism with unresponsive B-cells may lead to thymoma with hypogamma globulinemia (Good 1954; Korn et al. 1967).

Antigenic differences of cells including the expression of cell membrane receptors for regulative signals are programmed by the genome, i.e. by gene activation (Boveri 1914).

Thus unresponsiveness of cells or restricted cellular polymorphism excluding certain types of responsive cells can be of genetic origin. Viral infections and mutagens certainly can affect gene function. Essentially all mechanisms described before and discussed in the pathogenesis of malignant lymphomas can also act through production of unresponsive cells and secondary regulator cell hyperplasia. There are currently only sporadic investigations to this effect so that only suggestive evidence exists for this aspect of a dysregulative lymphoma theory. However, recent developments of a theoretical model of T cell functions and its disturbances may offer future insights into this phenomenon (Krueger et al. 2003b; Brandt et al. 2002, 2004).

Modeling of the T cell system and of lymphomagenesis

Theories of origins of malignant tumors including lymphomas and leukemias have changed according to scientific advancements and fashions. Genetic mutations were made responsible as early as 1914 by Boveri, and are still preferred today (Gauwerky and Croce 1995). Chemical carcinogens, ionizing radiation, and oncogenic viruses further contributed to understanding pathways of mutagenesis (Weinstein et al. 1995; Howley 1995). Despite of their logic and beauty, these concepts, as any, have their limitations. As known since long, there is no simple relationship of chemicals and radiation between mutagenic and carcinogenic effects (Auerbach 1940; Demerec et al. 1948; Burdette and Haddox 1954). Also, "oncogenic" viruses do not tend to be oncogenic in every *in vivo* case unless some other factors such as immune deficiency, integration into specific genomic sites, or disturbed repair mechanisms coincide with infection

(Klein 1989; Purtilo et al. 1985; Krueger and Ferrer Argote 1994; Hollander and Fornace 1995). Even the best known examples of cytogenetic abnormalities in malignant lymphomas and leukemias occur only in a certain percentage of cases, usually in <1–30%: the 95% c-myc activation through t(8:14)(q24;q32) translocation in Burkitt's lymphoma or B cell acute lymphocytic leukemia are rather exceptions (LeBeau and Larson 2000). Also, it has not always been clearly identified which cytogenetic abnormalities are primary (and thus of etiologic significance), which are secondary (and may rather influence the course of an existing neoplasia), and which are just "noise".

It appears thus, that although much is known about the potential etiologies of lymphomas and leukemias, their pathogenesis still needs further clarification.

Mimicking human pathology in transplant recipients, malignant lymphomas were induced in a number of experiments by applying the pathogenetic concept of immunosuppression and coincident persistent immunostimulation (Krueger 1972a, 1975; Krueger et al. 1971b, 1979, 1983; Kraus and Krueger 1981; Heine et al. 1983a, b). This pathogenetic principle, immune deficiency and coincident immune stimulation, appeared to prevail in all cases independent of their etiology. It lead to formulation of a *dysregulative lymphoma theory* (Krueger 1972b, 1975) and to the design of a computer model to simulate lymphoma pathogenesis (Krueger 1989). The conceptual design of this model was to manipulate factors which control either one of following functions: cell proliferation, cell differentiation, cell release, and cell inhibition (Figs. 8 and 9). Applying it to the T cell system, we were able to mimic the development of a rather large number of malignant lymphomas (Krueger 1989).

Subsequently, follow-up studies in human patients with persistent active infection by lymphotropic human herpesviruses, Epstein-Barr virus and human herpesvirus-6, and with HIV infection identified characteristic T- and B-cell changes in prelymphoma-tous conditions progressing to malignant lymphoma. Prelymphomatous changes were classified as "atypical polyclonal lymphoproliferation (APL), atypical lymphoproliferative disorders, or – independent of identified viral infection – as autoimmune lymphoproliferative syndrome (Krueger 1993; Krueger et al. 1988, 1989, 1992; Frizzera 1992; Lim et al. 1998).

The development of such changes was reminiscent to what was observed earlier in experimental mouse lymphomagenesis (Krueger 1972a, 1993).

It appears thus justified in further studying the concept of dysregulative lymphoma development as pathogenetic principle to review data of functional changes in human lymphoma entities.

Exemplary cases of lymphoproliferative diseases with genetic defects causing immune dysregulation were recently summarized (Table 7). All show a combination of some functional defect with enhanced stimulation and thus support the concept of dysregulative lymphomagenesis. With the multitude of possible pathways, as shown in the table, conceptual theories such as the dysregulative lymphoma theory (Krueger 1989) become necessary in order to construct computer-assisted models for further investigation. Otherwise, the abundance of individual data available can not be put into a realistic context to each other. This has become dramatically apparent recently, when genetically engineered corrections of SCID (severe combined immune deficiency) cases developed unexpected leukemias (Marshall 2003). One specific genetic defect in these patients was corrected without apparent consideration of side effects

$$\dot{w} = \mu_w + w\left(\sum_{h=1}^{H}P_{w_h} \cdot \sum_{j=1}^{J}D_{w_j} \cdot \sum_{k=1}^{K}I_{w_k} + ax + by + cz\right),$$

$$\dot{x} = \mu_x + x\left(\sum_{l=1}^{L}P_{y_l} \cdots \sum_{m=1}^{M}D_{x_m} - \sum_{n=1}^{N}I_{x_n} + dy - ez\right) + fw,$$

$$\dot{y} = \mu_y + y\left(\sum_{q=1}^{Q}P_{y_q} - \sum_{r=1}^{R}D_{y_r} - \sum_{s=1}^{S}I_{y_s} + gz\right) + ux,$$

$$\dot{z} = \mu_z + z\left(\sum_{\theta=1}^{\Theta}P_{z_\theta} - \sum_{\omega=1}^{\Omega}D_{z_\omega} - \sum_{\gamma=1}^{\Gamma}I_{z_\gamma}\right) + \delta y.$$

Fig. 9 Initial model equations. Notation: w,x,y,z cell pools (see Fig. 8: bone marrow, thymic cortex, thymic medulla, periphery); dot notation represents rate of change (e.g. dw/dt); μ average regeneratory potential of each compartment; P, D, I sum of proliferation, differentiation and inhibition factors at work within each compartment; a to δ various feedback and feedforward values (see Krueger et al. 2004)

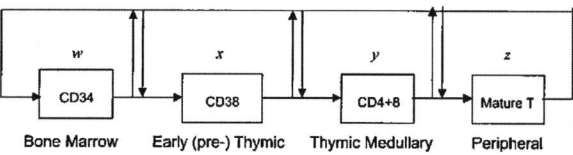

Fig. 8 Conceptional block diagram of the T cell model with feed-forward and feed-backward regulation

Table 7 Functional activity of CD ligands, interleukins and genomic alterations in NHL (From Krueger et al. 2001)

Maturation stage	B Immature	B Intermediate	B Mature	T Immature	T Intermediate	T Mature
Entity	B-ALL	B-CLL, DLCL, MCL	LPL, MM	T-ALL LGCL	T-CLL	MF, PTCL
CD Markers	CD19/10 (Bartlett 1972)	CD5/19/20/22/24/25	CD19/20/22/ 23/43/79a	CD1/2/5/7/34/99 16/56/25 (NK)	CD2/3/5/7/25 dp8+4+	CD2/3/5/7 8 <u>or</u> 4
Stimulation CD (by ligands)	Assoc. CD81/21 B cell activation, growth regulation development	IL-2, C-type lectin, P-selectin, IL-2 cell-cell cooperation activation	Assoc. CD54/21 ICAM, integrins cell–cell cooperation C'receptor stimul	LFA-3, BLAST-1 protectin, selectins NCAM, IL-2 cell–cell cooperation	LFA-3, BLAST-1 same as T Immat. in addition TCR stimulation by Aantigen	LFA-3, BLAST-1 same as T Immature in addition TCR stimulation by Aantigen
	In addition Ig-Receptor stimulation by antigen			activation	MHCII/I cell–cell cooperation	
Stimulation by IL	IL-7, IL-6, TNFα B cell growth and differentiation	IL-2, IL-6, IL-10, IL-15 B cell growth, activ. and differentiation	IL-6, IL-15 B cell growth and angiogenesis	IL-2, TNFα T cell activation	IL-2, IL-7 T cell proliferation and activation	IL-2, IL-4, IL-6, IL-10 T cell proliferation and activat., TH2 stimul TH1 inhibtion
Stimulation by cytogenetics	cyclinD1 cell prolifer.	cyclinD1, c-myc cell prolifer.	cyclinD1, ras cell prolifer.	tcl-1,2, rhom/ ttg-1,2 cell prolifer.	rhom/ttg-1,2 cell prolifer.	Signal transduction
Inhibition by IL		IL-4, IL-10, IL-8 Inhibit. of inflamm. inhibit. of apoptosis		IL-10 NK cell downreg.	IL-4 inhib. inflamm.	IL-4, IL-10 inhib. inflamm. NK cell downreg
Inhibition by cytogenetics	p53, IgH apoptosis specif. response	p53, bcl-2, IgH apoptosis specif. response	PAX-5, IgH specif. response	TCR specif. stimul.	TCR specif. stimul.	p53, TCR apoptosis

by interfering with other immunoregulatory cycles in the patients.

Since the presentation of a primitive computational model for simulation of lymphomagenesis in the first edition of this book (Krueger 1989), modeling studies of immune responses are actively performed over the past years. In particular, some works on T helper cell differentiation (into Th1/Th2 subsets) have been done (Fishman and Perelson 1999). Modeling of thymocyte differentiation was first studied by Mehr et al. (1995a, b, 1996). They used a set of four ordinary differential equations (ODEs) to describe numerical changes in different T cell pools. Some significant conclusions were drawn based on their model. Using similar approaches, we constructed model TCM-1 to validate the theory of "dysregulative lymphoma pathogenesis" (Brandt et al. 2002, 2004; Krueger et al. 2003b) by studying the dynamics of stem cells, early and late thymocytes as well as peripheral blood lymphocytes with viral infection as the cause of immunologic dysregulation. Although this model (see Figs. 8, 9) enabled us to realistically simulate cellular events following the infection with T lymphotropic viruses HHV-6, HIV-1 and HTLV-1 and apoptosis blockage related to Canale–Smith syndrome (Krueger et al. 2002a, 2003b), it afforded a rather large number of factors and parameters with designing of a somewhat detailed search algorithm (Krueger et al. 2004). Several attempts to simplify the computational modelling process finally led to Wang's technique of using mouse models and partial differential equations (Wang et al. 2003, 2004). This procedure describes T cell maturation processes as continual motion like moving fluids with the dimensions of time t and

maturity θ. The total cell number spreads along θ and forms a density distribution $\rho\,(\theta,\,t)$. The detailed model is described in the original publication (Wang et al. 2003, 2004).

Epilogue

It is difficult to comprehensively discuss all aspects of immune dysregulation as related to cancer. There are more than 100 factors in control of immune regulatory cycles including cytokines, chemokines, growth factors, neuropeptides and hormones (Krueger et al. 2002b, 2003a). There are more than 200 loci in the human genome which can be switched on or off and which are coding for immunoregulatory proteins (Venter et al. 2001). This does not include loci coding for general cell functions such as enzymes, modulators of cell cycle and signal transduction as well as many others. Computational models, as we have briefly discussed above, are an essential tool for our future understanding of all available data, their functional activities and their biological interactions.

Unfortunately, the limitation of knowledge and current scientific fashions in combination with pride and power of the proponents of such fashions are a major obstacle to freely and openly develop such new computational models. Biomedical data are often collected in a way that is not useful for computer simulation studies. Computational scientists, consequently, rather "play" with abstract data with limited application to a living individual. The "famous" biomedical scientists, on the other hand, know it all already and do not need mathematical approaches.

In our present review, we have attempted to openly discuss the older literature together with the newer one, and to draw the attention to the potentials of mathematical modeling. We hope that this may stimulate openness and curiosity in the younger generation for resuming their research independent of current fashions and in the classical Aristotelean sense of the natural sciences.

References

Abramova J, Majsky A, Koutecky J (1981) HLA Antigene bei germinativen Tumoren von verschiedener Lokalisation. Onkologie 4:19

Actor JK (2007) Elsevier's integrated immunology and microbiology. Mosby Elsevier, Philadelphia, pp 45–53

Ahmed A, Smith AH, Folks T (1979) Murine T cell heterogeneity. Cancer Treat Rep 63:613

Alexander P (1975) Escape from immune control by the shedding of membrane antigens: influence on metastatic behavior of tumor cells. In: Gottlieb AA et al. (eds) Fundamental aspects of neoplasia. Springer, NY, pp 101–108

Alexander P, Eccles SA, Gauci CLL (1976) The significance of macrophages in human and experimental tumors. Ann N.Y.A.S 276:124

Allison AC (1970) Tumor development following immunosuppression. Proc Roy Soc Med 63:1077

Armstrong MYK, Gleichmann E, Gleichmann H, Beldotti L, Andre-Schwartz J (1970) Chronic allogeneic disease. II. Development of lymphomas. J Exp Med 132:417

Armstrong MYK, Ruddle NH, Lipman MB, Richards FF (1973) Tumor induction by immunologically activated murine leukemia virus. J Exp Med 137:1163

Arora PK, Miller HC, Aronson LD (1978) Alpha-1-antitrypsin is an effector of immunological stasis. Nature 274:589

Arsenau JC, Sponzo R, Lewin RW, Schnipper LE, Bonner H, Young RC, Canellos GP, Johnson RE, DeVita VT (1972) Nonlymphomatous malignant tumors complicating Hodgkin's disease. Possible association with intensive therapy. N Engl J Med 287:1119

Askenase PW, Hayden BI, Gershon RK (1975) Augmentation of delayed-type hypersensitivity by doses of cyclophosphamide which do not affect antibody responses. J Exp Med 141:697

Astaldi A, Astaldi GCB, Wijermans P, Dagna-Bricarelli F, Kater L, Stoop JW, Vosasen JM (1978) Experiences with thymosin in primary immunodeficiency diseases. Cancer Treat Rep 62:1779

Atkinson K, Biggs JC, Ting A, Dodds AJ, Pun A (1982) Cyclosporin A is associated with faster engraftment and less mucositis than methotrexate after allogeneic bone marrow transplantation. Br J Haematol 53:265

Auerbach C (1940) Tests of carcinogenic substances in relation to production of mutations in drosophila melanogaster. Proc Roy Soc Edinburgh Sect B Biology 60:164

Bacacs T, Gergely P, Cormain S, Klein E (1977) Characterization of human lymphocyte subpopulations for cytotoxicity against tumor-derived monolayer cultures. Int J Cancer 19:441

Bach J-F, Carnaud C (1976) Thymic factors. Prog Allergy 21:342

Baldwin RW (1973) Immunological aspects of chemical carcinogenesis. Adv Cancer Res 18:1

Ball JK (1970) Immunosuppression and carcinogenesis: contrasting effects with 7,12 dimethylbenz-α- anthracene, benz-α-pyrene and 3-methylcholanthrene. J Nat Cancer Inst 44:1

Bar RS, Delor CJ, Clausen KP, Hurtbise P, Henle W, Hewetson JF (1974) Fatal infectious mononucleosis in a family. N Engl J Med 290:363

Bartholomaeus WN, Bray AE, Papadimitriou JM (1974) Immune response to a transplantable malignant melanoma in mice. J Nat Cancer Inst 53:1065

Bartlett GL (1972) Effect of host immunity on the antigenic strength of primary tumors. J Nat Cancer Inst 49:493

Bast RC, Zbar B, Borsos T, Rapp HJ (1974) BCG and cancer. N Engl J Med 290:1413

Belluco C, Nitti D, Frantz M (2000) Interleukin-6 blood levels is associated with circulating carcinoembryonic antigen and prognosis in patients with colorectal cancer. Ann Surg 7:133

Bendinelli M, Nardini L (1973) Immunosuppression by Rowson-Parr virus in mice. II. Effect of Rowson-Parr virus infection on the antibody response to sheep red cells in vivo and in vitro. Infect Immun 7:160

Bergholz M, Brehler R, Schauer A, Krueger G, Fischer R, Herbst EW (1983) Die SJL-Lymphadenopathie als Tiermodell. Verh Dtsch Ges Path 67:594

Blohm U, Roth E, Brommer K, Dumrese T, Rosenthal FM, Pircher H (2002) Lack of effector cell function and altered tetramer binding of tumor-infiltyrating lymphocytes. J Immunol 169:5522

Borzy MS, Hong R, Horowitz SD, Gilbert E, Kaufman D, De Mendoca W, Oxelius V-A, Dictor M, Pachman L (1979) Fatal lymphoma after transplantation of cultured thymus in children with combined immunodeficiency disease. N Engl J Med 301:565

Boveri TH (1914) Zur Frage der Entstehung maligner Tumoren. Fischer, Jena

Broder S, Edelson RI, Lutzner MA, Nelson DL, Meade BD, Waldman TA (1976) The Sezary syndrome. A malignant proliferation of helper T-cells. J Clin Invest 58:1297

Brandt ME, Wang G, Krueger GRF, Buja LM (2002) A biodynamical regulatory model of the human T-cell system. Eng Med Biol (EMBS) Conf Oct.20–23, Houston, Texas

Brandt ME, Krueger GRF, Wang G, Buja LM (2004) Feedforward and feedback mechanisms in a dynamical model of human T cell development and regulation. In Vivo 18:465

Broder S, Uchiyama T, Waldman TA (1979) Current concepts in immunoregulatory T cell neoplasms. Cancer Treat Rep 63:607

Bubenik J, Koldovsky P, Svoboda J, Adamova B (1967) Induction of tumors in mice with three variants of Rous sarcoma virus and studies on the immunobiology of these tumors. Folia Biol (Praha) 13:29

Buja LM, Eigenbrodt ML, Eigenbrodt EH (1993) Apoptosis and necrosis. Basic types and mechanism of cell death. Arch Path Lab Med 117:1208

Burnet MF (1970) Th concept of immunological surveillance. Progr Exp Tumor Res 13:1

Burdette WJ, Haddox CH Jr (1954) Mutation rate following treatment of neurospora with 20-methylcholanthrene and 1,2,5,6-dibenzanthracene in tween-80. Cancer Res 14:163

Bystryn JC, Smalley JR (1977) Identification and solubilization of iodinated cell surface human melanoma-associated antigens. Int J Cancer 20:165

Cabrera T, Lopez-Nevot MA, Gaforio JJ, Ruiz-Cabello F, Garrido F (2003) Analysis of HLA expression in human tumor tissues. Cancer Immunol Immunother 52:1

Camiener GW, Wechter WJ (1972) Immunosuppression – agents, procedures, speculations and prognosis. Progr Drug Res 16:67

Cantor H, Boyse EA (1975) Functional subclasses of T-lymphocytes bearing different Ly antigens. I. The generation of functionally distinct T-cell subclasses is a differentiative process independent of antigen. N Engl J Med 141:1376

Cantor H, Boyse EA (1977) Regulation of cellular and humoral immune responses by T cell subclasses. Cold Spring Harb Symp Quant Biol 41:23

Cerottini JC, Brunner KT (1974) Cell-mediated cytotoxicity, allograft rejection and tumor mimmunity. Adv Immunol 18:67

Chan SH, Day NE, Khor TH, Kunaratnam N, Chia KB (1981) HLA markers in the development and prognosis of NPC. Chinese Cancer Campaign 5:205

Chau GY, Wu CW, Lui WY (2000) Serum interleukin-10 but not interleukin-6 is related to clinical outcome in patients with respectable hepaocellular carcinoma. Ann Surg 231:552

Chiari H (1951) Ueber das feingewebliche Bild der bei Mesanthoinbehandlung zu beobachtenden Lymphknotensch wellung. Wien Klein Wschr 63:77

Chieco-Bianchi L, Sendo F, Aoki T, Barrera OL (1974) Immunologic tolerance to antigens associated with murine leukemia viruses: T-cell unresponsiveness? J Nat Cancer Inst 52:1345

Chism SE, Burton RC, Warner NL (1976) In vitro induction of tumor-specific immunity. II. Activation of cytotoxic lymphocytes to murine oncofetal antigens. J Nat Cancer Inst 57:377

Cohen AM (1976) Host immunity to growing sarcomas: tumor-specific serum inhibition of tumor-specific cellular immunity. Cancer 31:81

Coltman CA, Dixon DO (1982) Second malignancies complicating Hodgkin's disease: a southwest oncology group 10-year followup. Cancer Chemother Rep 66:1023

Coombs MM (1980) Chemical carcinogenesis: a view at the end of the first half century. J Path 130:117

Cooperband SR, Badger AM, Davis RC, Schmid K, Mannick JA (1972) The effect of immunoregulatory globulin (IRA) upon lymphocytes in vitro. J Immunol 109:154

Cooperband SR, Nimberg R, Schmid K, Glasgow AH, Mannick JA (1976) Humoral immunosppressive factors. Transplant Proc 8:225

Costanzi JJ, Gagliano RG, Delaney F, Harris N, Thurman GB, Sakai H, Goldstein AL, Loukas D, Cohen GB, Thomson PD (1977) The effect of thymosin on patients with disseminated malignancies. Cancer 40:14

Cremer NE (1967) Selective immunoglobulin deficiencies in rats infected with Moloney virus. J Immunol 99:71

Cremer NE (1973) Immunosuppression mediated by Moloney leukemia virus infection. In: Ceglowski S, Friedman H (eds) Virus tumorigenesis and immunogenesis. Academic, New York

Cremer NE, Taylor DON, Lenette EH (1971) Quantitation of immunoglobulin- and virus-producing cells in rats infected with Moloney leukemia virus. Immunol 107:689

Cunningham-Rundles C, Cunningham-Rundles S, Iwata T, Incefy G, Garofalo JA, Menendez-Botest C, Lewis V, Towmey JJ, Good RA (1981) Zinc deficiency, depressed thymic hormones, and T lymphocyte dysfunction in patients with hypogammaglobulinemia. Clin Immunol Immunopathol 21:287

Currie GA (1973) The role of circulating antigen as an inhibitor of tumour immunity in man. Br J Cancer 28:153

Currie GA, Alexander P (1974) Spontaneous shedding of TSTA by viable sarcoma cells: its possible role in facilitating metastatic spread. Br J Cancer 19:72

Davey GC, Currie GA, Alexander P (1976) Spontaneous shedding and antibody-induced modulation of histocompatibility antigens on murine lymphomata: correlation with metastatic capacity. Br J Cancer 33:9

Dawe CJ (1969) Phylogeny and ontogeny. Natl Cancer Inst Monogr 31:1

DeBruyns PP (1948) The effect of X-rays on the lymphatic nodule, with reference to the dose and relative sensitivities of different species. Anat Rec 101:373

DeLarco JE, Reynolds R, Carlberg K, Engle C, Todaro GJ (1980) Sarcoma growth factor from mouse sarcoma virus transformed cells. Purification by binding and elution from epidermal growth factor receptor-rich cells. J Biol Chem 255:3685

Demerec M, Wallace B, Witkin EM, Bertaini G (1948) The gene. Carnegie Institution of Washington Yearbook, Washington 48: 154,

Dent PB, Peterson RD, Goor RA (1965) A defect in cellular immunity during the incubation period of passage A leukemia in C3H mice. Proc Soc Exp Biol Med 119:869

Dent PB, Louis JA, McCulloch DB, Dunnett CW, Cerottini JC (1980) Correlation of elevated C1q binding activity and carcinoembryonic antigen levels with clinical features and prognosis in bronchogenic carcinoma. Cancer 45:130

Dixon FJ, Talmage W, Maurer PH (1952) Radiosensitivity and radioresistant phases in the antibody responses. Immunol 68:693

Domzig W, Lohmann-Matthes ML (1979) Antibody-dependent cellular cytotoxicity against tumor cells. II. The promonocyte identified as an effector cell. Euro J Immunol 9:267

Engers HD, Louis J (1979) Dissociation of the humoral and cell-mediated responses to allogens in mice by subl;ethal whole-body irradiation. Scand J Immunol 10:509

Editorial (1981) Immunocompromised homosexuals. Lancet II: 328

Fahey JL (1971) Cancer in the immunosuppressed patient. Ann Intern Med 75:310

Fauci AS, Lane HC (2000) Human immunodeficiency virus (HIV) disease: AIDS and related disorders. In: Harrison's principles of internal medicine. 14th edn. McGraw-Hill, NY

Fayad L, Keating MJ, Reuben JM (2001) Interleukin-6 and interleukin-10 levels in chronic lymphocytic leukemia: correlation with phenotypic characteristics and outcome. Blood 97:256

Felzmann T, Gardner H, Holter W (2002) Dendritic cells as adjuvants in antitumor immune therapy. Onkologie 25:456

Fenyoe EM, Klein E, Klein G, Swiech K (1968) Selection of an immunoresistant Moloney lymphoma subline with decreased concentration of tumor-specific surface antigen. J Natl Cancer Inst 40:69

Fernandez G, Nair M, Onoe K, Tanaka T, Floyd R, Good RA (1979) Impairment of the cell-mediated immunity functions by dietary zinc deficiency in mice. Proc Natl Acad Sci USA 76:457

Fidler IJ (1973) In vitro studies of cellular mediated immunostimulation of tumor growth. J Natl Cancer Inst 50:1307

Fishman MA, Perelson AS (1999) TH1/TH2 differentiation and cross regulation. Bull Math Biol 61:403

Fogel M, Gorelik E, Segal S, Feldman M (1979) Differences in cell surface antigens of tumor metastases and those of the local tumor. J Natl Cancer Inst 62:585

Friedman H, Kateley JR (1975) Lymphocyte surface receptors and leukemia-induced immunosuppression. Am J Clin Pathol 63:735

Friedman H, Thompson G, Halper IP, Knowles DM (1982) 0T cell: a human T-cell chronic lymphocytic leukemia that produces IL-2 in high titer. J Immunol 128:935

Frizzera G (1992) Atypical lymphoproliferative disorders. In: Knowles DM (ed) Neoplastic hematology. Williams & Wilkins, Baltimore, pp 459–495

Fudenberg HH (1971) Genetically determined immune deficiency as the predisposing cause of autoimmunity and the lymphoid neoplasia. AmJ Med 51:295

Gabrielsen AE, Good RA (1967) Chemical suppression of adaptive immunity. Adv Immunol 6:91

Gallin JI, Snyderman R (1999) Inflammation – basic principles and clinical correlates, 3rd edn. Lippincott, Williams & Wilkins, Philadelphia

Gatti RA, Good RA (1970) The immunological deficiency diseases. Med Clin North Am 54:281

Gauwerky CE, Croce CM (1995) Molecular genetics and cytogenetics of hematopoietic malignancies. In: Mendelsophn J, Howley PM, Israel MA, Liotta LA (eds) The molecular basis of cancer. WB Saunders Co, Philadelphia pp 19–37

Geldhof AB, van Ginderachter JA, Lin Y, Noel W, Raes G, de Baetselie P (2002) Antagonistic effect of NK cells on alternatively activated monocytes: a contribution of NK cells to CTL generation. Blood 100:4049

Gelehrter TD, Collins PS (1997) Principles of medical genetics, 2nd edn. Williams & Wilkins, Baltimore

Gershon RK, Birnbaum-Mokyr M, Mitchell MS (1974) Activation of suppressor T cells by tumour cells and specific antibody. Nature 250:594

Gilette RW, Fox A (1977) Changes caused in homing patterns of chromium 51-labeled lymphoid cells by Moloney sarcoma virus infection. J Natl Cancer Inst 58:1621

Gleichmann E, Gleichmann H (1976) Graft-versus-host reaction: a pathogenetic principle for the development of drug allergy, autoimmunity, and malignant lymphoma in nonchimeric individuals. Z Krebsforsch 85:91

Gleichmann E, Gleichmann H (1980) Spectrum of diseases caused by alloreactive T-cells, mode of sensitization to the drug diphenylhydantoin, and possible role of SLE-typical self- antigens in B-cell triggering. In: Krakauer RS, Cathcart MK (eds) Immunoregulation and autoimmunity. Elsevier North Holland, Amsterdam pp 73–83

Gleichmann H, Gleichmann E (1973) Immunosuppression and neoplasia. I. A critical review of experimental carcinogenesis and the immunosurveillance theory. Klin Wschr 51:255

Glicksman AS, Pajak TF, Gottlieb A, Nissen N, Stutzman L, Cooper MR (1982) Second malignant neoplasms in patients successfully treated for Hodgkin's disease: a cancer and leukemia group B study. Cancer Chemother Rep 66:1035

Goldstein G (1975) The isolation of thymopoietin (thymosin). Ann NY Acad Sci 249:177

Goldstein G (ed) (1977) Molecular control of proliferation and differentiation. Academic, NewYork

Golger D, Jones E, Powrie F, Elliott T, Gallimore A (2002) Depletion of CD25 + regulatory cells uncovers immune responses to shared murine tumor rejection antigens. Eur J Immunol 32:3267

Good RA (1954) Agammaglobulinemia: a provocative experiment of nature. Bull Univ Hosp Minn Med Found 26:1

Good RA, Finstad J (1968) The association of lymphoid malignancy and immunologic functions. Zorafonetis CID (ed) Proceedings of the International Conference on Leukemia/ Lymphoma. Lea & Febiger, Philadelphia, pp 175–197

Good RA, Finstad J (1969) Neoplasia and primitive vertebrate phylogeny: echinoderms, prevertebrates and fishes – a review. Natl Cancer Inst Monogr 31:41

Greenberg AH, Greene M (1976) Non-adaptive rejection of small tumour inocula as a model of immune surveillance. Nature 264:356

Greene MI, Pierres A, Dorf ME (1977) The I-J subregion codes for determinants on suppressor factor(s) which limit the contact sensitivity response to picryl chloride. J Exp Med 146:293

Griscelli C, Durandy A, Virelizier JL, Ballet JJ, Daguillard F (1978) Selective defect of precursor T-cells associated with apparently normal B-lymphocytes in severe combined immunodeficiency disease. J Pediat 93:404

Gruenwald HW, Rosner F (1970) Acute leukemia and immunosuppressive drug use. Ann Int Med 139:461

Grundmann E (ed) (1980) Metastatic tumor growth. Cancer Campaign 4:1

Gryboski J (1975) Gastrointestinal problems in the infant (acrodermatitis enteropathica). In: Major problems in clinical pediatrics, Vol XIII. Saunders, Philadelphia, pp 628–30

Hadden JW, Stewart WE (eds) (1981) The lymphokines. Humana, Clifton, NJ,

Hanto DW, Frizzera G, Purtilo DT, Sacramento K, Sullivan JL, Saemundsen AK, Klein G, Simmons RI, Najarian JS (1981) Clinical spectrum of lymphoproliferative disorders in renal transplant recipients and evidence for the role of Epstein-Barr virus. Cancer Res 41:4253

Harada S, Tatsumi E, Lipscomb H, Purtilo DT (1984) Responses to Epstein-Barr virus in immune deficient patients. In: Purtilo DT (ed) Immune deficiency and cancer. Plenum, New York, pp 123–141

Heimer R, Klein G (1976) Circulating immune complexes in sera of patients with Burkitt's lymphoma and nasopharyngeal carcinoma. Int J Cancer 18:310

Heine UI, Krueger GRF, Karpinski A, Munoz E, Krueger MB (1983a) Quantitative light and electron microscopic changes of thymic reticular epithelial cells during Moloney virus-induced lymhpoma development. J Cancer Res Clin Oncol 106:102

Heine UI, Krueger GRF, Munoz E, Karpinski A (1983b) Altered thymic epithelial cells may be decisive for Moloney virus-induced lymphoma development. EMSA meeting. Phoenix, Arizona (book of abstracts) p 41

Hellstroem I, Hellstroem KE (1969) Studies on cellular immunity and its serum-mediated inhibition in Moloney virus-induced mouse sarcomas. Int J Cancer 4:587

Hellstroem I, Hellstroem KE (1974) Cell mediated immune reactions to tumor antigens with particular emphasis on immunity to human neoplasms. Cancer 34:1461

Hellstroem I, Hellstroem KE, Pierce GE (1968) In vitro studies of immune reactions against autochthonous and syngeneic mouse tumors induced by methylcholanthrene and plastioc discs. Int J Cancer 3:467

Hellstroem KE, Hellstroem I (1976) Cell mediated immunity to mouse tumors: some recent findings. Ann NY Acad Sci 276:176

Henderson ES, Krueger GRF (1977) Medications and their toxicity. In: Altmann HW et al. (eds) Handb Allg Path. Springer Berlin, pp 745–791

Himmelweit F (ed) (1957) The collected papers of Paul Ehrlich. Pergamon, London

Hirsch MS (1973) Immunological activation of oncogenic viruses. In: Ceglowski CS, Friedman H (eds) Virus tumorigenesis and immunogenesis. Academic, New York, pp 131–140

Hirsch MS, Black PH, Tracy GS, Leibowitz S, Schwartz RS (1970) Leukemia virus activation in chronic allogeneic disease. Proc Nat Acad Sci USA 67:1914

Holland JM, Mitchell TJ, Gipson LC, Whitaker MS (1978) Survival and cause of death in aging germfree athymic nude and normal inbred C3H/He mice. J Natl Cancer Inst 61:1357

Hollander C, Fornace AJ (1995) Cell cycle checkpoints and growth arrest genes activated by genotoxic stress. In: Vos JMH (ed) DNA repair mechanisms: impact on human dsiseases and cancer. Springer, NY, pp 219–237

Hogg N, Koszinowski K, Mitchinson NA (1977) Principles governing the response to the cell surface antigens. Progr Immunol 3:532

Hoover RT, Fraumeni JF (1973) Risk of cancer in renal transplant recipients. Lancet II: 55

Howley PM (1995) Viral carcinogenesis. In: Mendelsohn J, Howley PM, Israel MA, Liotta LA (eds) The molecular basis of cancer. WB Saunders Co, Philadelphia, pp 38–58

Ioachim HL, Dorsett B, Sabbath M (1972) Loss and recovery of phenotypic expression of Gross leukemia virus. Nature New Biol 237:215

Ioachim HL, Keller S, Sabbath M (1974) Antigenic expression as a determining factor of tumor growth in Gross leukemia virus leukemia. Progr Exp Tumor Res 19:284

Iwata T, Incefy GS, Cunningham Rundles S, Cunningham-Rundles C, Smithwick E, Geller N, O'Reilly R, Good RA (1981) Circulating thymic hormone activity in patients with primary and secondary immune deficxiency diseases. Am J Med 71:385

Jerry LM, Rowden G, Cano PO, Phillips TM, Deutsch GF, Capek A, Hartman P, Lewis MG (1976) Immune complexes in human melanoma, a condsequence of deranged immune regulation. Scand J Immunol 5:845

Jones E, Dahm-Vicker M, Gogher D, Gallimore A (2003) CD25(+) regulatory T cells and tumor immunity. Immunol Lett 85:141

Jose DG, Seshadri R (1974) Circulating immune complexes in human neuroblastoma: direct assay and role in blocking specific cellular immunity. Int J Cancer 13:824

Kahn MF, Ariel J, Bloch-Michel H, Caroit M, Chanoat Y, Renier JC (1971) Le rique du leucose aigue apres traitement des rheumatisme inflammatoire chronique et des connectivities par les cytotoxiques a visee immunosuppresive. Rev Rheumatisme 46:163

Kall MA, Hellstroem I (1975) Specific stimulatory and cytotoxic effects of lymphocytes sensitized in vitro to either alloantigens or tumor antigens. Immunology 114:1083

Kaplan AM, Morahan PS (1976) Macrophage mediated tumor cell cytotoxicity. Ann NY Acad Sci 267:134

Karamaju LS, Levine PH, Sundar SK, Ablashi DV, Faggioni A, Armstrong GR, Bertram G, Krueger GRF (1983) Epstein-Barr virus related lymphocyte stimulation inhibitor: a possible prognostic tool for undifferentiated nasopharyngeal carcinoma. J Natl Cancer Inst 70:643

Kaufman WK, Kaufman DG (1993) Cell cycle control, DNA repair and initiation of carcinogenesis. FASEB J 7:1188

Kaur P, Schulof RS, Miller DR, Steinherz P, Good RA, Gupta S (1982) Immunoregulation by blasts from null cells and T cell leukemias. Cancer 49:43

Kenady DE, Chretien PB, Potvin C, Simon RM (1977) Thymosin reconstuction of T-cell deficits in vitro in cancer patients. Cancer 39:575

Kiessling R, Petranyi G, Klein G, Wigzell H (1976) Non T-cell resistance against a mouse Moloney lymphoma. Int J Cancer 17:275

Kim U, Baumler A, Carruther M (1975) Immunological escape mechanisms in spontaneous metastasizing mammary tumors. Proc Nat Acad Sci USA 72:1012

Klein G (1989) Epstein-Barr virus and its association with human disease: an overview. In: Ablashi DV, Faggioni A, Krueger GRF, Pagano JS, Pearson GR (eds) Epstein-Barr virus and human disease. Humana, Clifton, NJ, pp XVII–XXVII

Klein E, Klein G (1965) Antibody response and leukemic development in mice inoculated neonataly with the Moloney virus. Cancer Res 25:851

Klein E, Klein G (1966) Immunological tolerance of neonatally infected mice to the Moloney leukemia virus. Nature 209:163

Kleindienst P, Brocker T (2003) Endogenous dendritic cells are required for amplification of T cell responses induced by dendritic cell vaccines in vivo. J Immunol 170:2817

Kleist S von (1980) Immune surveillance and tumor defense mechanisms. Path Res Pract 170:289

Korn D, Gelderman A, Cage G, Nathanson D, Strauss AJL (1967) Immune deficiencies, aplastic anemia and abnormalities of lymphoid tissue in thymoma. N Engl J Med 276:1333

Kosmidis PA, Axelrod AR, Palacas C, Stahl M (1978) Angioimmunoblastic lymphadenopathy: a T-cell deficiency. Cancer 42:447

Kraus M (1981) Immunzytologische und immunfunktionelle Untersuchungen des lymphatischen Systems der C57Bl Maus bei kontinuierlicher Applikation von N-Nitrosobutyl-Harnstoff in leukaemogener und nicht-leukaemogener Dosierung. Thesis, Cologne University Medical Faculty, Hundt, Cologne

Kraus M, Krueger GRF (1981) T- and B-cell determination in various lymphoid tissues of mice during N-Nitrosobutylurea (NBU) leukemogenesis. J Cancer Res Clin Oncol 100:149

Kripke ML, Borsos T (1974) Immunosuppression and carcinogenesis. Isr J Med Sci 10:888

Krueger G (1970a) Versuch einer immunologischen Deutung der Lymphomentstehung. Dtsch Med J 21:28

Krueger GRF (1970b) Zur Pathogenese von Tumoren des lymphatischen Gewebes bei Transplantatempfaengern. Verh Dtsch Ges Path 54:175

Krueger GRF (1970c) Effect of dilantin in mice. I. Changes in the lymphoreticular tissue after acute exposure. Virchows Archiv A Path Anat 349:297

Krueger GRF (1971) Impaired immunologic reactivity as pacemaker for lymphoma development. Verh Dtsch Ges Path 55:200

Krueger GRF (1972a) Chronic immunosuppression and lymphomagenesis in man and mice. Natl Cancer Inst Monogr 35:183

Krueger GRF (1972b) Morphology of chemical immunosuppression. Adv Pharmacol Chemother 10:1

Krueger GRF (1972c) Host response during the latent period and the growth of an immunologically induced mouse leukemia. Beitr Path 142:132

Krueger GRF (1973) Immunosuppression and experimental lymphoma. J Natl Cancer Inst 50:1412

Krueger GRF (1974) Lymphoreticular neoplasia in immunosuppression. Facts and fancies. Beitr Path 151:221

Krueger GRF (1975) The significance of immunosuppression and antigenic stimulation in the development of malignant lumphomas. Recent Results Cancer Res 52:88

Krueger GRF (1979) Experimentelle Modelle immunologisch induzierter Tumoren. Verh Dtsch Ges Inn Med 85:1460

Krueger GRF (1985) Klinische immunpathologie. W.Kohlhammer Verlag, Stuttgart

Krueger GRF (1989) Abnormal variation of the immune system as related to cancer. In: Kaiser HE (ed) Cancer growth and progression. Kluwer Academic, Dordrecht, NL, pp 139–161

Krueger GRF (1993) Pathology of lymphoproliferative disorders in HIV infection. In: Schrappe M, Mauff G (eds) AIDS-SIDA. A comparison between Europe and Africa. Ediones Roche, Basel, pp 239–255

Krueger GRF, Bedoya V (1978) Hydantoin-induced lymphadenopathies and lymphomas: experimental studies in mice. Recent Results Cancer Res 64:265

Krueger GRF, Ferrer Argote V (1994) A unifying concept of viral immunopathogenesis of proliferative and aproliferative diseases (working hypothesis). In Vivo 8:493

Krueger GRF, Heine UI (1972) Morphogenesis of two immunologically induced mouse lymphomas. Cancer Res 32:573

Krueger GRF, Konorza G (1977) Angioimmunoblastic lymphadenopathy in persistent virus infection? Lancet II: 1135

Krueger GRF, Bedoya V, Grimley PM (1971a) Lymphoreticular tissue lesions in Steinbrinck-Chediak-Higashi syndrome. Virchows Archiv Abt A Path Anat 353:273

Krueger GRF, Malmgren RA, Berard CW (1971b) Malignant lymphomas and plasmacytosis in mice under chronic immunosuppression and persistent antigenic stimulation. Transplant 11:138

Krueger GRF, Fischer R, Flesch HG (1978) Sequential changes in T- and B-cells, virus antigen expression and primary histologic tumor diagnosis in virus-induced lymphomagenesis in mice. Z Krebsforsch 92:41

Krueger GRF, Bergholz M, Bartsch H-H, Fischer R, Schauer A (1979) Rubella virus antigen in lymphocytes of patients with angioimmunoblastic lymphadenopathy (AIL). J Cancer Res Clin Oncol 95:87

Krueger GRF, Karpinski A, Haas W, Lennert K (1981) Angioimmunoblastic lymphadenopathy in persistent virus infection and hyperimmunization syndrome: clinical and experimental data. XI Triennial World Congress Path, Jerusalem (book of abstracts) p 160

Krueger GRF, Karpinski A, Heine UI, Koch B (1983) Differentiation block of prethymic lymphocytes during Moloneyvirus induced lympohoma development associated with thymic epithelial defect. J Cancer Res Clin Oncol 106:102

Krueger GRF, Ablashi DV, Salahuddin SZ, Josephs SF (1988) Diagnosis and differential diagnosis of progressive lymphoproliferation and malignant lymphoma in persistent active herpesvirus infection. J Virol Methods 21:255

Krueger GRF, Manak M, Bourgeois N, Ablashi DV, Salahuddin SZ, Josephs SF, Buchbinder A, Gallo RC, Berthold F, Tesch H (1989) Persistent active herpesvirus infection associated with atypical polyclonal lymphoproliferation (APL) and malignant lymphoma. Anticancer Res 9:1457

Krueger GRF, Ablashi DV, Josephs SF, Balachandran N (1992) HHV-6 in atypical polyclonal lymphoproliferation (APL) and malignant lymphoma (Chapter 15). In: Abalshi DV, Krueger GRF, Salahuddin SZ (eds) Human herpesvirus-6. Elsevier Science, Amsterdam, pp 185–207

Krueger GRF, Nguyen AN, Uthman M, Brandt ME, Buja LM (2001) Dysregulative lymphoma theory revisited: wehat can we learn from cytokines, CD classes and genes? Anticancer Res 21:3653

Krueger GRF, Brandt ME, Wang G, Berthold F, Buja LM (2002a) A computational analysis of Canale-Smith syndrome: chronic lymphadenopathy simulating malignant lymphoma. Anticancer Res 22:2365

Krueger GRF, Marshall GR, Junker U, Schroeder HJ, Buja LM, Wang G (2002b) Growth factors, cytokines, chemokines and neuropeptides in the modeling of T-cells. In Vivo 16:365

Krueger GRF, Marshall GR, Junker U, Schroeder HJ, Buja LM (2003a) Growth factors, cytokines, chemokines and neuropeptides in the modeling of T-cell. Part II: Data tables of normal values in man. In Vivo 17:105

Krueger GRF, Brandt ME, Wang G, Buja LM (2003b) TCM-1: A nonlinear dynamical computational model to simulate cellular changes in the T-cell system: comceptional design and validation. Anticancer Res 23: 123, Krueger GRF, Brandt ME, Wang G, Buja LM (2004) Computational simulation of chronic persistent virus infection: Factors determining differences in clinical outcome of HHV-6, HIV-1 and HTLV-1 infections including aplastic, hyperplastic and neoplastic responses. Anticancer Res 24:187

Laguens G, Coconato S, Laguens R, Portiansky E, di Girolamo V (2002) Human regional lymph nodes draining cancer exhibit a profound dendritic cell depletion as comparing to those from patients without malignancies. Immunol Lett 84:159

Lai PK, Mackay-Scollay M, Fimmel PJ, Alpers MP, Keast D (1974) Cell-mediated immunity to Epstein-Barr virus and a blocking factor in patients with infectious mononucleosis. Nature 252:608

Lakshmi PM, Kolbe NP, Johri GN (1982) Ancylostoma caninum: transfer of lymphoid cells from thymus and adoptic immunity in mice. Int J Parasitol 12:227

Latner AL, Sherbet GV (1979) Aprotinin induces surface changes in malignant cells in culture: a possible model of antitumor action. Exp Cell Biol 47:392

Law IP, Hollinshead AC, Whang-Peng I, Dean JH, Oldham RK, Herberman RB, Rhode MC (1977) Familial occurrence of colon and uterine carcinoma and of lymphoproliferative malignancies. II. Chromosomal and immunologic abnormalities. Cancer 39:1229

LeBeau MM, Larson RA (2000) Cytogenetics and neoplasia. In: Hoffman R, Benz EJ, Schattil SJ, Furie B, Cohen HJ, Silberstein LE, McGlave P (eds) Hematoplogy, basic principles and practice, 3rd edn. Churchill Livingstone, NY, pp 848–870

Leventhal BG, Talal N (1970) Response of NZB and NZB-NZW spleen cells to mitogenic agents. J Immunol 104:918

Lim MS, Straus SE, Dale JK, Fleisher TA, Stetler-Sevenson M, Strober W, Sneller MC, Puck JM, Lenardo MJ, Kojo S, Elenitoba-Johnson J, Lin AJ, Raffeld M, Jaffe ES (1998) Pathological findings in human autoimmujne lymphoproliferative syndrome. Am J Pathol 153:1541

List AF, Greco FA, Vogler LB (1987) Lymphoproliferative diseases in immunocompromised hosts: the role of Epstein-Barr virus. J Clin Oncol 5:1673

Lombeck I, Schnippering HG, Kasparek K, Ritzl F, Kaestner H, Feinendegen LE, Bremer HJ (1975) Akrodermatitis enteropathica – eine Zinkstoffwechselstoerung mit Zinkmalabsorption. Z Kinderheilk 120:181

Look AT, Naegele RF, Callihan T, Herrod HG, Henle W (1981) Fatal Epstein-Barr virus infection in the child with acute lymphoblastic leukemia in remission. Cancer Res 41:4280

Louie S, Schwartz RS (1978) Immunodeficiency and the pathogenesis of lymphoma and leukemia. Semin Hematol 15:117

Low TLK, Thurman GB, Chincarini C, McClure JE, Marshall GD, Ha SK, Goldstein AL (1979) Current status of thymosin research: evidence for the existence of a family of thymic factors that control T-cell maturation. Ann NY Acad Sci 332:33

Malmgren RA, Bennison BE, McKinley TW (1952) Reduced antibody titers in mice treated with carcinogenic and cancer chemotherapeutic agents. Proc Soc Exp Biol Med 79:484

Mareel M, Leroy A (2003) Clinical, cellular and molecular aspects of cancer invasion. Physiol Rev 83:337

Marshall E (2003) Gene therapy a suspect in leukemia-like disease. Science 298,34, 2002 and Second child in French trial is found to0 have leukemia. Science 299:320

Marshall GD, Low TLK, Thurman GB, Hu SK, Rossio JL, Trivers G, Goldstein AL (1978) Overview of thymosin activity. Cancer Treat Rep 62:1731

Mehr R, Globerson A, Perelson AS (1995a) Modeling positive and negative selection and differentiation processes in the thymus. J Theor Biol 175:103

Mehr R, Fridkis-Harell M, Abel L, Segel L, Globerson A (1995b) Lymphocyte development in irradiated thymuses. Dynamics of colonization by progenitor cells and regeneratyion of resident cells. J Theor Biol 177:181

Mehr R, Perelson AS, Fridkis-Harell M, Globerson A (1996) Feedback regulation of T cell development in the thymus. J Theor Biol 181:157

Melief CJM, Schwartz RS (1975) Immunocompetence and malignancy. In: Becker FF (ed) Cancer Vol. I. Plenum, New York, pp 121–159

Melnick JF, Adam E, Rawls WE (1974) The causative role of herpes virus type 2 in cervical cancer. Cancer 34:1375

Mertens H, Krueger GRF (1976) Percent distribution of T- and B-lymphoid cells in spleen and lymph nodes of Moloney virus infected mice. Z Krebsforsch 85:169

Metcalf D (1971) The nature of leukemia. Neoplasm or disorder of hemopoietic regulation? Med J Australia 2:739

Miller AM, Walma EP, Klapwijk DE, van Bekkum DW (1982) The effect of cyclosporin-A on host-versus-graft disease in canine bone marrow transplantation. Exp Hematol 10:(Supl 11):36

Mills KH, Cawley JC (1982) Suppressor T-cells in chronic lymphocytic leukaemia: relationship to clinical stage. Leukemia Res 6:653

Mills KH, Cawley JC (1983) Abnormal monoclonal antibody-defined helper/suppressor T-cell subpopulations in multiple myeloma: relationship to treatment and clinical stage. Br J Haematol 53:271

Moretta L, Mingari MC, Webb MC, Pearl EA, Lydyard PM, Crossi CE, Lawton AR, Cooper MD (1977) Imbalance in T-cell subpopulations associated with immunodeficiency and autoimmune syndrome. Eur J Immunol 7:696

Moriya N, Miyawaki T, Ueno Y, Koizumi S, Taniguchi N (1981) Humoral helper activity of B-cell differentiation released from non-Hodgkin's lymphoma cells having both SRBC and complement receptors in the pokeweed mitogen system. Blood 57:1057

Moss DJ, Scott W, Pope JH (1977) An immunologic basis for inhibition of transformation of human lymphocytes by EB virus. Nature 268:735

Muller CP (1979) Wirkung der Zelldichte auf die Membranfluiditaet und deren Einfluss auf die Modulation spezifischer Membranrezeptoren. Thesis, University of Cologne, Hansen, Cologne

Nakano K, Cinader B (1980) Accelerated age-dependent decline in the T-suppressor cell capacity of SJL-mice. Eur J Immunol 10:309

Naor D (1980) Unresponsiveness to modified self antigen – a censorship mechanism controlling autoimmunity? Immunology 50:187

Nathanson N (2002) Viral pathogenesis and immunogenesis. Lippincott, Williams & Wilkins, Philadelphia, pp 130–157

Nelson K, Pollack S, Hellstroem KE (1975) Specific anti-tumor response of cultured spleen cells. III. Further characterization of cells which synthesize factor with blocking and antiserum-dependent cellular cytotoxic (ADC) activities. Int J Cancer 16:539

Nichols PW, Koss M, Levine AM, Lukes RJ (1982) Lymphomatoid granulomatosis: a T-cell disorder? Am J Med 72:467

Nicolson GL (1976) Trans-membrane control of receptors on normal and tumor cells. II. Surface changes associated with transformation and malignancy. Biochim Biophys Acta 458:1

Notkins AL, Mergenhagen SE, Howard RJ (1970) Effect of virus infections on the function of the immune system. Ann Rev Microbiol 24:525

O'Conor GT (1970) Persistent immunologic stimulation as a factor in oncogenesis, with special reference to Burkitt's tumor. Am J Med 48:279

Oppenheim JJ, Nordhoff H, Greenhill A, Mathieson BJ, Smith KA, Gills S (1980) Properties of human monocyte-derived lymphocyte-activating factor (LAF) and lymphocyte-derived mitogenic factor (LMF). In: de Weck A (ed) Biochemical characterization of lymphokines. Academic, New York, pp 399

Orsini F, Eppolito C, Ehrke MJ, Mihich E (1980) Inhibition by selected anticancer agents of the development of primary cell-mediated immunity against allogeneic tumor cells in culture. Cancer Treat Rep 64:211

Owen-Schaub L, Chan H, Cusack JC, Roth J, Hill LL (2000) Fas and Fas ligand interactions in malignant disease. Int J Oncol 17:5

Page AR, Hansen AE, Good RA (1963) Occurrence of leukemia and lymphoma in patients with agammaglobulinemia. Blood 21:197

Pardoll D (2003) Does the immune system see tumors as foreign or self? Annu Rev Immunol 21:807

Parker CW, Vavra JD (1969) Immunosuppression. Progr Hematol 6:1

Paul W (1998) Fundamental immunology, 4th edn. Lippincott Raven, Philadelphia

Peng L, Kjaergaard J, Plautz GE, Awad M, Drazba JA, Shu S, Cohen P (2002) Tumor-induced L-selectinhigh suppressor T cells mediate potent effector T cell blockade and cause failure of otherwise curative adoptive immunotherapy. J Immunol 169:4811

Penn I (1974) Occurrence of cancer in immune deficiencies. Cancer 34:858

Penn I (1978) Tumors arising in organ transplant recipients. Cancer Res 28:31

Penn I, Starzl TE (1972) Malignant tumors arising de novo in immunosuppressed organ transplant recipients. Transplant 14:407

Peters AM, Kohfink B, Martin H, Griesinger F, Wormann B, Gahr M, Roesler J (1999) Defective apoptosis due to a point mutation in the death domain of CD95 associated with autoikmmune lymphoproliferative syndrome, T-cell lymphoma and Hodgkin's disease. Exp Hematol 27:868

Peterson RD, Kelly WD, Good RA (1964) Ataxia teleangiectasia. Its association with a defective thymus, immunological deficiency disease and malignancy. Lancet I:1189

Pizon I (1955) Risques et dangers des radiations: radiations electromagnetiques et corpuscules de haute energie. Presse Med 63:1158

Plescia OJ, Grinwich K, Plescia AM (1976) Subversive activity of syngeneic tumor cells as an escape mechanism from immune surveillance and the role of prostaglandins. Ann NY Acad Sci 276:455

Prehn RT (1963) Function of depressed immunologic reactivity during carcinogenesis. J Natl Cancer Inst 31:791

Prehn RT (1976) Immunostimulation of highly immunogenic target tumor cells by lymphoid cells in vitro. J Natl Cancer Inst 56:833

Prehn RT, Lappe MA (1971) An immunostimulation theory of tumor development. Transplant Rev 7:26

Purtilo DT (1980) Epstein-Barr virus-induced oncogenesis in immune-deficient individuals. Lancet I: 300

Purtilo DT, Sakamoto K, Saemundsen AK, Sullivan JL, Synnerholm A-A, Anvret M, Pritchard J, Rich K, Sloper C, Sieff C, Pincott J, Pachman L, Knight J, Sandstedt B, Klein G (1981) Documentation of Epstein-Barr virus infection in immunodeficient patients with life-threatening lymphoproliferative diseases by clinical, virological and immunopathological studies. Cancer Res 41:4226

Purtilo DT, Sakamoto K, Barnabei V, Seeley J, Behctold T, Rogers G, Yetz J, Harada S (1982) Epstein-Barr virus-induced diseases in boys with X-linked lymphoproliferative syndrome (XLP): update on studies of the registry. Am J Med 73:49

Purtilo DT, Tatsumi E, Manolov G, Manolova Y, Harada S, Lipscomb H, Krueger GRF (1985) Epstein-Barr virus as an etiological agent in the pathogenesis of lymphoproliferative and aproliferative diseases in immune deficient patients. Rev Exp Path 27:113

Rapp F, Reed CL (1977) The viral etiology of cancer – a realistic approach. Cancer 40:419

Redlinger Jr., RE, Shimizu T, Remy T, Alber S, Watkins SC, Barksdale Jr., I (2003) Cellular mechanisms of interleukin-12 mediated neuroblastoma regression. J Pediat Surg 38:199

Reinherz EL, Rosen FS (1981) New concepts of immunodeficiency. 71: 511,

Reinherz EL, Schlossman SF (1980) The differentiation and function of human T-lymphocytes. Cell 19:821

Reinherz EL, Kung PC, Goldstein G, Schlossman SF (1979a)Further characterization of the human inducer T-cell subset defined by monoclonal antibody. J Immunol 123:2894

Reinherz EL, Parkman R, Rappaort J, Rosen FS, Schlossman SF (1979b)Aberrations of suppressor T-cells in human graft-versus-host disease. N Engl J Med 300:1061

Reinherz EL, Cooper MD, Schlossman SF (1981) Abnormalities of T-cell maturation and regulation in human beings with immunodeficiency disorders. J Clin Invest 68:699

Robins RA, Rees RC, Brooks GG, Baldwin RW (1979) Spontaneous development of cytotoxic activity in cultured lymph node cells from tumour-bearing rats. Br J Cancer 39:659

Rosenthal CJ, Noguera CA, Coppola A, Kapelner SN (1982) Pseudolymphoma with mycosis fungoides manifestations: hyperresponsiveness to diphenylhydantoin and lymphocyte dysregulation. Cancer 49:2305

Rosenstreich DL, Mitzel SB (1978) The participation of macrophage cell lines in the activation of T-lymphocytes by mitogens. Immunol Rev 40:102

Rowland GF, Hurd CM (1970) Target lymphoid cell population of carcinogen induced immunodepression in mice. Nature 227:167

Rubin BA (1971) Alteration of the homograft response as a determinant of carcinogenicity. Progr Exp Tumor Res 14:138

Russell SW, Gillespie GY, Pace JL (1980) Comparison of responses to activating agents by mouse peritoneal macrophages and cells of the macrophage line RAW 264. J Reticuloendothel Soc 27:607

Rygaard J, Poulsen CO (1976) The nude mouse vs. the hypothesis of immunological surveillance. Transplant Rev 28:43

Sakamoto K, Seeley J, Lindsten T, Sexten J, Yetz J, Bellow M, Purtilo DT (1982) Abnormal anti-Epstein-Barr virus antibodies in carriers of the X-linked lymphoproliferative syndrome and in females at risk. J Immunol 128:904

Saltzstein SL, Ackerman LV (1959) Lymphadenopathy induced by anticonvulsant drugs and mimicking clinically and pathologically malignant lymphomas. Cancer 12:164

Saxon A, Stevens RH, Golde DW (1979) Helper and suppressor T-lymphocyte leukemia in ataxia teleangiectasia. N Engl J Med 300:700

Schirrmachert V, Boslet K, Shantz G, Clauer K, Huebsch D (1975) Tumor metastases and cell mediated immunity in a model system in DBA/2 mice. IV. Antigenic differences between a metastasizing variant and the parental tumor line revealed by cytotoxic T-lymphocytes. Int J Cancer 23:145

Schwartz RS (1972) Immunoregulation, oncogenic viruses and malignant lymphomas. Lancet I: 1266

Schwarz RS, Andre-Schwarz J (1968) Malignant lymphoproliferative diseases. Interaction between immunological abnormalities and oncogenic viruses. Ann Rev Med 19:269

Schwartz RS, Beldotti L (1965) Malignant lymphomas following allogeneic disease: transition from immunological to a neoplastic disorder. Science 149:1511

Seeley JK, Sakamoto K, Ip S, Hansen P, Purtilo DT (1981) Abnormal lymphocyte subsets of X-linked lymphoproliferative syndrome. J Immunol 127:2618

Shankaran V, Ikeda H, Bruce AT, White JM, Swanson PE (2001) IFN gamma and lymphocytes prevent primary tumour development and shape tumour immunogenicity. Nature 410:1107

Shearer WT, Parker CW (1975) Humoral immunostimulation. V. Selection of variant cell lines. J Exp Med 142:1133

Sherbet GV (1982) The biology of tumor malignancy. Academic, New York

Shinitzky M (1976) Membrane changes in malignant cells – modulation of receptors and antigens by lipids. Bull Schweiz Aced Med Wiss 12:203

Siegal FP, Siegal M, Good RA (1976) Suppression of B-cell differentiation by leukocytes from hypogammaglobulinemic patients. J Clin Invest 58:109

Siegal FP, Siegal M, Good RA (1978) Role of helper, suppressor and B-cell defects in the pathogenesis of the hypogammaglobulinemias. N Engl J Med 299:172

Sjoegren HO (1973) Blocking and unblocking of cell-mediated tumor immunity. Meth Cancer Res 10:19

Sjoegren HO, Bansal SC (1971) Antigens in virally induced tumors. Progr Immunol 1:921

Sjoegren HO, Hellstroem I, Bansal SC, Hellstroem KE (1971) Suggestive evidence that the blocking antibodies of tumor-bearing individuals may be antigen-antibody complexes. Proc Nat Acad Sci USA 63:1372

Spreafico F, Anaclerio A (1977) Immunosuppressive agents. In: Hadden JW (ed) Comprehensive immunology vol 3 Immunopharmacology. Plennum, New York, pp 245–278

Stjernswaerd J (1965) Depressive effect of 3-methylcholanthrene. Antibody formation at the cellular level and reaction against weak antigenic homografts. J Natl Cancer Inst 35:885

Stjernswaerd J (1969) Immunosuppression by carcinogens. Antibiot Chemother 15:213

Stutman O (1969) Carcinogen-induced immune depression: absence in mice resistant to chemical oncogenesis. Science 166:620

Sundar SK, Ablashi DV, Karamaju LS, Levine PH, Faggioni A, Armstrong GR, Pearson GR, Krueger GRF, Hewetson JF, Bertram G, Sesterhenn K, Menezes J (1982) Sera from patients with undifferentiated nasopharyngeal carcinoma contain a factor which abrogates specific Epstein-Barr virus antigen-induced lymphocyte response. Int J Cancer 29:407

Szakal AK, Hanna MG (1972) Immune suppression and carcinogenesis in hamsters during topical application of 7,12-dimethylbenz-(-anthracene. J Natl Cancer Inst 35:173

Tada T, Taniguchi M, David CS (1976a) Properties of the antigen-specific suppressor T-cell factor in the regulation of antibody response in the mouse. IV. Special subregion assignment of the gene(s) that code(s) for the suppressive T-cell factoring the H-2 histocompatibility complex. J Exp Med 144:213

Tada T, Taniguchi M, Takemori T (1976b) Properties of primed suppressor T-cells and their products. Transplant Rev 26:106

Tamerius J, Hellstroem I, Hellstroem KE (1975) Evidence that blocking factors in the sera of multiparous mice are associated with immunoglobulins. Int J Cancer 16:456

Tamerius J, Nepom J, Hellstroem I, Hellstroem KE (1976) Tumor-associated blocking factors: isolation from sera of tumor-bearing mice. J Immunol 116:724

Ten Bensel RW, Stadlan GM, Krivit W (1966) The development of malignancy in the course of Aldrich syndrome. J Pediatr 68:761

Thomas JA, Janossy G, Graham-Braun JAC, Kung PC, Goldstein G (1982) The relationship between T-lymphocyte subsets and Ia-like antigen positive nonlymphoid cells in early stages of cutaneous T cell lymphomas. J Natl Cancer Inst 78:169

Ting CC, Rogers MJ (1977) Inhibition by sera and soluble antigens of T-cell-mediated cytotoxicity against leukemia-associated antigens. Nature 266:727

Treves AJ, Carnaud C, Trainin N, Feldman M, Cohen IR (1974) Enhancing T lymphocytes from tumor-bearing mice suppress host resistance to a syngeneic tumor. Eur J Immunol 4:722

Turk JL, Pareker D, Poulter LW (1972) Functional aspects of the selective depletion of lymphoid tissue by cyclophosphamide. Immunol 23:493

Uchiyama T, Yodoi J, Sagawa K (1977) Adult T-cell leukemia: cclinical and hematological features of 16 cases. Blood 50:481

Umiel T, Training N (1974) Immunological enhancement of tumor growth by syngeneic thymus-derived lymphocytes. Transplant 18:244

Venter CJ, Adams MD, Myers EW (& 267 additional co-authors)(2001) The sequence of the human genome. Science 291: 1304

Wainwright WH, Veltri RW, Sprinkle PM (1979) Abrogation of cell-mediated immunity by a serum blocking factor isolated from patients with infectious mononucleosis. J Infect Dis 140:22

Walder BK, Robertson MK, Jeremy D (1971) Skin cancer and immunosuppression. Lancet II: 1282

Waldmann TA, Durm M, Broder S, Blaese RM (1974) Role of suppressor T-cells in pathogenesis of common variable hypogammaglobulinemia. Lancet II: 609

Waldmann TA, Broder S, Durm M (1975) Suppresor T-cells in the pathogenesis of hypogammaglobulinemia associated with a thymoma. Trans Assoc Am Physicians 88:120

Waldmann TA, Blaese RM, Broder S, Krakauer RS (1978) Disorders of suppressor immunoregulatory cells in the pathogenesis of immunodeficiency and autoimmunity. Ann Int Med 88:226

Waldmann TA, Strober W, Blaese M (1980) T- and B-cell immunodeficiency diseases. In: Parker CW (ed) Clinical immunology, Vol I. Saunders, Philadelphia, pp 314–375

Waller EK, Ernstoff MS (2003) Modulation of antitumor immune responses by hematopoietic cytokines. Cancer 97:1797

Wang G, Krueger GRF, Buja LM (2003) A simplified and comprehensive computational model to study the behavior of T cell populations in the thymus during normal maturation and in infection with mouse moloney leukemiavirus. In Vivo 17:225

Wang G, Krueger GRF, Buja LM (2004) A continuous model studying T cell differentiation and lymphomagenesis and its distinction with discrete models. Anticancer Res 24:1813, (in print)

Warnatz H (1979) Cell mediated immune reactions in patients with colon carcinoma. In: Flad HD (ed) Immunodiagnosis and immunotherapy of malignant tumors. Springer, Berlin, pp 122–128

Wedderburn N, Salaman MH (1968) The immunodepressive effect of Friend virus. II. Reduction of haemolysin-producing cells in primary and secondary responses. Immunology 15:439

Weinstein IB, Carothers AM, Santella M, Perrera FP (1995) Molecular mechanisms of mutagenesis and multistage carcinogenesis. In: Mendelsohn J, Howley PM, Israel MA, Liotta LA (eds) The molecular basis of cancer. WB Saunders Co, Philadelphia, pp 59–85

West WH, Cannon GB, Kay D, Bonnard BD, Herberman RB (1977) Natural cytotoxic reactivity of human lymphocytes against a myeloid cell line: characterization of effector cells. J Natl Cancer Inst 58:155

Williams RM (1977) Experimental models with possible implications for the role of HLA in malignancy. In: HLA and malignancy. Alan Riss, NY, pp 21–28

Yokoe T, Iino Y, Morishita Y (2000) Trends of IL-6 and IL-8 levels in patients with recurrent breast cancer: preliminary report. Breast Cancer 7:187

Yoo BK, Cassin M, Lessin SR, Rook AH (2001) Complete molecular remission during biologic response modifier therapy for Sezary syndrome is associated with enhanced helper T type 1 cytokine production and natural killer cell activity. J Am Acad Dermatol 45:208

Yu DTY, Paulus HE, Peter JB (1975) Human lymphocyte subpopulations. Effect of immunosuppressive therapies. In: Williams RC (ed) Lymphocytes and their iunteractions. Raven, New York, pp 157–168

Zembala M, Mytar B, Popiela T, Asherson GL (1977) Depressed in vitro peripheral blood lymphocyte response to mitogens in cancer patients. The role of suppressor cells

Zollinger HU (1960) Radio-Histologie und Radio-Histopathologie. In: Buechner F (ed) Handb Allg Path. Springer, Berlin, pp 127–287

Zur Hausen H (1975) Oncogenic herpes viruses. Biochim Biophys Acta 417:25

H.E. Kaiser and A. Nasir (eds.), Selected Aspects of Cancer Progression:
Metastasis, Apoptosis and Immune Response, 223–246.
© *Springer Science + Business Media B.V.* 2008

CHAPTER THIRTEEN

A biodynamical model of human T-cell development and pathology: design, testing and validation

Michael E. Brandt[*]**, Gerhard R. F. Krueger, and Guanyu Wang**

Abstract: We describe a coupled ordinary differential equation model of human T-cell proliferative disorders based upon documented changes in various pools such as the bone marrow, thymic compartments and peripheral blood. The conceptual design of the model is based upon previously collected experimental data, its testing and validation by comparing with normal human cell pool data at various ages as well as their changes in response to HTLV-1, HHV-6 and HIV-1 viral infections. These viruses were chosen because they all target the same CD4 lymphocyte, yet produce different response patterns such as hyperplasia, aplasia and neoplasia. They were also selected because respective cell pool data were available for comparison with detailed human studies. The ultimate task of this modeling effort is to simulate the development of T-cell lymphomas and other immunoproliferative or aproliferative (i.e. aplastic) abnormalities reported in the literature.

Keywords: Immune system, T-lymphocytes, Proliferative diseases, Biocomputational modeling

Introduction

Normal and atypical cell proliferation in the T cell system and its differentiation entail a series of membranal events and interactions of the cell membrane

with cytokines and related substances that serve to communicate environmental influences into the cell. Such environmental signals serve to activate respective genetic codes that modulate cellular reactions. Disturbances in the control of this delicate network of cell stimulation and inhibition can easily imbalance the system resulting in pathologic over- or under-representation of certain control parameters; blocking of cell differentiation or of clonal expansion may result in clinical immune deficiency or in progressive cell proliferation with ultimate tumor development, as demonstrated in early experiments (Krueger et al. 1971, 1979, 1983, 1987a, Krueger 1972; Kraus and Krueger 1981; Heine et al. 1983a; Daefler and Krueger 1989a, b).

This project focuses on the concept of immunologic dysregulation preceding the development of atypical lymphoproliferative diseases and malignant lymphomas. It is based on extensive experimental studies and further confirmed by respective observations in human patients (Krueger et al. 1971, 1987b, 1988, 2001c; Krueger 1972, 1993; Purtilo et al. 1985; Krueger and Ferrer Argote 1994; Schonnebeck et al. 1991; Feaux de Lacroix et al. 1981; Haas et al. 1982; Heine et al. 1983b; Krueger 1989a; Krueger). According to this concept, diseases such as malignant lymphomas, aplasias or autoimmune disorders result from a disturbed balance of factors regulating cell proliferation, cell differentiation/function, and

University of Texas-Houston Health Science Center – Medical School and [*]School of Health Information Sciences cell inhibition or physiological cell death (apoptosis). Predominance of cell proliferation factors will ultimately cause lymphomas, predominance of apoptosis factors will cause aplasia, and unbalanced differentiation factors may contribute to autoimmune disorders.

On the other hand, a change in compartmental immune cell counts (or in the cells themselves) may perturb the balance between growth, differentiation and apoptosis factors; the network may absorb this perturbation and regain its original balance, or fail to do so, thereby exacerbating the alteration in cells and possibly leading to further imbalance. We may be able someday to better understand certain immune disorders if we can learn why the network fails to absorb the perturbations in some cases.

This "dysregulation hypothesis" functions independently of

1. The initial cause of the disturbance,
2. Transformation and atypia of certain cells,
3. Toxic loss of certain regulatory cells.

The designed model should accommodate, respectively, all such potential etiological factors.

We have developed a computer-implemented biomathematical model that is based upon the above-described concept of dysregulative lymphoma development. We found that simulated data from this model closely resembles data observed in previous human experimental studies of lymphomagenesis. In these studies, quantitative and functional changes were documented in various T cell pools such as the stem cell pool, the thymic compartments and peripheral lymphoid tissues or blood (Krueger 1989b).

The model we describe here is a significant update/revision of our previous model (Brandt et al. 2002) and has taken into account the exponential increase of the volume of data in recent years concerning cellular regulation and proliferation of the immune system. The results of computer simulations by this new comprehensive model of T cell proliferative changes will serve to (1) better target the kinds of clinical diagnostic investigations to be performed, and (2) decrease the number of investigations that are currently necessary to perform, thereby significantly reducing the amount of time required for obtaining additional pathogenetic data

of lymphomagenesis. The model is not designed to supplement or to replace current concepts of T cell development as such.

The specific aims of the study are to

1. More fully specify the network regulation of cell proliferation, differentiation, and apoptosis and to design a biologically realistic computational model of it.
2. Understand the mechanisms of uncontrolled cell proliferation and thus of tumor development, immune deficiency and aplasia.
3. Provide additional bases for diagnostic and therapeutic planning of lymphoproliferative and aproliferative disorders.

Conceptual basis of the immune system model

The present computational model deals with quantitative changes in defined cell pools of the T cell system including bone marrow stem cells, thymic cortical and medullary compartments and mature peripheral T lymphocytes. (Dunon et al. 1997; Liu and Auerbach 1991; Eren et al. 1987; Correla-Neves et al. 2001; Sen 2001; Chen et al. 1996; Zajac et al. 1998, 1999; Mollet et al. 2000; Spiegel et al. 2000; Appay et al. 2000; Champagne et al. 2001; Janossy et al. 1989; Leclercq and Plum 1996; Sanchez et al. 1998; Pawelec et al. 1998, 1999; Greenberger 1991; Suzuki et al. 2000; Kim et al. 2001; O'Sullivan et al. 2001; Montecino-Rogriguez et al. 2001). Within these pools, the processes of cell proliferation, differentiation, movement and death are under critical control of the respective local microenvironment including such components as reticular epithelial cells, macrophages and fibroblasts (Stutman 1978; Atkins et al. 1987; Lobach and Haynes 1987; Dappen et al. 1982; Surh and Sprent 1999; Crouse et al. 1980; Rothenberg and Lugo 1985; Allison 1987; Miescher et al. 1988; Kronenberg et al. 1986; Dardenne et al. 1977; Goldstein 1977; Crouse et al. 1985; Thiele et al. 1995; Wognum et al. 1996; de la Hera et al. 1989; Sarun et al. 1998; Bodey et al. 1999, 2000).

Such microenvironmental cells which influence the proliferation and maturation of T lymphocytes by various secretory products are unequally distributed in the organ stroma separating it into several poorly

defined compartments (Bodey et al. 1999; Wognum et al. 1996; Marrack et al. 1988; Peled et al. 1999). Therefore, we designed the basic computational model as presented in such a manner as to be capable of accommodating influences by such defined factors.

In each cell pool, T lymphocytes of different phenotypes can be identified, although the transition from one phase to the other is rather fluent. In addition, T-cell antigen receptor rearrangements characterize various stages of cell maturation (Eren et al. 1987; Sen 2001; Petri et al. 1995; Davis and Chien 1999; Korsmeyer 1987) and enable antigens to participate in selective instruction of respective cells. Current concepts of T cell maturation as proposed in major textbooks of immunology (e.g. Paul 1999) were included in our model design (Krueger et al. 2003; see specifically Figs. 1–3 there for details).

Cytokines and chemokines in T cell development

The orderly progress of cells through the various stages of T cell maturation is controlled by a number of microenvironmental factors provided by the local stroma and its cellular components. Such factors include substances controlling lymphocyte traffic and homing, cell proliferation and differentiation, as well as cell function and death (i.e. apoptosis).

Their recognition is brought about through cellular receptors whose availability also varies (Krueger et al. 1987a; Daefler and Krueger 1989a, b; Atkins et al. 1987; Lobach and Haynes 1987; Wognum et al. 1996; Inghirami and Knowles 1992; Bondurant and Koury 1999; Paraskevas and Foerster 1999; Sasada and Reinherz 2001; Kong et al. 1998, 1999).

Soluble substances are secretory products of cells in the microenvironment including lymphoid cells themselves. They include thymic epithelial factors such as thymopoietins, T-cell growth factors such as interleukins, interferons and other cytokines from macrophages and fibroblasts (Krueger et al. 1988; Dappen et al. 1982; Surh and Sprent 1999; Crouse et al. 1980; Rothenberg and Lugo 1985; Allison 1987; Miescher et al. 1988; Kronenberg et al. 1986; Dardenne et al. 1977; Goldstein 1977; Crouse et al. 1985; Thiele et al. 1995; Harris et al. 1994; Singh et al. 1998;

Weber et al. 1999; Theodor et al. 1997; Zevin-Sonkin et al. 1992; Szewczuk et al. 1997; Romagnani et al. 2000; Kondo et al. 2000; Arzt et al. 2000; Azuara et al. 2001; Robetamanith et al. 2001; Nakauchi et al. 2001; Fiorini et al. 2000; Dieu-Nosjean et al. 2001; Varas et al. 1998; Guerin et al. 1997; Appasamy 1999; Plum 1999; Le et al. 2001).

Chemotaxis of cells during their wandering through the virtual compartments, their homing and release from specific sites is controlled by a number of additional factors such as chemokines, integrins, and selectins a.o. (Thiele et al. 1995). Finally, additional pre-thymic and post-thymic regulatory influences result from intrathymic major histocompatibility complex (MHC) and antigen-restricted clonal development, activity of signal transduction mechanisms and NF-kB binding, lymphocyte hormone receptors and activity of neuroendocrine circuits, down-regulation by chalones, idiotypic network control and other lymphokine and "notch" mechanisms. A detailed review of all these various influences can be found in a separate paper (Krueger et al. 2002c).

In order to obtain an initial and workable computational model, this large number of diverse factorial influences on T cell maturation was narrowed down to a few primary ones, which were identified as proliferation factors (P), differentiation factors (D), and inhibition factors (I). Once validated, however, the model was designed in a way to permit the addition of specific factors if so suggested by the immunologic disorder under study.

Base model design

Different cell pools of the T cell lineage are identified by representative cell types in these pools, which are also reflected by respective changes in peripheral blood lymphocyte populations responding to disturbances and changes in said pool sizes (Krueger 1989b; Krueger 1985; Ogawa et al. 2000).

Such changes in the peripheral blood are a result of respective feedback and feed-forward mechanisms known since Rudolf Virchow's "Zellularpathologie" (1868) and are the basis for such common clinical diagnoses as "shift to the left" when less mature cell populations occur in the

blood in certain infections or leukemias. Over 100 years later, Metcalf (1971) discussed such mechanisms in myeloproliferative diseases. Our model is a predator–prey type implemented as a coupled set of continuous ordinary differential equations with both linear and simple nonlinear interaction terms. It was implemented in Matlab software for PC. The four compartments are:

1. C1: the bone marrow (stem cells).
2. C2: the pre-thymic circulating stem cell pool settling in the thymic subcortical space.
3. C3: the thymic medullary pool.
4. C4: the periphery (blood with or without lymphoid organs).

In order to conceptually consider "shift to the left" conditions of clinical medicine, each compartment communicates information regarding its current state to its forward neighbor in the T cell developmental chain from C1 to C4, and in turn compartments C2 to C4 then communicate their state information back to all the other compartments (C1 to C3). In addition, all the compartments are "aware" of their own state at any given time (they have memory) and regulate themselves to a certain degree using negative self-feedback. A simplified schematic diagram of the system base model is shown in Fig. 1. As can be seen from the figure, the model functions in a ring-type network configuration with feed-forward and feedback connections as shown and described

above. The base model ordinary differential equations (ODE's) are

$$\dot{w} = \mu_w + w\left(\sum_{h=1}^{H} P_{w_h} - \sum_{j=1}^{J} D_{w_j} - \sum_{k=1}^{K} I_{w_k} + ax - by + cz\right),$$

$$\dot{x} = \mu_x + x\left(\sum_{l=1}^{L} P_{x_l} - \sum_{m=1}^{M} D_{x_m} - \sum_{n=1}^{N} I_{x_n} - dy + ez\right) + fw,$$

$$\dot{y} = \mu_y + y\left(\sum_{q=1}^{Q} P_{y_q} - \sum_{r=1}^{R} D_{y_r} - \sum_{s=1}^{S} I_{y_s} + gz\right) + ux, \quad (1)$$

$$\dot{z} = \mu_z + z\left(\sum_{\theta=1}^{\Theta} P_{z_\theta} - \sum_{\omega=1}^{\Omega} D_{z_\omega} - \sum_{\gamma=1}^{\Gamma} I_{z_\gamma}\right) + \delta y.$$

The dot notation represents time rate of change, e.g. $\dot{w} = dw/dt$ and the following variable descriptions apply:

1. w represents the number of stem cells within C1 at time t.
2. x represents the number of stem cells within C2 at time t.
3. y represents the number of cells within C3 at time t.
4. z represents the number of fully developed T-cells within C4 at time t. It is the summation of the number of CD4 and CD8 cells. In simulations of chronic human herpes virus type 6 (HHV-6), human immunodeficiency virus type 1 (HIV-1) and human T-cell leukemia virus type 1 (HTLV-1)

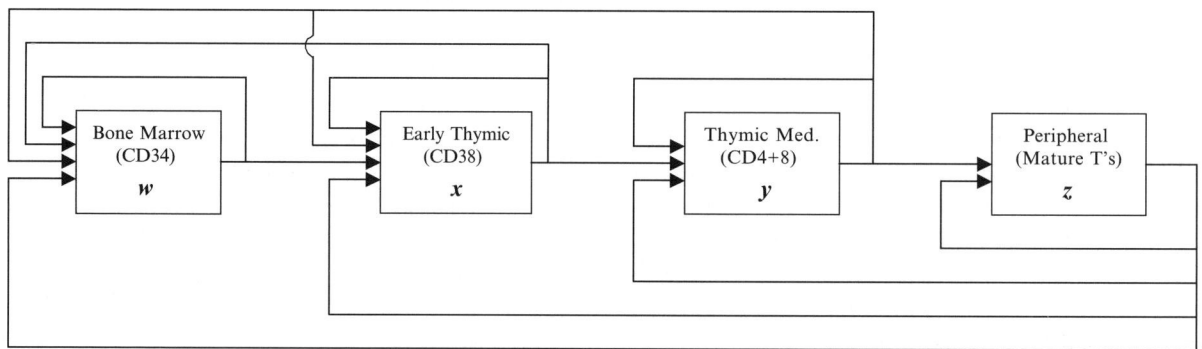

Fig. 1 Simplified block diagram of stepwise T cell differentiation with identified compartments and feedback controls. Identification of virtual compartments was done according to easily definable cell populations that can be determined in the peripheral blood. Feedback mechanisms enable changes in the individual cell compartments that are in balance with the peripheral blood

infections we model CD4 and CD8 counts separately (see results below).

5. $\mu_{[w\text{-}z]}$ represents the average regeneratory potential of each compartment [w-z].
6. $P_{[w\text{-}z]}$ are a group of proliferating factors at work within each compartment [w-z]. Each factor is represented as a real number.
7. $D_{[w\text{-}z]}$ are a set of differentiation factors at work within each compartment [w-z].
8. $I_{[w\text{-}z]}$ are a set of inhibition factors (apoptosis plus others) within each compartment [w-z].

Note that in this model currently only the state variables w, x, y, and z are functions of time t in which $\{w, x, y, z \geq 0 \text{ for } t \geq 0\}$ (use of the time variable t is implied in our notation). The remaining parameters are currently implemented as constants.

The feedback terms in the model are implemented as a multiplication of the state variable of the compartment itself, the inflow from a forward compartment, and a compartmental "cross-renewal" rate currently implemented as a scalar gain, e.g. gzy where g is the cross-renewal rate between compartments C3 and C4. The rate of change of cell numbers with time in each compartment is proportional to a summation of factors that decrease the rate of change, and/or factors that increase it. Several of these factors are in control of cell proliferation (P), differentiation (D), and inhibition (I) or apoptosis. For example for C1, the numbers of these factors are represented by H, J, and K, respectively. In simulations described here we used a single "pre-summated" factor for each P, D, and I in each of the four compartments in order to simplify the simulations and reduce the number of model parameters.

The model operates with textbook data for normal pool sizes (Krueger 1985, 1989b; Ogawa et al. 2000; Sing et al. 1988; Koury and Bondurant 1993; Pestano et al. 1999; Artavanis-Tsakonas et al. 1999; Kerre et al. 2001; MacDonald and Radtke 2001; Jaleco et al. 2001; Hochberg et al. 2001; Robey 1999) and is modulated according to data from human patients as published previously (Krueger et al. 2001a, b, 2002a, b). For further explanation of the biological (immunological) underpinnings of the model with extensive background information, please see Chapter 12 (Krueger and Buja) in this volume.

A caveat here: one should avoid taking the model as specified in Eq. 1 too literally with respect to its

parameters and its relationship with the biology of the human immune system. Simpler models are possible and may be worth further study (we have done some of this in fact but do not describe these here), yet what we are trying to do at this point is to capture some general features of the T cell development system under the hypothesis of regulation/dysregulation. Thus the model as specified is attempting to make use of two important system biology principles in network regulatory dynamics: Excitatory and inhibitory feedback and feedforward. It is noted that many biological systems utilize these two important tools for homeostatic self-regulation, especially in the neuroscience arena as well as in neuroimmunology.

On the other hand, one can imagine more complex models of the T cell system in which the individual cytokines are incorporated in the model individually, and/or the P, D, I factors are either a function of time or of the state variables themselves (w, x, y, z). Suffice to say that the type of modeling effort we describe here is still in its infancy. If we realize that not all the data and/or the factors that influence T cell development are currently available to us or known, then it becomes important for these and similar models to have predictive value in order to provide experimental investigators with clues about what factors to look for and where to look for them in the future. The model we describe here has already proven successful in this manner by demonstrating its qualitative similarity to data concerning the Canale–Smith syndrome (Krueger et al. 2002d). We continue refining the model as new data becomes available.

Simulations of the model were carried out using Matlab software we developed. The ODE's were solved using fourth-order Runge–Kutta integration. Each computer run was simulated continuously over a given time period depending upon the condition to be simulated (acute, chronic or late occurrence). Initial values at time $t_0 = 0$ (at birth) for cell pool numbers for all simulations discussed in the present manuscript were chosen as $w(t_0) = 100$, $x(t_0) = 10{,}000$, $y(t_0) = 1 \times 10^7$, and $z(t_0) = 1 \times 10^6$ based on standard textbook data.

The various parameters such as a, b, c, etc. are currently implemented as time static constants, but could be functions of time as well. Using both trial and error and a serial parameter search/optimization

procedure developed by us (see Appendix A) we found several sets of parameter values yielding suitably stable operating regions for Eqs. 1. Following this, we next attempted to determine a plausible working set of parameter values to simulate the immune system across a normal healthy human's lifespan (see next section). We later improved our search strategy – details of the current algorithm are provided in Appendix A.

In searching for parameters which will produce physiologically plausible model simulations, we took into account human clinical data we have collected and analyzed over the past ~25 years, available textbook data, and results from other studies in the literature. In the simulation studies described here time always begins at birth. Due to the present lack of various kinds of human immune system data, the computer modeling approach enables us to use *a priori* reasoning to explore a large space of "what if" scenarios.

Normative data across the human life span

After determining stable parameter regimes for the base model, we set out to find parameters which would result in an acceptable fit to the activity of the immune system across the lifespan of a "typical" normal-functioning human. Following use of both manual and semi-automated search procedures (see Appendix A) we arrived at the parameter values listed in Table 1, which accomplished our goal.

Table 1 Base model parameters for simulating a normal immune system across the lifespan

$w(t_0) = 100.0$	$x(t_0) = 1.0 \times 10^4$	$y(t_0) = 1.0 \times 10^7$	$z(t_0) = 1.0 \times 10^6$
$\mu_w = 50.0$	$\mu_x = 25.0$	$\mu_y = 2.5 \times 10^6$	$\mu_z = 1.0 \times 10^4$
$P_w - D_w - I_w$ $= -0.531$	$a = 5.0 \times 10^{-7}$	$b = 2.0 \times 10^{-7}$	$c = 5.0 \times 10^{-7}$
$P_x - D_x - I_x$ $= -0.04$	$d = 5.5 \times 10^{-8}$	$e = 3.6 \times 10^{-7}$	$f = 5.0 \times 10^{-7}$
$P_y - D_y - I_y$ $= -0.7$	$g = 1.0 \times 10^{-7}$	$u = 0.05$	$\delta = 0.001$
$P_z - D_z - I_z$ $= -0.023$			

Substituting these parameters into Eqs. 1, one obtains the following set of ordinary differential equations:

$$\dot{w} = 50.0 - 0.531w + (5.0 \times 10^{-7})wx - (2.0 \times 10^{-7})wy$$
$$+ (5.0 \times 10^{-7})wz,$$
$$\dot{x} = 25.0 - 0.04x - (5.5 \times 10^{-8})xy + (3.6 \times 10^{-7})xz$$
$$+ (5.0 \times 10^{-7})w,$$
$$\dot{y} = (2.5 \times 10^6) - 0.7y + 0.05x + (1.0 \times 10^{-7})yz, \qquad (2)$$
$$\dot{z} = (1.0 \times 10^4) - 0.023z + 0.001y.$$

We have also explored the robustness of these parameter values, which are reported in Appendix B. From Eqs. 2 we note the following salient system interactions for the base-level, normative model design, (1) there is negative self-feedback for each compartment, (2) there is positive feed-forward from compartment C1 to C4 (from $w \rightarrow x \rightarrow y \rightarrow z$), (3) compartment C4 ($z$) is in a positive feedback relationship with its previous compartments C3, C2 and C1 (y, x, w), and (4) compartment C3 (y) is in a negative feedback relationship with its previous compartments C2 and C1 (x, w).

Please note that these three characteristics of Eq. 2 shift the biological significance of the model specification in Eqs. 1 a bit. In practice there are three critical features of the model: self-feedback for each of the compartments, feed-forward to the next compartment in the chain, and cross-feedback from all forward compartments in the chain (e.g. a term such as gzy for feedback from compartment C4 back to C3). This may not exactly match what is currently understood of how the T cell development system functions biologically, yet it may do so in a qualitative functional sense.

Figure 2 shows a simulation run of normative developmental data using Eqs. 2. The main feature we observe is the slowly exponentially decreasing peripheral blood T cell count (z) across the lifespan which compares well with actual lymphocyte numbers in humans (Goldstein 1977) decreasing from 6,320 ± 3,000 per ml blood at age 1 month to 1,890 ± 830 per ml blood at age 80 years (T cells 78% ± 10% per ml blood and 61% ± 13% per ml blood respectively). The stem cell pool (w) remains fairly stable during this entire period, while early and late thymic cells (x, y) indicate physiologic involution of the organ. This is most prominent in the rapidly proliferating pool of early thymic cells.

Modeling and simulating a general viral challenge

We model and simulate an arbitrary viral (v) challenge to the base system by applying a stimulus at a time t_{vs} using a narrow impulse function. Virtual viral units in the model represent viral DNA or RNA load as determined in the peripheral blood of respective patients. First, we apply this stimulus to the mature T cell compartment (peripheral blood) only and study the behavior of the z-v subsystem using the following relations.

$$\dot{z} = \mu_z + z(P_z - D_z - I_z - k_2 v) + \delta y + k_1 v,$$
$$\dot{v} = v(l_1 - l_2 z), \qquad (3)$$
$$\delta = 0.0.$$

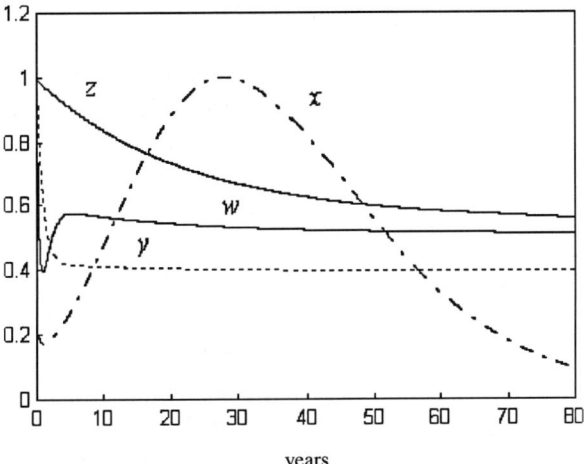

Fig. 2 Computer simulation of normative data over a human lifespan (birth to 80 years). Relative cell counts vs. time for CD34 (w), CD38 (x), CD4 + 8 (y), and relative cell counts vs. time for peripheral T cells (z). Absolute cell counts were normalized so as to fit the four curves onto a single plot

Here, $k_1 v$ represents the antigenic stimulation of the virus to the T cells. k_1 is always simulated with a positive sign as its factor mimics the virus' envelope glycoproteins stimulating T cell proliferation upon contact with the cell membrane. $k_1 vz$ represents T cell death due to viral replication and secondary to the cytotoxin of the T cells themselves. This factor acts on T cells following viral endocytosis. $l_1 v$ represents a virus proliferation factor, while $l_1 vz$ represents T cells destroying virus particles. Although we vigorously searched the parameter space of Eqs. 3, we were unable to find a reasonable fit of the model to real human data. The reason may relate to timing: that some processes occur earlier or later than others. We therefore decided to use "squashing" (weighting) functions with delays in Eqs. 3. We use the standard sigmoid function for this since its smooth shape can be easily controlled by changing its dual parameters. The modified model becomes

$$\dot{z} = \mu_z + z(P_z - D_z - I_z - k_2 \psi_2 v) + \delta y + k_1 \psi_1 v,$$
$$\dot{v} = v(l_1 \rho_1 - l_2 \rho_2 z), \qquad (4)$$
$$\delta = 0.0.$$

in which

$$\psi_{1,2} = \frac{1}{1 + \exp(\lambda_{1,2}(t - \tau_{1,2}))},$$
$$\rho_{1,2} = \frac{1}{1 + \exp(\lambda'_{1,2}(t - \tau'_{1,2}))} \qquad (5)$$

are functions of time t. Figure 3 shows plots of $\{\Psi_{1,2}, \rho_{1,2}\}$ for $\{\lambda, \lambda'\} > 0$ (Fig. 3a) and for $\{\lambda, \lambda'\} < 0$ (Fig. 3b).

(a)

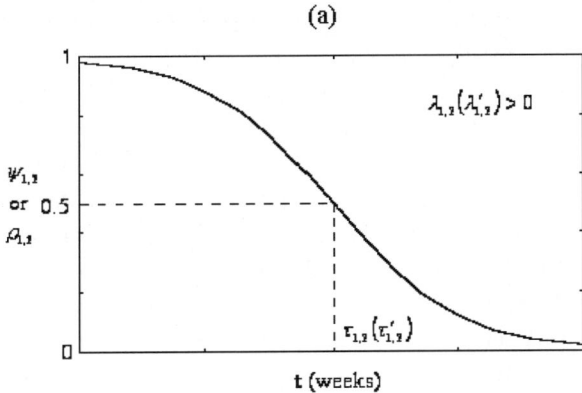

(b)

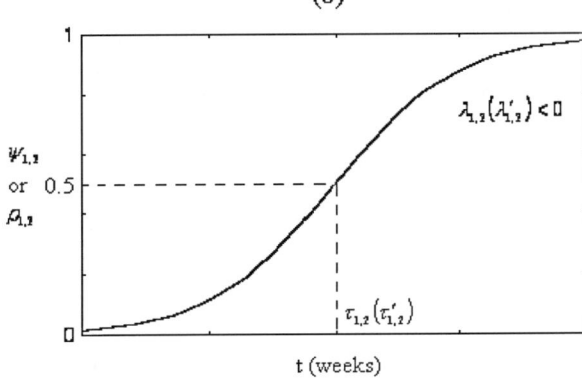

Fig. 3 Generalized plot of Eqs. 5 vs. time for (a) $\{\lambda, \lambda'\} > 0$, and (b) $\{\lambda, \lambda'\} < 0$

For example, for $\lambda' > 0$, Ψ_1 weights larger before time τ_1 than after time τ_1, meaning that the antigenic stimulation is greater prior to time τ_1 than afterward. Note that τ values can be negative (as well as positive) without leading to noncausality since simulated running time is always non-negative and $\tau_{vs} > 0$. We tested the stability of this system with success and were able to simulate a general viral challenge to the T cell system. For example, Fig. 4 shows simulated results for z and v as a function of number of weeks following viral stimulation using the following parameters in Eqs. 4–5:

$$k_1 = 2000.0, \qquad k_2 = 0.005,$$
$$\lambda_{1,2} = 0.7\,\text{week}^{-1}, \qquad \lambda'_{1,2} = 0.7\,\text{week}^{-1},$$
$$P_z - D_z - I_z = -0.014, \qquad v(t_{us}) = 100.$$
$$l_1 = 0.25, \qquad l_2 = 3.5 \times 10^{-8},$$
$$\tau_{1,2} = 12.0\,\text{weeks}, \qquad \tau'_{1,2} = 5.0\,\text{weeks},$$

Next we connected the remainder of the base system by making δ nonzero. The connected system is

$$\dot{w} = \mu_w + w(P_w - D_w - I_w + ax - by + cz),$$
$$\dot{x} = \mu_x + x(P_x - D_x - I_x - dy + ez) + fw,$$
$$\dot{y} = \mu_y + y(P_y - D_y - I_y + gz) + ux, \qquad (6)$$
$$\dot{z} = \mu_z + z(P_z - D_z - I_z - k_2\Psi_2 v) + \delta y + k_1\Psi_1 v,$$
$$\dot{v} = v(l_1\rho_1 - l_2\rho_2 z).$$

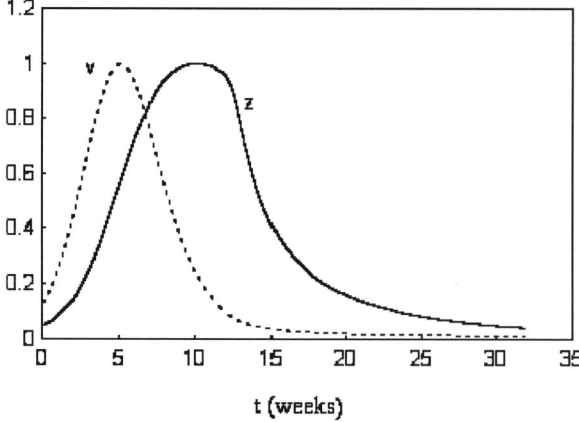

Fig. 4 Example plot of z and v vs. t (Eqs. 4–5). Data normalized to fit the plot

In this way the virus directly affects the system at compartment C4 (z) and indirectly affects compartments C1 to C3 through feedback from C4.

Data validation procedures and studies

Computer simulation runs were performed for validation purposes by initiating a simulated viral infection (DNA or RNA load = antigenic stimulus) of cells in pool C4. The viral stimulus effect on C4 propagates back through the remaining system compartments via feedback and feed-forward dynamics. The quantity of infected cells (e.g. 100,000–200,000/10^6 PBL) represents intensity of stimulus for pool initial expansion (e.g. mitotic activity: 10–20%) and differentiation (i.e. production of cytotoxic T cells ultimately lysing infected cells). Viral infection also induces apoptosis of infected cells.

Antigenic stimulation in combination with cellular apoptosis induces cell regeneration in pools C1, C2 and C3 via cytokine and growth factor production (i.e. proliferation factors from pools C3 and C4) with pool expansions of 10–20%. Cell differentiation factors from pools C3 and C4, targeting pools C2, C3 and C4 cause specific differentiation of cytotoxic T cells (Ttox) finally lysing infected cells. Ttox amounts to about 10–50% of the total expanded C4 pool. The simulation will come to an equilibrium when all HHV-6 infected cells are lysed.

In a young person, this may reinstitute the original state of cell numbers in the pools (i.e. 100% repair). In an adult, an age-dependent reduction of repair, i.e. in the production of thymus-primed T cells, must be introduced (e.g. beyond the age of 40 at 10%). Checkpoints in the patient for controlling the system's data are cells in pools C2, C3 (i.e. double positive CD4/CD8+ T cells in blood) and C4. Key cytokines and growth factors can also be measured in the blood similarly as for virus-infected, proliferating and apoptotic cells.

Representative data for validation studies were taken from human case studies. Infection with three different viruses were chosen, all of them targeting the same CD4 T helper cell population with different outcomes. Acute HHV-6 infection causes a reactive hyperplasia of T cell populations producing a clinical picture similar to infectious mononucleosis. All cell values return to normal after a given (short) time. Persistent active

infection with HHV-6 is accompanied by chronic fatigue syndrome-like changes with undulating cell values for a longer period of time, yet with the potential of return to normal and clinical recovery.

Infection with the human immunodeficiency virus type 1 (HIV-1) causes depletion of infected CD4 T helper cells with final aplasia and death of the patient. Opposite to this effect, human T cell leukemia virus type 1 (HTLV-1) also infects CD4 T helper cells, but after long latent periods will block death (apoptosis) in these cells permitting them unlimited growth. The final outcome is a T cell neoplasia. We have collected representative data from infections with these three viruses in such a way to be used in validating our computer simulation model (Clark and Brugge 1995; Tedder et al. 1995; Cyster 1999; Carrol and DeSousa 1983). Respective human clinical data for viral infections of HTLV-1, acute HHV-6, chronic HHV-6, and HIV-1 infection (respectively) are shown below in Figs. 5, 7, 9 and 11.

Computer runs were performed to simulate all separate conditions, i.e. normative (baseline) results across an individual's lifespan (see above), HTLV-1 infection (apoptosis blockade with concomitant viral stimulation), acute HHV-6 infection (Krueger et al. 2002d), chronic HHV-6 infection, and HIV-1 infection. Apoptosis blockade was accomplished by reducing the inhibition factor (I_z) for the peripheral T cell pool by 50% from its simulated normative value only (all other factors held constant).

Simulations of the model were produced which resemble the time courses of the above described viral infections in the patient data (described below). This was accomplished by inducing (simulated) cell proliferation secondary to viral stimulation (represented by viral DNA or RNA copy numbers in the blood) and simulated cell apoptosis as determined in the human patient. Dynamic changes in cell pool numbers then followed inbuilt feedback and feed-forward mechanisms.

HTLV-1 infection

Background

HTLV-1 causes a neoplastic response with development of adult T cell leukemia (ATCL) in about 1–2% of infected individuals (Krueger et al. 2002b).

It affects CD4+ T helper cells or their immature precursors and was the first human leukemogenic retrovirus discovered. ATCL was first described by Japanese authors (Uchiyama et al. 1977) with identification of the retrovirus as its causative agent by Poiesz et al. (1980).

HTLV-1 infections are endemic in restricted areas of southern Japan, the Caribbean islands, and certain parts of South America. The prevalence rate of HTLV-1 among volunteer blood donors in the USA is 0.02%, with increased prevalence in paid blood donors, African American health care clinics and Amerindians. The virus does not appear to be very virulent as only 1–2% of infected individuals develop clinical symptoms after long latent periods of up to 20 or 30 years (Ehrlich and Poiesz 1988). There are essentially two types of diseases which are causally related to HTLV-1 infection: ATCL/ cutaneous T cell lymphoma and, preferentially in the Caribbeans, tropical spastic paraparesis (TSP)/ HTLV-1 associated myelopathy (Gessain et al. 1985; Miyai et al. 1987). We focus on modeling of ATCL here only.

Human Data in infected individuals, virus replication remains at low levels for approximately 10 years following primary infection with average virus RNA copy loads in the peripheral blood of 1–10 copies per 10^3 peripheral blood lymphocytes (PBL). Thereafter, a logarithmic increase is observed in about 10% of infected individuals with between approximately 50 and 500 RNA copies per 10^3 PBL and a developing clinical picture of pre-leukemia and overt T cell leukemia. The latter coincides with a sudden drop in PBL apoptosis from initial levels of 20–30% to the almost normal rate of 5%, which is caused by apoptosis blockade in HTLV-1-transformed cells (Korsmeyer 1987; Inghirami and Knowles 1992).

Leukemic cell proliferation is characterized in the more rapidly progressing forms of ATCL by marked increases of immature T cells including up to a 160-fold increase in CD38 cells and a 119-fold increase in CD4+CD8+ double positive T cells. Figure 5 shows median values at each time point from measures taken from adult patients (males and females). Figure 5a shows normalized (by the maximum value) x-fold values (increases here) for CD38 (corresponding to x) and CD4+8 counts

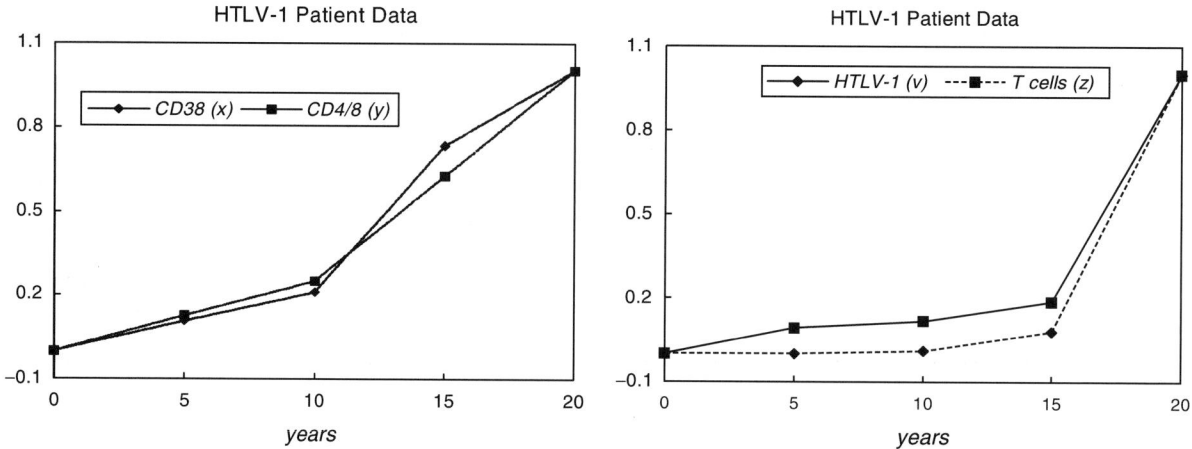

Fig. 5 Human cellular responses (median values at each time sample) in the T cell system following infection in individuals with HTLV-1, (left) normalized (by the data maximum) CD34 (w), CD38 (x) and CD4 + 8 (y) x-fold changes over a 20-year period. (Right) normalized (by the data maximum) peripheral T-cell (z) x-fold changes and viral load (v) over the same timespan

(corresponding to y). Figure 5b shows normalized x-fold values for T cell counts (corresponding to z) and normalized virus particle counts (genome copies/1,000 PBL, corresponding to v). See Krueger et al., (2002b) for a full description of the data sources used to construct Fig. 5.

Modeling and simulation

Using the model in Eqs. 6 with the set of normative parameter values in Table 1 we optimized the remaining parameters to arrive at

$$k_1 = 681.45, \qquad\qquad k_2 = 0.0069528,$$
$$\lambda_1 = -0.21009\,\text{year}^{-1}, \qquad \lambda_2 = 0.16195\ \text{year}^{-1},$$
$$\tau_1 = -7.692\ \text{year}, \qquad\qquad \tau_2 = -13.369\,\text{year},$$
$$v(t_{vs}) = 0.1,$$
$$l_1 = 0.8786, \qquad\qquad l_2 = -2.47\times10^{-8},$$
$$\lambda_1' = -0.04054\ \text{year}^{-1}, \qquad \lambda_2' = 0.32535\ \text{year}^{-1},$$
$$\tau_1' = 10.234\ \text{years}, \qquad\qquad \tau_2' = 9.888\ \text{years},$$

Figure 6 shows a typical simulation run with these parameter values and with the virus stimulation introduced at $t_{vs} = 30$ years of age. The slow exponential rise of both the viral load and total number of T cells observed in Fig. 6 functionally resembles that observed for the real data reported in Fig. 5. Note that in this

case l_2 itself is negative. Thus in this case total T-cell (z) counts have a proliferative effect on viral load.

Acute HHV-6 infection in adulthood

Background

In an infected individual, the virus DNA copies reside in the blood (i.e. serum + cells), thus copy numbers represent all available virus at any given time, intracellularly (probably three fourths), extracellularly (about one fourth) with and without bound antibodies. Only the intracellular virus multiplies with a generation time of about 48 h. HHV-6 binds easily to receptors of CD4 T cells (i.e. helper T cells), which are two thirds of all T cells in blood. Of these, virus readily infects about 60%. The virus also stimulates proliferation of CD8 T cells (i.e. cytotoxic T cells) with the CD4/CD8 ratio going down subsequently from 1.5 to 1.0 or even 0.75. It is the latter CD8 T cells which kill virus infected CD4 cells and thus inhibit indirectly virus proliferation (by killing host cells: virus can only proliferate intracellularly). One CD8 T cell can kill on average 8–10 virus infected CD4 cells before it will die itself. Specifically, the sequence of the immunological events can be described as follows:

Virus attaches to CD4 cell receptor and is internalized within about 6 h. Then this cell becomes antigenic and stimulates CD8 T cells to proliferate. Antigenic

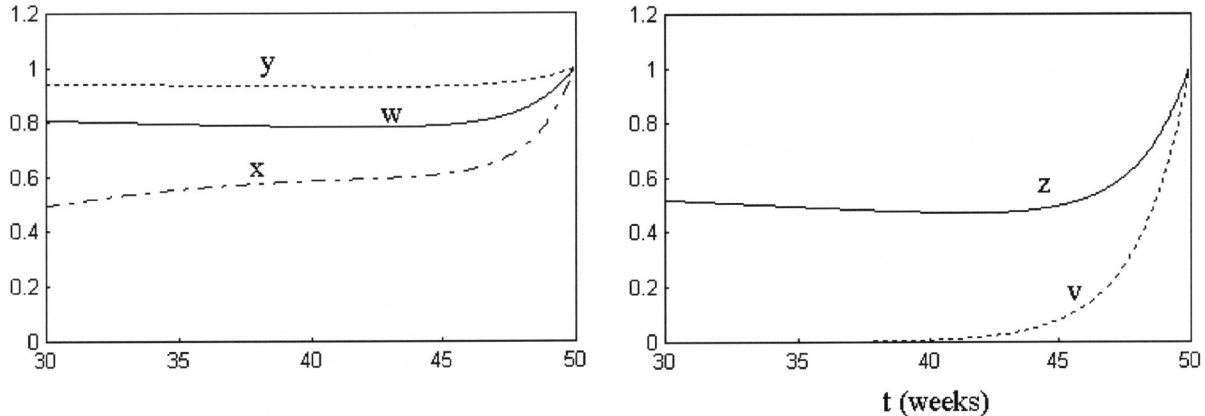

Fig. 6 Simulation of HTLV-1 infection initiated at 30 years of age in an otherwise healthy "virtual" adult. (Left) *w*, *x*, *y*, counts vs. time (20 years), (right) *z* and *v*. Data normalized to fit the plot. Note correspondence with patient data of Fig. 5

stimulation increases over the next 48 h as additional antigens become expressed in infected cells. As more CD4 cells become infected, antigenic stimulation and CD8 cell proliferation continue to increase. Infected CD4 cells die in response to virus replication itself and secondary to the effects of CD8 cytotoxic T cells. This may occur in measurable amounts as early as 4 days after infection, peaking after 10–14 days (depending on infectious dose). Thereafter, the numbers of infected cells decrease, i.e. intensity of antigenic stimulation decreases with subsequent decrease of CD8 cell proliferation and CD4 cell death. CD4 cells will regenerate to normal numbers that may take as long as 4–6 months.

Human data

First, we describe human data collected by our group to guide determination of the model parameters for simulation studies. Ten adult patients (males and females) with active HHV-6 variant A infections and clinical infectious mononucleosis-like disease (IM) were studied over a period of 32 weeks after onset of disease for their viral DNA load, changes in peripheral blood T-lymphocytes and subpopulations and frequency of cell death in peripheral blood cells (see Fig. 7).

Since the exact time of primary infection of the patients was unknown and thus no time relation of viral effects at cellular level were determined, we supplemented such data from separate tissue culture studies using HHV-6a infection of HSB2 cells. Patients with IM demonstrated an increase in HHV-6 DNA copies from 0 to 8.2 $\log_{10}$/5μL blood within 4 weeks that returned to normal by 16

weeks. Total T-lymphocytes increased by 20x normal levels following infection peaking at 8–10 weeks and then returning back to normal levels by 24–28 weeks.

Coincidently, less mature lymphoid cells carrying markers for stem cells, thymic cortical and medullary cells increase 8- to 10-fold indicating an enhanced mobilization of such cells from premature cell compartments. Cell death in peripheral mononuclear cells peaked at 30% at 8 weeks following onset of clinical disease and normalized by 24 weeks. HHV-6 replication in cell culture as determined by antigen expression, electron microscopy and harvesting of infectious virions indicated a complete cycle of virus infection and replication of at least 6 days. This data also compared well with others from the literature (Krueger et al. 2001a).

Modeling and simulation

Using the model in Eqs. 6 with the set of normative parameter values in Table 1 we optimized the remaining parameters to arrive at

$$k_1 = 60779.0, \qquad k_2 = 0.2258,$$
$$\lambda_1 = 35.6602 \, \text{year}^{-1}, \qquad \lambda_2 = -64.5812 \, \text{year}^{-1},$$
$$\tau_1 = 3.47 \, \text{months}, \qquad \tau_2 = 2.93 \, \text{months},$$
$$\upsilon(t_{\upsilon s}) = 1000,$$
$$l_1 = 24.879, \qquad l_2 = 1.1679 \times 10^{-6},$$
$$\lambda_1' = 35.8895 \, \text{year}^{-1}, \qquad \lambda_2' = -41.3943 \, \text{year}^{-1},$$
$$\tau_1' = 0.50 \, \text{month}, \qquad \tau_2' = 1.21 \, \text{months},$$

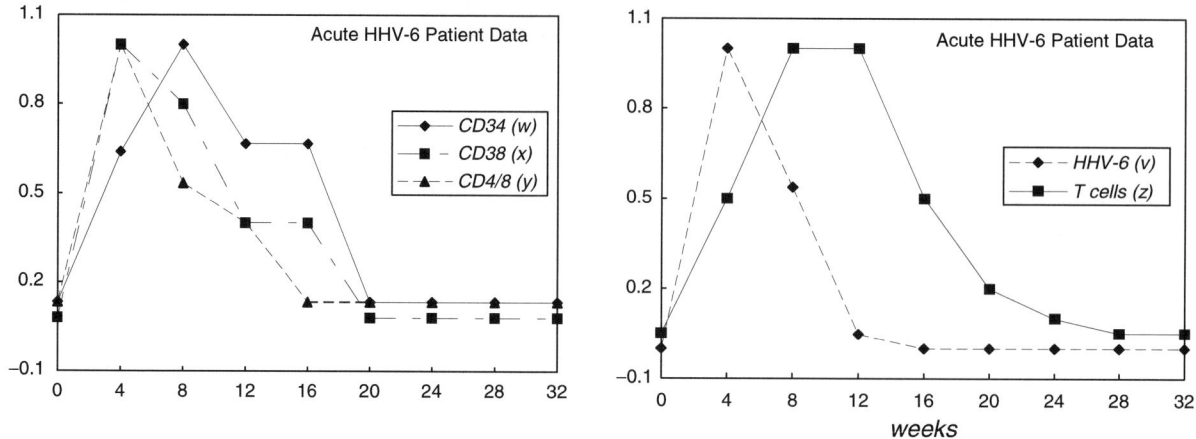

Fig. 7 Human cellular responses (median values at each time sample) in the T cell system following acute HHV-6 infection in 10 individuals, (left) normalized (by the data maximum) CD34 (*w*), CD38 (*x*) and CD4+8 (*y*) x-fold changes over a 32-week period. (Right) normalized (by the data maximum) T-cell (*z*) x-fold changes and viral load (*v*) over the same timespan

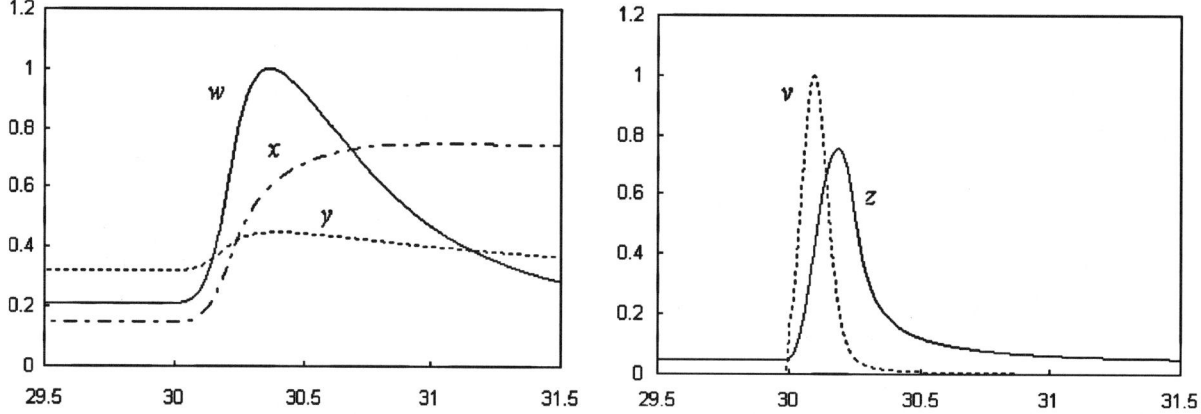

Fig. 8 Simulation of acute HHV-6 infection in infectious mononucleosis-like disease. Infection initiated at age $t = 30$ years in an otherwise typical healthy "virtual" adult. (Left) relative cell counts vs. time for CD34 (*w*), CD38 (*x*), and CD4+8 (*y*). (Right) relative counts vs. time for peripheral T cells (*z*) and for number of viral DNA copies (*v*). Data normalized to fit the plots. Note rapid return of *z* to near prestimulus baseline levels following termination of acute infection. Also note correspondence in time course of *z* and *v* with patient data shown in Fig. 7 (right panel)

Figure 8 shows a typical simulation run with these parameter values and with a viral stimulation introduced at $t_{vs} = 30$ years of age. From the figure it can be seen that immediately following viral stimulation the peripheral blood T cell count (*z*) ramps up rapidly and then returns to prestimulus baseline levels fairly quickly. Here, there is positive feedback of *v* to *z*, and negative feedback of *vz* to *z* at the same time. The I_z level is the same as in normative simulation runs (Fig. 2). The computational model thus simulates cellular changes that occur in the human patient following acute HHV-6 infection and the clinical picture of infectious mononucleosis (Krueger et al. 2001a). Interesting however, is the finding that immature CD38+ T cells remain elevated in the computational model for a longer period of time. We have not tested this in our patients with infectious mononucleosis, yet observed this phenomenon sporadically in other viral infections. We can observe a qualitative functional correspondence between the patient data (Fig. 7) and the simulated data from the model (Fig. 8).

Chronic HHV-6 infection in adults

Background

Due to increased biosystem complexity, in the case of modeling chronic HHV-6 infection it is necessary to consider the contributions of CD4 and CD8 T cells and model each one specifically. In this case the sequence of immunological events is quite different from acute infection given the much more prolonged combat among virus, CD4 T cells and CD8 T cells. The virus tends to curtail the ability of CD4 cells to multiply, which increases the former's numbers while decreasing the number of CD4 cells. The virus simultaneously stimulates CD8 cell production leading to increased destruction of virus. The prolonged combat continues leading to a longterm oscillatory system response.

Human data

Ten adult patients (males and females) with persistent active HHV-6 variant A infection and clinical chronic fatigue syndrome (CFS) were studied over a period of 24 months after initial clinical diagnosis (Krueger et al. 2001b). Results of the human data measures are shown in Fig. 9. CFS was diagnosed according to IIIP-revised CDC criteria as defined by the CFS Expert Advisory Group of the German Federal Ministry of Health in 1994. Changes in HHV-6 antibody titer, viral DNA load, peripheral blood T lymphocytes and subpopulations, as well as CD4/CD8 cell ratio and cell death (apoptosis) were monitored. Data were collected for comparison with respective changes in acute HHV-6 infection, and as a basis for future computer simulation studies. The results showed variable but slightly elevated numbers of HHV-6 DNA copies in the blood of patients with CFS, while PBL apoptosis rates were clearly increased. CD4/CD8 cell ratios varied from below 1 up to values as seen in autoimmune disorders. Contrary to acute HHV-6 infection, T lymphocytes do not exhibit the usual response to HHV-6 with elevation of mature and immature populations, suggesting a certain degree of unresponsiveness. The data suggest that persistent low dose stimulation by HHV-6 may favor imbalanced immune response rather than overt immune deficiency, which requires further confirmation by additional functional studies.

Modeling and simulation

The model relations used in this case are:

$$\dot{w} = \mu_w + w(P_w - D_w - I_w + ax - by + cz),$$

$$\dot{x} = \mu_x + x(P_x - D_x - I_x - dy + ez) + fw,$$

$$\dot{y} = \mu_y + y(P_y - D_y - I_y + gz) + hx,$$

$$\dot{z}* = \mu_{z*} + z*(P_{z*} - D_{z*} - I_{z*} - i\psi_2 \upsilon) + \delta* y + j\psi_1 \upsilon,$$

$$\dot{z}^@ = \mu_{z^@} + z^@(P_{z^@} - D_{z^@} - I_{z^@} + k\psi_4 \upsilon) + \delta^@ y - l\psi_3 \upsilon,$$

$$\dot{\upsilon} = \upsilon(m\rho_1 z* - n\rho_2 z^@ - o\rho_3),$$

$$z = z* + z^@,$$

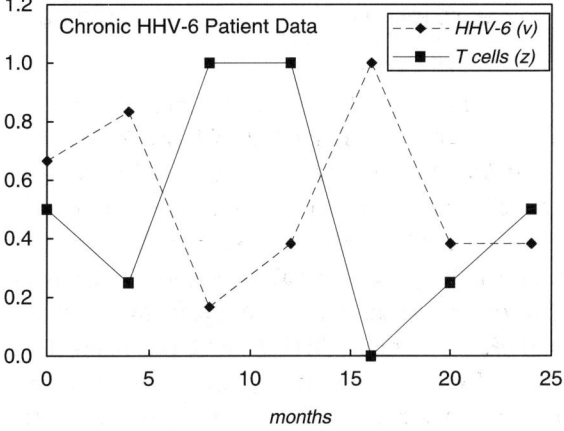

Fig. 9 Human cellular responses (median values at each time sample) in the T cell system following chronic HHV-6 infection in 10 individuals, (left) normalized (by the data maximum) CD34 (*w*), CD38 (*x*) and CD4+8 (*y*) x-fold changes over a 24-month period. (Right) normalized (by the data maximum) T-cell (*z*) x-fold changes and viral load (*v*) over the same timespan

where z^* are the number of CD4 T cells and $z^@$ are the number of CD8 T cells. $\delta^* = 0.6\delta$ and $\delta^@ = 0.4\delta$ represent two feed-forward (from bone marrow to the peripheral blood) gains for z^* and $z^@$, respectively, and Eqs. 7, 8 are a direct extension of Eqs. 6, 5 respectively. The parameters used are then as follows (w, x, and y subsystem parameters are same as those in Table 1):

$$\psi_{1,2} = \frac{1}{1 + \exp(\lambda_{1,2}(t - \tau_{1,2}))},$$

$$\psi_{3,4} = \frac{1}{1 + \exp(\lambda_{3,4}(t - \tau_{3,4}))}, \qquad (8)$$

$$\rho_{1,2,3} = \frac{1}{1 + \exp(\lambda'_{1,2,3}(t - \tau'_{1,2,3}))}.$$

$w(t_0) = 100.0,$ $x(t_0) = 1.0 \times 10^4,$ $y(t_0) = 1.0 \times 10^7,$ $z(t_0) = 1.0 \times 10^6$

$z^*(t_0) = 0.6z(t_0),$ $z^@(t_0) = 0.4z(t_0),$ $\mu_w = 50.0,$ $\mu_x = 25.0,$

$\mu_y = 2.5 \times 10^6,$ $\mu_{z^*} = 3.21 \times 10^5,$ $\mu_{z^@} = 1.27 \times 10^5,$

$P_w - D_w - I_w = -0.531,$ $a = 5.0 \times 10^{-7},$ $b = 2.0 \times 10^{-7},$ $c = 5.0 \times 10^{-7},$

$P_x - D_x - I_x = -0.04,$ $d = 5.5 \times 10^{-8},$ $e = 3.6 \times 10^{-7},$ $f = 5.0 \times 10^{-7},$

$P_y - D_y - I_y = -0.7,$ $g = 1.0 \times 10^{-7},$ $h = 0.05,$ $\delta = 0.001,$

$P_{z^*} - D_{z^*} - I_{z^*} = -0.478,$ $i = 3.09 \times 10^{-3},$ $k = 1.07 \times 10^{-3},$ $j = 0.01,$

$l = 5.6 \times 10^{-4}$ $m = 7.9 \times 10^{-5},$ $\lambda_1 = 0.0592\,\text{week}^{-1}$ $\lambda_2 = 0.0352\,\text{week}^{-1},$

$P_{z^@} - D_{z^@} - I_{z^@} = -0.463,$ $n = 5 \times 10^{-5},$ $o = 3.14 \times 10^{-7},$ $v(t_{vs}) = 3.8,$

$\lambda_3 = 0.028\,\text{week}^{-1},$ $\lambda_4 = -0.017\,\text{week}^{-1},$ $\lambda'_1 = -0.00045\,\text{week}^{-1},$ $\lambda'_2 = 0.0177\,\text{week}^{-1},$

$\lambda'_3 = 0.008\,\text{week}^{-1},$ $\tau_1 = -0.0178\,\text{week},$ $\tau_2 = 0.0122\,\text{week},$ $\tau_3 = -0.015\,\text{week},$

$\tau_4 = 0.037\,\text{week},$ $\tau'_1 = -0.0075\,\text{week},$ $\tau'_2 = -0.001\,\text{week},$ $\tau'_3 = -0.00944\,\text{week}.$

You will notice from the parameter signs and magnitudes that the net effect of HHV-6 over time is to reduce CD4 counts and increase CD8 counts. CD4 cells in turn tend to increase viral counts while CD8 cells tend to reduce viral counts. These correspond quite well to the virus' expected effects on the immune system. Figure 10 depicts the results of simulating a chronic HHV-6 infection using Eqs. (7–8). The results shown are in good agreement with the data from patients with chronic fatigue syndrome reported in Krueger et al. (2001b) (compare Figs. 9 and 10). We should point out that the waveform shapes (but not the time courses) shown

in Fig. 10 (right panel) are also very similar to the results reported in Brandt and Chen (2001) for modeling of HIV-1 infection.

Acute progressive HIV-1 infection

Background

While the previous cases of acute and chronic HHV-6 infection represent physiological and pathological hyperplastic cellular reactions to viral infection, HIV-1 infection provides data for an aplastic response to a viral infection targeting the same cell population as HHV-6. In this case HIV-1 infection with continuously replicating virus actively destroys the CD4+ T helper cell population and also interferes with their thymic regeneration. Since the latter also affects CD8+ cytotoxic T cells, host defense against the virus ultimately becomes deficient (among other mechanisms). HIV-1 infection, therefore, represents an ideal model for testing another pathological response of virus infection and for further validation of our model.

Human data

Nineteen adult patients (males and females) with persistent active HIV-1 infection and a rapidly

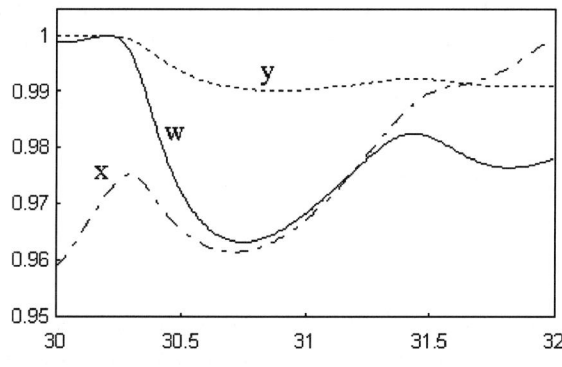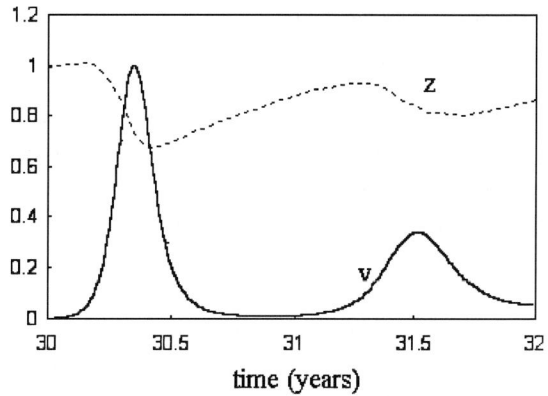

Fig. 10 Simulation of chronic HHV-6 infection in a typical, otherwise healthy "virtual" adult. Infection initiated at age $t = 30$ years. (Left) relative cell counts vs. time for CD34 (w), CD38 (x), and CD4+8 (y). (Right) relative counts vs. time for peripheral T cell counts ($z = z^* + z^@$) and for number of viral DNA copies (v). Data normalized to fit the plots. Note correspondence in undulating time course of z and v with patient data shown in Fig. 9 (right)

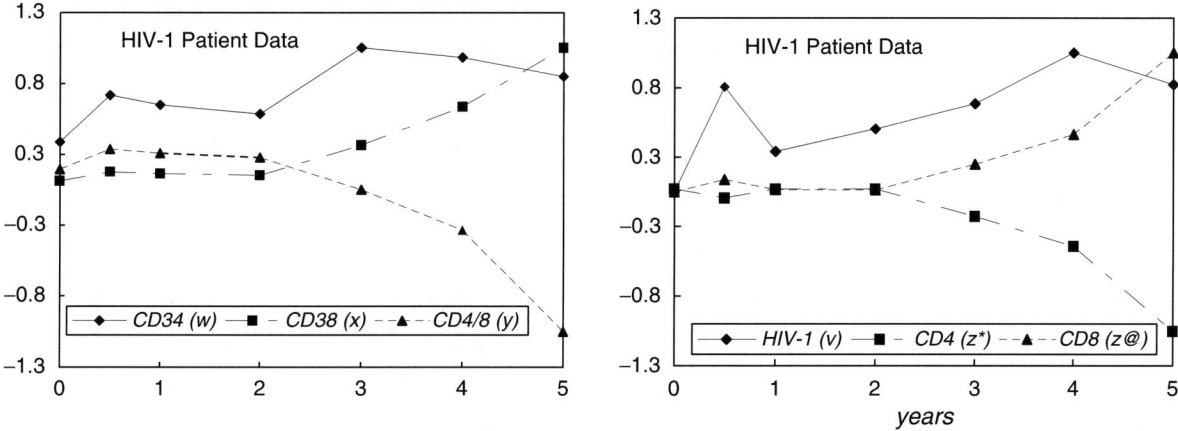

Fig. 11 Human cellular responses (median values at each time sample) in the T cell system following HIV-1 infection in 19 patients, (left) normalized (by the data maximum) CD34 (w), CD38 (x) and CD4+8 (y) x-fold changes over a 5-year period. (Right) normalized (by the data maximum) peripheral CD4 (z^*) and CD8 ($z^@$) x-fold changes and viral load (v) over the same timespan

progressive course without specific antiviral treatment were introduced in this study. HIV-1 infection was proven by two independent antibody ELISA tests and further confirmed by infection of tissue culture cells and demonstration of HIV antigen expression using Western blots (Krueger et al. 2002a).

Results of the human data measures are shown in Fig. 11. Similarly to HHV-6 infections, changes in RNA viral load, peripheral T cell populations and subpopulations, as well as CD4/CD8 cell ratio and cell death (apoptosis) were collected for subsequent computer simulation studies. The results show a progressive increase in PBL (peripheral blood

leukocyte) apoptosis paralleling the increase of viral RNA copies in patients' plasma and the decrease in CD4 T helper cells. Toward the end of the observation period (i.e. during years 3–5 post infection) numbers of immature CD38+ lymphoblasts increase signaling an intrathymic blockage of T cell differentiation. This rapidly progressive course of human AIDS represents the natural course of untreated HIV-1 infection as initially observed when HIV was detected. We selected these 19 patients for our study (out of a total of 200) in order to provide data for simulation studies that are representative of HIV-1 infection itself and not altered by additional influences through treatment regimens.

Modeling and simulation

The model equations used for simulation of HIV-1 infection are the same as those used for chronic HHV-6 (Eqs. 7–8). Here CD4 T cells (z^*) and CD8 T cells ($z^@$) are modeled separately as well. Note that although the virus directly impacts both CD4 and CD8 cell compartments it also indirectly impacts the bone marrow and thymus compartments through the feedback mechanism. The initial ratio of z^* to $z^@$ prior to infection was fixed at 1.8 (Krueger et al. 2002a). In accordance with this, $z(t_0)$ was partitioned satisfying $z^*(t_0)/z^@(t_0) = 1.8$, and δ was partitioned satisfying $\delta^*/\delta^@ = 1.8$. We then applied the search

algorithm to help determine the parameters to fit the data (see Appendix A).

Results

Figure 12 shows simulation results of Eqs. 7–8 using the set of parameter values just below (w, x, and y subsystem parameters are same as those in Table 1),

As in the case of chronic HHV-6 we note that based on the above parameter signs and magnitudes, the net effect of HIV-1 over time is to reduce CD4 cell counts and increase CD8 counts. CD4's in turn tend to increase viral counts while CD8 cells tend to reduce viral counts as we would expect. The general

$$
\begin{aligned}
&w(t_0) = 100.0, &&x(t_0) = 1.0 \times 10^4, &&y(t_0) = 1.0 \times 10^7, &&z(t_0) = 1.0 \times 10^6 \\
&z^*(t_0) = 0.6z(t_0), &&z^@(t_0) = 0.4z(t_0), &&\mu_w = 50.0, &&\mu_x = 25.0, \\
&\mu_y = 2.5 \times 10^6, &&\mu_{z*} = 5.46 \times 10^5, &&\mu_{z^@} = 4.92 \times 10^5, \\
&P_w - D_w - I_w = -0.531, &&a = 5.0 \times 10^{-7}, &&b = 2.0 \times 10^{-7}, &&c = 5.0 \times 10^{-7}, \\
&P_x - D_x - I_x = -0.04, &&d = 5.5 \times 10^{-8}, &&e = 3.6 \times 10^{-7}, &&f = 5.0 \times 10^{-7}, \\
&P_y - D_y - I_y = -0.7, &&g = 1.0 \times 10^{-7}, &&h = 0.05, &&\delta = 0.001, \\
&P_{z*} - D_{z*} - I_{z*} = -0.223, &&i = 2.35 \times 10^{-5}, &&k = 7.13 \times 10^{-8}, &&j = 9.0 \times 10^{-4}, \\
&l = 8.84 \times 10^{-4}, &&m = 6.7 \times 10^{-5}, &&\lambda_1 = 1.08 \, \text{week}^{-1} &&\lambda_2 = 0.04 \, \text{week}^{-1}, \\
&P_{z^@} - D_{z^@} - I_{z^@} = -0.511, &&n = 3.17 \times 10^{-5}, &&o = 1.476 \times 10^{-7}, &&\upsilon(t_{\upsilon s}) = 3.8, \\
&\lambda_3 = 0.75 \, \text{week}^{-1}, &&\lambda_4 = -0.24 \, \text{week}^{-1}, &&\lambda_1' = 0.2 \, \text{week}^{-1}, &&\lambda_2' = 0.6 \, \text{week}^{-1}, \\
&\lambda_3' = 0.6 \, \text{week}^{-1}, &&\tau_1 = 0.0106 \, \text{week}, &&\tau_2 = 0.004 \, \text{week}, &&\tau_3 = -0.0078 \, \text{week}, \\
&\tau_4 = -0.0094 \, \text{week}, &&\tau_1' = 0.0143 \, \text{week}, &&\tau_2' = -0.000947 \, \text{week}, &&\tau_3' = -0.0092 \, \text{week}.
\end{aligned}
$$

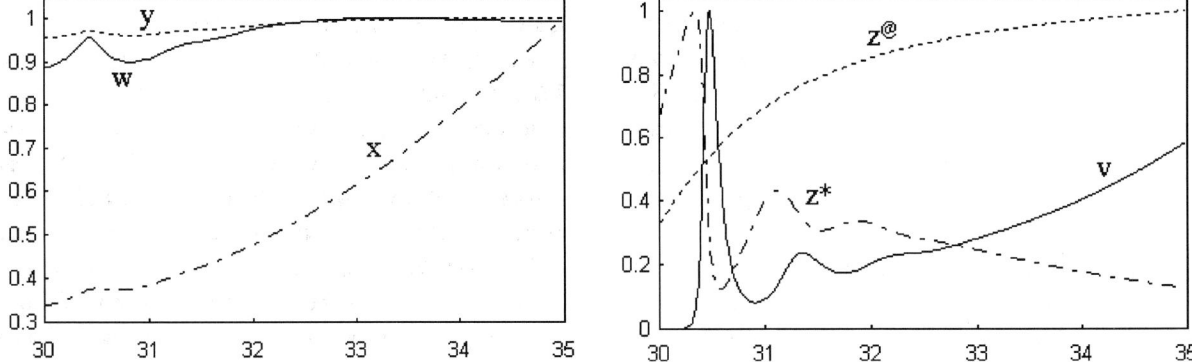

Fig. 12 Simulation of HIV-1 infection in a "virtual" adult patient. Infection initiated at age t = 30 years. (Left) relative cell counts vs. time for CD34 (w), CD38 (x), and CD4+8 (y). (Right) relative counts vs. time for peripheral CD4 (z^*) and CD8 ($z^@$) counts and for viral DNA copies (v). Data normalized to fit the plots. Note correspondence/resemblance with Fig. 11 patient data

tendency of these simulation results is similar to our human data (see Fig. 11), with an additional oscillatory response following the initial infection. There is eventual decline of z (not shown in Fig. 12) as would be expected in a patient with progressive infection leading to AIDS and decimation of the overall T-cell count (compare results in Figs. 11 and 12).

Discussion

We have developed a biodynamical model of human T cell proliferative activities for use in simulating pathologic conditions of cell proliferation as related to aplastic diseases or lymphoma development. Previous studies of the immune system using other computer modeling approaches (Bar-Or and Segel 1998; Segel and Bar-Or 1999; Kam et al. 2001; Forrest and Hofmeyr 2000; Warrender et al. 2001; Smith et al. 1998) focused on specific aspects of the peripheral cell pool only and its antigen/antibody/T cell regulation. They are therefore not useful for simulating quantitative and pathological cell pool changes which may ultimately lead to atypical (neoplastic) cell proliferation. This however, is the immediate task of our modeling endeavors in order to further elucidate the pathogenesis of malignant lymphomas.

Our basic idea, based on previous experimental data (Purtilo et al. 1985; Krueger and Ferrer Argote 1994; Schonnebeck et al. 1991; Feaux de Lacroix et al. 1981; Haas et al. 1982; Heine et al. 1983a; Krueger et al. 1987a, 1988, 2001; Krueger 1989a, 1993), was to identify certain pathological pathways from cell components in the peripheral blood of patients. PBL data derive from the activities of the immature cell pools (i.e. w, x, and y in our model) according to feedback and feedforward mechanisms described above. Considering the large number of known factors involved in the functional regulation of cell proliferation, differentiation, and cell death (Krueger et al. 2002c), our model was designed in an extensible, adaptive manner.

Previously, we successfully applied it to a rare human lymphoproliferative disease, the Canale–Smith syndrome (Krueger et al. 2002d). The rate of change with time in cell compartment numbers in the model is proportional to the summation of factors that decrease this rate, and/or factors which

increase it. Network regulation is brought about by feedback from other compartments and by regulation via negative self-feedback within each compartment. The latter represents a memory mechanism for cell numbers in each compartment at a given time. The model was shown to account for normative data across the lifespan.

Testing/validation of the model was carried out by simulating several different virus infections which target the same cell populations, then comparing these simulations with real patient data (Krueger et al. 2001a, b, 2002a). These patient data were part of controlled clinical trials initially not intended to be part of a computer simulation study, thus data collection was unbiased. They were the only ones available however where cell pool data were longitudinally monitored throughout the course of the disease. Such data unfortunately are widely missing in the literature. Only the HTLV-1 data were gleaned from a search of the literature reviewing some 1,000 publications (Krueger et al. 2002b).

The basic model (Eqs. 1) without viral infection was extended to include simulated infections and by incorporating a time delay mechanism using standard sigmoid functions. The modified system (Eqs. 6) thus enabled us to simulate some effects of external perturbations (viral infections) on the T cell development system and compartmental cell counts.

Stability testing of the system was carried out by keeping all parameters the same as in acute viral infection, except that both z and v were corrupted with noise. The system was able to withstand large noise perturbation, thus its robustness is good (see Appendix B). The model is also consistent with the expected effects of the various viruses on the human immune system with regard to the signs and magnitudes of its various parameters. The current model should form a framework for evaluating the effects of therapy (e.g. antiviral, gene therapies) or of other disorders of the immune system. It therefore appears to provide a useful basis for understanding regulatory disturbances of the T cell immune system and should assist in elucidating the pathogenesis of various diseases including malignant lymphomas. With continued development (e.g. by incorporating cytokine/chemokine factors) it very well may serve future clinical applications for diagnosis and treatment design.

Acknowledgement   Thanks to L. Maximilian Buja, M.D. of the UT Health Science Center Department of Pathology and Laboratory Medicine for his support of this research and helpful suggestions on this chapter.

Appendix A: parameter search algorithm

We describe our parameter search algorithm as applied (for simplicity) to Eqs. 3, the z-v subsystem (virus and C4 compartments only). There are 12 parameters then that need to be searched/optimized. We minimize the following objective function:

$$J(P) = \sum_{i}^{N} \left([z(P, ih) - \tilde{z}_i]^2 + [v(P, ih) - \tilde{v}_i]^2 \right),$$

where

$$P = (p_1, p_2, ..., p_{12}) = (k_1, k_2, \, l_1, l_2, \lambda_1, \lambda_2, \lambda'_1, \lambda'_2, \tau_1, \tau_2, \, \tau'_1, \, \tau'_2)$$

is the set of parameters to be determined, N is the total number of the samples, h is the numerical integration time interval between these samples, $z(P,t)$ and $v(P,t)$ are the concentrations of T-cells and virus as functions of time, which are generated by Eqs. 3 under the parameter set p. z_i and v_i (for $i = 0, 1,..., N$) are actual measured human data.

The aim is to minimize the objective $J(P)$ by changing the 12 parameters respectively. That is, we first increase p_1 by a value Δp_1. If this action makes the objective function smaller, we continue doing this until the objective function changes sign. We then divide Δp_1 in half and repeat the above-mentioned procedure. By doing this we reach a point that changing p_1 no longer affects the objective very much. We then in turn alter $p_2, p_3,..., p_{12}$. In this way, the objective function is minimized, at least to its local minima.

The detailed procedure is summarized as follows:

Step 1. Set $j = 0$.

Step 2. Arbitrarily choose initial values for the parameters, $P = (p_1,..., p_{12})$, with which Eqs. 3 are solved by the fourth order Runge–Kutta algorithm. $J(P)$ is obtained by Eq. 9 above.

Step 3. $j + 1 \rightarrow j$, if j reaches 13, stop.

Step 4. Increase p_j by a value m, i.e. $P_m = (p_1,..., p_{j+m}, ..., p_{12})$. Obtain $J(P_m)$ following the same procedure as Step 2.

Step 5. If $|J(P) - J(P_m)| < \varepsilon$ (a given small value), then $P_m \rightarrow P$, and then go to Step 3. Otherwise go to Step 6.

Step 6. If $|J(P) - J(P_m)| > 0$, then $P_m \rightarrow P$ and then go to Step 4. Otherwise make $-m/2 \rightarrow m$ (change sign and halve the step size), then $P_m \rightarrow P$ and then go to Step 4.

Steps 1–6 guarantee the algorithm to converge to a local minimal. If the result is still unsatisfactory, we re-run the optimization algorithm using another set of values as initial parameters. By doing so we may jump out of the previous local minimal and obtain a better result. Since the dimension of the z-v subsystem is 12, the optimization algorithm is time consuming. Furthermore, it is not easy to select the correct initial values. Usually it converges to the original local minimum after dozens of mutations.

Appendix B: studies of parameter robustness and sensitivity

Base normative model

We studied parameter robustness/sensitivity of the base normative model (Eq. 2). The open parameter range limits were determined in such a way that the basic "normative" pattern of the data shown in Fig. 2 was, for the most part, retained. Final decisions of "acceptability" were made by the clinical domain expert (GRFK). The parameter ranges were found in this way using a serial jack-knifing procedure of varying one parameter at a time while holding the other parameter values fixed according to Table 1 (we recognize this is a limited approach). We did this to provide some idea of the acceptable normal variation of the parameters and their robustness/sensitivity to noise.

Table 2 lists the parameters and their ranges, relative robustness, and polarity. Relative robustness was arbitrarily determined by dividing the range by the mid-value (similar to the measure of coefficient of variation): if the ratio is less than 0.5 we label it "low", if the ratio is between 0.5 and 1 we call it "med", and if greater than 1 we classify it as "high." Of the 17 parameters listed in Table 2, 11 are classified as medium to highly robust. The parameter polarity is determined by its dominant sign as being strongly positive or negative, positive or negative, or both positive and negative. Note that the average regeneratory potential (μ) of each compartment tends to be a (high) positive count.

Table 2 Normative Base model open parameter ranges

Parameter	Low	High	Relative robustness	Polarity
μ_w	40.0	80.0	Med	Strongly +
μ_x	−10.0	5.0×10^4	High	+
μ_y	2.3×10^6	2.7×10^6	Low	Strongly +
μ_z	0.7×10^4	1.1×10^4	Low	Strongly +
$P_w - D_w - I_w$	−1.0	−0.3	High	Strongly −
$P_x - D_x - I_x$	−0.07	−0.03	Med	Strongly −
$P_y - D_y - I_y$	−0.75	−0.65	Low	Strongly −
$P_z - D_z - I_z$	−0.025	−0.022	Low	Strongly −
a	-2.0×10^{-6}	3.0×10^{-6}	High	Both
b	2.0×10^{-7}	4.0×10^{-7}	Med	Strongly +
c	-6.0×10^{-7}	5.0×10^{-7}	High	Both
d	0.55×10^{-7}	0.65×10^{-7}	Low	Strongly +
e	3.4×10^{-7}	3.8×10^{-7}	Low	Strongly +
f	-2.0×10^{-6}	2.0×10^{-6}	High	Both
g	0.7×10^{-7}	1.8×10^{-7}	Med	Strongly +
u	−2.0	10.0	High	+
δ	0.0005	0.0012	Med	+

$k_1 = 1159.0,$ $k_2 = 0.003937,$ $l_1 = 0.5272,$ $l_2 = 2.547 \times 10^{-8},$
$\lambda_1 = 0.5951 \text{ week}^{-1},$ $\lambda_2 = -3.1056 \text{ week}^{-1},$ $\lambda_1' = 0.3845 \text{ week}^{-1},$ $\lambda_2' = -0.7825 \text{ week}^{-1},$
$\tau_1 = 21.3815 \text{ weeks},$ $\tau_2 = 12.7416 \text{ weeks},$ $\tau_1' = 2.1767 \text{ weeks},$ $\tau_2' = 2.14487 \text{ weeks},$

$P_z - D_z - I_z = -0.014, v(t_{vs}) = 100.$

All of the $P - D - I$ terms are strongly negative. This is the compartmental self-feedback term. None of the remaining parameters are purely negative although three of them (a, c, and f) can be either positive or negative. Parameters b and d, which are strongly positive, are actually negative feedback terms (refer back to Eq. 1).

Parameter variation and noisy measurements

We next performed studies to determine the amount of parameter variation that would allow for an approximately 10% variation in the relative amplitude and timing of peak maxima in the state variables. The 10% figure is a tolerance estimate of the amount of acceptable variation in a normal adult's immune system in which the system dynamics do not essentially change. The final decisions were made by visual inspection of the simulation runs by the co-author team. We explored this with virus compartment added in the model using Eqs. 4–5 with the parameter set shown in Table 2 (above):

Simulating and plotting this, we get results very similar to those shown in Fig. 4. Next, keeping all other parameters fixed we varied one parameter at

a time to achieve the 10% variation in the following features of the state variables v and z:

$$z_{max}/z(t_0) \in [18.0, 22.0], \ t_{zmax} \in [9.0, 11.0] \text{ weeks},$$
$$v_{max}/v(t_{vs}) \in [6.0, 8.0], \ t_{vmax} \in [4.5, 5.5] \text{ weeks},$$

where $z_{max}/(v_{max})$ is the peak value of $z(v)$ and t_{zmax} (t_{vmax}) is the time of occurrence of the peak value following the introduction of the viral stimulus to the system. The parameter ranges determined to ensure these conditions were found to be

$$k_1 \in [1138.0, 1326.0], \quad k_2 \in [0.00254, 0.018],$$
$$l_1 \in [0.460, 0.530], \quad l_2 \in [3.2 \times 10^{-8}, 2.5],$$
$$\lambda_1 \in [0.190, \infty] \text{ week}^{-1}, \quad \lambda_2 \in [-\infty, -1.312] \text{ week}^{-1},$$
$$\lambda_1' \in [0.374, 0.695] \text{ week}^{-1}, \quad \lambda_2' \in [-\infty, -0.589] \text{ week}^{-1},$$
$$\tau_1 \in [9.35, \infty] \text{ weeks}, \quad \tau_2 \in [11.45, 14.3] \text{ weeks},$$
$$\tau_1' \in [1.45, 2.207] \text{ weeks}, \quad \tau_2' \in [0.0, 2.32] \text{ weeks}.$$

It can be observed from this that there is considerable robustness in the parameter ranges of the system so as to keep the designated state characteristics within the 10% tolerances.

We also tested the stability of the overall system (Eqs. 6) by corrupting state variables z and v with additive

pseudorandom noise. The maximal values for the root-mean-square (RMS) amplitudes of these noises were σ $(v) = 1.0$ and $\sigma(z) = 4,000.0$, respectively. We used the parameter values for the normal immune system shown in Table 1 along with the following set:

$$k_1 = 60779.0, \qquad k_2 = 0.2258,$$
$$\lambda_1 = 35.6602 \, \text{year}^{-1}, \qquad \lambda_2 = -64.5812 \, \text{year}^{-1},$$
$$\tau_1 = 3.47 \, \text{months}, \qquad \tau_2 = 2.93 \, \text{months},$$
$$v(t_{vs}) = 100.$$

$$l_1 = 24.879, \qquad l_2 = 1.1679 \times 10^{-6},$$
$$\lambda_1^{'} = 35.8895 \, \text{year}^{-1}, \qquad \lambda_2^{'} = -41.3943 \, \text{year}^{-1},$$
$$\tau_1^{'} = 0.50 \, \text{month}, \qquad \tau_2^{'} = 1.21 \, \text{months},$$

Comparing Fig. B.1 with Fig. 4 we can see that the system is robust to fairly large noise perturbations in z and v without altering the essential system dynamic.

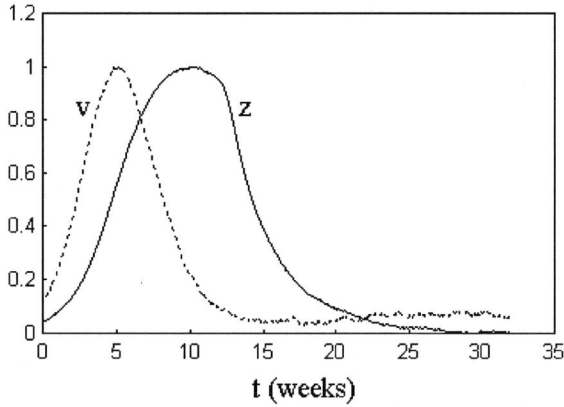

Fig. B.1 Stability testing of the system (Eqs. 4–5) by adding root-mean-square (RMS) noise for v ($\sigma = 1$) and for z ($\sigma = 4,000$). This demonstrates the robust response of the system to noise perturbation. Data normalized to fit the plot

References

Allison JP (1987) Structure, function and serology of the T-cell antigen complex. Annu Rev Immunol 5:503–540

Appasamy PM (1999) Biological and clinical implications of interleukin-7 and lymphopoiesis. Cytokines Cell Mol Ther 5:25–39

Appay V, Nixon DF, Donahoe SM, Gillespie GM, Dong T, King A, Ogg GS, Spiegel HM, Conlon C, Spina CA, Havlir DV, Richman DD, Waters A, Easterbrook P, McMichael AJ, Rowland-Jones SL (2000) HIV-specific CD8(+) T cells produce antiviral cytokines but are impaired in cytolytic function. J Exp Med 192:63–75

Artavanis-Tsakonas S, Rand MD, Lake RJ (1999) Notch signalling: cell fate control and signal integration in development. Science 284:770–776

Arzt E, Kovalovsky D, Igaz LM, Costas M, Plazas P, Refogo D, Paez-Pareda M, Reul JM, Stalla G, Holsboer F (2000) Functional cross-talk among cytokines, T-cell receptor, and glucocorticoid receptor transcriptional activity and action. Ann NY Acad Sci 917:672–677

Atkins B, Mueller C, Okada CY, Reinert RA, Weissman IL, Spangrude GJ (1987) Early events in T-cell maturation. Annu Rev Immunol 5:325–365

Azuara V, Grigoriadu K, Lembezat MP, Nagler-Anderson C, Perreira P (2001) Strain-specific TCR repertoire selection of IL-4-producing Thy-1 dull gamma delta thymocytes. Eur J Immunol 31:205–214

Bar-Or RL, Segel LA (1998) On the role of a possible dialogue between cytokine and TCR presentation mechanisms in the regulation of autoimmune disease. J Theor Biol 190:161–178

Bodey B, Bodey Jr., B, Siegel SE, Kaiser HE (1999) Molecular biological ontogenesis of the thymic reticulo-epithelial cell network during the organization of the cellular microenvironment. In Vivo 13:267–294

Bodey B, Bodey Jr., B, Siegel SE, Kaiser HE (2000) The role of the reticulo-epithelial (RE) cell network in the immuno-neuroendocrine regulation on intrathymic lymphopoiesis. Anticancer Res 20:1871–1888

Bondurant MC, Koury MJ (1999). Origin and development of blood cells. In: Lee GR, Foerster J, Lukens J, Paraskevas F, Rodger G (ed)Wintrob's clinical hematology 10th edn. Williams & Wilkins, Baltimore, pp 145–168

Brandt ME, Chen G (2001) Feedback control of a biodynamical model of HIV-1. IEEE T Bio-Med Eng 48:754–759

Brandt ME, Wang G, Krueger GRF, Buja LM (2002) A biodynamical regulatory model of the T-cell system. Proceedings of the Second Joint IEEE EMBS/BMES Conference.

Carrol AM, DeSousa M (1983) Thymus cell differentiation and in vivo T-cell migration: Migration of lectin-selected thymocytes. Cell Immunol 23:356–375

Champagne P, Ogg GS, King AS, Knabenhans C, Ellefsen K, Nobile M, Appay V, Rizzardi GP, Fleury S, Lipp M, Forster R, Rowland-Jones S, Sekaly RP, McMichael AJ, Pantaleo G (2001) Skewed maturation of memory HIV-specific CD8 T lymphocytes. Nature 410:106–111

Chen CH, Six A, Kubota T, Tsuji S, Kong FK, Gobel TW, Cooper MD (1996) T cell receptors and T cell development. Curr Top Microbiol Immunol 212:37–53

Clark EA, Brugge JS (1995) Integrins and signal transduction pathways: the road taken. Science 268:233–239

Correla-Neves M, Mathis D, Benoist C (2001) A molecular chart of thymocyte positive selection. Eur J Immunol 31:2583–2592

Crouse DA, Jordan RK, Sharp JG (1980). T-cell differentiation. In: Battisto and Knight (ed) Immunoglobulin genes and cell differentiation. Elsevier, North Holland/Amsterdam, The Netherlands, pp 65–78

Crouse DA, Turpen JB, Sharp JG (1985) Thymic non-lymphoid cells. Surv Immunol Res 4:120–134

Cyster JG (1999) Chemokines and cell migration in secondary lymphoid organs. Science 286:2098–2102

Daefler S, Krueger GRF (1989a) Expression of proliferation and differentiation antigens in response to modulation of membrane lipid fluidity in chronic lymphatic leukemia lymphocytes. Anticancer Res 9:501–506

Daefler S, Krueger GRF (1989b) Lack of dynamic lipid changes after binding of interleukin-2 in chronic lymphocytis leukemia lymphocytes indicates defective transmembrane signaling. Anticancer Res 9:743–748

Dappen GE, Crouse DA, Anderson RW, Jordan RK, Robinson JH, Sharp JG (1982) Morphological assessment of immunologically relevant cells in the thymus. Adv Exp Biol Med 149:389–399

Dardenne M, Pleau JM, Man NK, Bach JF (1977) Structure of circulating thymic factor: a peptide isolated from pig serum. I. Isolation and purification. J Biol Chem 252:8040–8044

Davis MH, Chien YH (1999). T-cell antigen receptors. In: Paul WE (ed) Fundamental immunology, 4th edn. Lippincott-Raven, Philadelphia, pp 341–366

de la Hera A, Marston W, Aranda C, Toribio ML, Martinez C (1989) Thymic stroma is required for the development of human T cell lineages in vitro. Int Immunol 1:471–478

Dieu-Nosjean MC, Massacrier C, Vanbervliet B, Fridman WH, Caux C (2001) Il-10 induces CCR6 expression during Langerhans cell development while IL-4 and IFN-gamma suppress it. J Immunol 167:5594–5602

Dunon D, Courtois D, Vainio O, Six A, Chen CH, Cooper MD, Dangy JP, Imhof BA (1997) Ontogeny of the immune system: gamma/delta and alpha/beta T cells migrate from thymus to the periphery in alternating waves. J Exp Med 186:977–988

Ehrlich GD, Poiesz BJ (1988) Clinical and molecular parameters of HTLV-1 infection. Clin Lab Med 8:65–84

Eren R, Zharhary D, Abel L, Globerson A (1987) Ontogeny of T cells: development of pre-T cells from fetal liver and yolk sac in the thymus microenvironment. Cell Immunol 108:76–84

Feaux de Lacroix W, Deying T, Krueger GRF (1981). Cell proliferation in the T lymphocyte lineage during the latent period of Moloney virus induced lymphomagenesis. Eleventh European Study Group for Cytokinetics in Pathology Meeting, Aarhus, Denmark (book of abstracts)

Fiorini E, Marchisio PC, Scupoli MT, Poffe O, Tagliabue E, Brentegnani M, Colombatti M, Santini F, Tridente G, Ramarli D (2000) Adhesion of immature and mature T cells induces in human thymic epithelial cells (TEC) activation of IL-6 gene transcription factors (NF-kappab and NF-IL6) and IL-6 gene expression: role of alpha3beta1 and alpha6beta4 integrins. Dev Immunol 7:195–208

Forrest S, Hofmeyr SA (2000). Immunology as information processing. In: Segel LA, Cohen IR (ed) Design principles for immune system and other distributed autonomous systems. Oxford University Press, New York, pp 361–387

Gessain A, Vernant JC, Maurs L (1985) Antibodies to human T-lymphotropic virus type 1 in patients with tropical spastic paraparesis. Lancet II 407–409

Goldstein G (1977) Molecular control of proliferation and differentiation. Ann NY Acad Sci 249:177–185

Greenberger JS (1991) The hematopoietic microenvironment. Crit Rev Oncol Hematol 11:65–84

Guerin S, Mari B, Fernandez E, Belhacene N, Toribio ML, Auberger P (1997) CD10 is expressed on human thymic epithelial cell lines and modulates thymopentin-induced cell proliferation. FASEB J 11:1003–1011

Haas W, Deying T, Krueger GRF, Feaux de Lacroix W (1982) Autoradiographic and immunocytologic identification of atypical cell proliferation during Moloneyvirus induced lymphoma development. In: Yoh DS, Blakeslee JR (eds) Advances in comparative leukemia research. Elsevier Biomedical, Amsterdam, The Netherlands, pp 241–243

Harris CA, Andryuk PJ, Cline S, Chan HK, Natarajan A, Siekierka JJ, Goldstein G (1994) Three distinct human thymopoietins are derived from alternatively spliced mrnas. Proc Natl Acad Sci USA 91:6283–6287

Heine UI, Krueger GRF, Karpinski A, Munoz E, Krueger MB (1983a) Quantitative light and electron microscopic changes in thymic reticular epithelial cells during Moloneyvirus induced lymphoma development. J Cancer Res Clin Oncol 106:102–111

Heine UI, Krueger GRF, Munoz E, Karpinski A (1983b) Altered thymic epithelial cells may be decisive for Moloneyvirus induced lymphoma development. In: Bailey GW (ed) Proceedings of the 41st Annual Meeting of the Electron Microscopic Society of America, pp 784–785

Hochberg EP, Chillemi AC, Wu CJ, Neuberg D, Canning C, Hartman K, Alyea EP, Soiffer RJ, Kalams SA, Ritz J (2001) Quantitation of T-cell neogenesis in vivo after allogeneic bone marrow transplantation in adults. Blood 98:1116–1121

Inghirami G, Knowles DM (1992) The immune system, structure and function. In: Knowles DM (ed) Neoplastic hematopathology Williams & Wilkins, Baltimore, pp 27–72

Jaleco AC, Neves H, Hooijberg E, Gameiro P, Clode N, Haury M, Henrique D, Parreira L (2001) Differential effects of notch ligands delta-1 and jagged-1 in human lymphoid differentiation. J Exp Med 194:991–1002

Janossy G, Campana D, Akbar A (1989) Kinetics of T cell development. Curr T Pathol 79:59–99

Kam N, Cohen IR, Harel D (2001) The immune system as a reactive system: modeling T cell activation with statecharts. Proceedings of Symposia Human-Centric Computing Languages and Environments. IEEE Computer Society Press, Stresa, Italy, pp 15–22

Kerre TCC, DeSmet G, DeSmedt M, Offner F, DeBosscher J, Plum J, Vanderkerckhove B (2001) Both CD34+38+ and CD34+38− cells home specifically to the bone marrow of NOD/ltsz scid/scid mice but sahow different kinetics in expansion. J Immunol 167:3692–3698

Kim JK, Takahashi I, Kai Y, Kiyono H (2001) Influence of enterotoxin in mucosal intranet: selective inhibition of extrathymic T cell development in intestinal intraepithelial lymphocytes by oral exposure to heat-labile toxin. Eur J Immunol 31:2960–2969

Kondo M, Scherer DC, Miyamoto T, King AG, Akashi K, Sugamura K, Weissman IL (2000) Cell fate conversion of lymphoid committed progenitors by instructive action of cytokines. Nature 407:383–386

Kong F, Chen CH, Cooper MD (1998) Thymic function can be accurately monitored by the level of recent T cell emigrants in the circulation. Immunity 8:97–104

Kong FK, Chen CL, Six A, Hockett RD, Cooper MD (1999) T cell receptor gene deletion circles identify recent thymic emigrants in the peripheral T cell pool. Proc Natl Acad Sci USA 96:1536–1540

Korsmeyer J (1987) Immunoglobulin and T-cell receptor genes reveal the clonality, lineage and translocations of lymphoid neoplasms. In: DeVita VT, Hellman S, Rosenberg (ed)

Important advances in oncology. Lippincott, Philadelphia, pp 3–25

Koury MJ, Bondurant MC (1993) Prevention of programmed death in hematopoietic progenitor cells by hematopoietic growth factors. News Physiol Sci (NIPS) 8:170–174

Kraus M, Krueger GRF (1981) T- and B-cell determination in various lymphoid tissues of mice during N-nitrosobutylurea (NBU) leukemogenesis. J Cancer Res Clin Oncol 100:149–165

Kronenberg M, Siu G, Hood LE, Shastri N (1986) The molecular genetics of the T-cell antigen receptor and T-cell antigen recognition. Annu Rev Immunol 4:529–591

Krueger GRF (1972) Chronic immunosuppression and lymphomagenesis in man and mice. Natl Cancer I Monogr 35:183–190

Krueger GRF (1985) Klinische immunpathologie. W. Kohlhammer, Stuttgart, Germany

Krueger GRF (1989a) The pathology of diphenylhydantoin-induced lymphoproliferative reactions in animals. In: Kammueller ME, Blocksma N, Deinen W (ed) Autoimmunity and toxicology Elsevier, Amsterdam, The Netherlands, pp 391–413

Krueger GRF (1989b) Abnormal variation of the immune system as related to cancer. In: Kaiser HE (ed) Cancer growth and progression Kluwer, Dordrecht, The Netherlands, pp 139–161

Krueger GRF (1993) Pathology of lymphoproliferative disorders in HIV infection. In: Schrappe M, Mauff G (ed) AIDS-SIDA, a comparison between Europe and Africa. Ed Roche, Basel pp 239–253

Krueger GRF, Ferrer Argote V (1994) A unifying concept of viral immunopathogeneis of proliferative and aproliferative diseases (working hypothesis). In Vivo 8:493–500

Krueger GRF, Malmgren RA, Berard CW (1971) Malignant lymphomas and plasmacytosis in mice under chronic immunosuppression and persistent antigenic stimulation. Transplantation 11:128–144

Krueger G, Fischer RM, Flesch HG (1979) Sequential changes in T- and B-cells, virus antigen expression and primary histologic diagnosis in virus-induced lymphomagenesis in mice. Zeitschrift fuer Krebsforschung 92:41–54

Krueger GRF, Karpinski A, Heine UI, Koch B (1983) Differentiation block of prethymic lymphocytes during Moloneyvirus induced lymphoma development associated with a thymic epitheliel defect. J Cancer Res Clin Oncol 106:153–157

Krueger GRF, Stolzenburg T, Muller C (1987a) Cell membrane lipid fluidity and receptor expression in Moloney- and Friendvirus transformed cells. In Vivo 1:343–246

Krueger GRF, Papadakis T, Schaefer HJ (1987b) Persistent active Epstein-Barr virus infection and atypical lymphoproliferation. Am J Surg Pathol 11:972–981

Krueger GRF, Ablashi DV, Salahuddin SZ, Josephs SF (1988) Diagnosis and differential diagnosis of progressive lymphoproliferation and malignant lymphoma in persistent active herpesvirus infection. J Virol Methods 21:255–264

Krueger GRF, Bertram G, Ramon A, Koch B, Ablashi DV, Brandt ME, Wang G, Buja LM (2001a) Dynamics of infection with human herpesvirus-6 in EBV-negative infectious mononucleosis: data acquisition for computer modeling. In Vivo 15:373–380

Krueger GRF, Koch B, Hoffmann A, Rojo J, Brandt ME, Wang G, Buja LM (2001b) Dynamics of chronic active herpesvirus-6 infection in patients with chronic fatigue syndrome: data acquisition for computer modeling. In Vivo 15:461–466

Krueger GRF, Nguyen A, Uthman M, Brandt ME, Buja LM (2001c) Dysregulative lymphoma theory revisited: what can we learn from cytokines, CD classes and genes? Anticancer Res 21:3653–3662

Krueger GRF, Koch B, Deninger Weldner J, Tymister G, Ramon A, Brandt ME, Wang G, Buja LM (2002a) Dynamics of active progressive infection with HIV1: data acquisition for computer modeling. In Vivo 15:513–518

Krueger GRF, Brandt ME, Wang G, Buja LM (2002b) Dynamics of HTLV-1 leukemogenesis: data acquisition for computer modeling. In Vivo 16:87–92

Krueger GRF, Marshall GR, Junker U, Schroeder H, Buja LM, Brandt ME, Wang G (2002c) Growth Factors, Cytokines, Chemokines and Neuropeptides in the Modeling of T-Cells, In Vivo 16:365–386

Krueger GRF, Brandt ME, Wang G, Berthold F, Buja LM (2002d) A computational analysis of Canale-Smith syndrome: chronic lymphadenopathy simulating malignant lymphoma. Anticancer Res 22:2365–2372

Krueger GRF, Brandt ME, Wang G, Buja LM (2003) TCM-1: A nonlinear dynamical computational model to simulate cellular changes in the T cell system; Conceptional design and validation, Anticancer Res 23:123–136

Le PT, Adams KL, Zaya N, Mathews HL, Storkus WJ, Ellis TM (2001) Human thymic epithelial cells inhibit IL-15 and IL-2 driven differentiation of NK cells from the early human thymic progenitors. J Immunol 166:2194–2201

Leclercq G, Plum J (1996) Thymic and extrathymic T cell development. Leukemia 10:1853–1859

Liu CP, Auerbach R (1991) Ontogeny of murine T cells: thymus-regultade development of T cell receptor-bearing cells derived from embryonal yolk sac. Eur J Immunol 21:1849–1855

Lobach DF, Haynes BF (1987) Ontogeny of the human thymus during fetal development. J Clin Immunol 7:81–97

MacDonald WA, Radtke F (2001) Notch 1-deficient common lymphoid precursors adopt a B cell fate in the thymuc. J Exp Med 194:1003–1012

Marrack P, Lo D, Brinster R, Palmiter R, Burkly L, Flavell RH, Kappler J (1988) The effect of thymus environment on T cell development and tolerance. Cell 53:627–634

Metcalf D (1971) The nature of leukemia: neoplasm or disorder of haematopoietic regulation? Med J Australia 2:739–746

Miescher C, Howe RC, Budd RC, MacDonald HR (1988) Expression of T-cell receptors by functionally distinct subsets of immature adult thymocytes. Ann NY Acad Sci 532:8–17

Miyai I, Saida T, Fujita M, Kitahara Y, Hirono N (1987) Familial cases of HTLV-1 associated myelopathy. Ann Neurol 22:601–605

Mollet L, Tai-Sheng L, Samri A, Tournai C, Tubiana R, Calvez V, Debre P, Katlama C, Autran B (2000) Dynamics of HIV-specific CD8+ T lymphocytes with changes in viral load. J Immunol 165:1692–1704

Montecino-Rogriguez E, Truong LT, Henderson AJ (2001) Long-term bone marrow cultures provide access to early lymphoid progenitors. J Hematother Stem Cell Res 10:107–114

Nakauchi H, Sudo K, Ema H (2001) Quantitative assessment of the stem cell self-renewal capacity. Ann NY Acad Sci 938:18–24

Ogawa T, Kitagawa M, Hirokawa K (2000) Age-related changes of human bone marrow: a histometric estimation of proliferative cells, apoptotic cells, T cells, B cells and macrophages. Mech Aging Dev 117:57–68

O'Sullivan NL, Skaandera CA, Montgomry PC (2001) Development of T cell lineages in rat lacrimal glands. J Curr Eye Res 22:375–383

Paul W (1999) Fundamental immunology. Lippincott-Raven, Philadelphia

Paraskevas F, Foerster J (1999) The lymphatic system (Chapter 18). In: Lee GR, Foerster J, Lukens J, Paraskevas F, Rodger G (ed) Wintrob's clinical hematology, 10th edn. Williams & Wilkins, Baltimore pp 430–463

Pawelec G, Muller R, Rehbein A, Hahnel K, Ziegler BL (1998) Extrathymic T cell differentiation in vitro from human CD34 + stem cells. J Leukoc Biol 64:733–739

Pawelec G, Muller R, Rehbein A, Hahnel K, Ziegler BL (1999) Finite life spans of t cell clones derived from CD34+ human haematopoietic stem cells in vitro. Exp Gerontol 34:69–77

Peled A, Petit I, Kollet O, Magid M, Ponomaryov T, Byk T, Nagler A, Ben-Hur H, Many A, Shultz L, Lider O, Alon R, Zipori D, Lapidot T (1999) Dependence of human stem cell engraftment and repopulation mice on CXCR4. Science 283:845–848

Pestano GA, Zhou Y, Trimble LA, Daley J, Weber GF, Cantor H (1999) Inactivation of misselected CD8 T cells by CD8 gene methylation and cell death. Science 284:1187–1191

Petri HT, Livak F, Burtrum D, Mazel ST (1995) T cell receptor gene recombination patterns and mechanisms: cell death, rescue, and T cell production. J Exp Med 182:121–127

Plum J (1999) Can T cell immunity in an adult be regenerated? Verhandlingen af Koninglike Akademie for Geneeskunde Belgie 61:457–464

Poiesz BJ, Ruscetti FW, Gazdar AF, Bunn PA, Minna JD, Gallo RC (1980) Detection and isolation of type C retrovirus particles from fresh and cultured lymphocytes of a patient with cutaneous T cell lymphoma. Proc Natl Acad Sci USA 77:7415–7419

Purtilo DT, Tatsumi E, Manolov G, Manolova Y, Harada S, Lipscomb H, Krueger GRF (1985) Epstein-Barr virus as an etiological agent in the pathogenesis of lymphoproliferative and aproliferative diseases in immune deficient patients. Int Rev Exp Pathol 27:113–183

Robetamanith T, Schroder B, Bug G, Muller P, Klenner T, Knaus R, Hoelzer D, Ottmann OG (2001) Interleukin 3 improves the ex vivo expansion of primitive human cord blood progenitor cells and maintains the engraftment potential of scid repopulating cells. Stem Cells 19:313–320

Robey E (1999) T cell fate by notch. Annu Rev Immunol 17:283–295

Romagnani P, Annunziato F, Piccini MP, Maggi E, Romagnani S (2000) Cytokines and chemokines in T lymphopoiesis and T cell effector function. Trends Immunol Today 21:416–418

Rothenberg E, Lugo JP (1985) Differentiation and cell division in the mammalian thymus. Dev Biol 112:1–17

Sanchez M, Alfani E, Visconti G, Passarelli AM, Migliaccio AR, Migliaccio G (1998) Thymus-independent T-cell differentiation in vitro. Brit J Haematol 103:1198–1205

Sarun S, Dalloul AH, Laurent C, Blanc C, Schmitt C (1998) Human CD34+ thymocyte maturation: pre-T and NK cell differentiation on neonatal thymic stromal cell culture. Cell Immunol 190:23–32

Sasada T, Reinherz EL (2001) A critical role for CD2 in both thymic selection events and mature T cell function. J Immunol 166:2394–2403

Schonnebeck M, Krueger GRF, Braun M, Fischer M, Koch B, Ablashi DV, Balachandran N (1991) Human herpesvirus-6 infection may predispose cells for superinfection by other viruses. In Vivo 5:255–264

Segel LA, Bar-Or RL (1999) On the role of feedback in promoting conflicting goals of the adaptive immune system. J Immunol 163:1342–1349

Sen J (2001) Signal transduction in thymus development. Cell Mol Biol 47:197–215

Sing GK, Keller JR, Ellingsworth LR, Ruscetti FW (1988) Transforming growth factor-a selectively inhibits normal and leukemic human bone marrow cell growth in vitro. Blood 72:1504–1511

Singh VK, Biswas S, Mathur KB, Haq W, Garg SK, Agarwal SS (1998) Thymopentin and splenopentin as immunomodulators; Current status. Immunol Res 17:345–368

Smith DJ, Forrest S, Ackley DH, Perelson AS Smith (1998) Using lazy evaluation to simulate realistic-size repertoires in models of the immune system. Bull Math Biol 60:647–658

Spiegel HM, Ogg GS, DeFalcon E, Sheehy ME, Monard S, Haslett PA, Gillespie G, Donahoe SM, Pollack H, Borkowsky W, McMichael AJ, Nixon DF (2000) Human immunodeficiency virus type 1- and cytomegalovirus-specific cytotoxic T lymphocytes can persist at high frequency for prolonged periods in the absence of circulating peripheral CD4(+) T cells. J Virol 74:1018–1022

Stutman O (1978) Intrathymic and extrathymic T-cell maturation. Immunol Rev 42:138–184

Surh CD, Sprent J (1999). The thymus and T-cell development (Chapter 9) In: Gallin JI et al. (ed) Inflammation: basic principles and clinical correlates Lippincott Williams & Wilkins, Philadelphia pp 137–149

Suzuki K, Oida T, Hamada H, Hitotsumatsu O, Watanaba M, Hibi T, Yamamoto H, Kubota E, Kaminogawa S, Ishikawa H (2000) Gut cryptopatches: direct evidence of extrathymic anatomical sites for intestinal T lymphopoiesis. Immunity 13:691–702

Szewczuk Z, Wieczorek Z, Stefanowicz P, Wilczynski A, Siemion IZ (1997) Further investigations on thymopentin-like fragments of HLA-DQ and their analogs. Arch Immunol Exp Ther (Warsaw) 45:335–341

Tedder TF, Steeber DA, Chen A, Engel P (1995) The selectins: vascular adhesion molecules. FASEB J 9:866–873

Theodor L, Shoham J, Berger R, Gokkel E, Trachtenbrot L, Simon AJ, Brok-Simon F, Nir U, Ilan E, Zevin-Sonkin D, Friedman E, Rechavi G (1997) Ubiquitous expression of a cloned murine thymopoietin cdna. Acta Haematol 97:153–163

Thiele J, Wickenhauser C, Baldus SE, Kuemmel T, Zirbes TK, Drebber U, Wirtz R, Thiel A, Hansmann ML, Fischer R (1995) Characterization of CD34+ human hematopoietic progenitor cells from the peripheral blood: enzyme-, carbohydrate- and immunocytochemistry, morphometry, and ultrastructure. Leukemia Lymphoma 16:483–491

Uchiyama T, Yodoi J, Sagawa K, Takatsuki K, Uchino H (1977) Adult T cell leukemia: clinical and hematological features of 16 cases. Blood 50:481–492

Varas A, Vicente A, Sacedon R, Zapata AG (1998) Interleukin-7 influences the development of thymic dendritic cells. Blood 92:93–100

Warrender C, Forrest S and Segel L (2001) Effective feedback in the immune system. In: Genetic and Evolutionary Computation Conference Workshop Program. Morgan & Kaufmann, pp 329–332

Weber PJ, Eckhard CP, Gonser S, Otto H, Folkers G, Beck-Sickinger AG (1999) On the role of thymopoietins in cell proliferation. Immunochemical evidence for new members of the human thymopoietin family. Biol Chem380:653–660

Wognum AW, DeJong MO, Wagemaker G (1996) Differential expression of receptors for hematopoietic growth factors on subsets of CD34+ hemopoietic cells. Leukemia Lymphoma 24:11–25

Zajac AJ, Blattman JN, Murali-Krishna K, Sourdive DJ, Suresh M, Altman JD, Ahmed R (1998) Viral immune evasion due to persistence of activated T cells without effector function. J Exp Med 188:2205–2213

Zajac AJ, Vance RE, Held W, Sourdive DJ, Altman JD, Raulet DH, Ahmed R (1999) Impaired anti-viral T cell responses due to expression of the Ly49A inhibitory receptor. J Immunol (Baltimore, MD.; 1950) 163:5526–5534

Zevin-Sonkin D, Ilan E, Guthmann D, Riss J, Theodor L, Shoham J (1992) Molecular cloning of bovine thymopoietin gene and its expression in different calf issues: evidence for a predominant expression in thymocytes. Immunol Lett 31:301–309

Index